T0301990

Edited by Mani Lakshminarayanan and Fanni Natanegara

Bayesian Applications in Pharmaceutical Development

Chapman & Hall/CRC Biostatistics Series

Shein-Chung Chow
Duke University School of Medicine

Byron Jones
Novartis Pharma AG

Jen-pei Liu
National Taiwan University

Karl E. Peace
Georgia Southern University

Bruce W. Turnbull
Cornell University

Analysis of Incidence Rates
Peter Cummings

Cancer Clinical Trials: Current and Controversial Issues in Design and Analysis
Stephen L. George, Xiaofei Wang, Herbert Pang

Data and Safety Monitoring Committees in Clinical Trials 2nd Edition
Jay Herson

Clinical Trial Optimization Using R
Alex Dmitrienko, Erik Pulkstenis

Mixture Modelling for Medical and Health Sciences
Shu-Kay Ng, Liming Xiang, Kelvin Kai Wing Yau

Economic Evaluation of Cancer Drugs: Using Clinical Trial and Real-World Data
Iftekhar Khan, Ralph Crott, Zahid Bashir

Bayesian Analysis with R for Biopharmaceuticals: Concepts, Algorithms, and Case Studies
Harry Yang and Steven J. Novick

Mathematical and Statistical Skills in the Biopharmaceutical Industry: A Pragmatic Approach
Arkadiy Pitman, Oleksandr Sverdlov, L. Bruce Pearce

Bayesian Applications in Pharmaceutical Development
Mani Lakshminarayanan, Fanni Natanegara

For more information about this series, please visit:
https://www.crcpress.com/go/biostats

Edited by Mani Lakshminarayanan and Fanni Natanegara

Bayesian Applications in Pharmaceutical Development

CRC Press is an imprint of the
Taylor & Francis Group, an **informa** business
A CHAPMAN & HALL BOOK

CRC Press
Taylor & Francis Group
6000 Broken Sound Parkway NW, Suite 300
Boca Raton, FL 33487-2742

First issued in paperback 2021

© 2020 by Taylor & Francis Group, LLC
CRC Press is an imprint of Taylor & Francis Group, an Informa business

No claim to original U.S. Government works

ISBN-13: 978-1-138-29676-3 (hbk)
ISBN-13: 978-1-03-217702-1 (pbk)
DOI: 10.1201/9781315099798

This book contains information obtained from authentic and highly regarded sources. Reasonable efforts have been made to publish reliable data and information, but the author and publisher cannot assume responsibility for the validity of all materials or the consequences of their use. The authors and publishers have attempted to trace the copyright holders of all material reproduced in this publication and apologize to copyright holders if permission to publish in this form has not been obtained. If any copyright material has not been acknowledged, please write and let us know so we may rectify in any future reprint.

Except as permitted under U.S. Copyright Law, no part of this book may be reprinted, reproduced, transmitted, or utilized in any form by any electronic, mechanical, or other means, now known or hereafter invented, including photocopying, microfilming, and recording, or in any information storage or retrieval system, without written permission from the publishers.

For permission to photocopy or use material electronically from this work, please access www.copyright.com (http://www.copyright.com/) or contact the Copyright Clearance Center, Inc. (CCC), 222 Rosewood Drive, Danvers, MA 01923, 978-750-8400. CCC is a not-for-profit organization that provides licenses and registration for a variety of users. For organizations that have been granted a photocopy license by the CCC, a separate system of payment has been arranged.

Trademark Notice: Product or corporate names may be trademarks or registered trademarks, and are used only for identification and explanation without intent to infringe.

Publisher's Note
The publisher has gone to great lengths to ensure the quality of this reprint but points out that some imperfections in the original copies may be apparent.

Visit the Taylor & Francis Web site at
http://www.taylorandfrancis.com

and the CRC Press Web site at
http://www.crcpress.com

Contents

Preface

Drug discovery and drug development have a long history with primitive stages dating back to the early days of human civilization. The long journey of the pharmacological treatment of disease began long ago with the use of herbs in Chinese medicine, Japanese traditional medicine, Ayurvedic medicine (traditional Indian medicine), and several other forms before venturing into synthetic drugs in the 1900s. The introduction of synthetic drugs led to the founding of pharmaceutical industries which continue to invest millions of dollars in the research and development of innovative treatment for both common and rare diseases. Key discoveries of the 1920s and 1930s, such as insulin and penicillin, were game changers during wartime in treating infections and resulted in mass manufacturing and distribution. Screening natural products and isolating active ingredients for treating diseases became part of drug discovery in the early 1930s. The active ingredients are normally the synthetic versions of the natural products. These synthetic versions, called New Chemical Entities (NCEs), have to go through many iterations and tests to ensure they are safe, potent, and effective.

Use of randomized clinical trials and rigorous statistical principles for establishing the safety and effectiveness of NCEs was first investigated in the 1948 evaluation of streptomycin for the treatment of tuberculosis by the British Medical Research Council under the direction of Sir Austin Bradford Hill, an epidemiologist and statistician. The concept of a double-blind placebo-controlled trial was introduced in the field trial of the Salk polio vaccine in 1954. Before World War II, no formal scientific rationale was required for conducting clinical trials before marketing a drug. In the 1960s, evidence from controlled clinical trials became necessary to get an approval for a new drug from the Food and Drug Administration (FDA). This scientific rationale was further enhanced through the efforts of the International Conference on Harmonization (ICH) in standardizing processes for drug approvals.

Statistical principles for clinical trials described in the ICH E-9 guideline (https://www.ich.org/fileadmin/Public_Web_Site/ICH_Products/Guidelines/Efficacy/E9/Step4/E9_Guideline.pdf, 1998) focused on giving directions to researchers in the design, conduct, analysis, and evaluation of trials using frequentist approach. The FDA's Critical Path Initiative in 2004 transformed the way FDA-regulated medical products are developed, evaluated, and manufactured. The Initiative's report recognized that there is a widening gap between scientific discoveries and their translation into medical treatments. It also positioned the Bayesian approach as a tool to create innovative and efficient clinical trials. In 2010, the Center for Devices and Radiological Health of the FDA issued an elaborate guidance for the use of Bayesian statistics in clinical trials involving medical devices.

This book presents a plethora of topics behind Bayesian applications in pharmaceutical development. It is an extension of some of the great efforts put forth by most of the authors in this book who participated in the Bayesian Scientific Working Group of the Drug Information Association. This working group, started in 2011, states its mission as follows: "to ensure that Bayesian methods are well-understood and broadly utilized for design and analysis throughout the medical product development process and to improve industrial, regulatory and economic decision making". The group is comprised of individuals from academia, industry, and regulatory authorities (http://www.bayesianscientific.org/). Authors who have contributed chapters to this book also participated as members of various subteams that were formed to deal with topics such as non-inferiority, safety, missing data, and others, most of which have resulted in publications.

Our intention is not only to provide a platform that will cover the entire spectrum of development of medical products, from pre-clinical to post-marketing, but also to take advantage of the great efforts put forth by these subteam members who have contributed to different topics in the use of Bayesian applications. It is our intention that this book would serve as a "go-to source" for the topics on the use of Bayesian methods in biopharmaceutical development. The principal audience for this book are statisticians, clinicians, pharmacologists, scientists, engineers, and epidemiologists who are typical members of any medical product development team. We believe that these members can use this book as a single source to find out much more about different topics than is currently available in multiple sources in the public domain. Besides providing theoretical development, the authors have also attempted to provide computer codes if and when required under each topic, in addition to case studies as a basis for describing some of the Bayesian solutions. We were also very conscious in ensuring that the principal emphasis of the book is to describe the problems that occur and the solutions that are used "in practice" rather than in theory.

This book has been designed to show many of the problems and solutions associated with applying Bayesian methods appropriately in the life of a medical product. The chapters in this book are written with the personal experience of the authors in developing and getting regulatory approval of such products. Several of our colleagues who have collaborated in the DIA Bayesian Scientific Working Group have also contributed tirelessly to various chapters in this book. First and foremost, we wish to thank all of the contributors for their efforts and patience as we strove toward completion of this undertaking. We hope that this book will pave way to the understanding and correct usage of Bayesian approaches and, as a result, will provide yet another statistical tool for strengthening regulatory applications. We are also thankful for the wonderful support of our family members. Finally, we take this opportunity to thank John Kimmel of Taylor and Francis for his support and patience throughout the development of this book.

Mani Lakshminarayanan
Statistical Consultant
Fanni Natanegara
Eli Lilly and Company, Indianapolis, IN

Editors

Dr. Fanni Natanegara is currently a Principle Research Scientist and a Group Leader for the Early Phase Neuroscience Statistics team at Eli Lilly and Company in Indianapolis, Indiana. She has 15+ years of pharmaceutical experience providing statistical leadership in designing, executing, and reporting clinical trials across drug development phases and therapeutic areas. She was instrumental in providing Bayesian education outreach and statistical consultation to statistical and non-statistical colleagues at Eli Lilly while working to accelerate the organization's clinical development transformation. She has published over 20 articles in peer-reviewed statistics and medical journals, as well as presented her work at multiple professional conferences. Dr. Natanegara is also actively engaged in external activities, including chairing the cross industry-regulatory-academic DIA Bayesian Scientific Working Group, serving as Vice-Chair and industry representative on the ASA statistical partnerships or collaborations between two or more entities across the academic, industry, and/or government sectors, and serving on multiple working groups in Pharmaceutical Research and Manufacturers of America (PhRMA) efforts to provide statistical influence on regulatory policies. She received her PhD in Statistics from Baylor University, Waco, Texas.

Mani Lakshminarayanan is a Statistical Consultant. He has over 30 years of experience working in the pharmaceutical industry. He has held several positions in management as well as in research during his tenure. He has provided statistical leadership across all drug development phases and therapeutic, including neuroscience, bone, respiratory, immunology, endocrinology, cardiovascular, and oncology. He was also instrumental in conceptualizing and creating the Early Clinical Development (Statistics) team at Centocor. Dr. Lakshminarayanan has published over 50 articles, technical reports, and book chapters, besides serving as a referee for several journals. He has volunteered his time to the American Statistical Association in various positions and is currently a core member of the Development Committee. He has also served as a core member of the cross industry-regulatory-academic DIA Bayesian Scientific Working Group for the past several years. Dr. Lakshminarayanan is also a part-time lecturer at the Department of Statistics and Biostatistics at Rutgers University. He has received several professional awards and holds a Six-Sigma Greenbelt for a Process Improvement Initiative. He has a PhD in Statistics from Southern Methodist University, Dallas, Texas and is a Fellow of the American Statistical Association.

Contributors

Simin Baygani
Eli Lilly and Company
Indianapolis, Indiana

Somer Blair
JPS Health Network
Fort Worth, Texas

Freda Cooner
Sanofi US
Bridgewater, New Jersey

Maria J. Costa
Novartis Pharma AG
Basel, Switzerland

Paul Faya
Eli Lilly and Company
Indianapolis, Indiana

Margaret Gamalo-Siebers
Eli Lilly and Company
Indianapolis, Indiana

A. Lawrence Gould
Merck & Co., Inc.
North Wales, Pennsylvania

Baoguang Han
Sarepta Therapeutics
Cambridge, Massachusetts

Cory R. Heilmann
Eli Lilly and Company
Indianapolis, Indiana

Lei Huang
Office of Biostatistics and Epidemiology CBER
Silver Spring, Maryland

Telba Irony
Office of Biostatistics and Epidemiology CBER
Silver Spring, Maryland

David Kahle
Department of Statistical Science Baylor University
Waco, Texas

Mani Lakshminarayanan
Statistical Consultant
Pennsylvania

Wen Li
Merck & Co., Inc.
North Wales, Pennsylvania

Ruitao Lin
Department of Biostatistics The University of Texas MD Anderson Cancer Center
Houston, Texas

Stacy Lindborg
Biogen
Cambridge, Massachusetts

G. Frank Liu
Merck & Co., Inc.
North Wales, Pennsylvania

Timothy H. Montague
GlaxoSmithKline
Brentford, United Kingdom

Melvin Munsaka
AbbVie, Statistical Sciences, R&D
Chicago, Illinois

Fanni Natanegara
Eli Lilly and Company
Indianapolis, Indiana

Karen L. Price
Eli Lilly and Company
Indianapolis, Indiana

Matilde Sanchez-Kam
SanchezKam, LLC

John W. Seaman, Jr.
Department of Statistical Science
Baylor University
Waco, Texas

John Sherington
UCB Pharma
Brussels, Belgium

Guochen Song
Biogen
Cambridge, Massachusetts

Michael Sonksen
Eli Lilly and Company
Indianapolis, Indiana

James D. Stamey
Department of Statistical Science
Baylor University
Waco, Texas

Phil Stanley
UCB Pharma
Brussels, Belgium

Qi Tang
Digital and Data Sciences Sanofi US
Bridgewater, New Jersey

Neal Thomas
Pfizer Inc.
New York

Ros Walley
UCB Pharma
Brussels, Belgium

J. Kyle Wathen
Janssen Research & Development, LLC
Titusville, New Jersey

Ying Yuan
Department of Biostatistics The University of Texas MD Anderson Cancer Center
Houston, Texas

Xin Zhao
Johnson & Johnson
New Brunswick, New Jersey

John Zhong
Biogen
Cambridge, Massachusetts

1

Introduction

Mani Lakshminarayanan
Statistical Consultant

Fanni Natanegara
Eli Lilly and Company

CONTENTS

1.1 Motivation

The development of medical products including drugs and medical devices follows a rigorous process governed by regulatory authorities to ensure the safety, potency, and efficacy of the final product. The process starts in the discovery phase, followed by testing in the preclinical stage, where basic questions about the safety of the drug are investigated, and then enters the clinical phase.

Little research has been done to understand the failure occurring in the preclinical or early stage of the drug development. Failures occurring at the later stages of the clinical phase are a well-recognized industry-wide problem with published data showing that the failure rate is more than 50% (Grignolo, 2016). In an analysis conducted by FDA, it was shown that out of 313 new molecular entity (NME) submissions received in 2013, only 151 were approved, and nearly half failed because they could not show efficacy (Shanley, 2016). Furthermore, Pharmaceutical Research and Manufacturers of America (PhRMA) reported that in 2013 alone, companies have spent nearly $10 billion to run 1,680 clinical trials involving 644,684 patients. A 2014 report published by the Tufts Center for the Study of Drug Development suggests that the current cost of bringing a new medicine to market, estimated to be as high as $2.6 billion, presents a major barrier to investment in innovation in drug development. Such failures are costly and wasteful, not only for the pharmaceutical industries but also for patients who participate in these studies with an ultimate hope that these will result in cures for their ailments.

There have been a plethora of investigations performed to identify possible reasons for these late-stage failures which include:

- Incomplete/inadequate characterization of relationship between exposure and treatment response in target population or indication
- Smaller observed effect sizes than the expected effect sizes which may be due to not having robust historical data needed to support the assumption
- Lack of experience with primary efficacy endpoints at the study sites
- Frequent changes in clinical study teams including high turnover of personnel
- Lack of interim analysis that could have prompted early termination for futility

One of the key considerations for avoiding potential failures especially at a later stage is to adopt some of the ideas put forth, for example, by Sheiner (1997). His suggestion was to use learning and confirmation cycles in that order, where progression of compounds is carried through a learning cycle that provides sufficient knowledge and can be used as a support to proceed to subsequent confirmatory stages. In the learning phase, it may be necessary to start with small early Phase II studies for proof of viability which will have the flexibility of stopping for futility with minimal spending. During this time, it is important for the teams to enhance their learning through available external data on similar indication or with the same mechanism of action to guide them in designing future trials. Such learning perspective may also include confirmation of disease activity, characterization of dose response, clear understanding of pharmacokinetic/pharmacodynamic (PK/PD), validation of endpoints and others (see Chapters 6 and 7). A composite look at the Phase II data and external information should be used to confidently drop the program if the results are not favorable. Such decisions will ensure that any unnecessary Phase III expenditure can be avoided. An overriding theme in this regard is to explore methodologies that can be used to synthesize all available data including those from previous trials, expert opinions and external data to guide the choice of trial design and increase efficiency. A Bayesian paradigm offers a natural platform for synthesizing current and historical data with prior beliefs.

In summary, there is an urgent need to make drug development less time-consuming and less costly. We need to use a framework that ultimately leads to a more timely and accurate decision-making process. Innovative trial designs/analyses, such as Bayesian approach, as laid out in the FDA's Critical Path Initiative list are essential to meet this need.

1.2 What Is Bayesian Method?

The original research by Rev Thomas Bayes was as a paper 'An essay towards solving a problem in the doctrine of chances' (Bayes, 1763) published posthumously by his friend Richard Price. Bayesian method is a concept based on conditional probability that most statistics students are taught as part of an early course in probability theory. From the perspective of statistical inference, Cox and Hinkley (1974) describe it as a method for combining a prior distribution for a parameter with current data or likelihood to extract a posterior distribution which represents all that you know for a given outcome of interest. Simply put, Bayesian method is a statistical framework for updating your belief as data and evidence accumulate. Mathematically, you can express the posterior distribution $p(\Theta|y)$ to be proportional to the product of likelihood $l(y|\Theta)$ and the prior distribution $p(\Theta)$

$$p(\Theta|\, y) \propto l(y|\, \Theta) \times p(\Theta)$$

where y is the data and Θ is a vector of parameters of interest. If a normalizing constant is included, then this mathematical expression can be written as an exact equation

$$p(\Theta|\, y) = (\text{normalizing constant}) \times l(\, y|\, \Theta) \times p(\Theta)$$

The mathematics of calculating a posterior distribution can be made tractable by using conjugate distributions, which have the property that for a particular distribution for the data, the distributional form of the conjugate prior distribution and the posterior distribution are the same, with updated parameters. All members of the exponential family of distributions have conjugate priors. Two examples are presented later in Section 1.5, one with closed form solution using conjugate priors and another example which would require an algorithm to sample from a distribution of random quantities.

1.3 Why Do We Need Bayesian Method in Pharmaceutical Development?

Consideration for using a Bayesian approach in a regulatory pharmaceutical development dates back to 1997 led by Dr. Greg Campbell, who was a Director of Biostatistics, FDA (Campbell, 2017). The setting in medical devices lends itself to a natural platform for using Bayesian methods. There exists a great deal of prior information for a medical device in part because often the mechanism of action is well-known, physical as opposed to pharmacokinetic/pharmacodynamics, and local as opposed to systemic. Moreover, devices can go through changes over time from model to model in terms of mechanism and in overall functional aspects, whereas an NME can remain unchanged. The availability of prior information and the inherent time-dependent variations provide a natural background for the use of Bayesian methods, which can help the industries and regulators to bring good technology to market sooner or with less current data by leveraging prior information (Berry, 1997).

Spiegelhalter (2004) provided insights into the use of Bayesian approaches in healthcare evaluation. He mentioned three characteristics of a Bayesian approach: acknowledgment of subjectivity and context, simple use of Bayes theorem, and use of a "community" of prior distributions to assess the impact of new evidence. Bayesian approach is rooted in probability theory as it explicitly allows for the possibility that the conclusions of an analysis may depend on who is conducting it and their available evidence and opinions (see Chapters 2, 3, and 4).

In any applied research such as healthcare evaluation, we are constantly faced with external information during the conduct of current research. External information can come in many forms that may impact the overall hypothesis, models under consideration, decision criteria, and other relevant research topics. As practitioners, we should be ready to embrace new information, apply our subjective judgment in the choice of what methods to use, what assumptions to invoke, and what data to include in their analyses. Incorporating subjectivity is natural in a Bayesian paradigm, and it can be accomplished via Bayes theorem. As Spiegelhalter (2004) indicates, subjectivity in healthcare interventions can come from various sources such as sponsors of clinical trials, investigators, reviewers, policymakers, and consumers at any time during the conduct of clinical trials. Clinical trial process is essentially dynamic in nature, in which any individual study takes place in a context of continuously increasing knowledge. As emergence of new information from the current trial as well as other trials is inevitable, it is prudent not to expect that a single trial alone will provide definitive conclusion on a given research hypothesis. As a result, it is critical that one explores a method such as a Bayesian approach that is naturally dynamic where the prior belief as well as current information can be combined to generate new data. As medical product development is time consuming and costly, it is critical for a drug and medical device maker to decide in the middle of drug development whether it is worth to continue to proceed till the end (see Chapters 11). In a Bayesian setting, these decisions that are made in the interim or at the end of a trial can easily be communicated in a probabilistic way, which should make the interpretation straightforward. For example, a Bayesian probability of Go/No Go decisions based on interim look is a very popular presentation in adaptive designs.

Another advantage of the Bayesian method is the ease of interpretation in probabilistic statements. In Bayesian analyses, any inferences regarding the parameter of interest will be based on the posterior distribution. This enables the user to make a probabilistic statement such as "There is an 80% probability that the mean difference in clinical outcome between active drug and placebo is greater than 0" or "There is a 5% probability that the difference in baseline score is less than 7 units." These are true probabilistic statements, not to be confused with the concept of p-value as explained in Goodman (1999).

Although Bayesian approach is the central theme of this book, we acknowledge that the frequentist approach has its merits and should be considered as complementary to the Bayesian framework. We are in full agreement of Lee and Chu's (2012) statement regarding the two approaches: "the past was combative, the present is competitive, and the future will be cooperative." Little's (2006) calibrated Bayesian concept also promotes the idea of co-existence between the two paradigms.

In summary, advantages of using Bayesian methods can be achieved in the following areas/topics (Gupta, 2012):

- Incorporation of prior information: Bayes formula presents a natural paradigm for updating information by combining prior and current data

- Adaptations in trial design: use of predictive probability in stopping the trial, making decisions based on outcome (given the observed outcomes up to that point), and other aspects

- Phase I dose finding study: use of methodologies such as Continual Reassessment Method or Target Probability Interval Approach in Phase I dose escalation studies

- Phase II Proof-of-Concept Study: use of predictive probability in making Go/No-Go Decision and availability of Bayesian concepts in two-stage and optimal flexible designs

- Seamless Phase II/III trials: use of Bayesian predictive power in integrating Phase II and Phase III trials

- Easy decision making: provision of a continuous learning structure by Bayesian paradigm as data accumulate, use of hierarchical models in determination of borrowing of information, and direct estimation of evidence for the effect of interest using posterior probability

- Post-marketing surveillance: use of posterior distribution from a pre-marketing study as a prior for surveillance

- Meta-analysis: use of hierarchical models in incorporating random covariates

1.4 Regulatory Consideration

ICH E9 Statistical Principles of Clinical Trials (1998) opened the door to the possibility of Bayesian applications in clinical trials by stating the following: "Because the predominant approaches to the design and analysis of clinical trials have been based on frequentist statistical methods, the guidance largely refers to the use of frequentist methods (see Glossary) when discussing hypothesis testing and/or confidence intervals. This should not be taken to imply that other approaches are not appropriate: the use of Bayesian (see Glossary) and other approaches may be considered when the reasons for their use are clear and when the resulting conclusions are sufficiently robust."

In medical device development, Bayesian applications have had much success and acceptance by the US and EU regulatory agencies. FDA's Center for Device and Radiation Health (CDRH) created a Bayesian Guidance (2010) to provide direction on statistical consideration on design and analyses of medical device trials using Bayesian approach. This guidance discusses the extent of prior information available on medical devices from other devices, overseas trials, and other historical data, which can enhance borrowing strength, and provides insight to ease in updating posterior information for any modification in the device as the behavior may be very similar (Irony, 2018).

Bayesian applications in pharmaceutical research have gained modest popularity in recent years (Natanegara, 2013). There is much openness to the applications in exploratory stages of clinical trials but not so much in the confirmatory setting. Like medical device development, past and emerging data are bona fide currencies in pharmaceutical development. Therefore, we need to have a framework to synthesize data coming from various sources and be able to make decisions as data accumulates. The Bayesian approach provides a solution to this problem.

Recent legislations such as the 21st Century Cures Act and Prescription Drug User Fee Act (PDUFA) VI champion the use of innovative trial designs and analyses such as Bayesian methods. Either or both legislations support the conduct of public meetings and creation of guidance documents as well as hiring expertise to handle submissions of innovative trials such as those using Bayesian approach. Chapter 14 of this book is dedicated to further understanding the regulatory perspective on Bayesian applications in medical product development.

1.5 Examples of Applications of Bayesian Methodology

1.5.1 Bayesian Analysis of Beta-Binomial Model

The Beta-Binomial model is a foundational example of Bayesian model as it involves one parameter, can be computed analytically, and serves as an illustration of a conjugate prior. Consider the binomial likelihood with success probability θ and S number of successes out of n trials.

$$L(\theta|y,n) = \theta^S (1-\theta)^{n-S}$$

$$S = \sum_{i=1}^{n} y_i$$

As a function of θ, the likelihood above is a Beta distribution with parameters $\alpha = S + 1$ and $\beta = n - S + 1$, i.e., $L(\theta|y, n) \sim \text{Beta}(S + 1, n - S + 1)$.

The Beta distribution is a conjugate prior to the binomial likelihood above since the resulting posterior is also a Beta distribution. That is, if the prior θ is distributed as Beta(α_0, β_0), then the posterior distribution is of a Beta distribution.

$$p(\theta|\, y) \sim \text{Beta}\,(\alpha_0 + S,\, \beta_0 + n - S)$$

The posterior distribution as a Beta-Binomial has several interesting properties. The posterior mean can be written as a weighted average of the maximum likelihood estimate (MLE) of θ and the mean of the prior distribution as follows.

$$\begin{aligned}(S+\alpha_0)/(n+\alpha_0+\beta_0) = &[n/(n+\alpha_0+\beta_0)]\,[S/n] \\ &+ [(\alpha_0+\beta_0)/(n+\alpha_0+\beta_0)]\,[\alpha_0/(\alpha_0+\beta_0)]\end{aligned}$$

In a Beta(α_0, β_0) prior, α_0 is the number of 1's (or successes) in a binomial trial, and (α_0 + β_0) is the prior's sample size. As a result, "weight" of the prior relates to information content of the prior. For example, if there are 20 observations, the prior has roughly two times the "weight" of the prior Beta(3,7) because the sum of the parameters of the Beta prior is an approximately equivalent sample size. Similar to a discussion like this, many of the aspects of a Bayesian analysis can be easily explained with an analysis on a binomial data with a conjugate Beta prior.

As a simple illustration, suppose the parameter of interest θ is the proportion of clinical trial participants who experienced nausea as a treatment emergent adverse event (TEAE) for a given compound. A previous study (study 1) involving the compound suggests that 10% out of 30 participants experienced nausea as a TEAE. The current study (study 2) with $n = 60$ having similar study design and patient population compared to study 1 observed nausea in 20% of patients. Given the information from these two studies, what is our best estimate of θ? Using the model set up previously, we have the following.

Study 1: prior $\theta \sim \text{Beta}(\alpha_0, \beta_0)$ where $\alpha_0 = 3$ and $\beta_0 = 27$

Study 2: Likelihood $L(\theta|y, n) \sim \text{Beta}(S + 1, n - S + 1)$ where $S = 12$ and $n = 60$

Posterior: $p(\theta|y) \sim \text{Beta}(\alpha_0 + S, \beta_0 + n - S)$

Hence, assuming equal weighting of studies 1 and 2, the posterior distribution of θ is Beta(15, 75) with posterior mean 17% and standard deviation 3.9%. All inference on θ in this example will be based on this posterior distribution. The overlay plots of the prior, likelihood, and posterior distributions are presented in Figure 1.1.

In cases where the posterior distribution does not have a closed form solution such as the Beta-Binomial example, one must rely on Markov Chain Monte Carlo (MCMC) algorithm to sample from a distribution of random quantities. Monte Carlo is a technique that uses repeated pseudo-random sampling. It makes use of algorithm to generate samples. Markov Chain, on the other hand, is a random process with a countable state space with the Markov property. According to Chen et al. (2000), Markov property means that the future state is dependent only on the present state and not on the past states. The combination of Markov Chains and Monte Carlo techniques is commonly referred to as MCMC (Robert and Casella, 2004).

1.5.2 Bayesian Analysis of Logistic Regression

In a logistic regression, we have the probability of success which varies from one subject to another (depending on their covariates); the likelihood contribution from the ith subject is binomial:

$$\text{likelihood}_i = \pi(x_i)^{y_i} (1 - \pi(x_i))^{(1-y_i)}$$

where $\pi(x_i)$ represents the probability of the event for subject i who has covariate vector x_i and y_i indicates the presence, $y_i = 1$, or absence, $y_i = 0$, of the event for that subject. Note that we can write $\pi(x)$ using the logistic model as

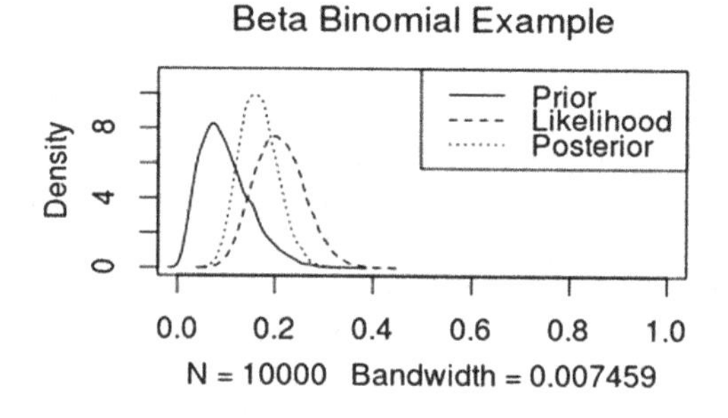

FIGURE 1.1
Prior, likelihood, and posterior distributions of a Beta-Binomial model.

$$\pi(x) = \frac{e^{\beta_0+\beta_1 x_1+\cdots+\beta_p x_p}}{1+e^{\beta_0+\beta_1 x_1+\cdots+\beta_p x_p}}$$

With associated likelihood written as

likelihood =

$$\prod_{i=1}^{n}\left[\left(\frac{e^{\beta_0+\beta_1 x_{i1}+\cdots+\beta_p x_{ip}}}{1+e^{\beta_0+\beta_1 x_{i1}+\cdots+\beta_p x_{ip}}}\right)^{y_i}\left(1-\frac{e^{\beta_0+\beta_1 x_{i1}+\cdots+\beta_p x_{ip}}}{1+e^{\beta_0+\beta_1 x_{i1}+\cdots+\beta_p x_{ip}}}\right)^{(1-y_i)}\right]$$

For the parameters, β_0, β_1, ..., β_{p}, any prior distribution, e.g., non-informative or informative, can be used, depending on the available prior information. If informative prior distributions are desired, it is often difficult to give such information on the logit scale, i.e., on the β parameters directly.

One may prefer to provide prior information on the Odds Ratio OR = $\exp(\beta)$ scale and mathematically transform back to the logit scale. Alternative approaches include deriving prior distributions for various categories or using a prior elicitation technique. Assuming a normal prior, $\beta_j \sim N(\mu_j, \sigma_j^2)$, Bayes theorem can be used to derive the posterior as follows.

posterior =

$$\prod_{i=1}^{n}\left[\left(\frac{e^{\beta_0+\beta_1 x_{i1}+\cdots+\beta_p x_{ip}}}{1+e^{\beta_0+\beta_1 x_{i1}+\cdots+\beta_p x_{ip}}}\right)^{y_i}\left(1-\frac{e^{\beta_0+\beta_1 x_{i1}+\cdots+\beta_p x_{ip}}}{1+e^{\beta_0+\beta_1 x_{i1}+\cdots+\beta_p x_{ip}}}\right)^{(1-y_i)}\right]$$
$$\times\prod_{j=0}^{p}\frac{1}{\sqrt{2\pi}\sigma_j}\exp\left\{-\frac{1}{2}\left(\frac{\beta_j-\mu_j}{\sigma_j}\right)^2\right\}$$

The above expression has no closed form expression, and even if it did, we would have to perform multiple integration to obtain the marginal distribution for each coefficient. We may use the Gibbs sampler as implemented by WinBUGS and other software (see Chapters 15 and 16) to solve approximately the properties of the marginal posterior distributions for each parameter.

The method of maximum likelihood and Bayesian analysis are closely related as can be seen in the equation relating posterior to likelihood (Berger, 1984). In fact, the Bayes theorem can be expressed, in terms of OR version, as posterior OR = likelihood ratio X prior OR. Under maximum likelihood estimation, we would compute the mode of the likelihood function and use the local curvature to construct confidence intervals. Hypothesis testing follows using likelihood-ratio (LR) statistics. Alternatively, using Bayes theorem, one can update the prior assumption of the parameter with data (likelihood) and develop an updated version, namely the posterior distribution. All inferences will use the posterior distribution. The dependence of the posterior on the prior provides an indication of how much information on the unknown parameter values is contained in the data. If the posterior is highly dependent on the prior, then the data likely has little signal, while if the posterior is largely unaffected under different priors, the data are likely highly informative.

Bayesian inference regarding the parameter and its uncertainty is based on the posterior distribution. Posterior distribution can be used to compute summary statistics and credible intervals. As noted above, there exists a close connection between Bayesian estimation and the maximum likelihood estimation. If the prior is flat, then the parameter estimate that maximizes the posterior (the mode, also called the maximum a posteriori estimate or MAP) is the same as the MLE. It is worth noting that uniform priors may not be appropriate in several situations because the underlying parameter space is unbounded, while an informative prior will make the MAP estimate different from that of the MLE.

The history of awareness and use of Bayesian methods in various applications including pharmaceutical development have shown some reluctance primarily for two reasons: philosophical differences, which has for the most part receded, and practical issues such as availability and ease of use of software to handle computational difficulties (Winkler, 2001). Inexpensive highspeed computing became widely available in the 1990s, and the surge in the development of iterative simulation techniques, such as MCMC methods, led to an explosion of Bayesian computational tools. More importantly, it helped Bayesians to focus more on the problem of modeling the data and not have to worry about computing. WinBUGS, as the Windows version of the original BUGS, played a key role in the growth of Bayesian statistics by making these methods more accessible computationally, and it is still being used in many applications currently amid other packages developed based on the freeware such as R.

1.6 Objectives and Organization of the Book

This book provides a comprehensive coverage of Bayesian applications across the span of drug development from discovery to clinical trial to manufacturing. In addition, it also covers broad topics such as prior distribution, regulatory perspective on Bayesian applications, and computational tools.

Central to Bayesian methods is the prior distribution specification; we start with three chapters covering the use of historical data in clinical trials and prior elicitation. Second, we have several chapters showcasing Bayesian applications in discovery and clinical phases as well as product development. In this regard, we also provide recommendations on best practices for simulation plan and report to assess the operating characteristics of Bayesian trial design/analyses as well as on appropriate Bayesian languages to be included in SAP and protocol. Third, we have a chapter to focus on regulatory perspectives on Bayesian applications in the drug development. A comprehensive list and review of computational tools is also another important topic that is discussed in this book. We conclude with two chapters on specialized topics in which Bayesian approach provides a natural framework on rare diseases and pediatrics. These chapters highlight practical examples as illustrations

TABLE 1.1
Title and Content of Each Chapter

Chapter	Title	Content
1	Introduction	Overview of Bayesian methods and motivation for Bayesian applications in drug development. This chapter provides the road map for the book.
2	On Incorporation of Historical Control Data in Clinical Trials	Reviews the rationale for incorporating historical information, and previous work for attenuating undue influence of historical information illustrated with real clinical trial examples.
3	Practical Considerations for Building Priors for Confirmatory Setting	A discussion on considerations of integrating historical information into confirmatory trials from a practical perspective, using the development of treatment for spinal muscular atrophy as an example.
4	The Practice of Prior Elicitation	Focuses on methodologies and challenges of prior elicitation, a process that enables one to translate expert knowledge and judgment into a probability distribution, a quantitative belief distribution.
5	Bayesian Examples in Preclinical In-Vivo Research	Bayesian methods appear to be rarely used in the analysis of preclinical pharmacology data. Issues relating to the introduction, implementation of Bayesian methods in preclinical research, and applications in preclinical in vivo pharmacology studies are discussed.
6	Planning a Model-Based Bayesian Dose Response Study	Recent meta-analyses combined with information specific to a new compound can be applied in an empirically based Bayesian design and analysis of a parametric Emax dose response phase II study. R package implementing these methods allows rapid implementation.
7	Novel Designs for Early Phase Drug Combination Trials	Reviews novel designs for phase I drug combination trials to identify the maximum tolerated dose combination(s) with a particular focus on the model-assisted drug combination designs along with freely available software for implementation and case examples.

(*Continued*)

TABLE 1.1 (*Continued*)
Title and Content of Each Chapter

Chapter	Title	Content
8	Executing and Reporting Clinical Trial Simulations: Practical Recommendations for Best Practices	The increase in use and acceptance of Bayesian methodology in clinical trials has led to a need for guidance on how to report and document such methodology. Recommendations on the level of detail to include in simulation plan, simulation report, protocols, and analysis plan are presented.
9	Reporting of Bayesian Analyses in Clinical Research: Some Recommendations	A checklist considering several key aspects of Bayesian analysis is proposed that can be useful in ensuring standardization of the reporting of Bayesian analyses, including design, interpretation, analyses, documentation, and regulatory setting. A hypothetical example is used to illustrate.
10	Handling Missing Data in Clinical Trials with Bayesian and Frequentist Approaches	Missing data is common in clinical trials but is often inadequately handled and could have an impact on estimation of treatment effects. Several case studies are included to demonstrate how to utilize Bayesian, multiple imputation, and some unconventional approaches for missing data analysis.
11	Bayesian Probability of Success for Go/No-Go Decision-Making	Discusses the framework for Go/No-Go decision making where probability of success is assessed utilizing all relevant information, sensitivity of probability of success is examined against different priors, and the risks associated with each decision made are evaluated using receiver operating characteristics (ROC) curves. A due diligence and an end of phase II Go/No-Go examples are given to illustrate this framework.

(*Continued*)

TABLE 1.1 (*Continued*)
Title and Content of Each Chapter

Chapter	Title	Content
12	Simulation for Bayesian Adaptive Designs Step-by-Step Guide for Developing the Necessary R Code	Reviews necessary steps and skills required to develop a custom simulation package for a Bayesian outcome adaptive randomization where randomization probabilities are altered prior to each patient enrolling and allow for the possibility of early stopping for superiority or futility. Complete designs are available in a GitHub repository and freely available to the reader.
13	Power Priors for Sample Size Determination in the Process Validation Life Cycle	Drug quality, safety, and efficacy are dependent on thoroughly validated manufacturing processes which span process design, qualification, and verification. Critical to this decision is the number of batches to be examined for the qualification stage. Bayesian assurance and sample size determination method incorporating prior knowledge of process performance, using both expert opinion and data, are presented.
14	Bayesian Approaches in the Regulation of Medical Products	Provides examples of current use of Bayesian approaches in the regulation of medical products, focusing on the use of prior information to increase the power of clinical trials, the use of Bayesian adaptive designs, and the use of simulations to assess the operating characteristics of Bayesian designs.
15	Computational Tools	An overview of the concepts underlying and motivating modern Bayesian computation highlighting statistical programing language R and the probabilistic programing language JAGS (just another Gibbs sampler). Implementations of three commonly used models in pharmaceutical drug development are presented including illustration of prior elicitation and sample size determination.

(Continued)

TABLE 1.1 (*Continued*)
Title and Content of Each Chapter

Chapter	Title	Content
16	Software for Bayesian Computation – An Overview of Some Currently Available Tools	Aims to provide a brief introductory overview to some selected commercial and open source tools for Bayesian analysis. To help elucidate the functionality of the software tools considered, we will describe each software with regard to its functionality and provide some illustration in some instances of usage via some applications as well as provide graphical view point along with some recommendations where appropriate.
17	Considerations and Bayesian Applications in Pharmaceutical Development for Rare Diseases	Introduces the background of rare disease definition, its regulations, and unique clinical development considerations for rare diseases and small sample size trials. Examples of innovative trial designs and novel analytical methodologies, especially Bayesian statistics, are presented to illustrate their implementation in real clinical trial settings.
18	Extrapolation Process in Pediatric Drug Development and Corresponding Bayesian Implementation for Validating Clinical Efficacy	Discusses the process of extrapolation to harmonize expectations from different regulatory authorities toward a single extrapolation strategy as well as relevant case examples. Bayesian statistical paradigm is an appropriate analytical tool in this setting to improve pediatric development programs that use all available information in the most efficient manner.
Appendix 1	A Brief Guide to Bayesian Model Checking	Covers critical features of Bayesian model building. Checking and documenting such features facilitate evaluation of the models, rendering potential problems with assumptions, convergence, and inferential accuracy.

along with programing codes such that readers can implement. Finally, in the appendix, we include a guide to Bayesian model checking which would apply across all these chapters. The title and content of each chapter are listed in Table 1.1.

Many of the contributors of this book are members of the DIA Bayesian Scientific Working Group . This group was formed in 2011 with a vision to ensure that Bayesian methods are well-understood and broadly utilized for design, analysis and throughout the medical product development process and to improve industrial, regulatory, and economic decision making. The group is comprised of individuals from academia, industry, and regulatory authorities who have volunteered their time and effort to carry out this vision. As Bayesian applications in medical product development are wide ranging, several sub-teams were formed to focus on various Bayesian topics such as safety, non-inferiority, prior specification, comparative effectiveness, joint modeling, program wide decision making, tools, and education. Each sub-team have written numerous publications in various statistical journals and conferences.

We expect this book will have a wide appeal to statisticians, scientists, and clinicians working in drug development who are motivated to accelerate and streamline the drug development process as well students who aspire to work in this field. We also hope this book provides enough examples motivated by real case studies to be used as supplementary materials for teaching Biostatistics or Pharmacoepidemiology courses for example.

References

Bayes T. An essay towards solving a problem in the doctrine of chances. *Philosophical Transactions of the Royal Society* 1763; 53: 370–418.

Berger JO and Wolpert RL (1984). The Likelihood Principle. Institute of Mathematical Statistics Lecture Notes—Monograph Series V6, Hayward, CA.

Berry DA (1997). Using a Bayesian approach in medical device development, Technical Paper.

Campbell G. Bayesian methods in clinical trials with applications to medical devices. *Communications for Statistical Applications and Methods* 2017; 24(6): 561–581.

Chen MH, Shao QM, and Ibrahim JG (2000). *Monte Carlo Methods in Bayesian Computation.* Springer, New York.

Cox DR and Hinkley DV (1974). *Theoretical Statistics.* Chapman and Hall, London.

FDA (2010). Guidance for the use of Bayesian Statistics in Medical Device Clinical Trials, Center for Devices and Radiological Health. Available at: www.fda.gov/medicaldevices/deviceregulationandguidance/guidancedocuments/ucm071072.htm [Accessed 31 January 2019].

Goodman SN. Towards evidence-based medical statistics: 1. The P value fallacy. *Annals of Internal Medicine* 1999; 130:995–1004.

Grignolo A and Pretorius S. Phase III trial failures: Costly, but preventable. *Applied Clinical Trials* 2016; 25(8).

Gupta SK. Use of Bayesian statistics in drug development: Advantages and challenges. https://www.ncbi.nlm.nih.gov/pmc/articles/PMC3657986/ *International Journal of Applied and Basic Medical Research* 2012; 2(1): 3–6.

ICH E9 document (1998). Available at: www.fda.gov/downloads/Drugs/GuidanceComplianceRegulatoryInformation/Guidances/UCM073137.pdf [Accessed 31 January 2019].

Irony T (2018). www.bayesianscientific.org/wp-content/uploads/2018/01/Irony_BayesKOL_11918.pdf [Accessed 1 February 2019].

Lee JJ and Chu CT. Bayesian clinical trials in action. *Statistics in Medicine* 2012; 31: 2955–2972.

Little RJ. Calibrated Bayes, for statistician general, and missing data in particular (with discussion). *Statistical Science* 2006; 26(2): 162–186.

Natanegara F, Neuenschwander B, Seaman JW, Kinnersley N, Heilmann CR, Ohlssen D, and Rochester G. The current state of Bayesian methods in medical product development: Survey results and recommendations from the DIA Bayesian Scientific Working Group. *Pharmaceutical Statistics* 2013; 13(1): 3–12.

Robert CP and Casella G (2004). *Monte Carlo Statistical Methods.* Springer, New York.

Shanley A. Preventing phase III failures. *Pharmaceutical Technology* 2016; 2016(1): 24–27.

Sheiner L. Learning versus confirming in clinical drug development. *Clinical Pharmacology and Therapeutics* 1997; 61(3): 275–291.

Spiegelhalter DJ. Incorporating Bayesian ideas into health-care evaluation. *Statistical Science* 2004; 19: 156–174.

Winkler RL. Why Bayesian analysis hasn't caught on in healthcare decision making. *International Journal of Technology Assessment in Health Care* 2001; 17: 56–66.

2

Incorporation of Historical Control Data in Analysis of Clinical Trials

G. Frank Liu, Wen Li, and A. Lawrence Gould

Merck & Co., Inc.

CONTENTS

2.1 Introduction

Double-blinded, randomized, controlled clinical trials are the gold standard for demonstrating the efficacy and safety of new drugs and biologic products. Depending on the disease and the existence of standard therapy, either placebo or an existing active treatment may serve as a control. If it is available, historical information about the effect of the control treatment in a similar patient population may be used to improve the sensitivity of treatment comparisons that drive decisions about the benefits and risks of the test treatment.

Historical data may be useful when a concurrent control is not feasible for ethical or practical reasons or for active-controlled clinical trials. For placebo-controlled clinical trials, placebo response information may be available from previously completed placebo-controlled trials. Active-controlled trials are often used in the development of a combination vaccine, in which the test vaccine is compared to each of the component vaccines in terms of immunogenicity measures such as geometric mean antibody titers [1]. Information for component vaccines is commonly available from previous trials or from the component vaccine labels.

Data from previous studies may provide useful information about potential treatment effects and/or variability for the control group in a new study that can be used for sample size and power calculations. This historical control group information also could be incorporated into analyses of treatment group effects. For example, if the current study and one or more historical trials are very similar, the historical information may be pooled using a meta-analytic approach [2,3]. This approach has been used in the evaluation of clinical trials of medical devices and rare diseases, and can be used in the design of early-phase single-arm oncology trials to evaluate the response of a test drug relative to the response rate with current standard care in a specific cancer population.

Historical trials may differ from a current trial in many respects, such as patient population, evaluation tools, and evolving standards of medical care. Bayesian approaches provide a way to account for heterogeneity across trials by using information from previous studies to specify informative prior distributions for quantities of interest [4] and for the design and analysis of pediatric trials using data from trials in adults [5]. Recently, Viele et al. [6] evaluated pooling, pooling after testing, power prior, and hierarchical model approaches for borrowing information for trials with binary endpoints. They found that simply pooling historical data may result in type-I error inflation, especially when the true response in the current control group differs from the historical data, and recommended the use of other methods that allow the amount of borrowing to be adjusted by specifying the value of a suitable parameter.

This chapter describes approaches for incorporating historical information in such a way as to control the influence of this information on the analysis of clinical trials. Section 2.2 provides an overview of common methods including commensurate prior, power prior, meta-analytic predictive (MAP) prior, and the test-then-pool approach to borrow historical information with some adjustment for the heterogeneity between historical information and the current control. Case examples are presented in Section 2.3 including an application of MAP prior for an indirect comparison between the treatment in the current trial and a control agent available from previous trials but not included in the current trial and applications of various approaches in a regression model to illustrate how these methods may borrow information and alter the analysis results. A discussion of the implications of the findings is presented in Section 2.4.

2.2 Methods for Incorporating Historical Information in Clinical Trials

Many methods have been proposed for incorporating historical information into the analysis of a current trial. To simplify the exposition, we describe the methods for continuous endpoints; the conclusions can be easily extended for other endpoints with generalized regression models.

Suppose that there are H historical trials, with observations $\mathbf{x}_h = \{x_{hj}, j = 1, \ldots, m_h\}$ among m_h subjects in the control group for the hth trial. A current trial includes a total of n subjects, n_C on the same control treatment as the historical trials with observations $\mathbf{y}_\mathrm{C} = \{y_{\mathrm{C}j}, j = 1, \ldots, n_\mathrm{C}\}$, and n_T subjects on a test treatment with observations $\mathbf{y}_\mathrm{T} = \{y_{\mathrm{T}j}, j = 1, \ldots, n_\mathrm{T}\}$.

Let $f_0(\mathbf{x}_h; \mu_{0h}, \eta)$ denote the likelihood of the control group observations $\mathbf{x}_h$ from the hth historical trial, with parameter(s) (μ_{0h}, η), $h = 1, \ldots, H$; μ_{0h} denotes the expected value for $\mathbf{x}_h$, and η denotes a nuisance parameter such as the variance. Further, let $f_\mathrm{C}(\mathbf{y}_\mathrm{C}; \mu_\mathrm{C}, \theta)$ denote the likelihood for the control group observations from the current trial with parameter(s) (μ_C, θ), and let $f_\mathrm{T}(\mathbf{y}_\mathrm{T}; \mu_\mathrm{C} + \lambda, \theta)$ denote the likelihood for the test group observations from the current trial, where λ is the mean difference between test and control; μ_C denotes the expected value of y_C, and θ is a nuisance parameter. The historical and current trial nuisance parameters may differ.

Remark 1: If the observations are binary with individual Bernoulli probability functions, then $\mathbf{x}_h$ can be replaced with $x_h = \sum_j x_{hj}$, $\mathbf{y}_\mathrm{C}$ can be replaced with $y_\mathrm{C} = \sum_j y_{\mathrm{C}j}$, and $\mathbf{y}_\mathrm{T}$ can be replaced with $y_\mathrm{T} = \sum_j y_{\mathrm{T}j}$. The likelihood f_0 for the control group observations (x_h) is binomial with parameters(p_h, m_h), $h = 1, \ldots, H$ where $p_h = s(\mu_{0h})$ and $s(\cdot)$ denotes a convenient link function, e.g., logit or probit. The likelihood f_C for the current trial control group (y_C) is binomial with parameters $(p_\mathrm{C}, n_\mathrm{C})$, and the likelihood f_T for the test group (y_T) is binomial with parameters $(p_\mathrm{T}, n_\mathrm{T})$, where $p_\mathrm{C} = s(\mu_\mathrm{C})$ and $p_\mathrm{T} = s(\mu_\mathrm{C} + \lambda)$.

2.2.1 Commensurate Priors

Suppose $\mathbf{x}_h \sim f_0(\mathbf{x}; \mu_0, \eta)$ with the same mean μ_0 for all of the historical trials and that μ_C is related to μ_0 via a "commensurate prior" [7,8],

$\mu_C \sim g(\mu_C; \mu_0, \xi)$. The product of the joint likelihood of the historical and current data and the density of μ_C is

$$f(\mathbf{x}, \mathbf{y}_C, \mathbf{y}_T, \mu_C; \mu_0, \lambda, \xi, \theta, \eta) = f_C(\mathbf{y}_C; \mu_C, \theta) f_T(\mathbf{y}_T; \mu_C + \lambda, \theta) \times g(\mu_C; \mu_0, \xi) \prod_{h=1}^{H} f_0(\mathbf{x}_h; \mu_0, \eta)$$

where $\mathbf{x} = (\mathbf{x}_1, \ldots, \mathbf{x}_H)$ is the collection of observed data from the historical trials. Let $g_{\mu_0}(\mu_0)$, $g_{\eta_0}(\eta)$, $g_{\lambda_0}(\lambda)$, $g_{\theta_0}(\theta)$, and $g_{\xi_0}(\xi)$ denote, respectively, prior densities for μ_0, η, λ, θ, and ξ. A complete specification for Bayesian calculations is the product

$$f(\mathbf{x}, \mathbf{y}_C, \mathbf{y}_T, \mu_C; \mu_0, \lambda, \xi, \theta, \eta) = f_C(\mathbf{y}_C; \mu_C, \theta) f_T(\mathbf{y}_T; \mu_C + \lambda, \theta) \times g(\mu_C; \mu_0, \xi) \times g_{\mu_0}(\mu_0) g_{\lambda_0}(\lambda) g_{\theta_0}(\theta) g_{\eta_0}(\eta) g_{\xi_0}(\xi) \times \prod_{h=1}^{H} f_0(\mathbf{x}_h; \mu_0, \eta) \tag{2.1}$$

Integrating (2.1) with respect to μ_C and μ_0 leads to an expression which is proportional to the posterior density of λ, the treatment effect,

$$f_{\text{post}}(\lambda; \mathbf{x}, \mathbf{y}_C, y_T, \xi, \theta, \eta) = v(\mathbf{x}, \mathbf{y}_C, \mathbf{y}_T, \lambda, \xi, \theta, \eta)\, g_{\lambda_0}(\lambda) \Big/ \int_{\lambda \in L} v(\mathbf{x}, \mathbf{y}_C, \mathbf{y}_T, \lambda, \xi, \theta, \eta)\, g_{\lambda_0}(\lambda) d\lambda \tag{2.2}$$

where L is the domain for the mean difference and

$$v(\mathbf{x}, \mathbf{y}_C, \mathbf{y}_T, \lambda, \xi, \theta, \eta) = \int_{\mu_0 \in M} \times \left\{ \int_{\mu_C \in M} f_C(\mathbf{y}_C; \mu_C, \theta) f_T(\mathbf{y}_T; \mu_C + \lambda, \theta)\, g(\mu_C; \mu_0, \xi)\, d\mu_C \right\} \times g_{\mu_0}(\mu_0) \prod_{h=1}^{H} f_0(\mathbf{x}_h; \mu_0, \eta)\, d\mu_0$$

M is the domain for the mean parameters. If all of the densities are normal, the quadratic form in the exponent of the joint density is

$$n_C(\overline{y}_C - \mu_C)^2 + n_T(\overline{y}_T - \lambda - \mu_C)^2 + b(\lambda - \Lambda)^2 + a(\mu_C - \mu_0)^2 + m(\overline{\overline{x}} - \mu_0)^2 + c(\mu_0 - \zeta)^2$$

where $m \propto \sum_h m_h$, Λ, and ζ are prior mean values for λ and μ_0; a, b, and c are constants related to the variance parameters of the prior distributions of μ_C, λ, and μ_0; and

$$\overline{\overline{x}} = \frac{\sum_h m_h \overline{x}_h}{\sum_h m_h}$$

Expanding and integrating first with respect to μ_0 and then with respect to μ_C leads to expressions for the posterior expectation and precision for λ:

$$E_{\text{post}}(\lambda) = \frac{\begin{array}{c} n_T\left\{(m+c+a)n_{\text{C}}\left(\overline{y}_{\text{T}}-\overline{y}_{\text{C}}\right)+am\left(\overline{y}_{\text{T}}-\overline{\overline{x}}\right)+ac\left(\overline{y}_{\text{T}}-\zeta\right)\right\} \\ +b\Lambda\left\{(m+c+a)\left(n_{\text{T}}+n_{\text{C}}\right)+a(m+c)\right\} \end{array}}{(m+c+a)\left(n_{\text{T}}n_{\text{C}}+b\left(n_{\text{T}}+n_{\text{C}}\right)\right)+a(m+c)\left(n_{\text{T}}+b\right)}$$

$$\text{prec}_{\text{post}}(\lambda) \propto (m+c+a)\left(n_{\text{T}}n_{\text{C}}+b\left(n_{\text{T}}+n_{\text{C}}\right)\right)+a(m+c)\left(n_{\text{T}}+b\right) \quad (2.3)$$

The influence of the historical data on the estimation of λ is controlled by the values of a and c. If the prior information is ignored, so that $a = c = 0$, then the posterior expectation of λ becomes

$$E_{\text{post}}(\lambda) = \frac{n_{\text{T}}n_{\text{C}}\left(\overline{y}_{\text{T}}-\overline{y}_{\text{C}}\right)+b\Lambda\left(n_{\text{T}}+n_{\text{C}}\right)}{n_{\text{T}}n_{\text{C}}+b\left(n_{\text{T}}+n_{\text{C}}\right)}$$

which does not depend on m. If only $c = 0$ (corresponding to the use of a flat prior for μ_0), then

$$E_{\text{post}}(\lambda) = \frac{\begin{array}{c} n_{\text{T}}\left\{(m+a)\,n_{\text{C}}\left(\overline{y}_{\text{T}}-\overline{y}_{\text{C}}\right)+am\left(\overline{y}_{\text{T}}-\overline{\overline{x}}\right)\right\} \\ +b\Lambda\left\{(m+a)\left(n_{\text{T}}+n_{\text{C}}\right)+am\right\} \end{array}}{(m+a)\left(n_{\text{T}}n_{\text{C}}+b\left(n_{\text{T}}+n_{\text{C}}\right)\right)+am\left(n_{\text{T}}+b\right)}$$

The values of a, b, and c may be very close to 0 when vague or flat priors are used for μ_{C}, μ_0, and λ. In that case, the posterior mean becomes $E_{\text{post}}(\lambda) = \overline{y}_{\text{T}} - \overline{y}_{\text{C}}$, which is just an estimate based only on the current study with no borrowed historical information.

The calculations can be generalized to allow for different true means ($\mu_{0h} \neq \mu_0$) in the various historical trials when there is more than one historical study ($H > 1$). Suppose that the μ_{0h} values are exchangeable, i.e., drawn from a parent distribution with parameters μ_0 and τ, $\mu_{0h} \sim w(\mu_{0h};\mu_0,\tau)$, $h = 1, \ldots, H$ (usually mean and standard deviation (SD)). Let $\boldsymbol{\mu}_0 = (\mu_{01}, \ldots, \mu_{0H})$, then the joint likelihood (2.1) can be written as

$$\begin{aligned} f\left(\mathbf{x},\ \mathbf{y}_{\text{C}},\ \mathbf{y}_{\text{T}}, \mu_{\text{C}},\ \boldsymbol{\mu}_0, \lambda, \xi, \theta, \eta, \tau\ \right) &= f_{\text{C}}\left(\mathbf{y}_{\text{C}};\ \mu_{\text{C}}, \theta\right) f_{\text{T}}\left(\mathbf{y}_{\text{T}};\ \mu_{\text{C}}+\lambda, \theta\right) \\ &\times g\left(\mu_{\text{C}};\ \boldsymbol{\mu}_0, \xi\right)\ \times \prod_{h=1}^{H} f_0\left(\mathbf{x}_h;\ \mu_{0h}, \eta\right) w\left(\mu_{0h};\ \mu_0,\ \tau\right) g_{\mu_0}\left(\mu_0\right) \\ &\times g_{\lambda_0}(\lambda) g_{\theta_0}(\theta) g_{\eta_0}(\eta) g_{\xi_0}(\xi) g_{\tau_0}(\tau) \end{aligned} \quad (2.4)$$

The remaining calculations are nearly the same as for the case when $\mu_{0h} = \mu_0$.

2.2.2 Power Priors

One could assume that the same expected values can be applied in the historical and current trials (i.e., $\mu_{\text{C}} = \mu_{0h} \equiv \mu_0$), so that the likelihood of the historical trial(s) could be used as a prior distribution for the corresponding parameter(s) of the current distribution. However, this may lead to the historical information inappropriately influencing the conclusions resulting from the analysis of the current trial. One way to adjust for this influence is to attenuate the contribution of the historical trials by raising the historical likelihood to a power between 0 and 1 and using this quantity as the prior distribution of the parameters of interest [9]. As before, suppose first that $H = 1$ (one historical trial) and that the likelihood for the historical trial is $f_0(\mathbf{x};\ \mu_0, \eta)$. A power prior for μ_C is

$$f_{\text{prior}}(\mu_C;\ \eta, \delta) \propto f_0(\mathbf{x};\ \mu_C, \eta)^{\delta}\, g_{\mu C}(\mu_C)\, g_{\eta 0}(\eta) \tag{2.5}$$

The power prior parameter $\delta \in [0, 1]$ provides a way to control the amount of the historical trial data contributed to the prior. Thus, $\delta = 0$ corresponds to no incorporation of historical data, while $\delta = 1$ is equivalent to taking the complete historical likelihood. A lower value of δ will incorporate less information from the historical trial.

The normalizing constant needed to make $f_0(\ \mathbf{x}; \mu_C, \eta)^{\delta}$ a proper density clearly depends on δ [10–12]. If the value of δ amounts to a constant that has been specified based on an external (to the current analysis) assessment of the consistency of the historical and current control groups, then no adjustment to the usual calculations for determining the posterior distribution of the model parameters is needed. However, the value of δ in principle could be regarded as a parameter in an overall model with its own prior distribution $\pi(\delta)$ so that (2.5) becomes (with incorporation of a required normalizing constant $C(\delta)$ [10–12])

$$f_{\text{prior}}(\mu_C;\ \eta,\ \delta) \propto C(\delta)\, f_0(\ \mathbf{x};\ \mu_C, \eta)^{\delta}\, g_{\mu C}(\mu_C)\, g_{\eta 0}(\eta)\, \pi(\delta) \tag{2.6}$$

The power prior approach differs from the commensurate prior model in that the uncertainty about the relationships between the control groups for the current and historical trials is accounted for by attenuating the contribution of the historical information to the prior distribution for the control group as expressed by the parameter δ instead of having the control group mean drawn from a distribution with the same expectation as the historical controls but with a parameter ξ that expresses consistency of the historical information with the observed control group outcome. It also differs in that the historical information no longer is considered data as such but instead is incorporated directly into the determination of the prior.

If $H > 1$, so that there are several historical trials, or groups of historical trials, the calculations can be generalized to allow for different degrees of attenuation of the historical information for the various trials or groups of trials, that is, for different values of the discounting factor, δ. Let $\boldsymbol{\delta}$ denote the vector of discounting factors, $\boldsymbol{\delta} = (\delta_1, \ldots, \delta_H)$, and $\mathbf{x} = (\mathbf{x}_1, \ldots, \mathbf{x}_H)$ be the collection of observed data from historical trials. The power prior for the control group mean μ_C can be written as an extension of (2.6),

$$f_{\text{prior}}(\mu_C, \boldsymbol{\delta};\ \mathbf{x}, \eta) = C(\boldsymbol{\delta})\, g_{\mu c}(\mu_C) \prod_{h=1}^{H} f_0^{\delta_h}(\mathbf{x}_h; \mu_C, \eta)\, \pi(\delta_h) \tag{2.7}$$

The product of the joint prior and the likelihood given the observed historical information is similar to (2.6) except that there are multiple historical datasets $\mathbf{x}_h$ and power parameters δ_h, $h = 1, \ldots, H$.

Most descriptions of the power prior approach in the literature assume binomial likelihoods, for which the analyses are fairly straightforward. However, the examples considered here use continuous data with adjustment for a covariate, for which the analyses are more complex. The following brief description outlines the calculations for normally distributed outcomes from a single historical trial or from a collection of historical trials when a common value of the power parameter δ applies. The goal is to obtain a posterior distribution for the treatment effect parameter(s) λ and for the precision (1/variance).

The first step is to specify the historical and current likelihoods and the prior distributions of the parameters. Thus, given δ, the likelihood for the historical data, which also is a prior distribution for the current data, is proportional to

$$\left(\frac{\eta}{2\pi}\right)^{\delta m/2} \exp\left\{-\frac{\delta\eta}{2}\left(\mathbf{X} - \mathbf{X}_0\boldsymbol{\beta}\right)'\left(\mathbf{X} - \mathbf{X}_0\boldsymbol{\beta}\right)\right\}$$

where $\mathbf{X}$ is a vector of length m and $\mathbf{X}_0$ is an $m \times p$ matrix (allowing for general linear models).

The likelihood for the current data is proportional to

$$\eta^{n/2} \exp\left\{-\frac{\eta}{2}\left(\mathbf{Y} - \begin{pmatrix}\mathbf{Y}_0 & \mathbf{Y}_1\end{pmatrix}\begin{pmatrix}\boldsymbol{\beta}\\ \lambda\end{pmatrix}\right)'\left(\mathbf{Y} - \begin{pmatrix}\mathbf{Y}_0 & \mathbf{Y}_1\end{pmatrix}\begin{pmatrix}\boldsymbol{\beta}\\ \lambda\end{pmatrix}\right)\right\}$$

where $\mathbf{Y}$ is a vector of length n, $\mathbf{Y}_0$ is an $n \times p$ matrix, and $\mathbf{Y}_1$ is an $n \times q$ matrix (general treatment effects). The parameter $\boldsymbol{\beta}$ is assumed to have a multivariate normal prior distribution with mean $\boldsymbol{\beta}_0$ and covariance matrix $\eta^{-1}\mathbf{V}_\beta$ so that the prior density of $\boldsymbol{\beta}$ is proportional to

$$\eta^{p/2} \exp\left\{-\frac{\eta}{2}\mathbf{z}'\mathbf{V}_\beta^{-1}\mathbf{z}\right\}$$

where $\mathbf{z} = \boldsymbol{\beta} - \boldsymbol{\beta}_0$. The parameter $\boldsymbol{\lambda}$ is assumed to have a multivariate normal prior distribution with mean $\boldsymbol{\lambda}_0$ and covariance matrix $\eta^{-1}\mathbf{V}_\lambda$ so that the prior density of $\boldsymbol{\lambda}$ is proportional to

$$\eta^{q/2} \exp\left\{-\frac{\eta}{2}\mathbf{w}'\mathbf{V}_\lambda^{-1}\mathbf{w}\right\}$$

where $\mathbf{w} = \lambda - \lambda_0$. Finally, the prior density of η is assumed to be a gamma density with parameters (a, b). The objective is an expression for the joint posterior density of $\boldsymbol{\lambda}$ and η; the parameter(s) $\boldsymbol{\beta}$ are nuisance parameters in this application. Details are provided in Appendix 2.2.

The posterior distribution of $\mathbf{w}$ is a (multivariate) normal distribution with mean

$$\tilde{\mathbf{w}} = \mathbf{Q}_w^{-1}\left(\mathbf{M}_{Y10}\mathbf{d}_{\beta 0Y} + \mathbf{M}_{Y11}\mathbf{d}_{\lambda 0Y} - \mathbf{M}_{Y01}\mathbf{Q}_z^{-1}\mathbf{A}\right) \tag{2.8}$$

and precision matrix $\eta\mathbf{Q}_{\mathrm{w}}$ with

$$\mathbf{Q}_w = \mathbf{V}_\lambda^{-1} + \mathbf{M}_{Y11} - \mathbf{M}_{Y10}\mathbf{Q}_z^{-1}\mathbf{M}_{Y01} \tag{2.9}$$

In these expressions,

$$\mathbf{M}_{Y10} = \mathbf{Y'}_1\mathbf{Y}_0,\ \mathbf{M}_{Y01} = \mathbf{M'}_{Y10},\ \mathbf{M}_{Y11} = \mathbf{Y'}_1\mathbf{Y}_1,\ \mathbf{d}_{\beta 0Y} = \hat{\beta}_{\mathbf{Y}} - \beta_0,$$

$$\mathbf{d}_{\lambda 0Y} = \hat{\lambda} - \lambda_0,$$

$$\mathbf{Q}_z = \mathbf{V}_\beta^{-1} + \delta\mathbf{M}_X + \mathbf{M}_{Y00},\ \mathbf{A} = \delta\mathbf{M}_X\mathbf{d}_{\beta 0X} + \mathbf{M}_{Y00}\mathbf{d}_{\beta 0Y} + \mathbf{M}_{Y01}\mathbf{d}_{\lambda 0Y},$$

$$\mathbf{M}_{Y00} = \mathbf{Y'}_0\mathbf{Y}_0,\ \mathbf{M}_X = \mathbf{X'}_0\mathbf{X}_0,\ \text{and}\ \mathbf{d}_{\beta 0X} = \hat{\beta}_{\mathbf{X}} - \beta_0$$

The normalization constants for these densities are functions of δ as they should be. Posterior confidence intervals/regions for λ and η given δ can be obtained analytically. Realizations from the marginal posterior densities for λ and η can be obtained easily by simulation and can be used to construct credible intervals/regions for λ and η. This calculation is illustrated in Section 2.3.

Remark 2: The calculations just described also apply when there are $H > 1$ historical trials. If a common value (or distribution) of δ is realistic, then no changes are required other than having the $\mathbf{X}$ vector and the $\mathbf{X}_0$ array consist of the stacked values for the individual studies, i.e., $\mathbf{X'} = (\mathbf{X'}_1, \ldots, \mathbf{X'}_H)$ and likewise for $\mathbf{X}_0$. When each historical trial has its own δ value or distribution, so that δ is replaced with $(\delta_1, \ldots, \delta_H)$, the calculations just described proceed as follows:

$$\delta\mathbf{M}_X \text{ is replaced with } \sum_{h=1}^{H} \delta_h\mathbf{M}_{Xh} \text{ where } \mathbf{M}_{Xh} = \mathbf{X'}_{0h}\mathbf{X}_{0h}$$

$$\delta\mathbf{M}_X\mathbf{d}_{\beta 0X} \text{ is replaced with } \sum_{h=1}^{H} \delta_h\mathbf{M}_{Xh}\mathbf{d}_{\beta 0Xh} \text{ where } \mathbf{d}_{\beta 0Xh} = \left(\hat{\boldsymbol{\beta}}_h - \boldsymbol{\beta}_0\right)$$

$$\delta S_{\mathbf{X}} \text{ is replaced with } \sum_{h=1}^{H} \delta_h S_{Xh} \text{ where } S_{Xh} = \left(\mathbf{X}_h - \mathbf{X}_{0h}\hat{\boldsymbol{\beta}}_h\right)'\left(\mathbf{X}_h - \mathbf{X}_{0h}\hat{\boldsymbol{\beta}}_h\right)$$

$$\delta\mathbf{d'}_{\beta 0X}\mathbf{M}_X\mathbf{d}_{\beta 0X} \text{ is replaced with } \sum_{h=1}^{H} \delta_h\mathbf{d'}_{\beta 0Xh}\mathbf{M}_{Xh}\mathbf{d}_{\beta 0Xh}$$

Remark 3: In the derivations, we have set the variance parameter η to be the same across the historical and current studies. Differing values of η across the historical trials and the current study can be accommodated by adjusting the δ_h values.

2.2.3 MAP Priors

The MAP approach is similar to the commensurate prior approach except that (1) the nuisance parameter η might differ among the historical trials

$(\eta_1, \ldots, \eta_H)$ and (2) the quantities μ_C and $\mu_{01}, \ldots, \mu_{0H}$ are exchangeable, i.e., realizations from a common distribution with expectation μ_0 and nuisance parameter(s) ξ. A commensurate prior assumes only that the historical trial parameters $(\mu_{01}, \mu_{02}, \ldots, \mu_{0H})$ are drawn from the same parent distribution. Specifically, the MAP assumes that historical data $x_h \sim N\left(\mu_{0h}, \eta_h^2\right), h = 1, \ldots, H$, and $\mu_{01}, \mu_{02}, \ldots, \mu_{0H}, \mu_C \sim N\left(\mu_0, \xi^2\right)$.

The fundamental assumption of the MAP approach is that the information from the historical trial control groups can be used to determine a predictive distribution for the value of μ_C, say $h\left(\mu_C; x_1, x_2, \ldots, x_H\right)$. This can be accomplished in various ways [2,13], in a frequentist setting [14] or in a fully Bayesian framework [15].

This chapter focuses on the Bayesian approach because it is easy to implement and interpret under the assumption of normal distributions with known standard errors. Thus, suppose that the control group likelihood for the hth historical trial is a normal distribution with mean μ_{0h} and known standard error η_h. Assume further that the historical and current trial control group means are realizations from a normal distribution with mean μ_0 and known SD ξ.

Integrating the product of the historical control group likelihoods and the prior densities of the historical control group means (i.e., $\mu_{01}, \mu_{02}, \ldots, \mu_{0H}$) with respect to these means yields the posterior density of μ_0, which is normal with

$$\begin{aligned} E\left(\mu_0 \middle| \overline{\overline{x}}\right) &= \overline{\overline{x}} = \sum w_h x_h / \sum w_h \text{ and } \mathrm{V}\left(\mu_0\right) \\ &= 1/\sum w_h \quad \text{where } w_h = \left(\eta_h^2 + \xi^2\right)^{-1} \end{aligned} \tag{2.10}$$

The product of the joint density of μ_C and the historical control means reduces to the product of a normal density with mean μ_0 and variance ξ^2 (for μ_C) and a normal density (for μ_0, conditional on $\overline{\overline{x}}$) with mean and variance given by (2.10). Integrating this product with respect to μ_0 gives the predictive density of μ_C given the historical control group data, which is normal with mean $\overline{\overline{x}}$ and variance $V(\mu_0) + \xi^2$.

A key assumption of this approach is that the between-trial variance ξ^2 is known. When it is unknown, it can be estimated, e.g., using a conventional random-effect meta-analysis [16–20]. However, the estimate will be imprecise unless there is a sufficient number of historical studies (large H), which rarely happens. When the number of historical studies is small, a fully Bayesian meta-analysis can be undertaken with different priors for ξ^2 [2, 21–25]. The MAP approach also assumes exchangeability among $\mu_{01}, \mu_{02}, \ldots, \mu_{0H}$ and μ_C. Full exchangeability requires the trials to be very similar, a judgment that generally is based on clinical and design considerations. This requirement can be loosened to partial exchangeability under certain circumstances [2].

Bayesian inference on the treatment effect using the MAP-type priors proceeds as implied by Equation (2.1) or (2.4), in which the likelihood of the historical data and the related prior settings are replaced by the MAP-type priors.

Application of the MAP approach to continuous data with covariate adjustments can be complex. The MAC (meta-analytic combined) approach, performing a meta-analysis with combined data from all historical control data and the current trial control data, may provide a practicable alternative because the analysis with an MAP prior is equivalent to an MAC approach [3]. For example, we can have the following setting for a regression model using the MAC approach:

$$\mathbf{x_h} \sim N\left(\mathbf{x_{0h}}\beta_h, \eta\right) \text{ for } h = 1, 2, \ldots, H$$
$$\mathbf{y_C} \sim N\left(\mathbf{y_{0C}}\beta_C, \theta\right)$$
$$\left\{\beta_h\right\}_{h \geq 1}, \beta_C \sim N\left(\beta_0, V_0\right)$$

where $\mathbf{x_h}$ and $\mathbf{x_{0h}}$ are observations from the historical trials and $\mathbf{y_C}$ and $\mathbf{y_{0C}}$ are observations from the current control. Some applications of this MAC approach are illustrated in Section 2.3.

2.2.4 Test-Then-Pool Approach

The test-then-pool approach is an empirical method for incorporating historical data [6]. Assume that there is just one historical trial and that normal likelihoods apply. Let μ_h and μ_c denote the means for the historical data and the current control, respectively. The test-then-pool approach first tests a null hypothesis $H_0\colon\ \mu_h = \mu_c$, e.g., with a t-test

$$T = \frac{\left|\hat{\mu}_h - \hat{\mu}_c\right|}{s_p\sqrt{\frac{1}{n_h} + \frac{1}{n_c}}},\ s_p = \sqrt{\frac{\left(n_h - 1\right)s_h^2 + \left(n_c - 1\right)s_c^2}{n_h + n_c - 2}}$$

where n, $\hat{\mu}$, and s^2 denote the sample size, sample mean, and sample variance, with subscripts h and c denoting the historical and current controls. If the test is *not significant* with a pre-specified significance level, e.g., 5% or 10%, that is, $T < t_{1-\alpha/2}\left(n_h + n_c - 2\right)$, then the consistency assumption is accepted and the historical data are pooled with the current control for further inference; otherwise, the historical data are discarded, and the analyses proceed using only the data from the current trial.

This testing approach may be a poor choice, however, because the aim essentially is to assess similarity rather than difference. A better approach is to frame the objective in terms of equivalence testing, that is, testing an equivalence hypothesis, $H_0\colon\ \left|\mu_h - \mu_c\right| > \Delta$ where Δ is a pre-specified margin for equivalence, which can be tested using the two one-sided hypothesis testing strategies familiar in bioequivalence literature [26]. This means testing two null hypotheses,

$$H_{0a}\colon\ \mu_h - \mu_c < -\Delta \text{ and } H_{0b}\colon\ \mu_h - \mu_c > \Delta$$

The null hypothesis of *non-equivalence* is rejected if the larger of the p values corresponding to these two tests falls below a critical value (e.g., 0.05). Rejecting non-equivalence means that the historical control data can be pooled with the current control data in carrying out the usual analyses. How should the value of Δ be specified? Typical choices in practice might be the SD s_h or perhaps $s_h/2$. Rejection of non-equivalence would mean that the difference between the current and historical control mean is less than a full or half-historical SD.

Extensions to multiple historical trials are straightforward. A simple approach is to perform a test for each historical trial against the current control, and pool all studies that meet the equivalence criterion with the current control.

2.3 Application Examples

2.3.1 Historical Placebo Control

A trial evaluating a new angiotensin converting enzyme inhibitor (ACEI) for treatment of congestive heart failure (CHF) used an active standard ACEI as a control instead of a placebo. The endpoint for the trial was percentage improvement from baseline in the duration of exercise on treadmill after 3 months of treatment. A total of 174 patients were randomized with 86 in the new ACEI group and 88 in the standard ACEI group. The observed mean (SD) of the log of percentage improvement were 0.269 (0.277) for the new ACEI and 0.215 (0.297) for the standard ACEI. This was, not surprisingly, a much smaller difference than would have been expected with an inactive control. However, how much of a difference might have occurred if the trial had included a placebo control arm, and would the difference have been large enough to reach statistical significance?

If a placebo group had been included in the trial, then the null hypothesis of no difference between the expected responses to the new ACEI and placebo might be tested using a t-statistic,

$$T = (\overline{x}_\mathrm{A} - \overline{x}_\mathrm{p}) \Big/ \sqrt{\frac{s_\mathrm{A}^2}{n_\mathrm{A}} + \frac{s_\mathrm{p}^2}{n_\mathrm{p}}} \tag{2.11}$$

where $\overline{x}_\mathrm{A}$, s_A^2, and n_A denote the sample mean, the sample variance, and the sample size for the new ACEI group, and $\overline{x}_\mathrm{p}$, s_p^2 and n_p denote the sample mean, the sample variance, and the sample size for the placebo group included in the trial. The null hypothesis would be rejected at the $100\alpha\%$ level if $T > t_{1-\alpha/2}\ (n_\mathrm{A} + n_\mathrm{p} - 1)$. But since no placebo group in fact was included, the best that can be done is to calculate the probability that the null hypothesis would have been rejected if one had been included. This means

TABLE 2.1
Exercise Tolerance Findings for Patients on Placebo from Eight Trials in CHF (w = watts, s = seconds)

Trial	Exercise Test type	Treatment Duration	n	Mean	SD
1	Treadmill (s)	12 weeks	42	0.048	0.39
2	Bike (w)	6 min	14	0.029	0.36
3	Treadmill (s)	10 weeks	13	0.21	0.30
4	Bike (s)	3 min	8	−0.01	0.39
5	Bike (s)	3 min	8	−0.01	0.52
6	Bike (w)	6 min	8	0.016	0.35
7	Bike (s)	6 min	37	0.071	0.28
8	Bike (s)	12 weeks	11	0.098	0.50

that $\overline{x}_{\text{p}}$ and s_{p}^2 are random variables instead of fixed observed quantities, and their distributions need to be determined. Previous trials of ACEI inhibitors in patients with CHF provide a basis for determining the distributions of $\overline{x}_{\text{p}}$ and s_{p}^2. Table 2.1 provides summary statistics of exercise tolerance findings for patients on placebo in eight prior trials.

Although the trials used different ways to measure exercise tolerance (bicycle or treadmill) and made the measurements after different durations of treatment (10 weeks to 6 months), the means and variances are reasonably consistent. Consequently, it is sensible to combine the findings using a Bayesian random effects model to get posterior distributions for the true mean μ_{p} and variance σ_{p}^2 of observations in the placebo group. Specifically, we assumed $\overline{x}_{\text{p}} \sim N\left(\mu_{\text{p}}, \frac{\sigma_{\text{p}}^2}{n_h}\right)$ and some flat prior for μ_{p} and σ_{p}^2. We can obtain the posterior distributions for μ_{p} and σ_{p}^2 by drawing samples from these posterior distributions to simulate placebo group outcomes. Then, we apply these samples in (2.11) to evaluate whether the null hypothesis would be rejected or not. With repeated samples for μ_{p} and σ_{p}^2, we can estimate the predictive probability of rejection.

The computations are entirely conventional and follow the model in Ref. [27]. The key finding is that the predictive probability of rejecting a null hypothesis of no difference from placebo turned out to be 0.78 for the standard ACEI and 0.92 for the new ACEI. The conclusion therefore is that the new ACEI is very likely different from placebo, and the standard ACEI is also quite likely different from placebo.

Remark 4: The predictive prior for this example was constructed from the historical trials without discounting. If there is reason to believe that the current trial differed materially from the historical trials, then the predictive prior could be discounted as described above.

2.3.2 Non-Inferiority Trials for a Vaccine Program

We consider two phase III double-blind randomized multicenter non-inferiority (NI) clinical trials evaluating concomitant use of Zostavax®, one with influenza virus vaccine [28] (Study I) and one with Pneumovax® (Study II). Subjects in both of these studies were randomly assigned to concomitant use (both vaccines together) or non-concomitant use (Zostavax® first, then the other vaccine about a month later).

A primary objective was to show that the antibody response to the test vaccine in the concomitant use group was non-inferior to that in the non-concomitant group. The statistical hypotheses are H_0: $\theta \leq 0.67$ and H_1: $\theta > 0.67$, where θ is the geometric mean titer (GMT) ratio between the test vaccine in concomitant and non-concomitant groups. The NI margin 0.67 corresponds to a no more than 1.5-fold decrease in the GMT between the groups.

The analysis that follows illustrates the application of the various historical control incorporation approaches using an analysis of covariance model that adjusts only for baseline (the original analyses used more elaborate models). Let y_{i0} and y_{i1} denote the observed log-titers for subject i at baseline and post-vaccination, respectively; the analysis model is

$$y_{i1} = \beta_0 + \beta_1 y_{i0} + \lambda z_i + \in_i, \quad \in_i \sim N\left(0, \sigma^2\right)$$

where $z_i = 1$ if the patient was in the concomitant use group or 0 if in the non-concomitant use group. All of the parameters are on the log-transformed titer scale so that the geometric mean titer ratio $\gamma = \exp(\lambda)$.

A placebo-controlled phase III efficacy trial was completed in the development program for Zostavax® before the two concomitant use studies were completed. The immunogenicity data from this phase III trial provided historical information for the two concomitant vaccine trials. Table 2.2 summarizes the log-titer of baseline and post-vaccination for Studies I and II as compared to the historical trial. The means and SDs are fairly similar between Study I and historical trial at both baseline and post-vaccination. The baseline mean for Study II is slightly lower than that of the historical trial, and the SDs for

TABLE 2.2
Summary of Log-Titer by Study and Treatment

		Control			Treatment		
Study		***N***	**Mean**	**SD**	***N***	**Mean**	**SD**
I	Baseline	380	5.6	1.19	375	5.6	1.11
	Post-vaccination	366	6.4	0.99	361	6.3	1.00
II	Baseline	234	5.0	0.93	235	5.3	0.99
	Post-vaccination	225	6.1	0.86	217	5.9	0.92
Historical	Baseline	655	5.6	1.04			
	Post-vaccination	667	6.2	0.96			

both baseline and post-vaccination are also slightly lower than those of the historical trial.

Conventional frequentist and Bayesian regression models (the latter using the rstanarm package [29]) were fit initially to the data from the current trials (Studies I and II). The results are displayed in the first section of Table 2.3. The results from analyses incorporating historical data using various approaches are discussed in the subsequent sections.

2.3.2.1 Test-Then-Pool Analysis

We use the equivalence test $H_0\colon\ |\mu_C - \mu_H| > \Delta$, where Δ is an equivalence margin. From the historical trial, the SD for log-titer at baseline and post-vaccination was about 1.0, so we take half of this SD as equivalence margin, that is, $\Delta = 0.5$. The equivalence testing results for each study are summarized in Table 2.4.

The equivalence test results imply that the historical trial outcomes are equivalent to the current trial outcomes for Study I, so that the historical control data can be pooled with the current trial control data. However, this is not true for Study II, so that the control group information from the historical trial cannot be pooled with the control group information for Study II. The results of the analyses are displayed in Table 2.5. There is strong evidence for NI on the basis of the Study I results, with the evidence being stronger (i.e., the lower credible interval has higher value) when the historical data are included. However, NI cannot be concluded for

TABLE 2.3
Summary of Analysis Results for Studies I and II

Method		Study I		Study II	
		GMTR[a]	LB_{025}	GMTR[a]	LB_{025}
Regression		0.93	0.84	0.70	0.61
Bayes regression		0.94	0.83	0.70	0.60
Commensurate prior	Flat prior for ξ, κ	0.94	0.84	0.70	0.60
	$\xi = 0.14$, $\kappa = 0.025$	0.95	0.85	0.71	0.61
	$\xi = 0.03$, $\kappa = 0.005$	1.04	0.95	0.82	0.72
Power prior	Fix $\delta = 0.20$	1.01	0.91	0.79	0.69
	Random δ without normalization factor	0.94	0.84	0.70	0.61
MAC	Flat prior for ξ, κ	0.94	0.85	0.7	0.61
	$\xi = 0.344$, $\kappa = 0.035$	0.94	0.84	0.7	0.61
	$\xi = 0.03$, $\kappa = 0.004$	0.99	0.89	0.77	0.69

[a]GMTR = Geometric mean titer ratio.
LB_{025} = 2.5% lower bound.

TABLE 2.4
Equivalence Test Conclusion between Current Control and Historical Data at Baseline and Post-Vaccination

Study	Endpoint	Equivalence test ($\Delta = 0.5$)
I	Baseline	Equivalent
	Post-vaccination	Equivalent
II	Baseline	Not equivalent
	Post-vaccination	Equivalent

TABLE 2.5
NI Analysis with Test-Then-Pool Approach

	Analysis	GMTR[a]	LB_{025}	p-value for NI
Study I	Current	0.93	0.84	<0.001
	Pooled	1.12	1.03	<0.001
Study II	Current	0.70	0.61	0.26

[a]GMTR = Geometric mean titer ratio.
LB_{025} = 2.5% lower bound.

Study II because the lower bound of 95% CI of 0.61 is lower than the NI margin of 0.67. The historical trial data cannot be pooled with the control group for this study because the equivalence is not concluded for baseline values.

2.3.2.2 Commensurate Prior

The prior densities of the parameters from the current trial (β_{0C}, β_{1C}) are assumed to be normally distributed with respective means β_{0H} and β_{1H} corresponding to the historical trial and respective variances ξ^2 and κ^2,

$$\beta_{0C} \sim N\left(\beta_{0H},\ \xi^2\right),\ \beta_{1C} \sim N\left(\beta_{1H},\ \kappa^2\right)$$

The posterior distributions of the historical control group parameters are obtained via MCMC assuming vague priors (Cauchy (0, 2.5) for the intercept and slope parameters, and half-Cauchy priors for the variances, ξ^2 and κ^2).

The analysis results using the commensurate prior approach with these flat priors were similar to the results using only the current data from Studies I and II (see the second section in Table 2.3). This implies that vague priors lead to little information being borrowed from the historical data, even though it is clear from Table 2.2 that the historical control results were very similar to the control results from Study I.

What happens when informative priors for ξ and κ are used? One possibility is to carry out a regression analysis of the control group from the historical trial, obtain values of the standard errors of the estimated values of β_{0H} and β_{1H}, and then assume that ξ and κ are proportional to these standard errors.

For this example, the regression analysis for the historical trial resulted in $\text{se}\left(\hat{\beta}_{0H}\right) = 0.14$ and $\text{se}\left(\hat{\beta}_{1H}\right) = 0.025$. When ξ and κ were set to these values, borrowing from the historical trial was minimal because the findings from applying the commensurate prior approach remained similar to the results based only on the current data for both studies (see the second section in Table 2.3). More borrowing from the historical control groups was observed when ξ and κ were set to about a fifth of these values ($\xi = 0.03$, $\kappa = 0.005$), and in fact the lower bound of the 95% CI for the GMTR exceeded 0.67 for Study II, implying that NI would be concluded.

2.3.2.3 Power Prior

As a first step, the calculations were carried out using the STAN package for the MCMC calculations [30, 31] with fixed values of δ, to see how little control information might be needed to bolster the findings from the current trial. We use vague priors for β_{0C}, β_{1C}, and σ_C^2, specifically, $\beta_{0C}, \beta_{1C} \sim \text{Cauchy}(0, 2.5)$ and $\sigma_C^2 \sim$ half Cauchy$(0, 2.5)$. Table 2.6 summarizes the findings. Increasing the contribution of the historical data to the evaluation of the control group outcomes caused the log likelihoods to decrease fairly substantially for both studies primarily because incorporating the historical data forces the estimates of the model parameters away from the values that would be obtained by maximizing the likelihoods. The results for $\delta = 0.2$ are also presented in the third section of Table 2.3.

It is clear from Table 2.6A that no historical data would be needed to provide a lower credible bound exceeding 0.70 for the GMTR. On the other hand, it is clear from Table 2.6B that this certainly is not true for the findings from Study II, and support from the historical trial is needed to have the lower credible bound for the GMTR above the NI margin of 0.67. Figure 2.1 provides a graphical summary of the dependence of the credible bounds for the GMTR on the value of δ.

Table 2.6 provides insight into the effects of varying the contribution of the historical information but at the cost of requiring that a value of δ be specified. This may be difficult if not unreasonable for practitioners to do in practice. Another option is to treat δ as random and assume a flat prior $\pi(\delta)$. If an appropriate normalization factor is not included, the result is about the same as when no historical data are borrowed (Table 2.3). Determining an appropriate normalization constant can be computationally challenging. Even with the normalization constant, the random power parameter may not reflect the difference or similarity between historical and current control data, which may lead to poor performance for the power prior method [7,8,32]. We consider here a more practical approach that adds to the model specification a prior distribution for δ based on the external data. This does not require complex integration because the posterior densities (2.8) and (2.9) are proper densities whose normalization constants depend on δ. Duan et al. describe a similar approach [10,11].

TABLE 2.6
Results of Power Prior Calculations for Fixed Values of δ

A. Study I

	GMTR				$1/\sqrt{\tau}$		
δ	Mean	LB_{025}	β_{0C}	β_{1C}	$=\sigma_C$	λ	LogLik
0	0.94	0.84	3.07	0.60	0.72	−0.07	−786.1
0.1	0.98	0.88	3.01	0.60	0.72	−0.02	−786.7
0.2	1.01	0.91	2.96	0.60	0.71	0.01	−788.0
0.3	1.04	0.94	2.92	0.61	0.71	0.03	−789.7
0.5	1.07	0.98	2.86	0.61	0.71	0.07	−793.0
0.8	1.11	1.02	2.79	0.62	0.70	0.10	−797.2

B. Study II

	GMTR				$1/\sqrt{\tau}$		
δ	Mean	LB_{025}	β_{0C}	β_{1C}	$=\sigma_C$	λ	LogLik
0	0.70	0.61	3.38	0.55	0.72	−0.36	−482.5
0.1	0.75	0.66	3.34	0.54	0.72	−0.29	−483.4
0.2	0.79	0.69	3.28	0.54	0.71	−0.24	−485.1
0.3	0.81	0.72	3.22	0.55	0.71	−0.21	−487.1
0.5	0.84	0.75	3.11	0.56	0.70	−0.17	−490.7
0.8	0.88	0.79	3.00	0.58	0.70	−0.13	−495.3

It is clear from Figure 2.1 that the lower bound of the GMTR based on the Study I outcome would exceed 0.67 regardless of the value of δ. This is not true for Study II. While any value of δ greater than about 0.17 may lead to a lower GMTR bound exceeding 0.67, specifying particular δ values in this way may generate unnecessary debate. It may be better (potentially less controversial) to specify a distribution of potentially acceptable δ values and see if the marginal lower GMTR bound exceeds 0.67. This is what was done to produce Table 2.7. It is clear that the lower 2.5% marginal posterior quantile for GMTR will exceed 0.67 if δ is distributed between 0.05 and 0.35 and centered at 0.17. It is easy to explore the effect of other assumptions about the distribution of δ.

2.3.2.4 MAP Prior

Assume $\{\beta_{0h}\}_{h\geq1}$, $\beta_{0C} \sim N(\beta_0, \xi^2)$ and $\{\beta_{1h}\}_{h\geq1}$, $\beta_{1C} \sim N(\beta_1, \kappa^2)$. Specific values of the hyper-parameters of ξ and κ may be based on previous knowledge of the variation or assuming a flat hyper-prior and estimated from the data. We may use vague priors for all other parameters, e.g., $\beta_0, \beta_1, \lambda \sim$ Cauchy $(0, 2.5)$ and $\sigma_H^2, \sigma_C^2 \sim$ half Cauchy $(0, 2.5)$.

To illustrate the MAP method through MAC, we used the historical study and Study II as two historical trials to analyze the findings from Study I (current trial) and used historical study and Study I as two historical trials to

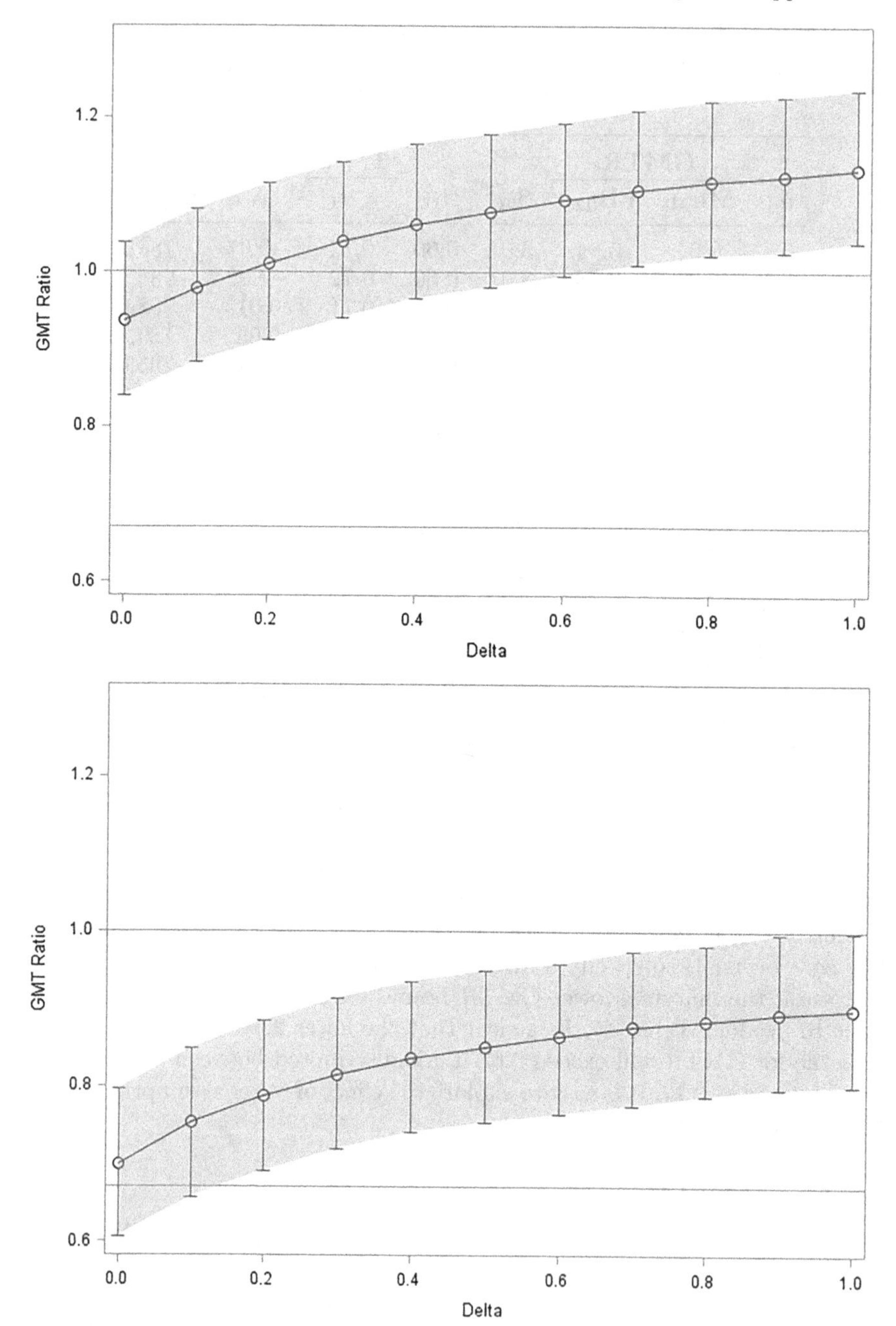

FIGURE 2.1
Posterior mean and 95% CI for GMT ratio as functions of δ.

TABLE 2.7
Posterior Quantiles of GMTR for Study II Given the Historical and Current Data as a Function of the Distribution of Values of the Power Parameter δ Used to Weight the Contribution of the Historical Data

						GMTR	
a	b	$E(\delta)$	δ_{50}	δ_{025}	δ_{975}	Mean	LB_{025}
1	4	0.2	0.16	0.01	0.60	0.776	0.654
2	8	0.2	0.18	0.03	0.48	0.779	0.664
3	12	0.2	0.19	0.05	0.43	0.781	0.676
4	16	0.2	0.19	0.06	0.40	0.781	0.677
5	20	0.2	0.19	0.07	0.37	0.781	0.679
6	24	0.2	0.19	0.08	0.36	0.783	0.683
1	5	0.167	0.13	0.01	0.52	0.766	0.650
2	10	0.167	0.15	0.02	0.41	0.765	0.659
3	15	0.167	0.15	0.04	0.36	0.768	0.666
4	20	0.167	0.16	0.05	0.34	0.772	0.669
5	25	0.167	0.16	0.06	0.32	0.768	0.669
6	30	0.167	0.16	0.07	0.30	0.773	0.674

Note: δ is assumed to have a beta $(a,\ b)$ distribution.

analyze the findings from Study II. The MAC analysis results are presented in the fourth section of Table 2.3. The results are similar to those from the regression model for both Study I and Study II when flat priors are used for ξ^2 and κ^2. This implies little information borrowing from the two historical trials for this MAC analysis.

The estimated SD values for the random intercept and slope were 0.344 and 0.035 from the meta-analysis model. The analysis results from the MAC approach were similar to the regression model results when $\xi = 0.344$ and $\kappa = 0.035$, so little information borrowed from the historical trial. However, the findings changed when the values of ξ^2 and κ^2 were reduced to about one-tenth of the observed values, i.e., $\xi = 0.03$ and $\kappa = 0.004$. The lower bound of the 95% CI for GMTR based on the Study II findings turned out to exceed 0.67, so that NI would be concluded.

2.4 Discussion

Bayesian methods provide a natural way to incorporating information from historical study control groups to improve the precision of estimation and sensitivity for detecting a treatment effect based on data from a current trial. However, these methods have had limited applications to the analysis of findings from actual clinical trials. The key problems are the determination of informative priors on the basis of historical data and, more importantly, determining how much information should be borrowed from historical data.

This chapter briefly reviews some currently available Bayesian approaches including test-then-pool, commensurate prior, power prior, and MAP prior. These methods are applied to real clinical trial examples to illustrate how historical data can be incorporated and the effects of different strategies for determining how much influence should be allowed for historical data.

It is attractive to let the data to determine how much information may be borrowed from historical studies, but the choice of parameters for prior distributions may not be a simple matter. Little information will be borrowed from historical studies when flat or non-informative priors are used. Informative priors for the parameters discounting the impact of historical data seem to be required in order to borrow any information at all. In effect, this means that little if any borrowing will be justified unless the historical and current control group findings are very similar.

A consequence of the analyses presented here is an emphasis on the criticality of the assumption that the historical data are "similar" to the current data in order for the historical data to have any effect on the final conclusions. Apparent similarity of studies, e.g., by virtue of their similarity of patient population, study design and procedures, and endpoints, cannot guarantee that they necessarily will have "similar" outcomes. Borrowing information from historical trials can inflate or decrease the Type I error rate when the control group outcomes from historical trials and a current trial differ [6].

A conventional hierarchical meta-analysis approach may be used to pool the data from the historical trials to form a prior for the control group of the current trial. This may constitute a "dogmatic" choice of prior, however, and could lead to the historical data unduly influencing the inference about the treatment effect [32]. When applying the power prior or commensurate prior or MAP/MAC prior described in this chapter, robust priors can help to down-weight the historical information according to its consistency with the current observations [33–35]. For the commensurate prior approach, a robust prior such as a Cauchy distribution instead of a Normal distribution could be assigned to the mean of the control group. Or, a robust component could be added to existing priors. When applying the MAP priors to binary observations, robust priors could be a weighted sum of an approximation of the MAP prior (through a mixture of conjugate priors) [3,21,36] and a robust component that accounts for prior-data conflict [3]. Gamalo-Siebers et al. described recently a Bayesian approach using an informative prior for the regression parameters that was constructed directly from the regression parameters estimated from the historical study [37].

When designing a trial, it is useful to express the prior information formally as a number of subjects, even though translating the historical information to a prior's effective sample size may not be straightforward. Most of the time, the derivation of the prior's effective sample size can be simplified if the normality assumptions about the historical information and the related priors are at least approximately true. For example, in the context of using a MAP prior, the prior's effective sample size can be expressed as the total sample size of history trials multiplied by the ratio of the predictive variance for no discounting historical data and the predictive variance for a between-trial variance [3].

Finally, a few possible designs have been discussed in literature for using historical trials [2,3]. One possible design for accommodating the potential inconsistency between the historical trials and the new trial is a two-stage design. This design gains some information about the control group of the new trial at the first stage and adapts the sample size at the second stage by considering the data-prior conflict information gained at the first stage.

Appendix 2.1: STAN Code for Calculations

Cmod.ComPr.stan COMMENSURATE PRIOR MODEL

```
data
{
  int ncasesC;
  int ncasesH;
  vector[ncasesC] Trt;
  vector[ncasesC] Y;
  vector[ncasesC] Y0;
  vector[ncasesH] X;
  vector[ncasesH] X0;
}
parameters
{
  vector[2] bH;
  vector[2] bC;
  vector[2] eta;
  real lambda;
  real<lower=0> sigmaC;
  real<lower=0> sigmaH;
}
transformed parameters
{
  vector[ncasesH]  EX = bH[1] + bH[2]*X0;
  vector[ncasesC]  EY = bC[1] + bC[2]*Y0 + lambda*Trt;
}
model
{
  sigmaC ~ cauchy(0,2.5);
```

```
  sigmaH ~ cauchy(0,2.5);
  eta ~ cauchy(0,2.5);
  bH ~ cauchy(0,2.5);
  lambda ~ cauchy(0,2.5);
  bC[1] ~ normal(bH[1],eta[1]);
  bC[2] ~ normal(bH[2],eta[2]);
  X ~ normal(EX,sigmaH);
  Y ~ normal(EY,sigmaC);
}
generated quantities
{
  real              GMTR = exp(lambda);
  real              LogLik;
  vector[ncasesC] LL;
  for (i in 1:ncasesC)
    LL[i] = normal_lpdf(Y[i]| EY[i], sigmaC);
  LogLik = sum(LL);
}
```

Appendix 2.2: Exact Posterior Distributions with Power Priors When the Data and Parameters Have Normal Distributions

Likelihood for historical data (also prior for current data – δ is a constant here):

$$\propto \left(\frac{\eta}{2\pi}\right)^{\delta m/2} \exp\left\{-\frac{\delta\eta}{2}(\mathbf{X}-\mathbf{X}_0\boldsymbol{\beta})'(\mathbf{X}-\mathbf{X}_0\boldsymbol{\beta})\right\} \quad \mathbf{X} \text{ is vector of length } m;$$

$\mathbf{X}_0$ is $m \times p$

Likelihood for current data:

$$\propto \eta^{n/2} \exp\left\{-\frac{\eta}{2}\left(\mathbf{Y}-(\mathbf{Y}_0 \quad \mathbf{Y}_1)\begin{pmatrix}\boldsymbol{\beta}\\ \boldsymbol{\lambda}\end{pmatrix}\right)'\left(\mathbf{Y}-(\mathbf{Y}_0 \quad \mathbf{Y}_1)\begin{pmatrix}\boldsymbol{\beta}\\ \boldsymbol{\lambda}\end{pmatrix}\right)\right\}$$

$\mathbf{Y}$ is a vector of length n;
$\mathbf{Y}_0$ is $n \times p$, $\mathbf{Y}_1$ is $n \times q$

$$\textbf{Prior for } \boldsymbol{\beta}: \propto \eta^{p/2} \exp\left\{-\frac{\eta}{2}\mathbf{z}'\mathbf{V}_\beta^{-1}\mathbf{z}\right\} \quad \mathbf{z} = \boldsymbol{\beta} - \boldsymbol{\beta}_0$$

$$\textbf{Prior for } \boldsymbol{\lambda}: \propto \eta^{q/2} \exp\left\{-\frac{\eta}{2}\mathbf{w}'\mathbf{V}_\lambda^{-1}\mathbf{w}\right\} \quad \mathbf{w} = \boldsymbol{\lambda} - \boldsymbol{\lambda}_0$$

Prior for η: $\propto \eta^{\frac{a}{2}-1} \exp\left(-\frac{b\eta}{2}\right)$

Product of likelihood and priors:

$$\propto (2\pi)^{-\delta m/2} \eta^{\frac{a+p+q+\delta m+n}{2}-1} \exp\left\{-\frac{\eta}{2} Z\right\}$$

where

$$Z = b + \mathbf{z}'\mathbf{V}_{\beta}^{-1}\mathbf{z} + \mathbf{w}'\mathbf{V}_{\lambda}^{-1}\mathbf{w} + \delta(\mathbf{X} - \mathbf{X}_0\beta)'(\mathbf{X} - \mathbf{X}_0\beta)$$
$$+ \left(\mathbf{Y} - (\mathbf{Y}_0 \quad \mathbf{Y}_1)\begin{pmatrix}\beta\\ \lambda\end{pmatrix}\right)' \left(\mathbf{Y} - (\mathbf{Y}_0 \quad \mathbf{Y}_1)\begin{pmatrix}\beta\\ \lambda\end{pmatrix}\right)$$

$$\beta = \hat{\beta}_{\mathrm{X}} - \left(\hat{\beta}_{\mathbf{X}} - \beta_0\right) + \mathbf{z} = \hat{\beta}_X - \mathbf{d}_{\beta 0X} + \mathbf{z}$$

$$\mathbf{X} - \mathbf{X}_0\beta = \mathbf{X} - \mathbf{X}_0\hat{\beta}_{\mathbf{X}} + \mathbf{X}_0\mathbf{d}_{\beta 0X} - \mathbf{X}_0\mathbf{z} \ \mathbf{X}'_0\left(\mathbf{X} - \mathbf{X}_0\hat{\beta}_{\mathbf{X}}\right) = 0 \, \mathbf{M}_X = \mathbf{X}'_0\mathbf{X}_0$$

So $(\mathbf{X} - \mathbf{X}_0\beta)'(\mathbf{X} - \mathbf{X}_0\beta) = \mathrm{S}_{\mathbf{X}} + \mathbf{d}'_{\beta 0X}\mathbf{M}_X\mathbf{d}_{\beta 0X} + \mathbf{z}'\mathbf{M}_X\mathbf{z} - 2\mathbf{d}'_{\beta 0X}\mathbf{M}_X\mathbf{z}$
where $\mathrm{S}_{\mathbf{X}} = \left(\mathbf{X} - \mathbf{X}_0\hat{\beta}\right)'\left(\mathbf{X} - \mathbf{X}_0\hat{\beta}\right)$

$$\begin{pmatrix}\beta\\ \lambda\end{pmatrix} = \begin{pmatrix}\hat{\beta}_{\mathbf{Y}}\\ \hat{\lambda}\end{pmatrix} - \begin{pmatrix}\hat{\beta}_{\mathbf{Y}} - \beta_0\\ \hat{\lambda} - \lambda_0\end{pmatrix} + \begin{pmatrix}\mathbf{z}\\ \mathbf{w}\end{pmatrix}$$
$$= \begin{pmatrix}\hat{\beta}_{\mathbf{Y}}\\ \hat{\lambda}\end{pmatrix} - \begin{pmatrix}\mathbf{d}_{\beta 0Y}\\ \mathbf{d}_{\lambda 0Y}\end{pmatrix} + \begin{pmatrix}\mathbf{z}\\ \mathbf{w}\end{pmatrix}$$

$$\mathbf{Y} - (\mathbf{Y}_0 \quad \mathbf{Y}_1)\begin{pmatrix}\beta\\ \lambda\end{pmatrix} = \left[\mathbf{Y} - (\mathbf{Y}_0 \quad \mathbf{Y}_1)\begin{pmatrix}\hat{\beta}_{\mathbf{Y}}\\ \hat{\lambda}\end{pmatrix}\right]$$
$$+ (\mathbf{Y}_0 \quad \mathbf{Y}_1)\begin{pmatrix}\mathbf{d}_{\beta 0Y}\\ \mathbf{d}_{\lambda 0Y}\end{pmatrix} - (\mathbf{Y}_0 \quad \mathbf{Y}_1)\begin{pmatrix}\mathbf{z}\\ \mathbf{w}\end{pmatrix}$$

$$(\mathbf{Y}_0 \quad \mathbf{Y}_1)'\left[\mathbf{Y} - (\mathbf{Y}_0 \quad \mathbf{Y}_1)\begin{pmatrix}\hat{\beta}_{\mathbf{Y}}\\ \hat{\lambda}\end{pmatrix}\right] = 0 \text{ so if } \mathbf{M}_{\mathrm{Y}}$$

$$= (\mathbf{Y}_0 \quad \mathbf{Y}_1)'(\mathbf{Y}_0 \quad \mathbf{Y}_1) = \begin{bmatrix}\mathbf{M}_{Y00} & \mathbf{M}_{Y01}\\ \mathbf{M}_{Y10} & \mathbf{M}_{Y11}\end{bmatrix}$$

and $S_Y = \left[\mathbf{Y} - (\mathbf{Y}_0 \quad \mathbf{Y}_1)\begin{pmatrix}\hat{\beta}_{\mathbf{Y}}\\ \hat{\lambda}\end{pmatrix}\right]'\left[\mathbf{Y} - (\mathbf{Y}_0 \quad \mathbf{Y}_1)\begin{pmatrix}\hat{\beta}_{\mathbf{Y}}\\ \hat{\lambda}\end{pmatrix}\right]$, then

$$\left[\mathbf{Y} - (\mathbf{Y}_0 \quad \mathbf{Y}_1)\begin{pmatrix}\beta\\ \lambda\end{pmatrix}\right]'\left[\mathbf{Y} - (\mathbf{Y}_0 \quad \mathbf{Y}_1)\begin{pmatrix}\beta\\ \lambda\end{pmatrix}\right] = S_Y + \begin{pmatrix}\mathbf{d}_{\beta 0Y}\\ \mathbf{d}_{\lambda 0Y}\end{pmatrix}'$$
$$\times \mathbf{M}_Y\begin{pmatrix}\mathbf{d}_{\beta 0Y}\\ \mathbf{d}_{\lambda 0Y}\end{pmatrix} + \begin{pmatrix}\mathbf{z}\\ \mathbf{w}\end{pmatrix}'\mathbf{M}_Y\begin{pmatrix}\mathbf{z}\\ \mathbf{w}\end{pmatrix} - 2\begin{pmatrix}\mathbf{d}_{\beta 0Y}\\ \mathbf{d}_{\lambda 0Y}\end{pmatrix}'\mathbf{M}_Y\begin{pmatrix}\mathbf{z}\\ \mathbf{w}\end{pmatrix}$$

So

$$
\begin{aligned}
Z =& b + \delta S_{\mathbf{X}} + S_Y + \mathbf{z}'\mathbf{V}_\beta^{-1}\mathbf{z} + \mathbf{w}'\mathbf{V}_\lambda^{-1}\mathbf{w} \ + \ \delta \mathbf{d}'_{\beta 0X}\mathbf{M}_X\mathbf{d}_{\beta 0X} \\
&+ \delta \mathbf{z}'\mathbf{M}_X\mathbf{z} - 2\delta \mathbf{d}'_{\beta 0X}\mathbf{M}_X\mathbf{z} + \begin{pmatrix}\mathbf{d}_{\beta 0Y}\\ \mathbf{d}_{\lambda 0Y}\end{pmatrix}' \mathbf{M}_Y \begin{pmatrix}\mathbf{d}_{\beta 0Y}\\ \mathbf{d}_{\lambda 0Y}\end{pmatrix} \\
&+ \begin{pmatrix}\mathbf{z}\\ \mathbf{w}\end{pmatrix}' \mathbf{M}_Y \begin{pmatrix}\mathbf{z}\\ \mathbf{w}\end{pmatrix} - 2\begin{pmatrix}\mathbf{d}_{\beta 0Y}\\ \mathbf{d}_{\lambda 0Y}\end{pmatrix}' \mathbf{M}_Y \begin{pmatrix}\mathbf{z}\\ \mathbf{w}\end{pmatrix} \\
=& b + \delta S_{\mathbf{X}} + S_Y + \mathbf{w}'\left(\mathbf{V}_\lambda^{-1} + \mathbf{M}_{Y11}\right)\mathbf{w} \\
&- 2\left(\mathbf{d}'_{\beta 0Y}\mathbf{M}_{Y01} + \mathbf{d}'_{\lambda 0Y}\mathbf{M}_{Y11}\right)\mathbf{w} + \mathbf{z}'\left(\mathbf{V}_\beta^{-1} + \delta\mathbf{M}_X + \mathbf{M}_{Y00}\right)\mathbf{z} \\
&+ \delta \mathbf{d}'_{\beta 0X}\mathbf{M}_X\mathbf{d}_{\beta 0X} + \begin{pmatrix}\mathbf{d}_{\beta 0Y}\\ \mathbf{d}_{\lambda 0Y}\end{pmatrix}' \mathbf{M}_Y \begin{pmatrix}\mathbf{d}_{\beta 0Y}\\ \mathbf{d}_{\lambda 0Y}\end{pmatrix} \\
&- 2\left(\delta \mathbf{d}'_{\beta 0X}\mathbf{M}_X + \mathbf{d}'_{\beta 0Y}\mathbf{M}_{Y00} + \mathbf{d}'_{\lambda 0Y}\mathbf{M}_{Y10} - \mathbf{w}'\mathbf{M}_{Y10}\right)\mathbf{z}
\end{aligned}
$$

First, complete the square on $\mathbf{z}$ which, in this context, is a nuisance parameter. Let

$$\mathbf{Q}_z = \mathbf{V}_\beta^{-1} + \delta\mathbf{M}_X + \mathbf{M}_{Y00} \text{ and } \tilde{\mathbf{z}} = \mathbf{Q}_z^{-1}\left(\mathbf{A} - \mathbf{M}_{Y01}\mathbf{w}\right)$$

where $\mathbf{A} = \delta\mathbf{M}_X\mathbf{d}_{\beta 0X} + \mathbf{M}_{Y00}\mathbf{d}_{\beta 0Y} + \mathbf{M}_{Y01}\mathbf{d}_{\lambda 0Y}$

Then the posterior density of $\mathbf{z}$ is multivariate normal with mean $\tilde{\mathbf{z}}$ and precision matrix $\tau\mathbf{Q}_z$

After integrating with respect to $\mathbf{z}$, the quadratic form reduces to

$$
\begin{aligned}
Z =& b + \delta S_{\mathbf{X}} + S_Y + \mathbf{w}'\left(\mathbf{V}_\lambda^{-1} + \mathbf{M}_{Y11}\right)\mathbf{w} - 2\left(\mathbf{d}'_{\beta 0Y}\mathbf{M}_{Y01} + \mathbf{d}'_{\lambda 0Y}\mathbf{M}_{Y11}\right)\mathbf{w} \\
&+ \delta \mathbf{d}'_{\beta 0X}\mathbf{M}_{\mathbf{X}}\mathbf{d}_{\beta 0X} + \begin{pmatrix}\mathbf{d}_{\beta 0Y}\\ \mathbf{d}_{\lambda 0Y}\end{pmatrix}' \mathbf{M}_Y \begin{pmatrix}\mathbf{d}_{\beta 0Y}\\ \mathbf{d}_{\lambda 0Y}\end{pmatrix} - \tilde{\mathbf{z}}'\mathbf{Q}_z\tilde{\mathbf{z}} \\
=& b + \delta S_{\mathbf{X}} + S_Y + \mathbf{w}'\left(\mathbf{V}_\lambda^{-1} + \mathbf{M}_{Y11}\right)\mathbf{w} - 2\left(\mathbf{d}'_{\beta oY}\mathbf{M}_{Y01} + \mathbf{d}'_{\lambda oY}\mathbf{M}_{Y11}\right)\mathbf{w} \\
&+ \delta \mathbf{d}'_{\beta oX}\mathbf{M}_X\mathbf{d}_{\beta oX} + \begin{pmatrix}\mathbf{d}_{\beta 0Y}\\ \mathbf{d}_{\lambda 0Y}\end{pmatrix}' \mathbf{M}_Y \begin{pmatrix}\mathbf{d}_{\beta 0Y}\\ \mathbf{d}_{\lambda 0Y}\end{pmatrix} - \mathbf{A}'\mathbf{Q}_z^{-1}\mathbf{A} \\
&+ 2\mathbf{A}'\mathbf{Q}_z^{-1}\mathbf{M}_{Y01}\mathbf{w} - \mathbf{w}'\mathbf{M}_{Y10}\mathbf{Q}_z^{-1}\mathbf{M}_{Y01}\mathbf{w} \\
=& b + \delta S_{\mathbf{X}} + S_Y + \mathbf{w}'\left(\mathbf{V}_\lambda^{-1} + \mathbf{M}_{Y11} - \mathbf{M}_{Y10}\mathbf{Q}_z^{-1}\mathbf{M}_{Y01}\right)\mathbf{w} \\
&- 2\left(\mathbf{d}'_{\beta 0Y}\mathbf{M}_{Y01} + \mathbf{d}'_{\lambda 0Y}\mathbf{M}_{Y11} - \mathbf{A}'\mathbf{Q}_z^{-1}\mathbf{M}_{Y01}\right)\mathbf{w} \\
&+ \delta \mathbf{d}'_{\beta 0X}\mathbf{M}_X\mathbf{d}_{\beta 0X} + \begin{pmatrix}\mathbf{d}_{\beta 0Y}\\ \mathbf{d}_{\lambda 0Y}\end{pmatrix}' \mathbf{M}_Y \begin{pmatrix}\mathbf{d}_{\beta 0Y}\\ \mathbf{d}_{\lambda 0Y}\end{pmatrix} - \mathbf{A}'\mathbf{Q}_z^{-1}\mathbf{A}
\end{aligned}
$$

The posterior distribution of $\mathbf{w}$ is normal with mean

$$\tilde{\mathbf{w}} = \mathbf{Q}_w^{-1}\left(\mathbf{M}_{Y10}\mathbf{d}_{\beta 0Y} + \mathbf{M}_{Y11}\mathbf{d}_{\lambda 0Y} - \mathbf{M}_{Y10}\mathbf{Q}_z^{-1}\mathbf{A}\right)$$

and precision matrix $\tau\mathbf{Q}_w$ where

$$\mathbf{Q}_w = \mathbf{V}_\lambda^{-1} + \mathbf{M}_{Y11} - \mathbf{M}_{Y10}\mathbf{Q}_z^{-1}\mathbf{M}_{Y01}$$

Integrating with respect to $\mathbf{w}$ yields the posterior density of τ, which is gamma with shape parameter $\tilde{a} = (a + \delta n_X + n_Y + p + q)/2$, where $p =$ length($\boldsymbol{\beta}$) and $q =$ length($\boldsymbol{\lambda}$) and scale parameter

$$\tilde{b} = \frac{1}{2}\left\{b + \delta S_{\mathbf{X}} + S_Y + \delta \mathbf{d}'_{\beta 0X}\mathbf{M}_X\mathbf{d}_{\beta 0X} + \begin{pmatrix}\mathbf{d}_{\beta 0Y}\\ \mathbf{d}_{\lambda 0Y}\end{pmatrix}' \right.$$
$$\left. \times\ \mathbf{M}_Y \begin{pmatrix}\mathbf{d}_{\beta 0Y}\\ \mathbf{d}_{\lambda 0Y}\end{pmatrix} - \mathbf{A}'\mathbf{Q}_z^{-1}\mathbf{A} - \tilde{\mathbf{w}}'\mathbf{Q}_w\tilde{\mathbf{w}}\right\}$$

We now have explicit posterior densities for $\mathbf{w}$ (i.e., for λ) and τ with correct normalization constants. These can be averaged over δ to give posterior expected values, quantiles, etc. for λ. They also can be used to simulate random draws from the posterior distributions so that posterior distributions of functions of λ can be obtained.

Remark: The calculations just described also apply when there are $H > 1$ historical trials. If a common value (or distribution) of δ is realistic, then no changes are required other than having the $\mathbf{X}$ vector and the $\mathbf{X}_0$ array consist of the stacked values for the individual studies, i.e., $\mathbf{X}' = (\mathbf{X}'_1, \ldots, \mathbf{X}'_H)$ (and likewise for $\mathbf{X}_0$). When each historical trial has its own δ value or distribution, so that δ is replaced with $(\delta_1, \ldots, \delta_H)$, the following changes are needed:

$$\delta\mathbf{M}_X \text{ is replaced with } \sum_{h=1}^{H} \delta_h \mathbf{M}_{Xh} \text{ where } \mathbf{M}_{Xh} = \mathbf{X}'_{0h}\mathbf{X}_{0h}$$

$$\delta\mathbf{M}_X\mathbf{d}_{\beta 0X} \text{ is replaced with } \sum_{h=1}^{H} \delta_h \mathbf{M}_{Xh}\mathbf{d}_{\beta 0Xh} \text{ where } \mathbf{d}_{\beta 0Xh} = \left(\hat{\boldsymbol{\beta}}_h - \boldsymbol{\beta}_0\right)$$

$$\delta S_{\mathbf{X}} \text{ is replaced with } \sum_{h=1}^{H} \delta_h S_{Xh} \text{ where } S_{Xh} = \left(\mathbf{X}_h - \mathbf{X}_{0h}\hat{\boldsymbol{\beta}}_h\right)'\left(\mathbf{X}_h - \mathbf{X}_{0h}\hat{\boldsymbol{\beta}}_{\text{h}}\right)$$

$$\delta\mathbf{d}'_{\beta 0X}\mathbf{M}_X\mathbf{d}_{\beta 0X} \text{ is replaced with } \sum_{h=1}^{H} \delta_h \mathbf{d}'_{\beta 0Xh}\mathbf{M}_X\mathbf{M}_{Xh}\mathbf{d}_{\beta 0Xh}$$

The calculations proceed as described above with these substitutions.

Appendix 2.3: R Code for Historical Prior Simulations

```
PostParams.fn <- function(dta, delta=0.3, dtacolsX=4:5, dtacolsY=1:3,
                          beta0=c(0,0), lambda0=0, vz=100, vw=100, tau_a=1, tau_b=1)
```

```
{
#  Let X denote a sample of observations on a control treatment from one or
#  more historical trials such that E(X) = beta2 + beta2*bsln and prec(X) =
#  tau*I. Let Y denote a sample of observations from a current trial such
#  that E(Y) = beta1 + beta2*bsln + z*lambda and prec(Y) = tau*I, where for
#  any element of Y, z = 0 if that element corresponds to an observation on
#  the control and z = 1 i that element corresponds to an observation on the
#  test treatment. Assume normality throughout.
#  Suppose that the prior expectation of beta = (beta1, beta2) is beta0 and
#  that the prior precision of beta is diag(1/vz). Also suppose that the
#  prior expectation of lambda is lambda0 and the prior precision of lambda
#  is diag(1/vw). Finally, suppose that the prior distribution of tau is
#  a gamma(tau_a, tau_b) distribution.
#  Assume that the historical control findings provide a prior distribution
#  for beta that, combined with the current data findings, leads to a
#  posterior distribution for beta, lambda, and tau. The contribution
#  corresponding to the historical data is the likelihood for the historical
#  data raised to the power delta. For the purpose of this calculation,
#  beta is assumed to be a nuisance parameter that is removed by integration
#  to give the posterior distribution of lambda with proper adjustment
#  for the value of delta (i.e., using the correct normalization constant).
#  This program returns the posterior mean and precision of lambda and the
#  parameters of the posterior distribution of tau given the historical and
#  current trial outcomes. The calculations actually are carried out for
#  general linear models, E(X) = Wx*beta and E(Y) = Wy*beta + U*lambda.
#  The key input is a text string (dta) giving the name of a list whose
#  elements are the values of X = observed historical outcomes, X01, X02, ...
#  = covariate values for historical outcomes (intercept constant is
#  assumed); Y = observed current outcomes, Y01, Y02, ... = covariate values
#  for current trial [same as historical trial]; and U1, U2, ... = covariate
#  values corresponding to the parameter vector lambda. Let H = the vector
#  of names of the elements of dta. The program identifies the various data
#  values by means of vectors of indices that assume a particular order
#  to the values:
#  dtacolsX = indices of H corresponding to (first) X, and (then) X01, X02, ...
#  dtacolsY = indices of H corresponding to (first) Y, and (then) Y01, Y02, ...,
#              U1, U2, ...
#                                        A. L. Gould  October 2017
 LikFn.fn <- function(dta, dtacolsX=4:5, dtacolsY=1:3, beta0=c(0,0),
                      lambda0=0)
 {
   likfnstep.fn <- function(dta, dtacols)
   {
     x <- paste0("c(", paste(dtacols, collapse=", "), ")")
     allvar <- eval(parse(text=paste0("names(", dta, ")[", x, "]")))
     depvar <- allvar[1]
     indvar <- allvar[-1]
     x <- paste0(depvar, "~", paste(indvar, collapse="+"))
     dta_array <- eval(parse(text=paste("as.data.frame(with(", dta, ",
```

```
                  cbind(", paste(allvar, collapse=", "), ")))", sep="")))
     cov_vals <- cbind(1, dta_array[, -1])
     xm <- as.matrix(cov_vals)
     M <- t(xm)%*%xm
     fit <- eval(parse(text=paste0("fit<-lm(", x, ", data=dta_array)")))
     S <- sum(fit$residuals^2)
     return(list(dta_array=dta_array, cov_vals=cov_vals, M=M, fit=fit,
                 l_par=length(coef(fit)), n_obs=nrow(dta_array), S=S))
   }

  XX <- likfnstep.fn(dta, dtacolsX)
  YY <- likfnstep.fn(dta, dtacolsY)
  lngthbeta <- XX$l_par
  lngthlambda <- YY$l_par - lngthbeta
  db0X <- coef(XX$fit) - beta0
  db0Y <- coef(YY$fit)[1:lngthbeta]
  dlam0Y <- coef(YY$fit)[-(1:lngthbeta)]-lambda0
  return(list(lngthbeta=lngthbeta, lngthlambda=lngthlambda, SX=XX$S, SY=YY$S,
              db0X=db0X, db0Y=db0Y, dlam0Y=dlam0Y, MX=XX$M, MY=YY$M,
              fitX=XX$fit, fitY=YY$fit, nX=XX$n_obs, nY=YY$n_obs))
}
###    PostParams.fn starts here  ###
  lik <- LikFn.fn(dta, dtacolsX=dtacolsX, dtacolsY=dtacolsY, beta0=beta0,
                  lambda0=lambda0)
  lngthbeta <- lik$lngthbeta
  lngthlambda <- lik$lngthlambda
  VbetaI <- diag(1/vz, lngthbeta)
  VlamI <- diag(1/vw, lngthlambda)
  db0Y <- lik$db0Y
  db0X <- lik$db0X
  dlam0Y <- lik$dlam0Y
  MX <- lik$MX
  MY <- lik$MY
  MY00 <- MY[1:lngthbeta, 1:lngthbeta]
  MY11 <- MY[-(1:lngthbeta), -(1:lngthbeta)]
  MY10 <- vec2mat.fn(MY[-(1:lngthbeta), 1:lngthbeta])
  MY01 <- t(MY10)
  A <- delta*MX%*%db0X + MY00%*%db0Y + MY01%*%dlam0Y
  Qz <- VbetaI + delta*MX + MY00
  QzI <- solve(Qz)
  Qw <- VlamI + MY11 - MY10%*%QzI%*%MY01
  QwI <- solve(Qw)
  wtilde <- QwI%*%(MY10%*%db0Y + MY11%*%dlam0Y - MY10%*%QzI%*%A)
  atilde <- (tau_a + delta*lik$nX + lik$nY + lngthbeta + lngthlambda)/2
  btilde <- 0.5*(tau_b + delta*lik$SX + lik$SY + delta*db0X%*%MX%*%db0X
            + c(db0Y, dlam0Y)%*%MY%*%c(db0Y, dlam0Y) - t(A)%*%QzI%*%A
            - t(wtilde)%*%Qw%*%wtilde)
  return(list(wtilde=wtilde, Qw=Qw, atilde=atilde, btilde=btilde))
}
```

```
PowPriorSim.fn <- function(dta, nrep_delta, nrep_tau, nrep_lam, delta_a=10,
                           delta_b=30, dtacolsX=4:5,
                           deltaprobs=c(.005,.025,.5,.975,.995),
                           dtacolsY=1:3, beta0=c(0, 0), lambda0=0, vz=100,
                           vw=100, tau_a=1, tau_b=1,
                           probs=c(.025,.05,.5,.95,.975))
{
  lambda <- array(0,nrep_delta*nrep_tau*nrep_lam)
  delta <- rbeta(nrep_delta, delta_a, delta_b)
  for (i in 1:nrep_delta)
  {
    params <- PostParams.fn(dta, delta=delta[i], dtacolsX=dtacolsX,
                            dtacolsY=dtacolsY, beta0=beta0, lambda0=lambda0,
                            vz=vz, vw=vw, tau_a=tau_a, tau_b=tau_b)
    tau <- rgamma(nrep_tau, params$atilde, params$btilde)
    for (j in 1:nrep_tau)
    {
      k <- nrep_lam*((j-1) + (i-1)*nrep_tau) + 1:nrep_lam
      sigma <- solve(params$Qw)/tau[j]
      wtilde <- rmvnorm(nrep_lam, params$wtilde, sigma)
      lambda[k] <- wtilde + lambda0
    }
  }
  GMTR <- exp(lambda)
  GMTRquants <- quantile(GMTR,probs)
  return(list(call=sys.call(), date=date(), dta=dta, delta_a=delta_a,
              delta_b=delta_b, nrep_delta=nrep_delta, nrep_tau=nrep_tau,
              nrep_lam=nrep_lam, lambda=lambda, GMTR=GMTR,
              GMTRmean=mean(GMTR), GMTRquants=GMTRquants,
              deltaquants=qbeta(deltaprobs,delta_a,delta_b)))
}
PowPriorSimSum.fn <- function(dta, nrep_delta, nrep_tau, nrep_lam, delta_ab,
                          dtacolsX=4:5, deltaprobs=c(.005,.025,.5,.975,.995),
                          dtacolsY=1:3, beta0=c(0,0), lambda0=0, vz=100,
                          vw=100, tau_a=1, tau_b=1,
                          probs=c(.025,.05,.5,.95,.975))
{
  delta_ab <- vec2mat.fn(delta_ab)
  ndelts <- nrow(delta_ab)
  result <- matrix(0,nrow=ndelts,ncol=11)
  for (i in 1:ndelts)
  {
    z <- PowPriorSim.fn(dta, nrep_delta, nrep_tau, nrep_lam,
                        delta_a=delta_ab[i,1], delta_b=delta_ab[i,2],
                        dtacolsX=dtacols, deltaprobs=deltaprobs, beta0=beta0,
                        lambda0=lambda0, vz=100, vw=100, tau_a=1, tau_b=1,
                        probs=probs)
    result[i,] <- c(i,delta_ab[i,], round(c(delta_ab[i,1]/sum(delta_ab[i,]),
                    z$deltaquants[3], z$deltaquants[2], z$deltaquants[4],
```

```
                  z$GMTRmean, z$GMTRquants[3], z$GMTRquants[1],
                  z$GMTRquants[2]),3))
  }
  colnames(result) <- c("Case", "del_a", "del_b", "E(delta)", "Med(delta)",
                        "delta025", "delta975", "E(GMTR)", "Med(GMTR)",
                        "GMTR025", "GMTR05")
  return(list(call=sys.call(),date=date(),result=result))
}
```

References

[1] Wang WWB, Mehrotra DV, Chan ISF, Heyse JF. Statistical considerations for noninferiority/equivalence trials in vaccine development. *Journal of Biopharmaceutical Statistics* 2006; **16**:429–441.

[2] Neuenschwander B, Capkun-Niggli G, Branson M, Spiegelhalter DJ. Summarizing historical information on controls in clinical trials. *Clinical Trials* 2010; **7**:5–18.

[3] Schmidli H, Gsteiger S, Roychoudhury S, O'Hagan A, Spiegelhalter D, Neuenschwander B. Robust meta-analytic-predictive priors in clinical trials with historical control information. *Biometrics* 2014; **70**:1023–1032.

[4] Berry DA. Bayesian clinical trials. *Nature Review Drug Discovery* 2006; **5**:27–36.

[5] Schoenfeld DA, Zheng H, Finkelstein DM. Bayesian design using adult data to augment pediatric trials. *Clinical Trials* 2009; **6**:297–304.

[6] Viele K, Berry S, Neuenschwander B, Amzal B, Chen F, Enas N, Hobbs B, Ibrahim JG, Kinnersley N, Lindborg S, Micallef S, Roychoudhury S, Thompson L. Use of historical control data for assessing treatment effects in clinical trials. *Pharmaceutical Statistics* 2014; **13**:41–54.

[7] Hobbs BP, Carlin BP, Mandrekar SM, Sargent DJ. Hierarchical commensurate and power prior models for adaptive incorporation of historical information in clinical trials. *Biometrics* 2011; **67**:1047–1056.

[8] Hobbs BP, Sargent DJ, Carlin BP. Commensurate priors for incorporating historical information in clinical trials using general and generalized linear models. *Bayesian Analysis* 2012; **7**:639–674.

[9] Ibrahim JG, Chen M-H. Power prior distributions for regression models. *Statistical Science* 2000; **15**:46–60.

[10] Duan Y, Smith EP, Ye K. Using power priors to improve the binomial test of water equality. *Journal of Agricultural, Biological, and Environmental Statistics* 2006; **11**:151–168.

[11] Duan Y, Ye K, Smith EP. Evaluating water quality using power priors to incorporate historical information. *Environmetrics* 2006; **17**:95–106.

[12] Neuenschwander B, Branson M, Spiegelhalter DJ. A note on the power prior. *Statistics in Medicine* 2009; **28**:3562–3566.

[13] Spiegelhalter DJ., Abrams KR, Myles JP. *Bayesian Approaches to Clinical Trials and Health-Care Evaluation,* John Wiley & Sons, New York, 2004.

[14] Tarone R. The use of historical control information in testing for a trend in proportions. *Biometrics* 1982; **38**:215–220.

[15] Dempster A, Selwyn M, Weeks B. Combining historical and randomized controls for assessing trends in proportions. *Journal of the American Statistical Association* 1983; **78**:221–227.

[16] DerSimonian R, Laird N. Meta-analysis in clinical trials. *Control Clinical Trials* 1986; **7**:177–188.

[17] DerSimonian R. Meta-analysis in the design and monitoring of clinical trials. *Statistics in Medicine* 1996; **15**:1237–1248.

[18] DerSimonian R, Kacker R. Random-effects model for meta-analysis of clinical trials: An update. *Contemporary Clinical Trials* 2007; **28**:105–114.

[19] Hardy RJ, Thompson SG. A likelihood approach to meta-analysis with random effects. *Statistics in Medicine* 1996; **15**:619–629.

[20] Paule RC, Mandel J. Consensus values and weighting factors. *Journal of Research of the National Bureau of Standards* 1982; **87**:377–385.

[21] Dallal S, Hall W. Approximating priors by mixtures of natural conjugate priors. *Journal of the Royal Statistical Society: Series B* 1983; **45**:278–286.

[22] Daniels MJ. A prior for the variance components in hierarchical models. *Canadian Journal of Statistics* 1999; **27**:569–580.

[23] DuMouchel W. Bayesian meta-analysis. *in* Berry, D. (ed.) *Statistical Methodology in the Pharmaceutical Sciences*, Marcel Dekker, New York, 1990: 509–529.

[24] Gelman A. Prior distributions for variance parameters in hierarchical models. *Bayesian Analysis* 2006; **1**:515–533.

[25] Polson NG, Scott JG. On the Half-Cauchy prior for a global scale parameter. *Bayesian Analysis* 2012; **7**:887–902.

[26] Schuirmann DJ. A comparison of the two one-sided tests procedure and the power approach for assessing the equivalence of average bioavailability. *Journal of Pharmacokinetics and Biopharmaceutics* 1987; **15**:657–680.

[27] Gould AL. Using prior findings to augment active-controlled trials and trials with small placebo groups. *Drug Information Journal* 1991; **25**:369–380.

[28] Kerzner B, Murray AV, Cheng E, Ifle R, Harvey PR, Tomlinson M, Barben JL, Rarrick K, Stek JE, Chung MO, Schodel FP, Wang WW, Xu J, Chan IS, Silber JL, Schlienger K. Safety and immunogenicity profile of the concomitant administration of ZOSTAVAX® and inactivated influenza vaccine in adults aged 50 and older. *Journal of the American Geriatric Society* 2007; **55**:1499–1507.

[29] STAN Development Team. *RStanArm: Bayesian applied regression modeling via Stan. R package version 2.15.3.* STAN Development Team, 2017 http://mc-stan.org.

[30] Carpenter R, Gelman A, Hoffman MD, Lee D, Goodrich B, Betancourt M, Brubaker M, Guo J, Li P, Riddell A. Stan: A probabilistic programming language. *Journal of Statistical Software* 2017; 76: DOI: 10.18637/jss/v076.i01.

[31] STAN Development Team. *RStan: the R interface to Stan Version 2.16.0* STAN Development Team, 2017 http://mc-stan.org.

[32] Gravestock I, Held L. Adaptive power priors with empirical Bayes for clinical trials. *Pharmaceutical Statistics* 2017; **16**:349–360.

[33] Cook J, Fúquene JA, Pericchi LR. Skeptical and optimistic robust priors for clinical trials. *Revista Colombiana de Estadistica* 2011; **34**:333–345.

[34] Fúquene JA, Cook JD, Pericchi LR. A case for robust Bayesian priors, with applications to clinical trials. *Bayesian Analysis* 2009; **4**:817–846.

[35] O'Hagan A, Pericchi LR. Bayesian heavy-tailed models and conflict resolution: A review. *Brazilian Journal of Probability and Statistics* 2012; **26**:372–401.

[36] Diaconis P, Ylvisaker D. Quantifying prior opinion. *in* Bernardo, J. M., DeGroot, M. H., Lindley, D. V., and Smith, A. F. M. (eds.) *Bayesian Statistics 2: Proceedings of the Second Valencia International Meeting September 6/10, 1983,* Elsevier, Netherlands, 1985: 133–156.

[37] Gamalo-Siebers M, Gao A, Lakshminarayanan M, Liu G, Natanegara F, Raikar R, Schmidli H, Song G. Bayesian methods for the design and analysis of noninferiority trials. *Journal of Biopharmaceutical Statistics* 2016; **26**:823–841.

3

Practical Considerations for Building Priors for Confirmatory Studies

Guochen Song, John Zhong, and Stacy Lindborg
Biogen Inc.

Baoguang Han
Sarepta Therapeutics

CONTENTS

3.1 Introduction

One of the main differences between the Bayesian approach and the frequentist approach is the way in which data from an external source can be incorporated into the analysis of a current trial and guide to update the current data. The use of external evidence in a confirmatory trial, however, also draws the most contentious criticism for the use of Bayesian analysis in drug development. One must examine the different data sources carefully and determine if the data are exchangeable with data from the current clinical trial. The ideal situation, where the historical data are exchangeable with data from the current trial at the individual patient level, is fairly rare. However, exchangeability at the aggregated data level or conditional on some other controllable situations may be realistic. If the exchangeability assumptions cannot be made, the analysis method needs to adequately reflect the problem, and the borrowing from historical data will be subjected to scrutiny. In this chapter, a Phase 3 drug development program for spinal muscular atrophy (SMA) will be used as an example to illustrate the technique of incorporating historical data from an observational study into the statistical analyses of an interventional clinical trial to improve its efficiency.

Before an effective treatment is developed, observational studies may have been conducted to evaluate the natural history of the disease. For example, the Pediatric Neuromuscular Clinical Research Network (PNCR) conducted a prospective, observational, cohort study of patients with SMA (Finkel et al, 2014); the data collected were very valuable in designing the clinical trials that studied new treatments for SMA.

A clinical trial usually has much stricter eligibility criteria than an observational study. These restrictions are in place not only to ensure the population under study is relatively homogeneous, but also to make sure the trial is safe and ethical in the sense that it will not enroll patients who should not be exposed to the experimental therapy. These restrictions can make the populations under study and natural history studies to be quite different. For example, the eligibility criteria for ENDEAR, a Phase 3, randomized, double-blind, sham-procedure controlled study to assess the clinical efficacy and safety of nusinersen administered intrathecally in patients with infantile-onset SMA, required patients to have two copies of the SMN2 gene and be younger than 210 days old at screening. In contrast, the PNCR study included all patients with SMA who were willing to participate; patients could also have been enrolled posthumously.

To enhance the validity of the exchangeability assumption between the data in the observational study and the clinical trial, one may apply the eligibility criteria used in the clinical trial to the observational study data. Pocock (1976) laid out six criteria that a historical control should satisfy. In practice, especially for observational data, it is impractical to retrospectively apply all inclusion/exclusion criteria to data already collected because the observational study may not have collected all the relevant variables. Some key disease characteristics, on the other hand, are more likely to be part of the dataset, and applying the eligibility criteria based on such characteristics might address the problem of validity to an extent. It is worth noting that sometimes a variable (i.e., specific, critical information) can be constructed from available data, even though it was not collected originally. For example, in the ENDEAR trial, disease duration, which was defined as the time between the onset of SMA and the starting of trial (soon after which treatment starts), was considered as a factor that might affect the treatment effect. As no treatment existed when the PNCR study started, this data point was not captured. When analyzing the data, the disease duration variable must be reconstructed to make the studies comparable.

This chapter presents the strategy used to integrate observational study data with clinical trial data to enhance the analyses supporting the development of Spinraza® (nusinersen).

3.1.1 Description of Clinical Trials in Infantile-Onset SMA

Two clinical studies were conducted in subjects with infantile-onset SMA. Both studies used dose levels that were the scaled equivalent, based on cerebrospinal fluid volume, to those that would be given to two-year-olds.

The CS3A study (Finkel et al., 2016) was a Phase 2, multicenter, open-label, multiple-dose, dose-escalation study to evaluate the safety, tolerability, and pharmacokinetics of nusinersen administered intrathecally to patients aged $\leq$7 months who had clinical signs and symptoms of SMA at $\leq$6 months of age at Screening. The original design of the study was to follow patients every 2–3 weeks for 36 weeks, and the protocol was amended to have the telephone follow-up up to 2.5 years.

The ENDEAR study (CS3B) was a Phase 3, multicenter, randomized, double-blind, parallel-group, sham-procedure controlled study to evaluate the efficacy, safety, tolerability, and pharmacokinetics of nusinersen administered to patients with two *SMN2* copies who were $\leq$7 months of age and had onset of clinical signs and symptoms of SMA at $\leq$6 months of age and at Screening (Finkel et al., 2017). There were 121 patients randomized 2:1 to receive a scaled equivalent 12 mg dose of nusinersen or undergo a sham procedure as control, respectively. The follow-up period in the study after initiation of treatment was 13 months. The final assessment was scheduled on Day 394 ($\pm$7 days).

3.1.2 Description of Observational Study

At the time of CS3A planning, after thorough review of recent published natural history studies, the SMA type I population enrolled in a study conducted by the Pediatric Neuromuscular Clinical Research Network for SMA (PNCR) was determined to be the most appropriate choice of control group (Finkel et al., 2014). That review revealed that both the PNCR study and Study CS3A utilized age at death or permanent ventilation (defined as tracheostomy or at least 16 h/day of ventilation support for at least 14 days in the absence of an acute reversible illness), motor function as assessed by the Children's Hospital of Philadelphia Infant Test for Neuromuscular Disorders (CHOP INTEND), and neuromuscular electrophysiology as measured by ulnar nerve compound muscle action potential (CMAP).

The PNCR study observed SMA Type I patients with characteristics like those in Studies CS3A and CS3B in terms of genetic confirmation of SMN1 deletion or homozygous mutation and known *SMN2* copy number. Moreover, the observational study and clinical trials used the same definition of SMA Type I (i.e., symptom onset within 6 months of age). Unlike the clinical trials, the PNCR study had no age limit at enrollment for its SMA Type I population, as expected given the goal of the PNCR registry. PNCR SMA Type I patient natural history study data included demographic information (age, sex, and ethnicity), SMA history (age at symptom onset, age at clinical diagnosis, and history of motor developmental milestones gained and/or lost), other relevant medical and surgical history, medication/supplements taken, physical examination findings, motor function test results (CHOP INTEND), ulnar nerve compound muscle action potential (CMAP) and motor unit number estimation (MUNE), ventilation information, and details on death.

3.2 Statistical Method

3.2.1 Data Preparation

The data from Study CS3B were considered the primary data. Data from the observational studies and Study CS3A and PNCR were considered supplemental, in other words, the contrast of analysis focused on model parameters for CS3B by borrowing information from CS3A and PNCR.

The endpoint of interest for this analysis was the time to death or permanent ventilation. For Study CS3B, the start date for calculation of day to death/permanent ventilation or day of censoring was the date of first dose/sham procedure, and if the date of first dose/sham procedure was incomplete, then the date of randomization was used. For the open-label Study CS3A, the date of first dose was used as the start date. Permanent ventilation was defined as $\geq$16 h ventilation/day continuously for $\geq$21 days in the absence of an acute reversible event or tracheostomy in the CS3B study protocol, but it was defined as $\geq$16 h ventilation/day continuously for >2 weeks in the absence of an acute reversible illness in the CS3A protocol and the PNCR study.

As mentioned previously, the patients' disease duration variable needed to be reconstructed in the PNCR study, but the start date in the PNCR study was not comparable to the clinical studies. To address this issue, the following exercise was performed: the median age (in days) of patients at start date was calculated for the combined CS3A and CS3B studies and was denoted as Med_CS3. The median age (in days) at enrollment for the natural history study was calculated and denoted as Med_NH. The start date of the natural history study then was calculated as follows:

$$\text{start date} = \text{study enrollment date} + \max\left(\text{Med\_CS3} - \text{Med\_NH}, 0\right).$$

In other words, because the observational study enrolled patients at a much younger age, adjustment was made to make the studies more comparable. If a patient experienced an event before or on the adjusted start date, the patient was considered as not eligible for this analysis and was excluded from the dataset.

Further, for the natural history study, data of each patient were reviewed by the clinicians and statisticians together, and those patients' baseline characteristics deemed not to meet the key inclusion/exclusion criteria of Study CS3B were excluded from the efficacy analysis.

3.2.2 Hierarchical Model Approach

A Bayesian hierarchical piecewise exponential model with adjusted covariates (Ibrahim et al, 2001; Spiegelhalter et al, 2004; Berry et al, 2011; Han et al, 2017) was used for this analysis. Based on Viele et al. (2014), comparing with other data combination methods, "borrowing behavior tends to be 'flatter' for hierarchical models, borrowing moderately over a long range, while still displaying dynamic borrowing (borrowing is reduced when the true control rate is far from the historical data)". In other words, the hierarchical model has a desirable feature that if the difference between the historical data and current study is big, borrowing will be minimal, while if the difference is small, borrowing will be moderate. The hierarchical model assumes exchangeability of parameters conditioning on the observed covariates. This may be a reasonable assumption given that the inclusion/exclusion criteria review has been applied to PNCR study to select similar patients to those in clinical trials.

For each study, let $D = \{t, \delta, \boldsymbol{X}, \boldsymbol{Z}\}$ denote the datasets, where (t, δ) was the observed time and the event indicator (i.e., for patients having an event, $\delta = 1$, otherwise, $\delta = 0$). $\boldsymbol{X}$ represented baseline characteristics shared among all the three datasets that were adjusted in the model. $\boldsymbol{X} = \{x_1, x_2\}$, with x_1 = age at onset, x_2 = disease duration at baseline. $\boldsymbol{Z} = \{I_{00}, I_{01}, I_{10}, I_{11}\}$ represented indicator variables to reflect the treatment received and the relevant study. The coding scheme of variable $\boldsymbol{Z}$ is listed in Table 3.1.

$\boldsymbol{\beta} = \{\beta_{00}\ \beta_{01}\ \beta_{10}\ \beta_{11}\}'$ were coefficients corresponding to $\boldsymbol{Z}$.

Then, $\gamma_1 = \beta_{11} - \beta_{10}$ was denoted as the contrast between patients treated with nusinersen and the sham control in Study CS3B, which was the main parameter that we were interested in. The 95% Bayesian credible interval (BCI) of γ_1 was reported. The parameter γ_1 was the log-hazard ratio (logHR) between the patients treated with nusinersen and those treated with a sham control in Study CS3B after adjusting baseline characteristics.

The time interval (maximum 401 days) was divided into $K = 5$ intervals, each of which contained roughly the same number of events.

Under the piecewise exponential hazard model framework, the hazard function for each individual i was expressed as

$$\lambda_i(t) = \lambda_0(t) \exp\left(\boldsymbol{x}_i'\boldsymbol{\alpha} + \boldsymbol{z}_i'\boldsymbol{\beta}\right)$$

TABLE 3.1
Coding Scheme of the Indicator Variables

			CS3B	
	PNCR (Natural History)	**CS3A (Nusinersen)**	**SHAM Control**	**Experiment (Nusinersen)**
I_{00}	1	0	0	0
I_{01}	0	1	0	0
I_{10}	0	0	1	0
I_{11}	0	0	0	1

where $\lambda_0(t)$ was the baseline hazard function for time t. The coefficients $\boldsymbol{\alpha}$ and $\boldsymbol{\beta}$ were assumed not to change over time.

Following the notation of Han et al. (2017), the likelihood function for the piecewise hazard model at each time interval was expressed as follows:

$$\prod_{i=1}^{N} \lambda_i(t)^{\delta_i} \exp\left\{-\int_0^t \lambda_i(u)\, du\right\}.$$

A gamma prior (0.01, 0.01) was assigned to the baseline hazard $\lambda_0(t)$ for each patient at each time interval. This was a non-informative prior in the sense that the mean was set as unity and the variance was very large. The rate parameter, however, had minimal effects on the analysis results in this setting.

A diffuse independent multinormal prior $\mathcal{N}\left(0, 10^2 I_2\right)$ was assigned to the regression coefficients $\boldsymbol{\alpha}$, with $I_2 = \begin{pmatrix} 1 & 0 \\ 0 & 1 \end{pmatrix}$ being the identity matrix and $\mathcal{N}(\cdot)$ denoting the normal density (Han et al, 2017). Next, it was assumed that

$$\beta_{00} \sim \mathcal{N}\left(\mu_0, \tau_0^2\right), \beta_{10} \sim \mathcal{N}\left(\mu_0, \tau_0^2\right),$$
$$\beta_{01} \sim \mathcal{N}\left(\mu_1, \tau_1^2\right), \beta_{11} \sim \mathcal{N}\left(\mu_1, \tau_1^2\right),$$

where μ_0 was the common mean to patients in natural history study or treated with a sham control in CS3B and μ_1 was the common mean for patients treated with nusinersen in CS3A or CS3B, both of which were assigned to a diffuse prior $N\left(0, 10^2\right)$. The parameters τ_0 and τ_1 reflected the heterogeneity between the studies, and they were assigned to a moderate prior of IG (1, 0. 1) (Viele et al, 2014). Note that exchangeability was assumed between the natural history data and data from the patients treated with a sham control in Study CS3B, and between data from patients in Study CS3A and patients treated with nusinersen in Study CS3B, after adjusting $\boldsymbol{X}$.

As a sensitivity analysis, the priors IG (1, 0.01) and IG (1, 1) were planned for the variance parameters τ_0 and τ_1.

3.3 Analysis Results and Operation Characteristics of Selected Scenarios

Table 3.2 summarizes the results of applying the approach discussed above on the final dataset after the trial was completed. The rjags package (Martyn Plummer, 2016) was used to fit these models, with 10,000 Markov Chain Monte Carlo (MCMC) updates and 1,000 burn ins. The BCI was calculated based on MCMC samples. The results reported hazard ratio (HR), using an exponential transformation. The standard deviation of the logHR was computed directly from the MCMC sample, which reflected the variability of the

TABLE 3.2
Analysis Results

HR	95% CI or BCI for HR	Method	Prior on τ	SD (log HR)	No. of Events
0.445	(0.201, 0.771)	Study CS3B, adjusting covariates	NA	0.276	59
0.404	(0.239, 0.687)	Hierarchical, adjusting covariates	IG(1, 0.1)	0.263	65[a]
0.37	(0.224, 0.621)	Hierarchical, adjusting covariates	IG(1, 0.01)	0.261	67[a]
0.433	(0.252, 0.739)	Hierarchical, adjusting covariates	IG(1, 1)	0.273	60[a]
0.344	(0.172, 0.560)	Pooled data, adjusting covariates	NA	0.248	77

[a]Calculated based on a linear interpolation of the SD(logHR) and the number of events between the no-borrowing case (first row) and the complete pooling case (last row).
BCI, Bayesian credible interval; CI, confidence interval; NA, not applicable; IG, inverse gamma; SD(logHR), standard deviation of logHR.

logHR of posterior distribution, same as the standard error in a frequentist approach.

The first row (in Table 3.2) is the result of using data from the Phase 3 clinical trial only, fitting a piecewise exponential adjusted for the covariates specified. The last row, on the other hand, pooled all available data and applied the same model. The three rows in the middle are the results using a Bayesian hierarchical model, but with different prior on the precision parameter τ, and the first of these rows is considered as the result of using primary approach and the other two as the results of sensitivity analysis. It can be seen that with the prior IG(1, 0.1), the standard deviation of the logHR is 0.263, which lies between the standard deviation calculated from the model using the dataset from the Phase 3 study only and the one using the pooled data. In Study CS3B, 59 events were observed. The pooled data added 18 more eligible events. As there are covariate adjustments, it is not possible to exactly calculate the actual information borrowed; however, a simple linear interpolation between the standard deviation and the number of events of the two extremes, though not completely accurate, can shed some light on this. For example, the model using IG(1, 0.1) on the precision can be roughly viewed as a fit from a dataset that contains 65 events, i.e., 6 events were borrowed from historical data.

The robustness of the Bayesian approach can be verified through examining the frequentist operating characteristics (e.g., type I and type II errors). Such validation is regarded as an essential element of validating a Bayesian inference (Gamalo-Siebers et al, 2017). Extensive simulation was carried out to examine the operating characteristics at the planning stage. Note that the CS3A data and the PNCR data were both available at the time when the simulation was performed, yet the median time to event was not observed in the CS3A study, but the patients in the PNCR data had a median time to event time approximately 23 weeks in the selected patients. We assumed these data as fixed. For the CS3B part, we simulated data assuming the median time to permanent ventilation of the treated and sham control group following an exponential distribution with different parameters. Table 3.3 describes some of the selected simulation scenarios for illustration purpose.

In Table 3.3, the last row assumed the true median time to permanent ventilation or death as 5.5 months regardless of treatment. Using the model described above with hierarchical borrowing, the estimated HR averaging over all the runs was about 0.97, and there were 3% of times the upper bound of the two-sided 95% Bayesian credible interval for the HR was less than 1. After study completion, the CS3B study reported the median time to permanent ventilation or death in the sham control group as 22.6 weeks, which was very similar to the selected PNCR patients (23.5 weeks) as well as the assumed value in this row (about 22 weeks). The median time to permanent ventilation or death was not reached for CS3A study at month 13. In the scenario described in this row, the PNCR data is similar to the sham control group in CS3B; hence, the model allows more information borrowing from PNCR

TABLE 3.3
Operation Characteristics from Simulation

Median Time (Months) to Event (TRT, SHAM)	True HR	Estimation from Simulation	Claiming Success Using Hierarchical Model with IG(1, 0.1) Prior Out of 10,000 Runs (%)	Claiming Success Using Simulated Data for CS3B Only (%)
(16.5, 5.5)	0.33	0.33	99.8	99.1
(11, 5.5)	0.5	0.48	89.3	82.9
(11, 11)	1	0.89	6.3	2.6
(5.5, 5.5)	1	0.97	3.0	2.6

The results above were based on 10,000 runs for each row; within each run, 10,000 MCMC simulations were performed. This table is for illustration purposes only, and in actual practice, a larger number of runs (e.g., 100,000) might be needed if the purpose is to show type I error control. It is time-consuming to run MCMC simulations like this even on high-capacity clusters; hence, careful planning is needed.

data. On the other hand, the treated arm would be very different from the CS3A study, and hence borrowing would be minimal from the CS3A study. The false positive rate was slightly inflated in this scenario compared to a true null case.

In the case that the median time to permanent ventilation or death was assumed to be 11 months in the nusinersen treated group and 5.5 months in the sham control group (second row), borrowing could occur in both groups. The simulated power was 89.3%. The CS3B trial was originally designed to detect a HR of 50%, with the designed power being around 80% assuming median time to permanent ventilation or death to be 5–6 months for patients treated with sham procedure. The proposed method was more powerful than the one without borrowing as the simulated power for the without borrowing method was 82.9%.

With the CS3A and PNCR data already available, a frequentist type I error type of assessment for the approach does not fit the situation. However, as stated in the Adaptive Designs for Clinical Trials of Drugs and Biologics Guidance for Industry (FDA, 2018), even though some null hypothesis space parameters may not be sensible, it might help to understand worst-case scenarios. The third row of Table 3.3 describes such a case. In the case that the median time to permanent ventilation or death was 11 months in both groups, the false positive rate was 6.29%. In this case, the model borrows some information from PNCR, but the patients assigned to SHAM do better than patients in PNCR; the model borrows some information from CS3A, yet patients treated with nusinersen in CS3B do worse than patients in CS3A. The false positive rate would be inflated as expected. It is worth emphasizing that the assumption in this scenario is not realistic and should only be treated as a hypothetical worst-case scenario.

3.4 Discussion

The Organ Drug Act (US Congress, 1983) defines rare diseases as those affects 200,000 persons in the United States. In drug development programs for rare diseases, enrollment is often very difficult, and data source is very valuable. In a disease area such as SMA, which is estimated to affect 1 in 11,000 live births (Finkel et al., 2017), utilizing existing data would reduce development time and deliver effective treatment to patients faster. Examining the eligibility criteria to decide which data should be properly included in the analysis has to be performed before the simulation exercise, as the data source cannot change. This effort requires clear communication among clinical development team members.

This example shows a way to incorporate historical observational data and open-label, noncontrolled data into a pivotal dataset analysis. The prior for the precision parameter governs how much information one can borrow from the historical data. Unless there are multiple historical studies, which is rare, this prior cannot be determined from data per se. Extensive simulation has to be done to examine the operating characteristics, especially the type I error rate in a frequentist sense, for this parameter. In addition to the current hierarchical model, alternative approaches such as power prior (Ibrahim and Chen, 2000) and commensurate prior (Murray et al, 2014) may also be explored. Application of such methods is ongoing and will be published in future publications.

In the draft guidance (FDA, 2018), FDA stated that one of the potential ways of applying Bayesian design is to "explicit borrowing of information from external sources, e.g., previous trials, natural history studies, and registries, via informative prior distributions to improve the efficiency of a trial". The example in this chapter exemplifies such a case. In designing such trials, we found that special attention needs to be paid to justify the exchangeability between external data and the current clinical trial, and simulations to demonstrate worst-scenario type I error rate under the null hypothesis space should be planned carefully.

References

Berry SJ, Carlin BP, Lee JJ, Mueller P. *Bayesian Adaptive Methods for Clinical Trials.* Boca Raton, FL: CRC Press; 2011.

FDA, 2018, Adaptive Designs for Clinical Trials of Drugs and Biologics Guidance for Industry. www.fda.gov/downloads/drugs/guidances/ucm201790.pdf.

Finkel RS, Chiriboga C, Vajsar J, et al. Treatment of infantile-onset spinal muscular atrophy with nusinersen: A phase 2, open-label, dose-escalation study. *Lancet.* 2016; 388(10063):3017–3026.

Finkel RS, McDermott MP, Kaufmann P, et al. Observational study of spinal muscular atrophy type I and implications for clinical trials. *Neurology.* 2014; 83:810–817.

Finkel RS, Mercuri E, Darras BT, et al. Nusinersen versus sham control in infantile-onset spinal muscular atrophy. *N Engl J Med.* 2017; 377(18):1723–1732.

Gamalo-Siebers M, Savic J, Basu C, et al. Statistical modeling for Bayesian extrapolation of adult clinical trial information in pediatric drug evaluation. *Pharm Stat.* 2017; 4:232–249.

Han B, Zhan, J, Zhong J, Liu D, Lindborg S. Covariate-adjusted borrowing of historical control data in randomized clinical trials. *Pharm Stat.* 2017; 16(4):296–308.

Ibrahim JG, Chen MH. Power prior distributions for regression models. *Stat Sci.* 2000; 15:46–60.

Ibrahim JG, Chen MH, Sinha D. Bayesian survival analysis. New York: Springer; 2001.

Martyn Plummer. rjags: Bayesian Graphical Models using MCMC. R package version 4-6. http://CRAN.R-project.org/package=rjags. 2016.

Murray, TA, Hobbs, BP, Lystig, TC Semiparametric Bayesian commensurate survival model for post-market medical device surveillance with non-exchangeable historical data. Biometrics 2014; 70:185–191.

Pocock SJ. The combination of randomized and historical controls in clinical trials. *J Chronic Dis.* 1976; 29:175–88.

Spiegelhalter DJ, Abrams KR, Myles JP. *Bayesian Approaches to Clinical Trials and Health-Care Evaluation.* Chichester, UK: John Wiley & Sons Ltd; 2004.

US Congress. Orphan Drug act. www.fda.gov/RegulatoryInformation/LawsEnforcedbyFDA/SignificantAmendmentstotheFDCAct/OrphanDrugAct/default.htm. 1983.

Viele K, Berry S, Neuenschwander B, et al. Use of historical control data for assessing treatment effects in clinical trials. *Pharm Stat.* 2014; 13:41–54.

4

The Practice of Prior Elicitation

Timothy H. Montague
GlaxoSmithKline

Karen L. Price
Eli Lilly & Company

John W. Seaman, Jr.
Baylor University

CONTENTS

4.1 Prior Elicitation in Drug and Medical Device Development

As the use of Bayesian approaches continues to expand in the field of drug development and health technology assessment, there is a need for tools to develop defensible informative prior distributions based on expert opinion. Ideally, relevant empirical data should provide the evidential basis for such priors. However, it is often the case that limitations in data, our understanding of the problem, or both may preclude direct construction of a data-driven prior. Even in cases where a substantial body of empirical evidence is available, there is very often a translational gap between the setting(s) to which the available data relate and the new setting under consideration. In such cases, how should we engage the wealth of expert knowledge and experience within a scientific community to develop informative priors that can bridge the gap between sparse or indirect evidence and the quantities of interest to decision makers?

While the goal of an elicitation is to provide a probability distribution for unknown parameters, the process creates transparency about not only what is known (summarized in an evidence dossier) and the rationale for experts' beliefs, but the gaps in knowledge as well as sources of uncertainty. Furthermore, the discussions can reveal aspects of the study design that may have not been considered which can help the drug development team optimize study protocols. The elicitation exercise has proven to increase understanding of the potential study outcomes and appreciation of the risks of the current study design. Dallow et al. (2018) provide some further commentary on the benefits of prior elicitation in decision making.

The focus of this chapter will be on the methodologies and challenges of prior elicitation, a process that enables translation of expert knowledge and judgment into a probability distribution—a quantitative belief distribution. The goal of this process is to capture the best estimate of what is known "right now" about the true value of an unknown quantity, such as the true treatment effect of a particular asset, as well as quantifying the uncertainty around that estimate. We assume that this prior distribution will be used in an associated Bayesian analysis, but this need not be the case if uncertainty modeling is desired for its own sake. We provide an extended example and discuss how these distributions can be utilized to improve the drug development process.

4.1.1 The Need for Expert Opinion

A key challenge when using Bayesian methods is the selection of the prior. In some instances, historical data in a similar setting can be used to construct a prior (see Chapters 2, 17, and 18). However, frequently this is not the case. In situations involving novel mechanisms of action, novel endpoints, or a new target population (e.g., pediatric indications), there may simply be no prior data relevant to the treatment effect of interest. Or it may be that the existing data cannot be used to predict the treatment effect in the next clinical study due to a lack of direct relevance, high levels of variability, or both. For example, there may be differences between the historical trials and the planned trial that cannot easily be modeled. The impact of differences in sampled populations, endpoints, durations, etc. can be difficult to quantify directly. Additional data may exist, perhaps from other compounds, other indications, or pre-clinical data. Such indirect sources of information can be difficult to incorporate into a prior. For example, a wealth of information may be available about safety and efficacy of a compound prescribed for indications in adults. If it is thought the same compound can be used for pediatric indications, how should the "adult data" be employed, if at all? Expert opinion will be needed in such considerations, both in the decision to use such data and in its "translation" for prior construction in the pediatric context.

The decision to use elicitation can be complex but always requires justification. For purposes of illustration, suppose that we have a binary event of interest. We wish to investigate the use of a novel compound (call it "study drug") for treating psoriasis. We will use a measure of the severity of psoriasis, the PASI score, as our primary efficacy measure of interest. It combines the severity (erythema, induration, and desquamation) and percentage of affected area (Fredriksson and Pettersson, 1978). Specifically, efficacy is defined as a 75% reduction in the PASI score—the so called PASI-75 score. Suppose we must design a clinical trial to assess the efficacy and safety of the study drug to treat this disease. There are several questions to consider when determining whether or not prior elicitation may be useful, including:

- What historical data is available for the study drug?
- What historical data is available for other compounds used to treat psoriasis? Of those, do any have a similar mechanism of action that may be used to possibly inform the likely effect for the study drug?
- What phase is currently being designed (e.g., dose finding, phase 3)?
- Is there any historical data on other endpoints of interest that may relate to PASI-75 (e.g., PASI-50 scores or improvement in Quality of Life (QoL) using the Dermatology QoL Index)?

- Is there any historical data for the study drug or key competitors in other indications that may be related to psoriasis?
- Are clinicians or other individuals with the necessary expertise available?

Answers to these questions may lead to exclusive use of historical data to construct needed prior distributions. But those answers may instead contribute to an evidence dossier facilitating expert elicitation (see Section 4.2).

4.1.2 Overview of Prior Elicitation

The use of expert opinion in the construction of a prior probability distribution—*prior elicitation*—is a multi-disciplinary process. Its purpose is to provide a quantitative summary of the current state of knowledge and uncertainty about an unknown quantity of interest in order to fill the data gaps where sufficient empirical evidence is not yet available. According to the United States National Academy of Sciences [NAS (2002)], the rigorous use of expert elicitation for the analyses of risks is considered to be quality science.

In the biomedical context, the elicitation process provides a formal link between scientific beliefs and clinical trial designs. Before the use of Bayesian methods, that link was often implicit (and unchallenged) in the trial design process. The foundation of an elicitation is an interview process in which experts are asked a series of questions about their beliefs regarding one or more uncertain quantities, such as a mean treatment effect or the probability of a specific adverse event. Based on the expert's responses to these questions, the statistician can derive a probability distribution for the quantity of interest. The probability distribution will reflect what the expert believes about the value of the quantity as well as the uncertainty of that belief. Prior distributions constructed in this way are sometimes called *belief distributions*.

Prior elicitation requires participation of several experts, ideally with a variety of backgrounds and beliefs. In our experience, this usually includes four to eight individuals. Subject-matter experts are clearly needed, and care must be taken in their selection. Duplication of expertise may assist in gauging differential opinions among experts but at the risk of overweighting opinions held in common. The latter is problematic if the experts are chosen with similar experience, for example, from the same research hospitals.

Statisticians typically provide probabilistic training for the experts prior to the elicitation. Statisticians must also "set the mathematical stage" on which the prior is considered, formulating the corresponding data model, prior probability model, and other technical aspects of the problem. This includes, crucially, a means of providing feedback to the expert, with subsequent adjustments of the belief distribution. The actual elicitation is managed by a *facilitator*, a person with expertise in the often subtle art of formulating and asking questions of the expert.

In the background is the decision maker or group that will ultimately use the information. Their needs should help formulate questions in the larger context, of course. But even in detail, the ultimate consumer of analyses dependent on the elicited prior may influence operational choices, such as the scale used for various questions.

Here is a brief sketch of the elicitation process:

- Preparation: Activities for the individuals involved, including those conducting the elicitation and the expert. This includes gathering background information, selecting one or more quantities of interest, as well as specification of related data models and prior families. It also involves choosing experts, preparing software, and logistical arrangements. A detailed protocol for elicitation should be written, as discussed further in Section 4.2.
- Elicitation: The actual exercise in which assessments are made of expert's knowledge and uncertainty.
- Fitting: Using information from the last step, construction of the required probability distribution(s).
- Summarization for feedback: Construction of summaries for feedback to the expert.
- Feedback and assessment: Is the expert satisfied with the probability distribution? If not, return to the elicitation step.

Elaborations on this scheme will be needed when more than one expert is involved, as we shall see.

4.2 A Framework for Eliciting Priors on a Simple Unknown Quantity

Once it has been decided to construct a belief distribution, the elicitation team must have a clear understanding of the statistical model and decision problem of interest. In general, this includes not just the quantities about which expert opinion is needed but the potential for dependencies among them. In this section, we will focus on the easier problem of a single, relatively simple unknown quantity to establish the basic ideas underlying the elicitation process. In the setting of clinical development, the parameter of interest will often be a measure of the treatment effect relative to a control. This could be the absolute difference between the treatment and control, a relative difference, or a hazard ratio (i.e., time to event outcome).

In most cases, an expert's judgment of a treatment effect will depend on what he or she believes about the control response. In some cases, the expert may be willing to assume that a treatment difference is independent of the value of the control response. One solution is to first elicit an independent prior for the control response and then elicit the treatment response conditional on assuming a particular value for the control response (typically the mean or mode from the elicited control prior). In this scenario, one could choose to elicit the treatment response and then convert the prior to one for the relative or absolute treatment difference (by simply dividing or subtracting by the fixed value). Or one could elicit the relative or absolute treatment difference conditional on a particular value for the control response (which can also be converted to prior of the treatment response).

If a data model and prior distribution families have been chosen, as we assume here, the parameters for which a joint prior is needed will be identified. The data model, for example, might be specified for a future clinical trial or a meta-analysis of existing trials. Often, elicitation will not be focused on the parameters of such models. Instead, the choice of what quantity to elicit should be operationally relevant to the expert, chosen in such a way as to allow priors on the needed parameters to be inferred. That is, the quantity should be defined in such a way that expert is able to apply his or her knowledge as directly and fully as possible without necessitating "mental gymnastics". The statistician then converts that information into a prior on model parameters. Thus, for example, eliciting a mode and a percentile is probably preferable to asking about a mean and variance. Both can be used to infer the location and scale of a prior, but the former specification is likely to be closer to an operational scale for the expert. We consider several examples of this later.

Questions for the experts should be carefully formulated, taking into account typical practice in a therapeutic area. For example, when answering questions about time to progression-free survival, a clinician may prefer to think about the number of treatment cycles to date rather than time since diagnosis. It is therefore important to identify such preferences before the elicitation.

4.2.1 SHELF

Prior elicitation is an exercise in measurement—the assessment of expert opinion and uncertainty. Good measurement practice requires a well-defined protocol. Thus, when carrying out an elicitation, it is advisable to use a formal, documented, framework or process. The framework should cover all aspects of the elicitation including the planning, execution, and documentation of the elicitation. The SHeffield ELicitation Framework (SHELF), proposed by Oakley and O'Hagan (2019), is one such process. This comprehensive framework covers many aspects of the elicitation process in a formal and consistent way. Some key features of SHELF include the role of the facilitator (who is an expert in the elicitation process), the use of "behavioral aggregation" to construct a single "consensus prior distribution", along with corresponding documentation. SHELF also includes software to provide real-time feedback to the experts regarding the way in which their individual beliefs are translated into probability distributions and the consensus prior. Here, we present SHELF to illustrate the value of a formal framework as well as key principles of elicitation. The basic steps in the SHELF framework are outlined in Table 4.1. We discuss them below. They can be modified to meet the needs of the specific situation.[1]

4.2.1.1 Step 1: Establish an Elicitation Protocol

This step should establish the questions to be asked and the order in which they are pursued. As noted above, this may require the selection of an underlying statistical model to identify parameters of interest. This model should be thoroughly described. Again, the questions asked of an expert will often be

TABLE 4.1
Main Steps in SHELF Elicitation Process

1. Establish a protocol	Specify the questions to be asked, including specific wording. This may require the selection of an underlying statistical model to identify parameters of interest
2. Select experts	These can be both internal and external to the organization and should involve only those that have a good understanding of the details that need to be elicited
3. Train experts	Provide experts with an overview of the elicitation process and the use of subjective probabilities and probability distributions
4. Evidence dossier	Prepare and review an evidence dossier that captures all pertinent information that the experts would rely upon to formulate their opinion
5. Elicit individual priors	Elicit, in a masked fashion, individual priors from each expert (i.e., experts are unaware of what other experts believe at this point)
6. Discuss individual priors	Share and review results from individual elicitations including each expert's rationale for their beliefs; discuss differences between experts
7. Agree consensus prior	Where possible, elicit a "consensus" prior from the experts which is based on what they collectively agree a "Rational Independent Observer" would determine after having observed the previous conversations
8. Documentation	Provide a written record of the elicitation session

[1]See Section 4.9 for a discussion of associated software to facilitate implementation of the SHELF protocol. A related tool, called MATCH, is also discussed there.

distinct from the parameters on which priors will be placed using the elicited information. The protocol should detail the transformations necessary for conversion of experts' answers to the necessary prior distributions. For example, suppose one is interested in the probability of an adverse event and a logistic regression model is selected to make use of several covariates. That model should be described in the protocol, along with the general form of joint prior structure. In this case, prior distributions on the regression coefficients will be needed. Instead of querying experts about regression parameters directly, questions about conditional means should be asked. The transformations from experts' answers to such questions back to the regression parameters of interest should be detailed in the protocol. See Section 4.4.2 for an illustration. Finally, the protocol should specify the methods to be used for obtaining a consensus distribution.

4.2.1.2 Steps 2 and 3: Selection and Training of Experts

One key aspect of a prior elicitation is the selection of experts. The group of experts chosen should cover the range of available knowledge and perspectives without duplication. Individuals with the relevant experience and expertise are often members of the drug development company and likely to be members of the project team, or they may be external key opinion leaders who have a vested interest in the proposed trial (e.g., unmet need for patients). Although there is nothing wrong with such individuals being experts for the elicitation process, there is a potential for this to lead to over-optimism (or potentially under-optimism). Documenting panel members' expertise, any potential conflicts of interest, as well as gaps in the areas of expertise covered by the panel creates transparency about any potential biases. There is no ideal solution to the balance between expertise and impartiality when selecting the experts, and having a skilled facilitator is critical to ensure experts' in-depth knowledge is brought to the prior elicitation process effectively in an unbiased fashion.

Prior to conducting the elicitation, it is important the experts be trained with the objective of becoming familiar with the process and concepts of subjective judgments, as well as to educate them on the common pitfalls of making judgments in the presence of uncertainty. See, for example, Chapter 3 of O'Hagan et al. (2006) for a thorough review.

4.2.1.3 Step 4: Evidence Dossier

A key part of the pre-elicitation preparation phase is the assembly of an evidence dossier summarizing all relevant historical data and biological or technical information pertaining to the quantities to be elicited. This might include quantitative summaries (e.g., meta-analysis) of previous internal or external studies of the molecule of interest (possibly for different populations, indications, or endpoints) or of a related molecule with a similar mechanism. Alternatively, the dossier may focus on the endpoint of interest just in control populations if no relevant historical data relating to the active molecule is available. In elicitations conducted during early phase development, there may be no relevant clinical data, in which case the evidence dossier will usually focus on summarizing the relevant biology and evidence supporting the potential mechanisms of action of the new molecular entity under consideration. While the evidence dossier should be comprehensive, it should also be as concise as possible to ensure that experts will have sufficient time to review and digest the data in advance of the session. The evidence dossier is typically constructed by the statistician and clinical lead of the project.

The evidence dossier should be sent to all experts in advance of the elicitation session. Experts should also be asked to provide any additional evidence (e.g., published or unpublished data) that they are aware of that is relevant to the elicitation problem prior to the session. During the elicitation session, but before constructing individual belief distributions, the authors of the dossier will present a review of its contents. This provides all experts with a common understanding of the relevant data and knowledge about the unknown quantities to be elicited.

During the review, experts will have the opportunity to critique the scientific merits and flaws of the studies in the evidence dossier. For example, the expert might point out differences in study populations or changes in medical practice since the conduct of a study included in the dossier. Here too, experts will be able to share any additional knowledge or data that was not captured in the evidence dossier. This could be a published or unpublished study unfamiliar to the authors of the dossier. However, the facilitator may want to advise the experts to carefully consider how much weight they want to give this new information, given its lack of thorough vetting.

While the experts are encouraged to critique the science of studies in the dossier during the review, it is important that they do not provide their opinions about the relative merit of a study and/or how they plan to weigh that evidence. In particular, this precludes a summary judgment about the inclusion or exclusion of a study by any one expert. Such judgments are considered as a group after individual belief distributions are constructed and are the focus of Step 6.

4.2.1.4 Steps 5–7: The Individual and Consensus Prior Elicitation

The elicitation session itself is conducted with all experts gathered in a single room or a virtual equivalent. The facilitator will lead the experts through steps 5–7 of the framework.

Step 5 is the elicitation of each expert's individual probability distribution. This is carried out in a blinded fashion designed so that experts are unaware of the judgments of the other experts. There are several techniques that can be used to elicit an expert's probability distribution (quartile/tertile, probability, and roulette) as discussed in Section 4.3.

Individual judgments are shared and discussed in Step 6. A graphical representation of each expert's individual belief distribution is shown to all the experts. Each expert is then asked to explain the rationale for their judgments (e.g., what data they put most weight on, what data they discounted). Other experts may comment on the rationale, either agreeing with or challenging it. It is important that all experts participate in this exchange and that no expert dominates the discussion.

Individual judgments are combined into a single "consensus" probability distribution in Step 7. This distribution should reflect the collective beliefs of the experts. There are several methods for doing this, as discussed in Section 4.5. In SHELF, this is achieved by employing a technique known as "behavioral aggregation", as detailed in Section 4.5.1. The facilitator is crucial in mediating this process and in keeping the discussion focused and relevant. The experts need not—indeed, should not—be required to accept the consensus prior as their own personal prior belief.

It is not always possible to gather experts in a single location, or even to have them available at the same time in different locations. In such cases, the elicitation can be carried out using one of the following approaches:

- Using communication technologies to convene the expert panel in a "virtual" room (e.g., video teleconference). In this setting, it is recommended that a facilitator be available in each room.

- Conduct one-on-one interviews to elicit individual priors from each expert. It is critical that the rationale for the expert's belief (Step 5) also be documented. It is recommended that the same facilitator conduct each interview.

- Elicit individual priors via questionnaire-based approach whereby experts initially independently produce a prior. This process is sometimes referred to as the *Delphi method.*

The Delphi method was originally developed by the RAND corporation in the 1950s for use in classified studies conducted for the United States Air Force on bombing requirements (Dalkey and Helmer, 1972). In the prior elicitation context, it is an iterative, questionnaire-based approach whereby experts initially independently produce a prior using a survey type tool, together with a brief written justification for their beliefs. Then, through a process of controlled interaction, each expert is given the opportunity to review the anonymized opinions from their peers and to update their prior distributions. This process continues iteratively until a consensus prior emerges or the facilitator decides to stop and mathematically aggregate the individual priors. Anonymity of experts is a specific feature of the Delphi protocol, as the technique is intended to reduce the social and political pressures to accept judgments that can arise from the dynamics within groups. By removing identifying information from feedback, experts are free to concentrate on the merits of the feedback information itself without being influenced by potentially irrelevant cues.

The original Delphi method was intended to capture point estimates only and not the uncertainty around these, and there are several online tools available to facilitate use of the Delphi approach in this context. However, an extension that captures uncertainty as well, called probabilistic Delphi, has been developed. For an extensive overview, see European Food Safety Authority (2014), Section 6.3.

An advantage of the Delphi method is that experts don't have to come together for a face-to-face or video conference meeting, and it eliminates the psychological issues of group discussion. However, there is a risk that experts do not fully engage in the procedure. Furthermore, careful training and feedback is required to ensure that each expert understands the process and the questions being asked. The limited interaction between experts also loses a principal benefit of the behavioral aggregation approach (see Section 4.5.1). Our experience of using prior elicitation to support internal decision-making in large drug development organizations suggests that the interaction among experts and exchange of rationale for the individual priors during the group stage of the elicitation is of value in its own right. Such an exchange allows stakeholders to see the diversity of opinions that experts hold about a potential drug effect and to uncover the reasons for heterogeneity, and potential resolution of these differences, in a totally transparent way.

As mentioned, the facilitator has a key role in the success of the elicitation. The facilitator guides the experts through the process, ensuring that all viewpoints are shared and debated and at the end presents the fitted probability distribution representing the experts' beliefs. The facilitator has a critical role in navigating the numerous challenges that can arise during an elicitation. In addition to the lead facilitator, it is helpful to have one or more assistants responsible for recording key details of the elicitation session and running the software to fit probability distributions using the values elicited from the experts.

4.3 Methodologies for Eliciting Priors for Simple Unknown Quantities

There are several methods for using expert opinion to construct priors and a vast literature on the subject. For a general overview, see Garthwait et al. (2005) and Oakley and O'Hagan (2019). See also Kadane and Wolfson (1998), Soares et al. (2011), and Kinnersley and Day (2013). Graphical methods emphasizing prior predictive distributions are also available, along with associated software. See Casement and Kahle (2017).

We now consider a few of the more commonly used methods for eliciting an unknown scalar quantity. This quantity may not—probably will not—be a parameter for the underlying data model but can be used to infer the needed prior on that parameter. All of these methods have the following foundation.

Begin by specifying the data model and associated prior families. In this way, the quantities to be elicited are formally identified. For example, suppose we are interested in the probability of an adverse event for a certain treatment in a well-defined population. We are planning a trial for which a binomial data model is appropriate. Then, we may wish to select a prior distribution from the beta family for the probability of an adverse event, call it θ. In this case, direct elicitation of a prior on θ is plausible, as we shall see in Section 4.3.4. Or perhaps we are interested in change from baseline for a continuous efficacy measurement whose distribution is thought to be well-approximated by a normal distribution with unknown mean, μ, and variance, σ^2. We might elicit information directly on μ, perhaps selecting a normal prior. However, direct elicitation on σ^2, or even σ, is inadvisable. Rather, it is better to elicit information about a mode and percentile and infer the mean and variance. We consider such issues in more detail in Section 4.4.

More generally, let θ be the unknown scalar quantity. For any elicitation method, the expert must provide a plausible range, R, for the values of θ. Here R should have a high prior probability (e.g., $> 99\%$) of containing θ. That is, the expert should be very surprised if θ is not in R. Knowing this as well as certain percentiles or "bin" probabilities can often facilitate the selection of a prior from a family of probability distributions.

4.3.1 The Probability Method

The probability method requires that the expert provide a probability for each of three intervals within R. Let θ be the unknown quantity. Let θ_1, θ_2 and θ_3 be fixed values in R such that $\theta_1 < \theta_2 < \theta_3$. The facilitator will ask experts to answer the following questions:

- What is your probability that θ is less than or equal to θ_2?
- What is your probability that θ is less than or equal to θ_1?
- What is your probability that θ is less than or equal to θ_3?

These probabilities can be used to help select a prior from an appropriate family of priors on θ.

The facilitator must pay careful attention to the selection of each fixed value or interval, to avoid biasing the experts' judgments by subconsciously drawing their attention to particular values (referred to as "anchoring bias"). Indeed, the SHELF authors do not recommend using the probability method for individual judgments due to the potential for anchoring bias. This method is, however, often used to elicit the consensus prior. In addition to SHELF, this method is also implemented in the MATCH software for elicitation. See Morris et al. (2014) and Section 4.9.

4.3.2 Quartile Method

The quartile method asks the expert to split the plausible range R into four intervals of equal probability by asking questions to obtain the expert's lower quartile, median, and upper quartile (25th, 50th, and 75th percentiles, respectively) for the value of θ. To determine the median, the expert is asked to choose a value, a, in R such that θ has the same probability of being above or below a. To do this, the value of a is adjusted until the expert is indifferent between the following gambles: Gamble 1 is that if $\theta < a$, then a large reward is received. In Gamble 2, a large reward is received if $\theta > a$. (There is no penalty for being wrong.)

To determine the lower quartile, the expert is asked to assume that θ is below the median elicited above (a) and is then asked to choose a value, b_{L}, such that $\theta < b_{\mathrm{L}}$ with the same probability that $b_{\mathrm{L}} < \theta < a$. The upper quartile is chosen in a similar fashion. The "tertile" method is very similar with R being split into three intervals of equal probability. A prior can be selected from a family of distributions to approximately match these percentiles. These methods can be used to elicit both individual and consensus priors. (They are implemented in SHELF and MATCH.)

4.3.3 Roulette Method

The roulette method requires that the expert build a histogram to represent the belief distribution (individual prior). The plausible range R is split into equally spaced intervals, and the expert is asked to place "chips" in the bins to create a histogram which represents his or her beliefs, as depicted in Figure 4.1. The width and number of intervals should be chosen so as

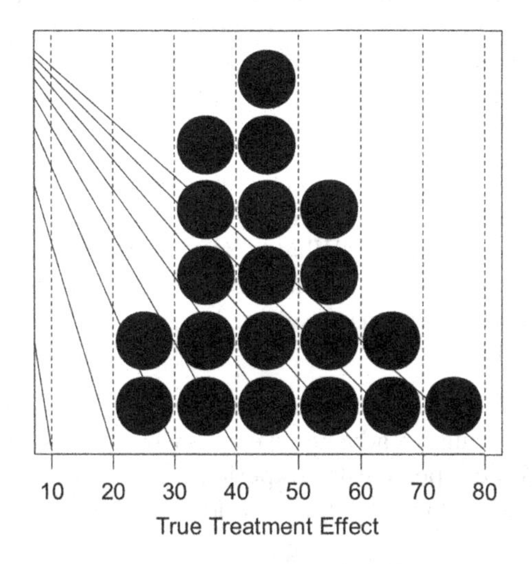

FIGURE 4.1
Example of a "bins and chips" (roulette) elicitation method.

to simplify the elicitation task for the expert (in terms of level of precision required) at the expense of accuracy in representing the expert's "true belief". Similar considerations should be given to the number of "chips". In general, it is recommended that R be split into ten bins and that experts are given 20 "chips" (corresponding to 5% probability each). A probability distribution can now be chosen that approximates the probabilities in these bins.

Our experience is that experts find roulette more intuitive to use when eliciting individual priors. However, the roulette method is generally not recommended for eliciting the consensus prior as it can be cumbersome to use in a group setting. (This method is implemented in both SHELF and MATCH.)

4.3.4 Mode–Percentile Method

Elicited values of a mode and a percentile can be used to select a member of a family of distributions suited to representing uncertainty about an unknown parameter. As an illustration, suppose θ is the probability of an adverse event. The familiar beta family of probability distributions is commonly used to model uncertainty about a probability. The beta density has the form

$$p(\theta|\alpha,\beta) = \frac{1}{B(\alpha,\beta)}\theta^{\alpha-1}(1-\theta)^{\beta-1}, \tag{4.1}$$

where $\alpha > 0, \beta > 0$, $0 < \theta < 1$, and

$$B(\alpha,\beta) = \int_0^1 t^{\alpha-1}(1-t)^{\beta-1}dt$$

is the beta function. We denote this distribution by Beta(α, β).

Clearly, the facilitator should not attempt to elicit values of α and β directly from the expert. As we have noted, it is always preferable to ask questions in a scale and context familiar to the expert. If θ is the probability of a certain adverse event for some treatment, then the facilitator might proceed as follows:

- First, the expert should be directed to imagine a group of 100 similar patients on treatment at a given dose and for a common length of time. Then, the expert is asked, "What is the most likely number of patients suffering an adverse event in that group?" This value is treated as the mode of the beta distribution, given by $(\alpha - 1)/(\alpha + \beta - 2)$ assuming $\alpha > 1$ and $\beta > 1$.

- Next the expert is asked, "What is the largest (or smallest) value that the number suffering an adverse event can be?" This response is treated as an upper (lower) percentile. Experience has shown that assignment of extreme percentiles to that "upper" ("lower") number should be avoided.

Now, with these values specified, an inversion process can be used to determine values of α and β. See Wu et al. (2008) for details. Similar schemes for other families of distributions can also be constructed.

4.3.5 Mixture Method

In many situations, there is a reasonable probability that the drug being tested will demonstrate no efficacy in the planned endpoint of interest. Furthermore, it may be scientifically implausible that the drug being tested will have a true negative effect. In these situations, eliciting a unimodal distribution may not accurately capture an expert's belief about the true effect since there may be insufficient probability around a near-zero effect.

To overcome this issue, one may choose to elicit a bimodal ("spike and smear") distribution. This is done by (1) eliciting the expert's probability, say w, that the drug has a true positive or favorable effect and (2) eliciting the distribution of this effect size under the assumption that the drug does have a favorable effect. A mixture distribution can then be formed to represent the overall prior for the treatment effect. For example, if the prior conditional on the drug having a favorable effect has been elicited on the scale of the treatment difference, this distribution can be weighted by a factor $1 - w$ and then a "spike" can be added with weight w at 0 (absolute difference) or 1 (relative difference) to represent the probability of no effect.

Figure 4.2a gives a hypothetical example to illustrate this approach. Note that, because of the difficulty in plotting a mixture of a discrete and continuous distributions (the height of the spike is ill defined), we follow Walley et al. (2015) and scale the height of the spike to equal its probability mass and scale the continuous conditional distribution so that its maximum equals its total probability. If a prior for the control response has also been elicited, this can be combined with the bimodal prior for the treatment difference to provide a mixture prior for the active response (black distribution in Figure 4.2b), which has weight w on the control response distribution (gray distribution in Figure 4.2 (b)) and weight $1 - w$ on the conditional distribution for the active response (dark gray distribution in Figure 4.2 (b)).

A useful source of information when eliciting bimodal priors are statistics on the success rates in drug development. See, for example, Biotechnology Innovation Organization (2018). Through the use of cross-industry benchmarking of success at each stage of drug development, experts can be provided with background success rates for many diseases under consideration for compounds at different stages of drug development. These statistics can help calibrate experts' opinions on the probability of a compound failing to demonstrate efficacy. Walley et al. (2015) used such a portfolio prior in a case study in which they elicited a prior for the treatment effect in a proof of concept study. Their approach was to construct a bimodal or spike and smear prior by eliciting a distribution on the basis of experts' beliefs about the treatment effect if the compound was "active" and then adding a spike of probability at 0 with weight 60%, which they argue is comparable with industry attrition rates at this stage of development. Alternatively, these portfolio success rates can be used as additional background evidence, to be taken together with all information on the specific compound of interest that the experts synthesize

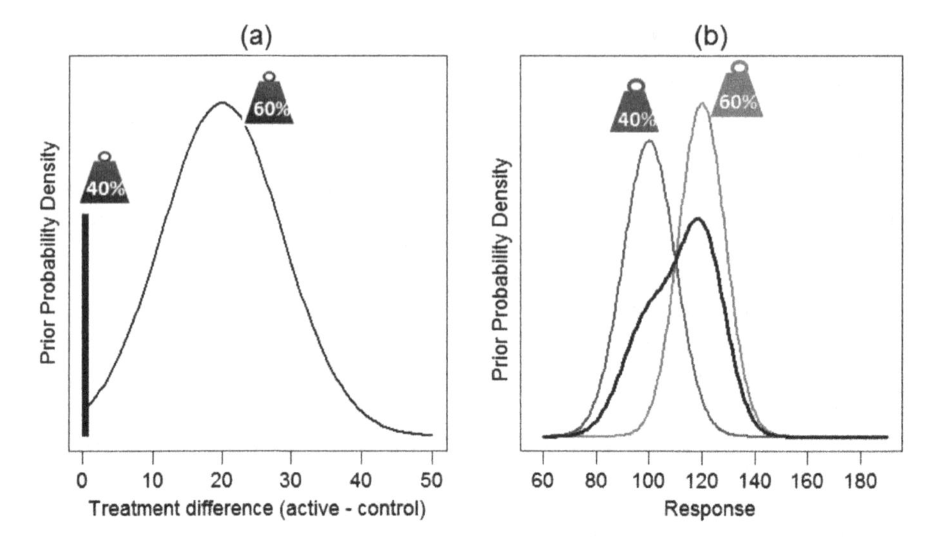

FIGURE 4.2
(a) Illustration of bimodal "spike and smear" distribution for a treatment difference where experts have given 40% probability that there is no true benefit of the drug (represented by spike at zero). (b) Illustration of mixture prior for response on active treatment.

to formulate their prior. This forces experts to robustly articulate and justify if and why they believe the probability of a true positive effect is higher for a specific compound than the overall portfolio success rate, which in turn is valuable information for decision makers and investment boards when balancing risks across the portfolio. More general mixture distributions can also be used, at the expense of adding additional prior parameters. For example, mixtures of two beta distributions might be used, for which interpretation is straightforward.[2] In some cases, one might construct a mixture of the control effect distribution (e.g., no effect) and treatment effect distribution. See Chapter 14 for further examples.

4.4 Indirect Elicitation of Priors

As noted earlier, to begin the elicitation process, a data model must be selected, along with a suitable family of prior distributions. The parameters indexing the family of priors need not correspond to potentially observable

[2]See, for example, Bernardo and Smith (1994), pp. 279–285, for an illustration with mixtures of conjugate priors.

quantities, that is, to a scale familiar to the expert—an "operational" scale. Again, a general principle of elicitation is that experts should be asked about observable quantities rather than abstract shape or scale parameters (Garthwaite et al., 2006, especially p. 689). So, a major feature of any elicitation of prior distributions is to translate questions about parameters into queries on an operational scale. For prior means and proportions, this is straightforward, as we have seen. Other parameters, such as variances and correlations, are problematic, as they can be difficult to assess directly. We illustrate this with two common problems in this section. As we shall see, the elicitation methods are highly problem dependent.

4.4.1 An Example: Eliciting Priors for a Correlation Coefficient

Our first example concerns construction of a prior on the correlation coefficient. Suppose we are studying a treatment for obesity and we are interested in modeling an expert's opinion about the relationship between reduction in A1c from baseline (Y, in mg/dL) to the reduction in weight from baseline (X, in kg) in a certain treatment group. As these values are likely to be dependent, our model must include an assessment of their association. If a bivariate normal prior is appropriate for this problem, this requires a prior on the correlation, ρ.

Direct elicitation for ρ has been studied but, like other scale parameters, can be problematic.[3] Instead, we can take advantage of a relationship between ρ and certain conditional probabilities (Gokhale and Press, 1982). Specifically, the expert can be asked questions about quadrant or "concordance" probabilities, and the results can induce a prior on the correlation. It is easier to assess an expert's opinions about probabilities than about correlation coefficients. To this end, suppose we elicit a beta prior on

$$p_C \equiv \Pr(Y > y_m | X > x_m), \tag{4.2}$$

where y_m and x_m are the medians of Y and X, respectively. Then, we can induce a prior on ρ using the identity

$$\rho = \sin(\pi(p_C - 0.5)). \tag{4.3}$$

Eliciting the beta distribution needed as a prior on p_C in this example might proceed as follows. We assume that marginal priors for the reduction in A1c and weight have already been elicited from the expert, yielding medians y_m and x_m, respectively. The following questions are asked of the same expert:

- First, think about patients in this treatment group with a reduction in weight exceeding the median reduction, x_m. What is the chance a patient

[3]See O'Hagan et al., 2006, Chapter 5 for an overview of such problems; see also the R function copulaSample in the package SHELF, Oakley and O'Hagan (2019).

chosen from *that* group will have a reduction in A1c larger than the median, y_m? (Denote the expert's answer by p_{50}. Treat this as the mode of the beta prior distribution for p_C.)

- Next, we need to assess your uncertainty about the value p_{50}. So, still conditioning on the reduction in weight exceeding the median for this treatment group, how much larger than p_{50} might that chance be? (Treat this answer as the 75th percentile of the beta on p_C, denoted by p_{75}.)

We can use p_{50} and p_{75} to determine the shape parameters of a beta distribution to serve as a prior on p_C.[4]

4.4.2 An Example: Eliciting Priors for Regression Parameters

For a second example, we turn to a common problem for the practicing Bayesian statistician: specifying priors on regression parameters. Here, we sketch a method by which priors for regression parameters can be determined via indirect elicitation. The priors produced by this method—induced priors on the regression parameters—are known as *conditional means priors* (CMPs).[5]

Eliciting prior information about regression coefficients can be problematic. Intercepts may not correspond to zero-values of covariates. Partial slopes can be difficult to interpret. Average response values, on the other hand, are easier to elicit from experts. This is the central idea behind CMPs. For instance, in logistic regression, success probabilities for various covariate settings are typically much easier to elicit than beliefs about changes in log-odds ratios.

The idea is easier to understand in the context of a specific problem. Suppose patients being treated with a certain drug may experience hypertension. Potential confounders for the group under study include obesity and smoking habits. Let y_i be an indicator for hypertension, and assume

$$y_i \sim \text{Bernoulli}(\theta_i),$$

where the probability, θ_i, depends on three binary covariates: smoking (S_i), obesity (O_i), and treatment (T_i). A logistic regression model for this has the form, for $i = 1, \ldots, n$,

$$\text{logit}(\theta_i) \equiv \log\left(\frac{\theta_i}{1-\theta_i}\right) = \beta_0 + \beta_S S_i + \beta_O O_i + \beta_T T_i, \tag{4.4}$$

[4]This is easy to do numerically using, for example, epi.betabuster, an R function in the package epiR. The package is described at https://CRAN.R-project.org/package=epiR.; see R Core Team (2018). This induces a prior on ρ through Equation (4.3).

[5]Thanks to James D. Stamey and David J. Kahle, both of Baylor University, for contributions on the topic of CMPs in this section. Conditional means priors are also known as BCJ priors (after Bedrick, Christensen, and Johnson) following Bedrick et al. (1996, 1997).

where, for simplicity, we have not included an interaction term. Here, regression coefficients are interpreted on a log-odds scale. For example, $\exp(\beta_S)$ is the multiplicative change in odds for smoking, with the other covariates fixed. Thus, for an obese patient under treatment, the odds of responding are changed by a factor of $\exp(\beta_S)$ compared to a non-smoker.

Although elicitation could use a change-in-odds scale, it would be better to ask questions in the scale of the expected response for fixed values of the covariates. This is likely closer to the expert's experience than thinking in terms of changes in odds. For example, suppose we ask the expert to think about an obese smoker under treatment, asking what is the most likely probability of hypertension, along with an extreme value for that probability. If we interpret these as, say, a mode and a percentile, we can select a beta distribution to represent the expert's uncertainty about that probability, as we did in Section 4.3.4.

This is depicted for several covariate configurations in Figure 4.3.[6] These prior distributions can be "plugged" into Equation (4.4), which can then be inverted to induce priors onto the regression coefficients, as shown in Figure 4.4. CMPs typically do not have closed form distributions but can be easily simulated. This is displayed in Figure 4.5. The entire sampling/inversion process can be done in a probabilistic programing language such as BUGS, JAGS, or Stan.[7]

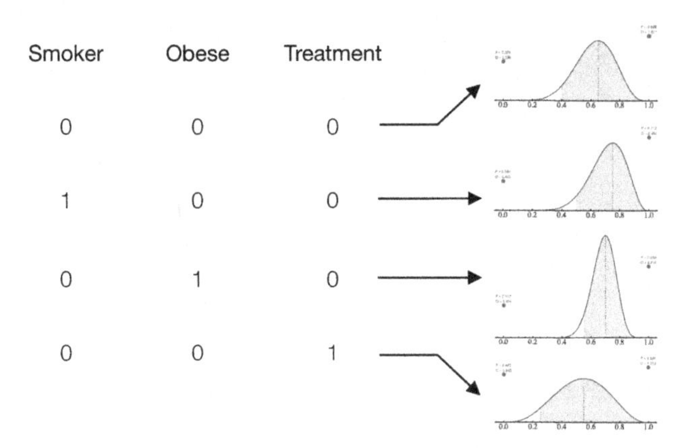

FIGURE 4.3
The first step of the CMP approach is to elicit prior distributions at various design configurations.

[6]In general, for p covariates, we construct K $p \times 1$ "configuration vectors" of covariate values. Note that this requires the design matrix associated with the scenario configurations to be invertible. The design matrix may have a vector of ones prepended to it representing the intercept term.

[7]See, for example, the R package indirect by G. Hosack, located at https://CRAN.R-project.org/package=indirect.

$$\begin{bmatrix} 1 & 0 & 0 & 0 \\ 1 & 1 & 0 & 0 \\ 1 & 0 & 1 & 0 \\ 1 & 0 & 0 & 1 \end{bmatrix} \begin{bmatrix} \beta_0 \\ \beta_S \\ \beta_O \\ \beta_T \end{bmatrix} = g(\begin{bmatrix} \\ \\ \\ \\ \end{bmatrix}) \longleftrightarrow \begin{bmatrix} \beta_0 \\ \beta_S \\ \beta_O \\ \beta_T \end{bmatrix} = \begin{bmatrix} 1 & 0 & 0 & 0 \\ 1 & 1 & 0 & 0 \\ 1 & 0 & 1 & 0 \\ 1 & 0 & 0 & 1 \end{bmatrix}^{-1} g(\begin{bmatrix} \\ \\ \\ \\ \end{bmatrix})$$

FIGURE 4.4
Inverting the regression equation to induce priors on the regression coefficients.

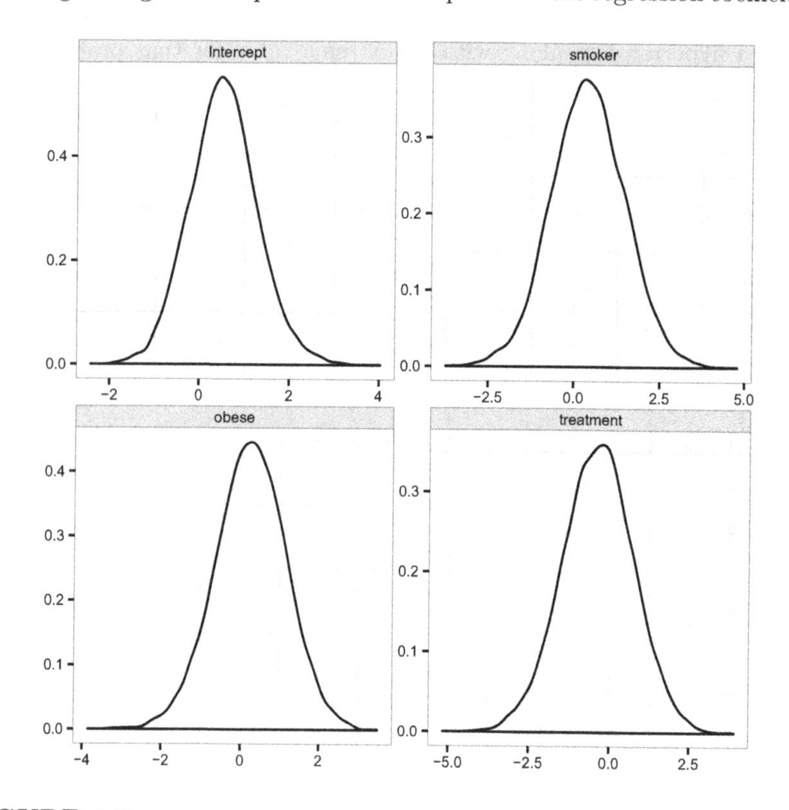

FIGURE 4.5
The induced priors on the regression coefficients can be sampled from at will and then visualized.

4.4.3 Feedback

Constructing a probability distribution to represent an expert's opinions requires feedback to the expert to assure fidelity to those opinions. Ideally, this should be done not just for the quantity of interest, but also using potentially observable consequences of using the distribution. Thus, for example, suppose the elicitation is on median progression-free survival, resulting in, say, a certain gamma distribution. The expert can be shown the gamma density with several features, such as quartiles, indicated. Or perhaps elicitation concerns the difference between a continuous efficacy measure and a baseline for a control and two treatments. The expert can be shown all three densities, with one or more quartiles marked for comparison.

In some cases, the elicitation will enable creation of a complete data model. Then, the prior predictive distribution can be used to generate hypothetical "observations". Descriptive statistics and graphical representations of this simulated "data" can be used for feedback to the expert. One can also provide a *prior effective sample size*—that is, loosely speaking, the number of observations that would be necessary to obtain the same amount of extra information about the parameters represented by the prior.[8] See, for example, Morita et al. (2008).

If, in addition to a data model, one has determined study success criteria, then the prior predictive distribution can also be used to compute the prior probability of study success. Evaluation of a prior with respect to this probability is recommended in the FDA Bayesian Guidance (2010):

> "This is the probability of the study claim before seeing any new data, and it should not be too high. What constitutes 'too high' is a case-by-case decision. In particular, we recommend the prior probability not be as high as the success criterion for the posterior probability." (p. 39)

It may not be appropriate to share this with the expert, but it is a vital aspect of evaluating the suitability of an prior distribution.

4.5 Consensus Priors vs. Mathematical Pools

Eliciting opinions from several experts will lead to multiple probability distributions. Unlike a traditional scientific experiment in which the selection of a technique for combining results can (and should) be made before any data is collected, with expert elicitation, it is necessary to see the results before determining if aggregation is even appropriate. There is a very rich literature on the aggregation of probabilistic representations of expert beliefs. For an extensive overview, see French (2011) or Clemen and Winkler (1999). A briefer review can be found in Chapter 9 of O'Hagan et al. (2006).

Expert opinions will differ for many reasons. Understanding the origin of differences among experts can lead to consensus or revision of the elicitation protocol—see Morgan and Henrion (1990). Indeed, this understanding about the source of the differences can be as, or more, valuable than any aggregate

[8]This is also recommended in the FDA Bayesian Guidance (2010).

finding. Nevertheless, it is usually desirable, where possible, to obtain a single distribution that encapsulates the beliefs of several experts and which can be used for analysis and decision making.

There are two options available for obtaining a combined single distribution. The first is to elicit a distribution from each expert individually and independently of the others and mathematically combine the resulting distributions into a single distribution. This approach is referred to as *mathematical aggregation*. The second is based on a guided interaction among the experts, through which a single distribution is elicited from the group as a whole. This is referred to as *behavioral aggregation*. As we have noted, SHELF uses the latter. In this section, we consider both approaches.

4.5.1 Consensus Priors Using Behavioral Aggregation

The method of behavioral aggregation assembles the experts in a group and elicits their beliefs as if they were a single expert. This is the approach adopted in SHELF to obtain a "consensus prior". To avoid "group think" mentality and specious consensus, it is crucial that behavioral aggregation follows a structured and well-documented process. In the SHELF protocol, individual beliefs are elicited first, followed by a facilitated discussion of these judgments, before asking experts to collectively agree on a consensus prior. Here, the experts are asked to make reasonable judgments about what would be the opinion of a Rational, Impartial Observer (RIO) who has witnessed the individual priors and subsequent discussion. This requires the experts to agree on how much weight should be given to each of their individual priors and supporting arguments. It does not (and should not) require each expert to accept the RIO consensus as their own personal prior belief. The result is called the *RIO prior*.

Having the ability to provide "real-time" feedback to experts through graphical displays of both their individual priors and the RIO consensus prior is critical to the process. When discussing these priors, it is important that the facilitator and experts not just focus on the mode of their distributions but also on the tails. If possible, it is helpful to provide feedback on specific quantiles (see Section 4.4.3).

In some instances, experts may not be able to put aside their beliefs, or there may be fundamental and valid differences in their opinions. Ultimately, failure to reach a consensus does not need to be a major issue, provided that the elicitation exercise can reveal why differences occur. In these situations, eliciting two (or more) separate priors on the basis of the differing opinions may be appropriate as the differing priors and their supporting rationale illuminate the uncertainty of their beliefs.

Alternatively, when there is a high degree of overlap of all the experts' individual priors, mathematical pooling (using a simple average of the individual priors—see below) can be an efficient and acceptable method to achieve a consensus prior. In such situations, after first explaining the purpose of the consensus prior to the experts, we may choose to show them the mathematically pooled prior and ask if they all agree that this provides a reasonable representation of what an independent observer might believe.

4.5.2 Consensus Priors Using Mathematical Aggregation

There are several formal mathematical methods of aggregation. Chief among these are Bayesian approaches, opinion pooling, and Cooke's method. The Bayesian approach treats the experts' opinions as "data", and a "supra-Bayesian" decision maker updates his or her own prior distribution with that information. For a review of this method, see Clemen and Winkler (1999).

Opinion pooling methods construct weighted averages of the distributions obtained using elicitation from experts. These relatively simple methods are widely used, although they have problems. The distributions that result need not represent the opinions of any one person, let alone a consensus opinion from the group of experts. There are also "coherency" issues, as we shall see. We provide this detail in part to illustrate the sometimes subtle pitfalls lurking within seemingly reasonable prior elicitation methodologies.

Suppose θ is a quantity of interest about which multiple experts have been queried. Let $\pi_i(\theta)$ be the probability distribution constructed using elicitation results from the ith expert, $i = 1, \ldots, n$. A *linear opinion pool* is a mixture distribution of the form

$$\pi_{\text{lin}}(\theta) \equiv \sum_{i=1}^{n} w_i \pi_i(\theta), \tag{4.5}$$

where the positive weights, w_i, $i = 1, \ldots, n$, sum to one.

Choosing the weights, w_i, is, of course, a critical task. Experts may be given equal weights $(1/n)$, if all are thought to have equal expertise. An approach based on gauging differential expertise among the elicited experts is provided by Cooke (1991) and extensively evaluated in Cooke and Goossens (2004). Other methods can be used to model experts beliefs as they change over time. See, for example, Satopää et al. (2014).

How a mixture like Equation (4.5) should be updated on receipt of new information—perhaps trial data—is problematic. One can, for example, update the individual π_i's, treating them as n prior distributions and obtaining n posterior distributions using the new data. Then a new linear pooled distribution can be constructed. Alternatively, one can take the original linear pooled distribution, $\pi_{\text{lin}}(\theta)$, and, using it as the prior, obtain a posterior with the new data. In any realistic setting, these two approaches will not yield the same result. Thus, linear opinion pooling is said to be not *externally Bayesian*, a term due to Madansky (1964).

A method that *is* externally Bayesian uses a weighted geometric average of the elicited probability distributions:

$$\pi_{\log}(\theta) \equiv k \sum_{i=1}^{n} \pi_i(\theta)^{w_i}, \tag{4.6}$$

where k is a normalizing constant and the weights, w_i, are defined as above. Unfortunately, although this *logarithmic opinion pool* is externally Bayesian, it has another, equally serious flaw. Suppose each expert is asked about mutually exclusive events, A and B. If C is the event "A or B", then coherency demands that $\Pr(C) = \Pr(A) + \Pr(B)$. There are two ways to obtain a pooled probability for C. We can compute the probability by adding $\Pr(A)$ and $\Pr(B)$ from each expert and pool the resulting sums, or we can pool the elicited probabilities for A and B first and then add the pooled results. With a logarithmic opinion pool, these approaches will not yield the same probability. We say logarithmic pooling fails the *marginalization* requirement.

It turns out that no mathematical aggregation method will satisfy both the external Bayesian and marginalization requirements. Despite these shortcomings, pooling is often employed, with the linear opinion pool being the most commonly used.

4.5.3 Discussion

There are many pros and cons to be balanced in choosing between mathematical and behavioral aggregation, and there is no conclusive evidence to consistently favor one method over the other. The process of behavioral aggregation can help to provide insights and resolve differences of opinion and build commitment to the eventual decision/analysis for which the prior is being elicited. On the other hand, there is a danger that minority or extreme opinions may be lost in building a common group view; the facilitator plays a crucial role in ensuring that the behavioral aggregation is not dominated by only one or two experts. For a planning or a regulatory decision with a wide variety of stakeholders, mathematical aggregation can be an advantage, both because it is explicit, auditable, and, in a sense, objective and also because it leaves all opinions and eventualities in the analysis. However, when individual priors are very diverse, mathematical pooling may create an average set of judgments that lacks any substantive meaning.

4.6 Biases, Heuristics, and Statistics

The goal of elicitation is to model an expert's knowledge and uncertainty about an unknown quantity—not to obtain an estimate close to some "true" value of that quantity. Although each expert is presented with the same evidence and has their judgments assessed by the same process, we don't expect elicitation from different experts to yield identical prior distributions. Experts differ in their training and experiences, in how much weight they place on different sources of information, and in the scientific paradigms they adhere to, and this leads to legitimate differences in their opinions and judgments. However, we do want to obtain an accurate representation of what each individual expert believes about the truth. There are heuristics, or rules of thumb, that most people (including experts) use to make judgments under uncertainty which can lead to prior distributions that fail to accurately represent the expert's beliefs. One common heuristic is anchoring and adjustment. This occurs when judgments under uncertainty are anchored at some starting value and adjusted outward, usually insufficiently. There is a diverse literature on commonly used heuristics and sources of bias that can affect judgment—see, for example, Chapter 3 of O'Hagen et al. (2006) and Kynn (2008). Anyone conducting an elicitation needs to be aware of the key heuristics and potential biases which affect human judgments and decision making and understand how he or she can reduce or avoid these.

The potential for introducing bias into an expert elicitation can be reduced by following a rigorous protocol for conducting the elicitation as described in Section 4.2. Training experts, which should be an integral part of any elicitation protocol, can include discussion of the common pitfalls that people encounter when making judgments under uncertainty. O'Hagan has developed an online training course that can be taken by experts before participating in an elicitation session conducted using the SHELF framework.[9] Most importantly, the facilitator needs to be aware of the key heuristics and biases which affect human judgments and decision making and understand how he/she can reduce or avoid these. Dallow et al. (2018) provide some further commentary on typical challenges encountered when conducting prior elicitations within a large drug development organization and propose some strategies to minimize or overcome these.

To run a successful elicitation process, the experts require a good understanding of both probability and common statistical terms (e.g., quartiles). For example, when using the quartile approach to elicit individual priors, experts are asked to provide their lower quartile, median, and upper quartile for the true value of the quantity of interest. Experts can struggle with the conceptualization of these terms, resulting in distributions which do not in any way represent their beliefs. Careful education is critical to avoid such misunderstandings. Alternatively, one may choose another elicitation method such as the roulette approach, which experts may find more intuitive.

Another issue is that experts sometimes struggle with the concept of the "true" treatment effect, not understanding that this refers to a population quantity, in effect, the result of an exhaustive sample. This misunderstanding risks injecting sampling uncertainty into the elicited prior. Kinnersley and Day (2013) note this potential problem. Once again, the training of experts is crucial to minimizing such risk.

[9]See www.tonyohagan.co.uk/shelf/ecourse.html

4.7 Example: Rhinitis Study

To illustrate several features of the prior elicitation process discussed in this chapter, we consider an example from Dallow et al. (2018). This elicitation focused on a fixed-dose combination (FDC) of two different drugs with different mechanisms of action. Positive data were reported in a phase II proof of concept (PoC) study in an allergen challenge model in which rhinitis patients were exposed to controlled amounts of allergen and the effect of the FDC was assessed using symptom scores. The next planned study was to assess the FDC in a phase III study assessing rhinitic patients in a real-world environmental setting. In addition to the PoC study, data were also available from similar in-house molecules assessed both in the challenge model as well as an environmental setting, plus summary results from a similar FDC in a series of phase III trials. This set of data allowed the degree of association between phase III response and PoC response to be assessed.

The team developing the phase III study were required to provide an estimate of assurance for two purposes. First, that estimate was to be used in seeking approval from internal governance boards to commit to phase III. Second, with that approval, the estimate of assurance would be used to optimally design the phase III trial. It was decided that although there was a wealth of available data, there were still uncertainties around what the effect of the FDC would be in the setting of phase III (i.e., environmental setting), so a prior elicitation session was conducted. A group of six experts was convened, with expertise spanning knowledge of the disease area, mechanisms of action, knowledge of previous development programs relating to the monotherapies, and statistical expertise within the disease area. It was felt that these six experts provided a balance between those who were or were not directly involved in the project (all were internal to the company) and had a wide enough coverage of relevant backgrounds and specific areas of expertise for a successful elicitation.

When planning the elicitation, one consideration was how to structure the quantities to be elicited. A key uncertainty was how the observed treatment difference between FDC and monotherapy in the PoC challenge model would translate to the environmental setting in phase III. One option was to break down the elicitation question into two parts: (1) elicit beliefs about the true treatment effect for the FDC in the challenge model in the phase III target population and (2) elicit beliefs about the relationship between the allergen challenge model treatment effect and the environmental treatment effect. However, after discussing different structuring options with the experts, they expressed a preference for directly eliciting the phase III environmental treatment effect, rather than separate elicitation of the different components.

The SHELF protocol was used for the elicitation, using the quartile method to elicit each expert's plausible range, median, upper quartile, and lower quartile for the true mean treatment difference between FDC and monotherapy in the phase III setting. Individual and consensus priors for the treatment difference between the FDC and the corticosteroid component are presented in Figure 4.6. Figure 4.6a illustrates one expert's quartile judgments together with several alternative fitted parametric distributions. Figure 4.6b shows the fitted prior distributions for all six experts, together with the final consensus prior which was obtained using the SHELF behavioral aggregation approach.

In parallel with the expert elicitation, a model-based prior was derived for comparison purposes. For the model-based prior, a logic similar to the two-step structuring discussed above was used: a linear relationship was assumed between the true treatment differences in the challenge model and environmental settings. A Bayesian hierarchical model was specified for the intercept and slope parameters and fitted to all available PoC (allergen challenge model) and phase III (environmental) study results from other compounds. The predicted phase III treatment difference based on the PoC data for the investigational product was then used as the model-based prior. This model-based prior is shown in Figure 4.6b. It is noticeable that this has considerably more uncertainty than the experts' elicited priors. The elicited priors place negligible probability on the event that the treatment effect is less than zero (for the mechanisms of action in this setting, it was considered to be scientifically implausible for the combination to be worse than monotherapy), whilst the model-based approach suggests this is plausible. Conversely, the model-based

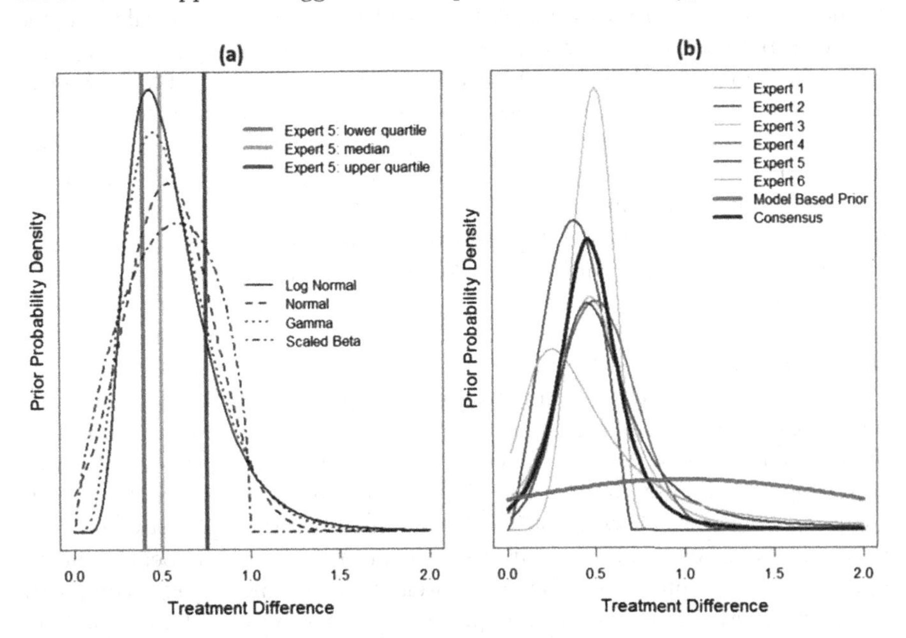

FIGURE 4.6
Rhinitis example from Dallow et al., 2018. (a) Quartile judgments for one expert and fitted parametric prior distributions and (b) best-fitting individual priors, consensus prior and model-based prior.

prior also has considerable probability of a treatment difference above 1.0, in contrast with the elicited priors. A benefit of the elicited priors is the fact that the experts are able to bring in the broader knowledge of the FDC mechanisms and published data on other molecules leading to belief that the effect would be very unlikely to be negative, while clinical expertise of the disease informed a maximal plausible efficacy threshold.

The elicited consensus prior was subsequently used to determine estimates for assurance (approximately 55%), which in turn informed the sample size of the phase III studies as well as the overall development strategy. As a result of quantifying the risk of failure, the development team favored a staggered approach to the two phase III studies to mitigate the cost and risk of a parallel phase III approach. The model-based assurance (approximately 80%) was higher than that based on the elicited prior and would have led to a less appropriate development plan. See Crisp et al. (2018) for further discussion of the assurance calculations for this case study.

4.8 Unexpected Benefits

While the goal of an elicitation is to provide a probability distribution for unknown parameters, the process creates transparency about not only what is known (summarized in the evidence dossier) and the rationale for experts' beliefs, but also the gaps in knowledge and sources of uncertainty. Furthermore, the decisions might reveal previously unexamined aspects of the study design which can help the team optimize study protocols. The elicitation exercise has proven to increase teams' understandings of the potential study outcomes and appreciation of the risks of the current study design.

Perhaps most importantly, the prior elicitation process facilitates a rich and scientifically driven review of evidence. This fosters a more robust collective understanding of the problem and the decisions to be made. Prior to the elicitation, teams must summarize all relevant existing information about the potential treatment that might inform the probability distribution of the quantities of interests. This forces the team to construct a concise summary of current knowledge. During the elicitation process, the experts will be able to review and discuss all relevant existing data and share additional knowledge and concerns. Furthermore, following the elicitation of individual priors, experts are encouraged to discuss the rationales for their beliefs and to understand why some may place more weight on certain evidence compared to other experts. This affords the opportunity for robust, systematic debate among the experts.

An additional benefit of the elicitation process is the opportunity to involve experts from "outside" the project team. This is something that project teams may not routinely consider in the absence of the elicitation process but is strongly recommended for any elicitation. Outside experts can provide an independent perspective. In some ways, the inclusion of these experts is analogous to "peer review" of the team's beliefs about the quantities of interest.

4.9 Software

To aid the facilitator in carrying out an elicitation, it is advisable to have a software package that can record the experts' opinions, convert them into parametric distributions and provide graphical representation of the experts' beliefs in real time. Being able to provide the graphical representation of the individual beliefs will aid the discussion of rationale for an expert's belief and help the facilitator to identify areas of agreement, disagreement, and uncertainty.

As mentioned, SHELF includes an R package (Oakley and O'Hagan, 2019) that provides graphical tools to support elicitation and recording of individual priors and consensus distributions. The software supports use of tertiles, quartiles, or the roulette method for eliciting individual judgments. Any of these methods can also be used to aid elicitation of the RIO consensus prior (although roulette is generally not recommended as it can be cumbersome to use with a group). The software also supports use of the probability method for this stage. The SHELF R package also implements methods to select and graphically display the parametric distribution (normal, Student-t, scaled beta, log normal, log Student-t, or gamma) that provides the best fit to each expert's values. An option is also available to mathematically aggregate several individual priors using the linear pool method and display this alongside all the individual priors.[10]

The most recent version of SHELF (version 4.0) has also been extended to include two multivariate elicitation methods (Gaussian copula method and a Dirichlet method), which can be used for eliciting individual priors for two or more correlated quantities when it is not possible to remove the dependence through structuring. See the SHELF documentation for more details.

Acknowledgements

The authors would like to acknowledge the contributions of Nicky Best (GSK), Nigel S. Dallow (GSK), David J. Kahle (Baylor University), Sam R. Miller

[10]A web-based implementation of some of the code within the SHELF R package, the MATCH Uncertainty Elicitation Tool, is available at http://optics.eee.nottingham.ac.uk/match/uncertainty.php.

(Exploristics), Michael D. Sonksen (Eli Lilly & Co.), and James D. Stamey (Baylor University).

References

Bedrick, E., Christensen, R., and Johnson, W. (1996) A new perspective on priors for generalized linear models. *Journal of the American Statistical Association*, **91**(436), 1450–1460.

Bedrick, E., Christensen, R., and Johnson, W. (1997) Bayesian binomial regression: Predicting survival at a trauma center. *The American Statistician*, **51**(3), 211–218.

Bernardo, J. and Smith, A. (1994) *Bayesian Theory*. Wiley: New York.

Biotechnology Innovation Organization (2018) Clinical development success rates 2005–2016. Available: www.bio.org/press-release/bio-releases-largest-study-ever-clinical-development-success-rates.

Casement, C. and Kahle, D. (2017) Graphical prior elicitation in univariate models. *Communications in Statistics–Simulation and Computation*, **47**, 2906–2924.

Clemen, R. and Winkler, R. (1999) Combining probability distributions from experts in risk analysis. *Risk Analysis*, **19**(2), 187–203.

Cooke, R. (1991) *Experts in Uncertainty*. Oxford University Press: Oxford.

Cooke, R. and Goossens, L. (2004) Expert judgement elicitation for risk assessments of critical infrastructures. *Journal of Risk Research*, **7**(6), 643–656.

Crisp, A., Miller, S., Thompson, D., and Best N. (2018) Practical experiences of adopting assurance as a quantitative framework to support decision making in drug development. *Pharmaceutical Statistics*, **17**(4), 317–328.

Dalkey, N. and Helmer, O. (1972) An experimental application of the Delphi method to the use of experts. RAND corporation, Santa Monica, CA, Report RM-727/1.

Dallow, N., Best, N., and Montague, T. (2018) Better decision making in drug development through adoption of formal prior elicitation. *Pharmaceutical Statistics*, **17**, 301–316.

European Food Safety Authority (2014) Guidance on expert knowledge elicitation in food and feed safety risk assessment. *EFSA Journal*, **12**(6), 3734 (274 pages Available: www.efsa.europa.eu/en/efsajournal/pub/3734).

Food and Drug Administration (2010) Guidance for the use of Bayesian statistics in medical device clinical trials. Available: www.fda.gov/RegulatoryInformation/Guidances/ ucm071072.htm.

Fredriksson, T. and Pettersson, U. (1978) Severe psoriasis–oral therapy with a new retinoid. *Dermatologica*, **157**, 238–244.

French, S. (2011) Aggregating expert judgement. *Revista de la Real Academia de Ciencias Exactas, Físicas y Naturales. Serie A. Matemáticas*, **105**, 181–206.

Garthwaite, P., Kadane, J., and O'Hagan, A. (2005) Statistical methods for eliciting probability distributions. *JASA*, **100**(470), 680–701.

Gokhale, D. and Press, J. (1982) Assessment of a prior distribution for the correlation coefficient in a bivariate normal distribution. *Journal of the Royal Statistical Society, A*, **145**(2), 237–249.

Kadane, J. and Wolfson, L. (1998) Experiences in elicitation. *The Statistician*, **47**(1), 3–19.

Kinnersley, N. and Day, S. (2013) Structured approach to the elicitation of expert beliefs for a Bayesian-designed clinical trial: A case study. *Pharmaceutical Statistics*, **12**, 104–113.

Kynn, M. (2008) The 'heuristics and biases' bias in expert elicitation. *Journal of the Royal Statistical Society A*, **171**, Part 1, 239–264.

Madansky, A. (1964) External Bayesian groups. The RAND Corporation.

Morgan, M. and Henrion, M. (1990) *Uncertainty: A Guide to Dealing with Uncertainty in Quantitative Risk and Policy Analysis.* Cambridge University Press: Cambridge, MA.

Morita, S., Thall, P., and Müller, P. (2008) Determining the effective sample size of a parametric prior. *Biometrics*, **64**, 595–602.

Morris, D., Oakley, J., and Crowe, J. (2014) A web-based tool for eliciting probability distributions from experts. *Environmental Modelling and Software*, **52**, 1–4.

NAS (2002) *Estimating the Public Health Benefits of Proposed Air Pollution Regulations.* The National Academies Press: Washington, DC.

O'Hagan, A., Buck, C., Daneshkhah, A., Eiser, J., Garthwaite, P., Jenkinson, D., Oakley, J., and Rakow, T. (2006) *Uncertain Judgements.* Wiley: New York.

Oakley, J. E. and O'Hagan, A. (2019) SHELF: The Sheffield elicitation framework (version 4.0), School of Mathematics and Statistics, University of Sheffield. Available: http://tonyohagan.co.uk/shelf.

R Core Team (2018) R: A language and environment for statistical computing. R Foundation for Statistical Computing, Vienna, Austria. Available: www.R-project.org/.

Satopää, V., Jensen, S., Mellers, B., Tetlock, P., and Ungar, L. (2014) Probability aggregation in time-series: Dynamic hierarchical modeling of sparse expert beliefs. *The Annals of Applied Statistics*, **8**(2), 1256–1280.

Soares, M., Bojke, L., Dumville, J., Iglesias, C., Cullum, N., and Claxton, K. (2011) Methods to elicit experts' beliefs over uncertain quantities: Application to a cost effectiveness transition model of negative pressure wound therapy for severe pressure ulceration. *Statistics in Medicine*, **30**, 2363–2380.

Walley, R., Smith, C., Gale, J., and Woodward, P. (2015) Advantages of a wholly Bayesian approach to assessing efficacy in early drug development: A case study. *Pharmaceutical Statistics*, **14**, 205–215.

Wu, Y., Shih, W., and Moore, D. (2008) Elicitation of a beta prior for Bayesian inference in clinical trials. *Biometrical Journal*, **50**(2), 212–223.

5

Bayesian Examples in Preclinical In Vivo Research

John Sherington, Phil Stanley, and Ros Walley

UCB Pharma

CONTENTS

5.1 Introduction

Bayesian methods appear to be rarely used in the analysis of preclinical pharmacology data. A literature search by Walley et al. [1] for "in vivo" and "Bayes*" (where the * means "any phrase beginning with Bayes...") covered approximately 270,000 papers in the period from January 2010 to December

2014 and resulted in "no papers disclosing *in vivo* lab data ... addressed with a univariate Bayesian analysis, nor of constructing priors or predictive distributions based on similar historical studies".

Possible reasons why Bayesian methods are not widely used in preclinical research may be that they are more statistically challenging and not available in standard software packages. They therefore require the support of statisticians to implement them. In contrast to clinical studies, data from preclinical studies are usually analysed by the scientists who run the study, typically with little or no input from professional statisticians. Scientists who perform their own data analysis often favour the traditional statistical approaches with an emphasis on t-tests or basic analysis of variance and p values. Even with more statistical support, if Bayesian analyses are to be used regularly, then the issues of which software, who will run the analysis, and in what timeframe all need to be addressed. This issue has been highlighted by Natanegara et al. [2] who give practical recommendations for how to grow the application of Bayesian methods.

In this chapter, we describe three applications of Bayesian methodology in preclinical pharmaceutical research, with the emphasis on *in vivo* pharmacology studies. This emphasis reflects our view that this area is a priority for implementing novel statistical approaches because of the inherent variability in most *in vivo* research. Our initial interest in applying such methods in this area was motivated by two factors. First, the increasing accessibility of Bayesian methods and their application in clinical research prompted us to ask where they could have a role in preclinical research. Second, the increasing importance of the 3Rs initiative in animal research (Replacement, Reduction, Refinement) [3] raises the question from scientists of why so many animals are needed in each control group in each experiment for repeated *in vivo* models and whether there is a way of making use of data or "borrowing information" from control animals from previous studies.

Bayesian methodology is an ideal approach for incorporating prior information from previous studies into the design and analysis of the latest study. The use of data from historical controls to replace or supplement control groups in regularly run studies is the focus of our first application. We then give two further examples of where we have applied Bayesian methods to preclinical data. Our second example gives a Bayesian method for detecting and downweighting outliers in the analysis of small data sets. The final example gives a method for analysing data with non-standard distributions, in this case, scores with many zeros.

5.2 Use of Historical Control Data

The use of Bayesian methods to incorporate data from historical controls is increasingly being used in early-phase clinical trials (see Chapter 2). There is considerable potential for the application of such methods in regularly repeated preclinical *in vivo* studies too, where each study has a number of control animals providing valuable data. This can result in a more effective use of animals, either by reducing the total number of animals used or by increasing the precision of the treatment differences. This will lead to clearer results and will support the ethical use of animals in research.

In repeated studies, such as in drug screening programmes, the protocol remains consistent throughout, with animals of the same strain, age, and body weight ranges. Even the same equipment, room, and personnel may be used for each study. As a result, the historical studies provide very useful information for the analysis of the current study, by providing data for a meta-analysis. So a Bayesian approach allowing for study-to-study variation has great potential in this setting. As the most frequently used animal models may be run several times a month, using such an approach could result in marked savings over the lifetime of the assay, even if the reduction in animal usage per study is only slight.

Preclinical studies differ from clinical studies in that many preclinical studies include more than one control group. In addition to the control group against which the treated groups are compared, additional control groups may be included to characterise the assay. For example, animals are commonly challenged with a known stimulus and then treated with experimental compounds to try to counteract the effect of the stimulus. As well as a vehicle treated control group, such studies will usually also include an unchallenged control group which may be treated with a vehicle or left completely untreated. The purpose of this group may be to establish the "window" between the challenged and unchallenged control groups, i.e. the magnitude of the challenge effect, or to convert response values to percentages, or to check consistency with previous studies, but it will not be used directly in the statistical comparison of the experimental drugs. Some repeated studies will include a marketed drug for reference, which may or may not be used in any formal statistical comparison with the experimental drugs. The historical data for control groups used as comparators can be utilised to create an informative prior in a Bayesian analysis. We refer to this as a "full Bayesian approach". But when a control group is not used for formal comparisons, we present the alternative of completely replacing it with predictive limits based on the posterior predictive distribution for its observed mean. We refer to this option as a "Replacement approach".

5.2.1 Statistical Methodology

In this section, we describe the Bayesian meta-analysis of the historical data and two alternatives for using predictions from the meta-analysis in the analysis of future studies.

5.2.1.1 Bayesian Meta-Analytical Approach

This approach is based on that of Neuenschwander et al. [4]. It assumes that a set of H studies has been observed, and we wish to make predictions about the parameters of the next study.

For each historical study, h ($h = 1, \ldots, H$), the number of animals in the control group is represented as n_h, with observed responses $Y_{ih}, i = 1, \ldots, n_h$, which have an observed mean $\overline{Y}_h$ and standard deviation (SD) s_h. The responses of each animal are assumed to be continuous, independent, and identically distributed following a Normal distribution with parameters θ_h and σ_a^2, where θ_h is the mean for study h and σ_a^2 is the between-animal variance, which is assumed to be the same for all studies:

$$Y_{ih}|\theta_h, \sigma_a^2 \sim N\left(\theta_h, \sigma_a^2\right), i = 1, \ldots, n_h; h = 1, \ldots, H$$

It is also possible to work with the sufficient statistics rather than the raw data, and this may be more convenient in some situations, depending on the availability of the raw data. This approach is used in the Neuenschwander paper [4], and the standard errors are assumed known. However, in our situation, the individual *in vivo* studies are typically quite small, so the standard errors will not be well estimated. So, we assume that the true between-animal variance, σ_a^2, is the same in each study and that the observed standard errors estimate the true standard errors. We therefore assume that the sample means and the sample SDs follow a Normal and Chi-squared distribution, respectively:

$$\overline{Y}_h|\theta_h, \sigma_a^2 \sim N\left(\theta_h, \frac{\sigma_a^2}{n_h}\right)$$

$$s_h^2 \sim \left(\frac{\sigma_a^2}{n_h - 1}\right) \cdot \chi^2_{n_h - 1}$$

It is more convenient to represent the Chi-squared distribution as a Gamma distribution for use in OpenBUGS, as described in Evans et al. [5]:

$$s_h^2 \sim \Gamma\left(\frac{2\sigma_a^2}{n_h - 1}, \frac{n_h - 1}{2}\right)$$

We assume that the data for the new study are Normally distributed, and we represent the sample mean as $\overline{Y}^*$, the number of animals as n^*, and the true mean as θ^*:

$$\overline{Y}^*|\theta^*, \sigma_a^2 \sim N\left(\theta^*, \frac{\sigma_a^2}{n^*}\right)$$

We use a hierarchical model assuming exchangeability between the true study means, $\theta_1 \ldots \theta_H$, θ^*. These parameters are also assumed to be Normally distributed, with parameters μ and σ_s^2:

$$\theta_1 \ldots \theta_H, \theta^* \sim N(\mu, \sigma_s^2)$$

A vague Normal prior is used for the population mean, μ; and a vague inverse Gamma prior is used for the between-animal variance, σ_a^2. We follow Gelman et al. [6], who advise using a vague half-Normal prior for the between-study SD, σ_s.

5.2.1.2 Using the Predictions from the Meta-Analysis

We can use the Bayesian meta-analytic predictive analysis in two different ways. First, we can obtain a predictive distribution for the ***true*** mean of the next study (θ^*). Alternatively, we can obtain a predictive distribution for the ***observed*** mean of the next study for a given sample size $\left(\overline{Y}^*\right)$.

The conventional approach in the literature [4] is to use the predictive distribution for θ^* as a prior for the control group mean in the next study, which we approximate with a Normal distribution. Other treatment groups will use a vague prior. We refer to this as the "full Bayesian approach", and we illustrate this in the Novel Object Recognition case study (Example 5.1b). We believe this will be the more appropriate approach when the control group is used for formal comparisons against the experimental groups. However, this forces the analysis of the next study to be Bayesian which is likely to be beyond the capability of many of the *in vivo* scientists to implement themselves. We have found the concept of effective sample size (ESS) a very helpful tool to communicate how the full Bayesian approach works. To estimate the strength of the prior, an effective sample size can be estimated to indicate the number of extra animals that the prior is equivalent to. If we can assume that the posterior is approximately Normal, then the effective sample size can be approximated by the ratio of the between-animal variance estimate for the current study $\left(s_a^2\right)$ divided by the variance of the estimate of θ^* from the meta-analysis.

$$\text{ESS} \approx \frac{s_a^2}{\text{Var}(\theta^*)}$$

Whilst in the case of unknown variance the effective sample size is only approximate, in our case study, assuming a known variance had little impact on the outcome. In some contexts, the concept of the prior probability of success may also be used to assess the value of prior information. However, this is not a useful metric in preclinical work because there may be little historical information on the treatment effects, so vague priors are typically used for these, and formal success criteria are not usually specified.

An alternative use of the Bayesian meta-analytic predictive analysis is where a control group is not used for formal statistical comparisons against experimental drugs but included in the study for other reasons such as establishing a background or "baseline" value or benchmark. In this case, we may rely solely on the predictive distribution for $\overline{Y}^*$, and this control group can then be excluded from future studies. We refer to this as the "Replacement approach". The data from the new study are analysed (probably by the scientist) using a conventional frequentist ANOVA, and the 95% credible interval for the predictive distribution is superimposed on the graphical presentation of the data. This has the advantage that it leaves the analysis in the biologists' hands except when the predictive distribution needs updating. We illustrate this in Example 5.1a, the lipopolysaccharide (LPS)-induced cytokine release model case study. With this approach, it may still be desirable to include animals in this control group at regular intervals or when one of the experimental factors changes in order to check that the assay remains stable.

5.2.2 Implementation

The first task is to create a single data set of all the historical data. This may not be straightforward, as combining raw data from different studies may not have been considered when the data were originally recorded. Little attention may have been given to ensuring consistency of data management and nomenclature across studies. If raw data are not readily available, sufficient statistics (mean, SD, and the number of observations) from the relevant control group for each study may be easier to obtain. Consideration also needs to be given as to which studies to include or exclude. Rather than using formal criteria, we have done this in consultation with the scientists with the assistance of quality control (QC) charts.

QC charts should be created first, for both mean responses and individual values for each control group. In our experience, although QC charts are not routinely used by *in vivo* scientists, they are routinely used in some of the other preclinical areas, so *in vivo* scientists may have had some exposure to them. These charts have several uses, e.g. to identify outlying studies or individual values and to check that the responses are stable over time. The latter is particularly important for the Bayesian analysis. In addition, we use them to illustrate the size of between-animal variation as compared to between-study variation, and this provides a useful demonstration of why animals from previous studies provide some relevant information but not as much as animals in the current study.

To implement the Bayesian meta-analytic approach, a meta-analysis is applied to the data from the control groups of each of the studies. We typically omit the last 1–3 studies and retain these as a "test set" to illustrate our recommended analysis compared to the traditional frequentist approach. Where it is proposed to reduce the size of a treatment group, we have simulated this by deleting some animals from the data set of these last studies to illustrate the effect of the Bayesian approach. This proved more accessible to scientists rather than discussing the theoretical properties of each approach. For brevity, we only present the analysis and results from one of the last three studies in each example.

Example 5.1a: Using the Replacement Approach

The LPS-induced model of endotoxaemia in mice is an *in vivo* pharmacology model used for assessing the effect of novel inhibitors on pro-inflammatory pathways.

Mice are given a prophylactic treatment of either a vehicle, a test compound, or an anti-inflammatory reference antibody and later challenged with

LPS to induce an inflammatory response. A further group of animals are dosed with a vehicle but do not receive the LPS challenge, and this is the unchallenged control group. After several hours, plasma is collected from all mice, and IL-6 levels are determined. The comparisons of interest are between the LPS-challenged vehicle group and compound treated groups. The purpose of the unchallenged group is to ascertain the IL-6 level of "normal" mice and determine a window compared to the LPS-challenged vehicle treated group. It is not involved in any formal statistical comparisons.

Raw data from 25 historical studies were combined. Each study was run to the same protocol and had an unchallenged group and several challenged groups including a vehicle treated group, a reference treatment, and typically 2–4 experimental treatment groups consisting of different novel test compounds or increasing doses of a single test compound. Between five and eight mice were allocated to each group. The data were analysed on the log scale to conform to the assumption of Normality, and the results were back-transformed and presented in terms of geometric means.

QC charts were created for the unchallenged control group, the challenged vehicle group and the challenged reference group, and for the assay window (defined as the ratio of the geometric mean of the challenged vehicle group to that of the unchallenged vehicle group). Some studies were excluded after discussion with the scientists because of known issues, e.g. a more variable assay in the initial development phase. These data were also excluded from the Bayesian meta-analysis. An example QC chart is given in Figure 5.1 showing the geometric mean IL-6 values of the unchallenged group for each study plotted against study number with 2-sigma control limits.

In this example, the unchallenged group was not used in the pairwise treatment comparisons but was used to demonstrate the size of the assay window, i.e. to give some idea of the lowest mean IL-6 obtainable. Thus, we explored whether predictive limits could be used in future studies to replace the animals used in this group.

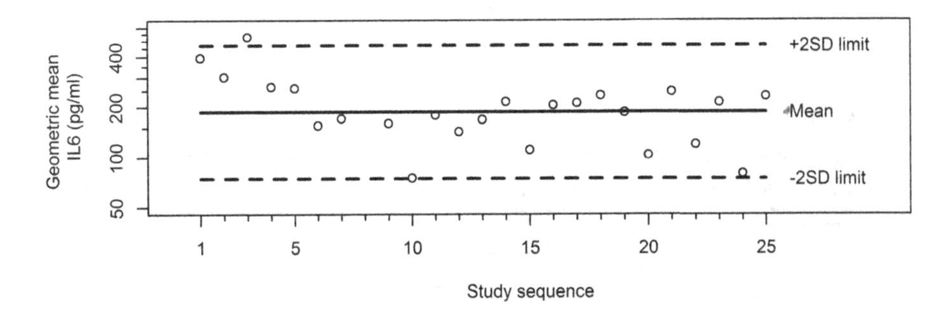

FIGURE 5.1
QC chart for the geometric mean of the unchallenged group from 24 historical LPS studies, presented on a log scale. The solid line represents the overall mean, and the dotted lines indicate 2-sigma control limits.

The results of the Bayesian meta-analysis of $\log_{10}$(IL-6) for the unchallenged group are given in Table 5.1. Four location parameters are estimated: the population mean (μ), the prediction for the "true" mean for a new study (θ^*), the prediction for the observed mean for a new study with $n = 8$ $\left(\overline{Y}^*\right)$, and the prediction for a single animal from a new study (Y^*). The posterior means of μ, θ^*, $\overline{Y}^*$, and Y^* were the same to two decimal places. As expected, the SD for θ^* was noticeably larger than that of μ, reflecting the additional study-to-study variation, σ_s^2. The SDs for the parameters relating to predictions of observed data, $\overline{Y}^*$ and Y^*, were larger still, reflecting the animal-to-animal variation, σ_a^2. The estimates of variance reflect the two levels of variation: σ_s^2 for the variation in the "true" study means (θ_i) and σ_a^2 the animal-to-animal variation within a study.

We analysed the next study using conventional analysis of variance on the full data set and also the data set excluding the unchallenged group. For the latter, we added two reference lines representing the 2.5% and 97.5% points of the predictive distribution of either $\overline{Y}^*$ or θ^* for the unchallenged group. The results are shown in Figure 5.2. The limits for $\overline{Y}^*$, the prediction for the *observed* mean of an unchallenged vehicle group in a study with $n = 8$ (Figure 5.2b), may be more appropriate on a scatterplot of individual data points, whereas the limits for θ^*, the prediction for the "true" mean of an unchallenged vehicle group (Figure 5.2c), are more appropriate for a plot of treatment means. In practice, only one of these would be required.

For this example, the approach of replacing the unchallenged group with reference lines from the predictive distribution was judged to be sufficient for future studies. This was because there were many historical studies; the QC charts suggested the assay was stable over time, giving confidence in the assumption of exchangeability; the unchallenged treatment group was not used for formal pairwise treatment comparisons; and the predictive distribution is precise relative to the large window between the challenged and unchallenged vehicle groups. The benefits of this approach are that this model will require fewer animals in future studies and the scientists can still analyse the future studies themselves, maintaining a quick turnaround.

TABLE 5.1
Summary of Estimation of Parameters from the Bayesian Meta-Analysis of $\log_{10}$-Transformed IL-6 Data from the Unchallenged Group

Parameter	Mean	SD
μ	2.26	0.05
θ^*	2.26	0.12
$\overline{Y}^*$	2.26	0.19
Y^*	2.26	0.45
σ_s^2	0.012	0.014
σ_a^2	0.184	0.025

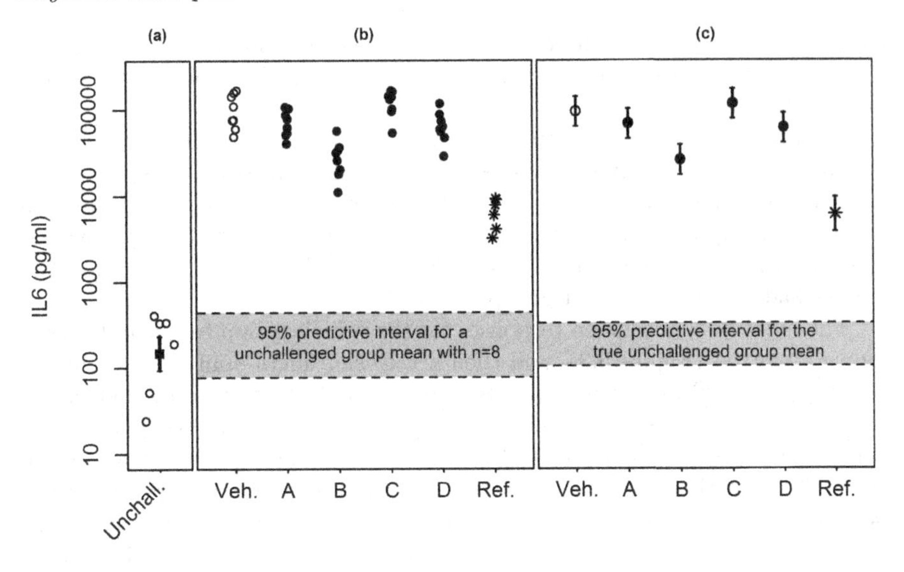

FIGURE 5.2
Analysis of the LPS study 25. (a) shows the data, geometric mean, and 95% confidence interval for the unchallenged group from a frequentist ANOVA of the full data set; (b) shows the data excluding the unchallenged group and overlaying a predictive distribution for a geometric mean of this group assuming $n = 8$ (based on $\overline{Y}^*$); and (c) shows the geometric means and 95% confidence intervals from an ANOVA of the data set omitting the unchallenged group and overlaying a predictive distribution for the "true" geometric mean of the unchallenged group (based on θ^*). The data are presented using a log scale. A, B, C, and D are the investigative treatments/compounds; Unchall, unchallenged; Veh, vehicle or control; Ref, reference group.

Example 5.1b: Using a Full Bayesian Approach

The Novel Object Recognition model in rats is a model of recognition memory, allowing the evaluation of short-, medium-, or long-term memory depending on the time-lag used. It is used to evaluate effects of drugs on cognitive function. Each experiment is performed in two trials. During the first trial (acquisition), the rat explores an arena containing two identical objects. During the second trial (retention trial), one of the objects is replaced by a novel object. It is expected that if the rat recognises the object from the acquisition trial, it will spend less time exploring that object in the retention trial, compared to the time spent exploring the novel object. An "index of differential exploration" is calculated as the difference in times spent exploring the new and familiar objects in the retention trial divided by the sum of both times. This index is regarded as representative of the functionality of recognition. In this example, the protocol used a delay of 24 h between acquisition and retention. With this delay, it was expected that untreated animals would have a low index reflecting their inability to distinguish between the familiar and novel objects. Drugs are tested for their ability to improve the index of differential exploration reflecting improved memory.

Data from 17 historical studies were combined. Each study was run to the same protocol and had a vehicle control group and two to six treatment groups consisting of different novel test compounds or increasing doses of a single test compound. There were typically between 10 and 15 rats allocated to each group; however, some early pilot studies had fewer, and some key studies had up to 24 rats per group.

The individual and mean responses for the vehicle group from the 17 studies were plotted (Figure 5.3), and then a Bayesian meta-analysis was carried out using the 226 untreated animals from the first 16 studies. In contrast to the previous example, the estimated between-study variance component was much smaller than the within-study variance component (by a factor of more than 200), and this can clearly be seen in Figure 5.3. The posterior distribution for θ^* from the Bayesian meta-analysis was used as the prior distribution for the true mean of the vehicle group in a full Bayesian analysis of the 17th study, which had a vehicle control group, three tests compounds (A, B, and C), and a reference compound. The very low between-study variance led to

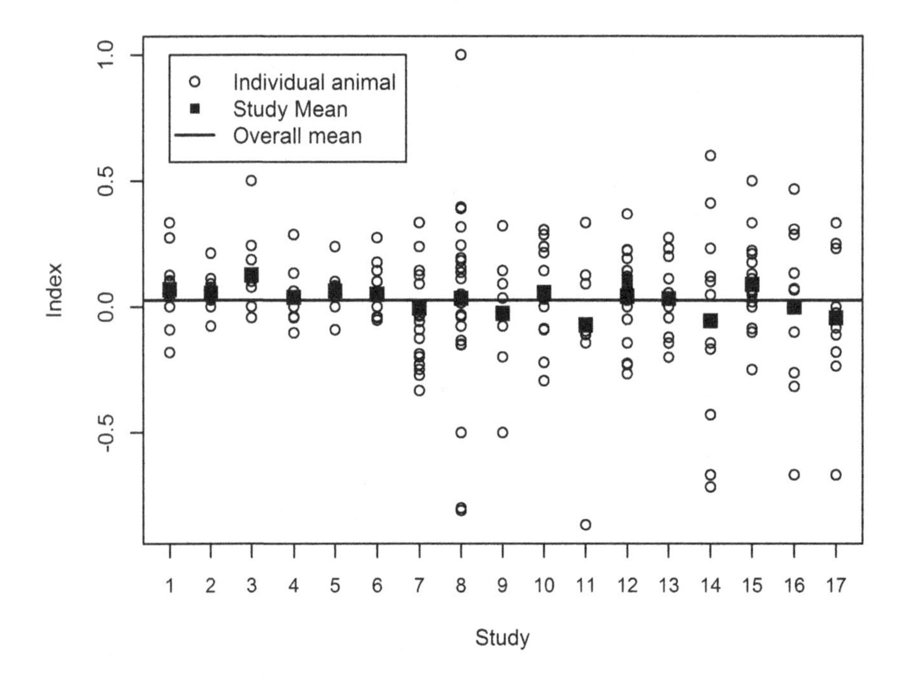

FIGURE 5.3
Scatterplot of the vehicle group data from 17 studies showing the individual animal data, the study means, and the overall mean.

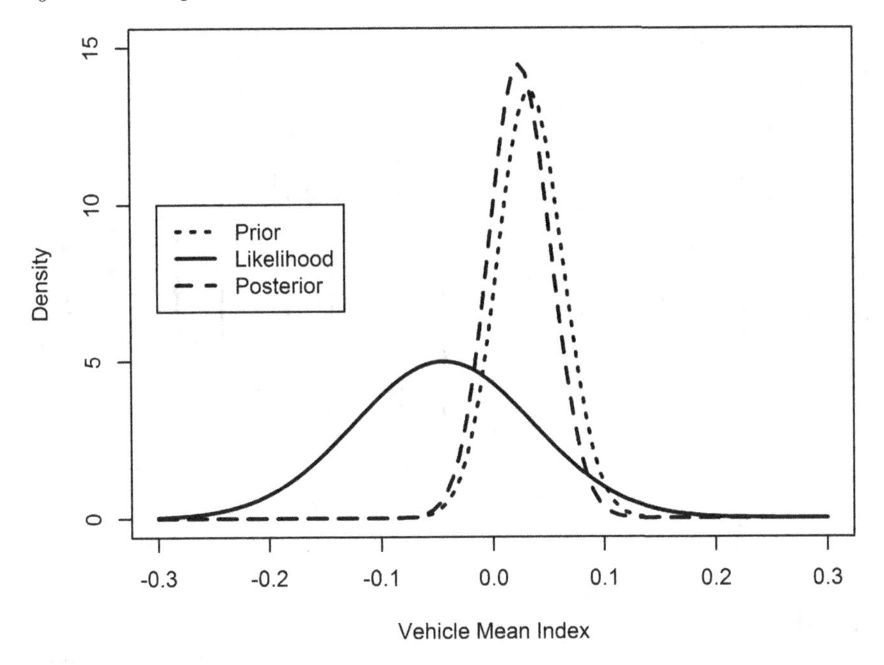

FIGURE 5.4
Plots of the prior and posterior densities and likelihood for the vehicle mean showing the strong influence the informative prior has on the posterior distribution.

a very informative prior for the analysis of the 17th study (Figure 5.4). The prior had an effective sample size of about 90 additional animals, and the impact of this is illustrated in Figure 5.5.

Figure 5.5a shows that the impact was particularly marked for the estimated vehicle mean (the mean estimated using the informative prior) and for its estimate of precision. Figure 5.5b shows the increase in precision in the estimated treatment differences due to the informative prior for the vehicle mean. The increase in precision of the difference was less marked than that for the vehicle mean as the precision of the comparator mean had been unaffected by the prior. Although frequentist confidence intervals and Bayesian credible intervals have a different interpretation, a frequentist analysis is equivalent to a Bayesian analysis with non-informative priors. Therefore, the 95% confidence intervals are equivalent to 95% credible intervals from such an analysis and can be used to demonstrate the impact of an informative prior. Although the interpretation of the data is unchanged for this study, the same style of plots (with real or hypothetical data) can be used to demonstrate how the additional precision could make results less ambiguous. Alternatively, one could use a similar plot to show equivalent precision between the standard frequentist approach with the full data set and a Bayesian analysis using a data set

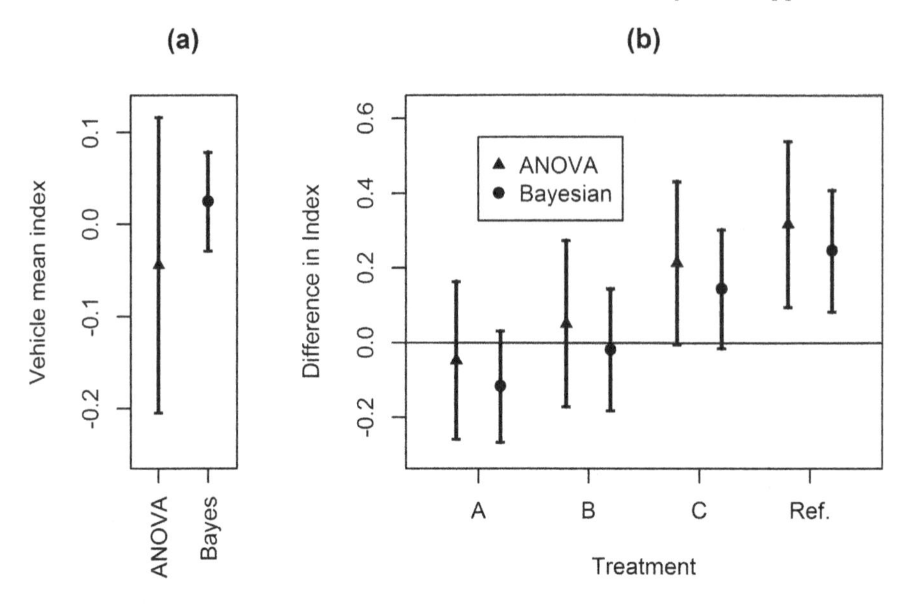

FIGURE 5.5
Analysis of the Novel Object Recognition study 17. (a) shows the vehicle group mean and interval estimates from the frequentist ANOVA and the full Bayesian analysis with the informative prior, and (b) shows estimates of the treatment differences and 95% confidence intervals for the comparisons of the treated groups to the vehicle group from the frequentist ANOVA contrasted with the treatment differences and 95% credible intervals from the full Bayesian analysis with the informative prior.

with an appropriately reduced number of vehicle control animals. We have used these plots in discussions with scientists to show the beneficial value of the full Bayesian approach, either in terms of the increase in precision or in reducing the number of vehicle control animals.

Figure 5.5a also shows that the vehicle mean from the Bayesian analysis was higher than that from the conventional ANOVA. This can be explained using a QC chart of vehicle means, which shows that the new study (study 17) had a lower-than-average mean (as seen in Figure 5.3). The informative prior from the historical vehicle data had the effect of “adjusting” the mean of the new study in the direction of the historical mean.

To estimate the strength of the prior, we can compare the point estimate of the animal-to-animal variance from this analysis with the variance of the prior for the mean and so obtain an effective sample size to indicate the number of extra animals that the prior is equivalent to. In this example, the prior distribution was estimated to be equivalent to approximately 90 additional animals (compared to the 226 animals in the historical data set). As this prior is so informative, it will have a large effect on the results of any future study. Before using this methodology in earnest, one would need to confirm with all stakeholders that it is appropriate to put this much emphasis on the historical data. Indeed, this is an extreme illustrative example, and before routine use, we would down-weight the prior by inflating its variance and thereby reducing the effective sample size. If there is a concern that there is the possibility of a mismatch between an informative prior and the data, i.e. a so-called prior-data conflict, then before running the study, either a rule of when to discard the prior could be agreed or a robust prior using a mixture of Normal distributions could be used, as in Schmidli et al. [7].

Figure 5.6 shows the posterior probability density for the difference between the vehicle group and Compound C. This demonstrates how a Bayesian approach allows flexible probability statements. For example, this posterior density could be used to estimate the probability that Compound C increases the index by a biologically relevant value, e.g. more than 0.2.

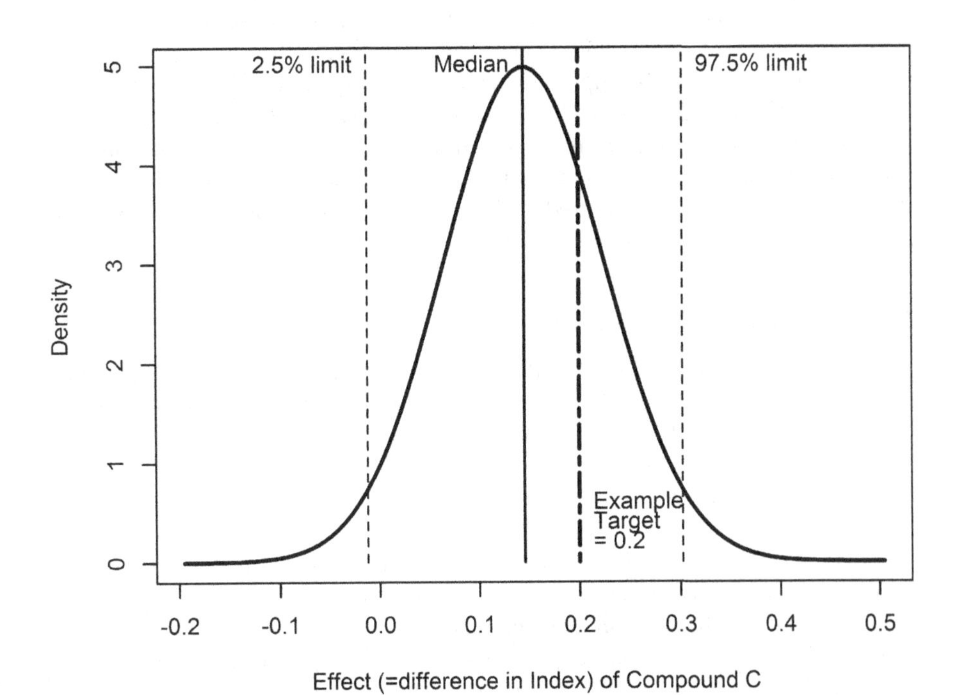

FIGURE 5.6
Posterior probability density for the difference between vehicle and compound C from the full Bayesian analysis with the informative prior illustrating the calculation of additional probability statements.

5.3 Mixture Distributions for Robust Handling of Outliers

Outlying observations are not uncommon in preclinical studies, and they can be particularly influential in small studies, such as most animal experiments. In the absence of a good justification for their omission, the standard dilemma arises: omitting an outlier may bias the results, whereas leaving the outlier in the analysis may invalidate the required assumptions of Normality and variance homogeneity, rendering confidence intervals and p values unreliable.

A pragmatic solution frequently adopted in preclinical studies is to analyse the data with and without the suspected outlier and hope that the two analyses give similar conclusions. However, if the outlier is influential and the conclusions of the two analyses differ markedly, then the transparent option is to present both analyses and the resulting uncertainty.

A more formal approach is to assume a mixture likelihood where most of the observations come from one distribution and a small minority come from a different distribution. Such an approach allows any outliers to be included in the analysis, but down-weights their influence [8].

We demonstrate this approach using data which have been artificially "spiked" with one or more outliers. We then apply the method to a genuine data set which contains several suspected outliers. In both examples, comparisons are made between the treatment groups, and the effect of down-weighting the potential outliers is compared with the results from a conventional Normal model (ANOVA) fitted using a Bayesian approach.

5.3.1 Statistical Methodology

The analysis involves fitting a model which is a mixture of two Normal distributions. Rather than use a mixture of Normal distributions comprising of two distributions with different means and a common variance, in our example, we use a mixture of two Normal distributions which have a common mean and different variances. We do not assume that outliers will be in the same direction (i.e. they could be high or low values).

Observation j from the ith treatment group, y_{ij}, is assumed to be drawn from one of two categories: either non-outlier or outlier. $T_{ij} = 1, 2$ is the true category for observation ij, where category k for treatment group i has a Normal distribution with mean μ_i and variance σ_k^2. The mean parameter, μ_i, does not depend on the non-outlier/outlier category. We assume an unknown proportion p of the population are in category 2 (outliers) and $1 - p$ in category 1. The model is thus

$$y_{ij} \sim N\left(\mu_i, \sigma^2_{T_{ij}}\right)$$

$$T_{ij} \sim \text{Bernoulli}(p)$$

In our examples, we have assumed that the variance of the outliers is five times that of non-outliers, i.e. $\sigma_2^2 = 5 \times \sigma_1^2$. In other words, the non-outlier distribution for treatment i is $N(\mu_i,\ \sigma^2)$, and the outlier distribution is $N(\mu_i,\ 5\sigma^2)$.

For comparison, we fitted a conventional Normal model (one-way ANOVA) with a Bayesian approach.

$$y_{ij} \sim N\left(\mu_i, \sigma^2\right)$$

For both models, we used vague priors for the means and variances: Normal(0, 100^2) for the means and inverse Gamma(0.01, 0.01) for the variance.

For the Bernoulli probability, p, the probability of an observation coming from the "outlier" distribution, we used a reasonably informative Beta(1,19) prior distribution. The Beta distribution was chosen as it is the conjugate prior distribution for the Bernoulli, and the particular values of the parameters, $\alpha = 1$ and $\beta = 19$, were chosen without reference to the data, to reflect our subjective prior beliefs about the probability of an observation being an outlier. We chose these parameter values as they put realistic limits on p, with a mean of 0.05 (5%), a median of 0.035 (3.5%), and an upper 95th percentile of 0.15 (15%). Based on our experience with the *in vivo* models used, these seemed reasonable characteristics for a prior. In different areas, a different prior may be more appropriate. This prior is plotted in Figure 5.7.

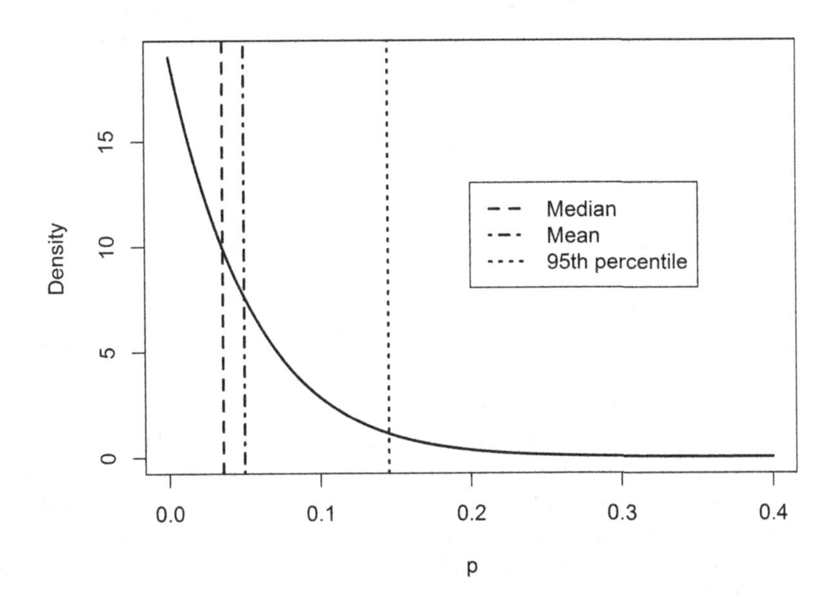

FIGURE 5.7
The Beta(1,19) prior density for the Bernoulli probability, p, of an observation being from the "outlier" distribution.

A Bayesian approach to fitting such models is convenient as it has the advantage of allowing a reasonably informative prior to be placed on the mixture parameter, p, the probability of an observation being an outlier (i.e. coming from the "outlier" distribution). The Bayesian approach with a reasonably informative prior may overcome any difficulties in reliably estimating all the model parameters from the small data sets which are typical of *in vivo* studies.

Example 5.2a: Artificially Spiked Data

The first example takes data from a study with five groups: an untreated control, three test compounds (A–C), and a reference compound. The experiment was designed with eight animals per group, except for the reference compound with six animals. From the original data set (data set 1), three further data sets were created by replacing chosen observations with more extreme values to represent outliers whilst keeping the group sizes the same. The four data sets (one original and three modified) are:

1. Original data without modification (no outliers);
2. Medium outlier: Replacing one observation for compound A with an outlier 3.5*SD below the mean;
3. Large outlier: Replacing one observation for compound A with an outlier 5*SD below the mean;
4. Several outliers: Replacing four observations across the groups with outliers of various magnitudes between 2.5*D and 5*SD below the group mean.

The data are shown in Figure 5.8.

The Bayesian analysis provided the posterior median of the Bernoulli parameter, p, for each of the data sets. For the original data set, p was 0.052, which was higher than the prior median of 0.035 suggesting the possibility of outlying values even in the original data set. The value of p increased as the number and size of the outliers increased, and it was estimated as 0.084 for data set 4, which contained the most outliers.

Figure 5.9 shows the mean and 95% credible intervals for compound A from the mixture model and, for comparison, from a Bayesian conventional Normal model (ANOVA). For the unmodified data set, both models gave identical means and similar credible intervals. In contrast, for data set 3 which had one extremely low observation, the mixture model down-weighted the outlier and so gave a noticeably higher mean compared with the conventional model. For data set 4, with several outliers, the mixture model gave noticeably narrower credible intervals. This is because for the conventional model, the estimate of the variance, σ^2, has been inflated by the presence of extreme observations, whereas the mixture model deems extreme observations to come from a distribution with a variance of $5\sigma^2$, thereby reducing their impact on the variance estimate. Table 5.2 gives the posterior medians for the variance

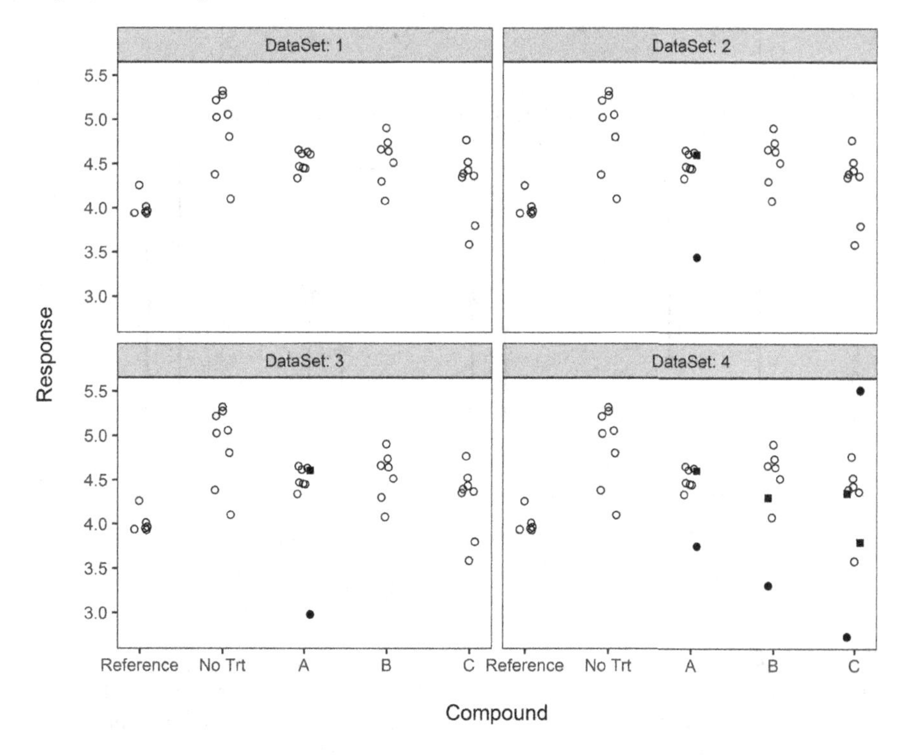

FIGURE 5.8
Scatterplots of the "spiked" data sets. The open circles show the original data. For data sets 2, 3, and 4, the filled squares are replaced by the filled circles to represent the potential outliers.

estimates from the two models for each of the four data sets. The estimate of σ^2 from the original data was 0.099. It is clear to see that with the conventional analysis, the estimates of σ^2 increased with the number and size of "spiked" outliers. The use of the mixture model, however, maintained the estimate of σ^2 to around that of the conventional estimate for the original data set (approximately 0.1). The estimate for data set 4, with several outliers, was inflated but to a much lesser extent than with the conventional analysis. The reduction in the estimated variance will have the knock-on effect of reducing the standard errors and therefore increasing the precision of any comparisons made between the treatment groups.

While the main objective of the mixture model is to down-weight any outliers rather than just identify them, we can obtain the posterior probability of being an outlier for each observation. From each iteration of the MCMC simulation, we obtain a Bernoulli estimate (0 or 1) for T_{ij}, i.e. whether the data point comes from the outlier or the non-outlier distribution. By taking the mean of these over all the iterations, we obtain the estimated probability that the individual observation is an outlier.

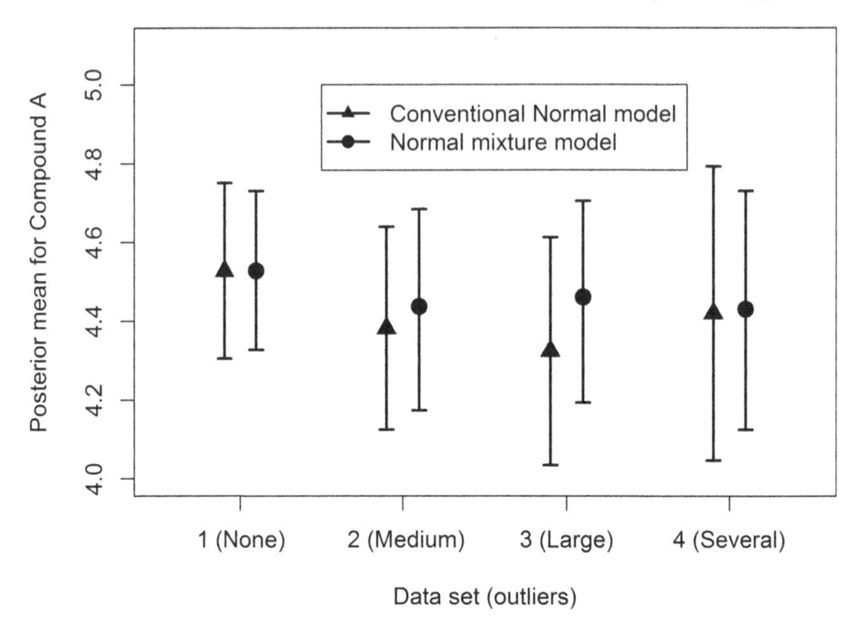

FIGURE 5.9
Posterior means and 95% credible intervals for Compound A from each of the four data sets from the conventional linear model and the mixture model.

TABLE 5.2
Median Values of the Posterior Distributions for the Original and Three Spiked Data Sets

Parameter	Data Set 1 Original Data	2 Medium Outlier	3 Large Outlier	4 Several Outliers
Variance (σ^2)—Conventional	0.099	0.131	0.165	0.276
Variance (σ^2)—Mixture model	0.078	0.100	0.096	0.159

The variance estimates from the conventional analyses increase with the number and magnitude of outliers. The Bayesian analysis using the mixture model has lower and more consistent variance estimates.

Figure 5.10 gives the scatterplots of the data with the spiked observations annotated with these probabilities. It can be seen that the single large outlier in data set 3 and also the high outlier in compound C for data set 4 have high posterior probabilities of coming from the $N(\mu,\ 5\sigma^2)$ "outlier" distribution (0.92 and 0.83, respectively). Conversely, the outlier in Compound A from data set 4 only has a probability of 0.19 of being an "outlier".

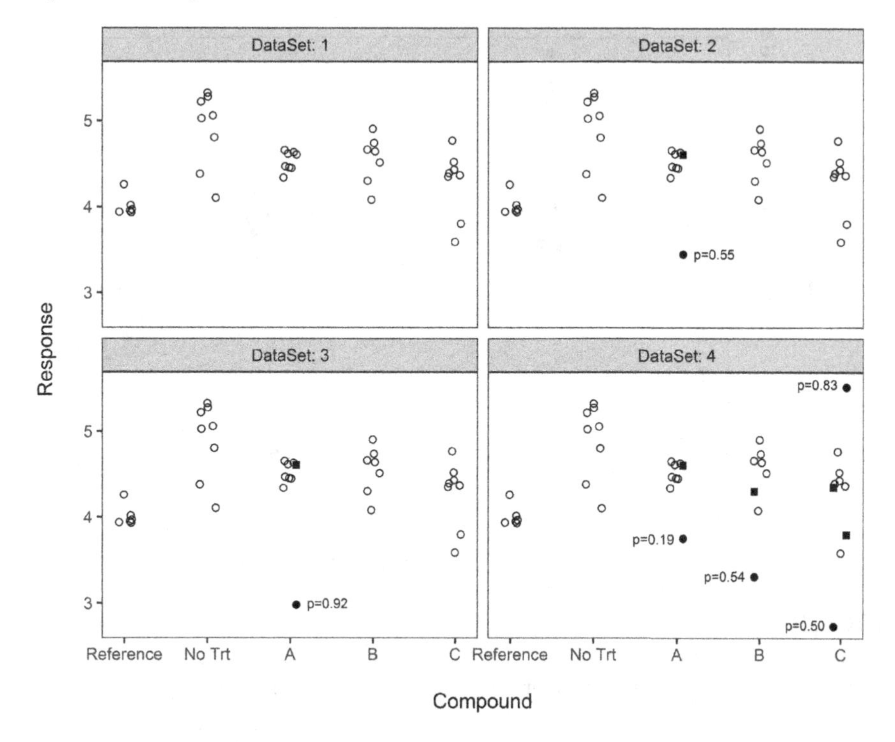

FIGURE 5.10
Scatterplots of the data with the probability of being an outlier indicated for each of the "spiked" data points. The open circles show the original data, and the filled squares are replaced by the filled circles to represent the potential outliers (as in Figure 5.8).

Example 5.2b: Flow Cytometry Outlier Example

For the second example, the method was applied to a study where outliers were suspected. In the inflammatory mediator-induced cell recruitment model, male mice are dosed with either control or an active treatment, and 1 h post dose, animals receive an inflammatory mediator injected into the peritoneal cavity. To assess mediator-induced cell recruitment, animals are sacrificed 3 h later and peritoneal lavage fluid assessed by flow cytometry. This *in vivo* model is known to have issues with occasional outliers, and the example shown is an extreme case. There were five groups: three doses of a test compound, a reference compound, and an untreated control group with eight animals per group and one missing value. Log_{10}(Neutrophilia count) was analysed.

The observed data are shown in Figure 5.11. It is clear that the two highest neutrophilia counts in the no treatment group would have an influence on that group mean and consequently on any treatment comparisons involving that group. Similarly, the two high counts in the reference group would affect its mean, and the two low counts in treatment group B would reduce its mean.

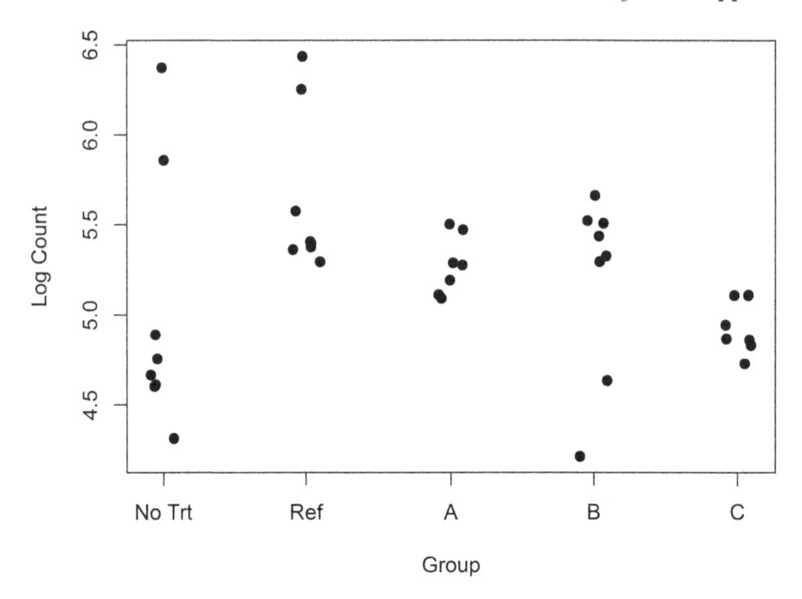

FIGURE 5.11
Scatterplot of the $\log_{10}$(neutrophilia count) data showing the potential outliers in the No Treatment group, Reference group, and Group B.

The objective is to be able to use all the observed data, but to down-weight the highly influential extreme observations in the analysis. Figure 5.12 shows the differences between the four treated groups versus the no treatment group from both the conventional Normal model and the Normal mixture model. For all groups, the Normal mixture model gave a larger estimate of the difference compared to the conventional Normal model due to down-weighting the influence of the two outliers in the No Treatment group.

The median of the posterior distribution of σ^2 was reduced from 0.21 for the conventional Normal model to 0.11 for the Normal mixture model. This means that for the mixture model, most of the observations came from a distribution with $\sigma^2 = 0.11$, but a significant minority of observations came from a distribution with σ^2 five times that, i.e. 0.55. As a result, the overall widths of the credible intervals from the Normal mixture model and the conventional linear model end up being very similar in this example, but those from the mixture model tend to be slightly asymmetric.

Figure 5.13 shows the prior and posterior densities for the Bernoulli parameter, p. The posterior median was 0.10, a marked increase from the prior median of 0.035. This indicates that our updated belief in the probability of an observation coming from the "outlier" distribution is now around 10% with a 95th percentile of 24%.

Both examples show the benefits of the Bayesian mixture model. In these cases, we used a mixture of two Normal distributions, with one variance five

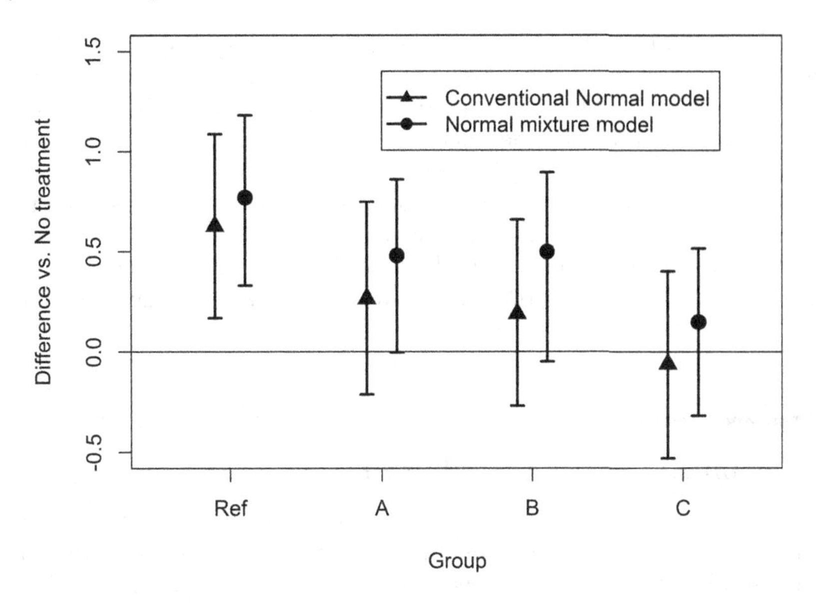

FIGURE 5.12
Posterior means and 95% credible intervals for the differences of each group compared to the No Treatment group from the conventional linear model and the mixture model.

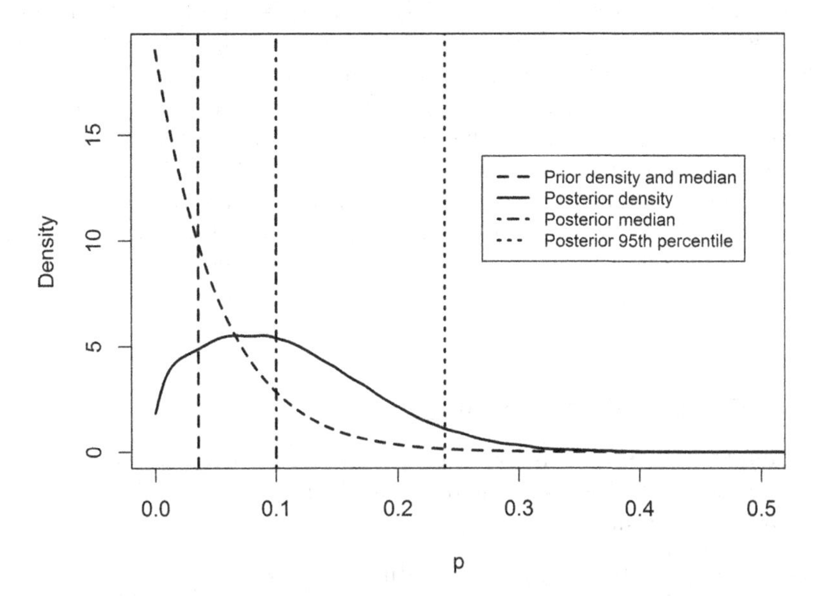

FIGURE 5.13
The prior and posterior densities for the Bernoulli probability, p, of an observation being from the "outlier" distribution. The prior is a Beta(1,19) distribution.

times the size of the other. Some such constraint is necessary as there is insufficient data to estimate two independent variances. The ratio of 5 between the two variances is admittedly somewhat arbitrary but seems to work well with this data. In practice, one should test the sensitivity of the results to this assumption.

Alternative mixtures may be appropriate in other situations and should depend on a plausible biological mechanism. For example, if outliers are only expected in one direction due to failure of an artificial challenge, then a mixture of two Normal distributions with a shift in means may be useful.

5.4 Non-Standard Distributions

A Bayesian approach can often be useful when analysing data with an unusual distribution. While it is often possible to use frequentist methods (e.g. maximum likelihood, generalised linear models) for analysing data with non-Normal distributions, the availability of software now makes the Bayesian approach an attractive option for such data.

The example we give involves data from the collagen antibody-induced arthritis model in mice. Typically, this model is run with a vehicle control group and two to five test groups with eight mice per group. For each animal, each paw is scored for arthritis on a 0–3, scale and the scores are summed to give a score of 0–12 on each day for eight consecutive days. This type of scoring is not uncommon in *in vivo* work despite the fact that the basic 0–3 scale is an ordinal scale rather than an interval scale, so a sum of four such scores can be ambiguous. For example, a score of 3/12 could arise from one paw with a score of 3 with the other paws all scoring zero or three paws each scoring 1. It is not obvious that these are equivalent. However, scoring such as this is largely accepted among researchers. It was decided that the most appropriate summary for each animal was the area under the curve (AUC) of the scores over the 8 days.

5.4.1 Statistical Methodology

An example of data from one typical experiment with a control group and five ascending doses of a test compound is given in Figure 5.14 and demonstrates some awkward features:

- Variance appears to increase with the mean level,
- A large number of zero values.

Analysing such data using a log transformation with a suitable offset (e.g. $\log_{10}(x + 1)$) is one pragmatic approach, but it did not work well for this model due to the excessive number of zeros.

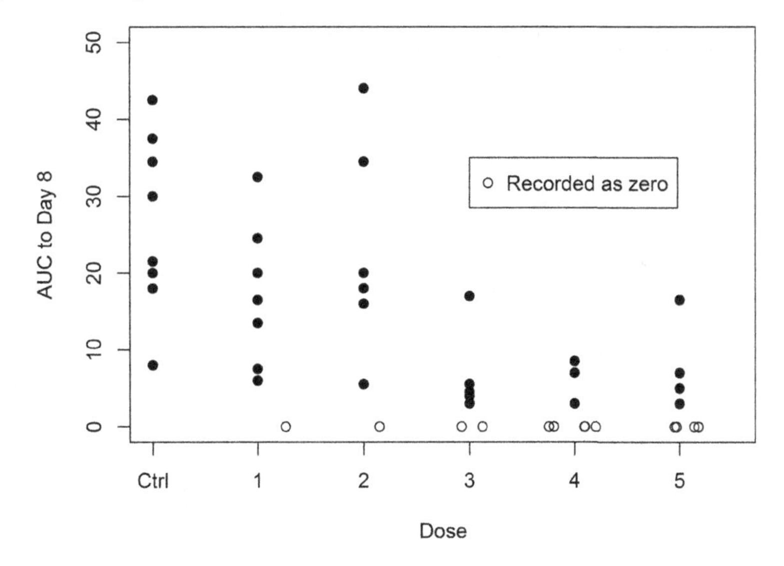

FIGURE 5.14
Scatterplot of AUC data showing the heterogeneous group variances and a large number of zero values.

In this example, we analysed the data using a left-censored Normal distribution, with a different variance for each group. This assumes that there is some underlying latent variable, z, with a Normal distribution where negative values get recorded as zero.

Figure 5.15 demonstrates the idea of a latent variable. For this plot, the latent variable ($n = 20$) was randomly generated from a Normal distribution with a mean of 5 and an SD of 7. Only values above zero can be observed, giving rise to the censored data. Any latent values less than zero are observed as zero.

To demonstrate the statistical approach used, the example data were analysed with three different models, all using a Bayesian approach:

A. Standard linear model with equal variances (ANOVA). This was included for comparison as it is the analysis that may be typically carried out by a scientist.

B. Linear model with different variances for each group, including the zero values.

C. Linear model with different variances for each group, with zero values regarded as left-censored, i.e. treated as ≤ 0 and modelled with a left-censored Normal distribution.

Model A: $y_{ij} \sim N(\mu_i, \sigma^2)$
Model B: $y_{ij} \sim N(\mu_i, \sigma_i^2)$
Model C: $z_{ij} \sim N(\mu_i, \sigma_i^2)$; $y_{ij} = z_{ij}$ if $z_{ij} > 0$; $y_{ij} = 0$ if $z_{ij} \leq 0$

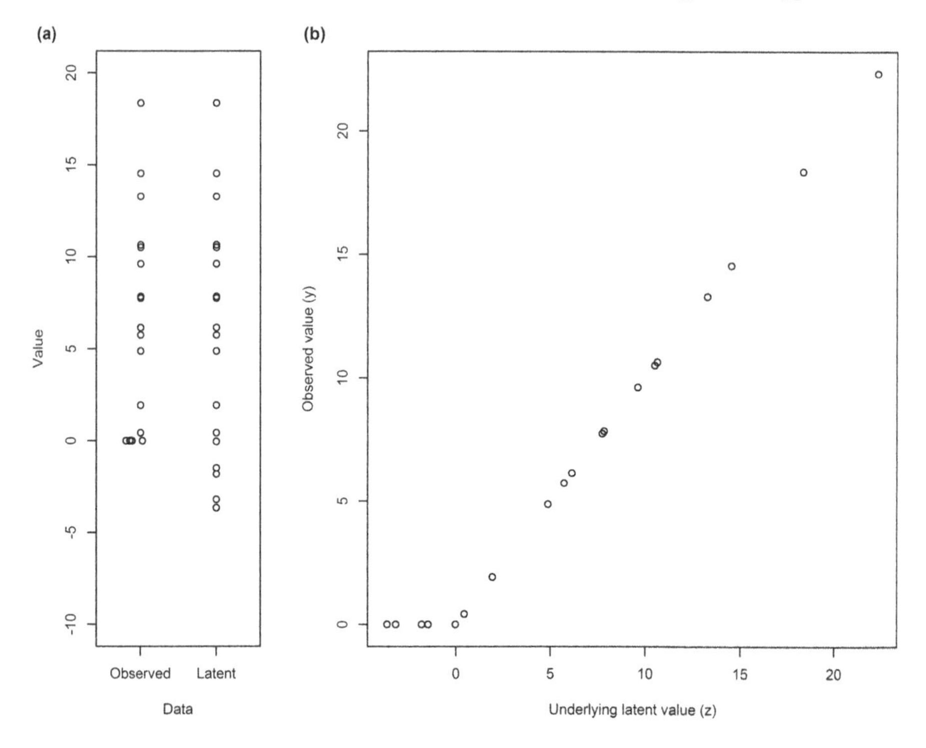

FIGURE 5.15
Data simulated from a Normal(5, 7^2), demonstrating the effect of censoring. (a) gives scatterplots for both the observed (censored) data and the underlying latent variable. (b) shows the relationship between the latent variable and the observed data.

All models used vague priors for the group means (Normal(0, 100^2)) and variances (inverse Gamma(0.01, 0.01)).

Figure 5.16 summarises the results and gives the median and 95% credible intervals for the posterior distribution of the estimates of the mean differences versus the control group from all three models.

As expected, the means from models A and B, where the zero values were included, were similar, but the credible intervals differed due to model B allowing for the unequal group variances. Model C however gave markedly different estimates of the treatment differences versus the control group. This model estimated the means of the underlying latent variable (z) rather than the AUC (y). This is particularly noticeable for doses 4 and 5 which had large numbers of zero/censored observations. The estimated differences versus the control were larger in magnitude for these two groups because regarding the zeros as unknown, negative values had a large influence on the estimated group means of the latent variable (z). The posterior medians for the mean of the latent variable for both of these groups were negative, reflecting the fact that

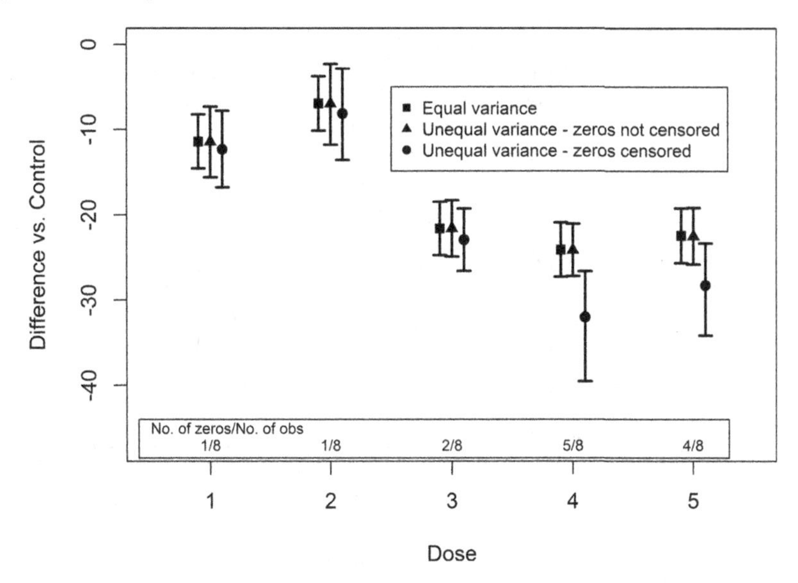

FIGURE 5.16
Medians and 95% credible intervals from the posterior distributions of the estimates of the mean differences between each treatment and the vehicle control for each of the three models.

at least half of the animals had latent scores ≤ 0. Also, for doses 4 and 5, the credible intervals from model C were wider than those from model B as the variance estimates from model B regard all the zero observations as clumped together with no variation, whereas model C allowed for extra variation among the values recorded as zero.

An illustration of the posterior densities for the means of the control group and dose 4 and their difference (dose 4 – Control) is given in Figure 5.17. Note that the posteriors for the dose 4 mean and the difference between dose 4 and Control are noticeably skewed, resulting in the asymmetric credible interval seen in Figure 5.16 for the difference of dose 4 versus control. Also, the median of the posterior for dose 4 is negative, reflecting the fact that five out of the eight animals in this group had scores of zero.

The above results and plots refer to the posterior distribution of the means of the latent variable, z. However, it is also possible to obtain analogous results for the expectations of the observed values, y, if this is thought to be more relevant. This will depend on the context and the relevance or interpretability of the underlying latent variable. In some applications of censored data distributions, such as survival analysis or assays with a lower limit of quantification, the latent variable parameters are usually the most relevant. In other applications, the decision may be less clear-cut.

This example demonstrates the use of a Bayesian approach for fitting non-standard models to data. In this case, the approach allows flexible distributions

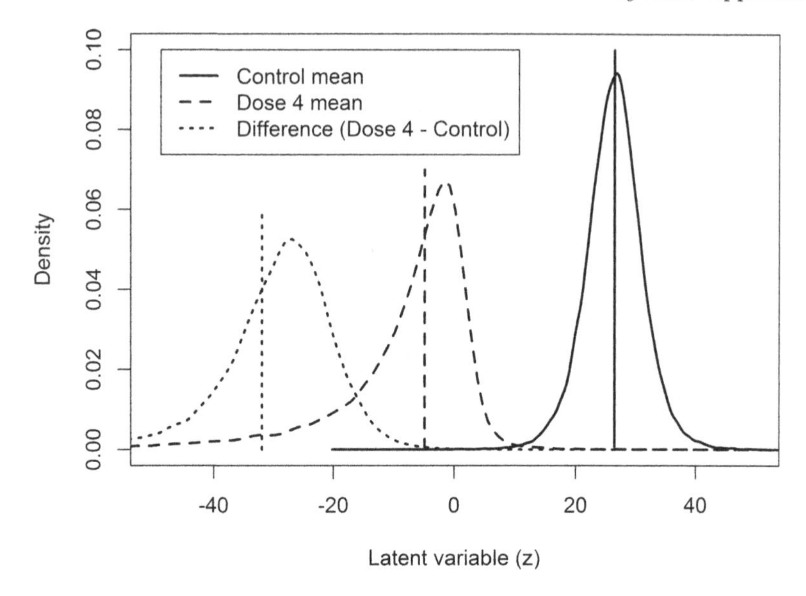

FIGURE 5.17
The posterior densities for the control group mean, the dose 4 group mean, and the difference between them.

to be fitted in a straightforward manner. Additionally, as in Example 5.1b, the Bayesian methodology allows probability statements about the size of effects to be made.

5.5 Discussion

We have described some applications of Bayesian methodology in preclinical research, specifically to *in vivo* testing, initially driven by the possibility of using historical control data to reduce animal usage in models used repeatedly over time.

We have demonstrated that Bayesian methods may be used effectively in the analysis of stable, repeated *in vivo* studies, and indeed we have demonstrated that this is a natural home for their use because the historical data is extremely compatible with the next study. We hope that as the value of accessing historical control data is appreciated, both in terms of QC charts and Bayesian methods, more attention will be given to optimising the storage of data for cross-study meta-analyses. This would reduce the data management aspect of future projects and further ease the implementation of these methods. Clearly, for these methods to have greatest impact, one should focus on studies that are repeated frequently. We believe that using Bayesian methods in stable, repeated *in vivo* studies will result in a more effective use of animals, either by reducing the total number of animals used or by increasing the precision of key treatment differences, thereby leading to clearer results.

Although we have focussed on Normal data in Section 5.2, the meta-analytic approach can be applied to non-Normal data within the GLM context [9,10]. This could also be extended to an approach robust to outliers by replacing the Normal likelihood with a mixture of two Normal distributions, as in Section 5.3, or a heavy tailed distribution [8, pp. 49–52]. It would also be possible to extend the methodology in terms of the construction of informative priors. If the between-animal variance appears to be stable across studies, a potential extension to the full Bayesian approach would be to use an informative prior for this variance. One may also wish to consider using an informative prior for the treatment effect, obtained via a review of the historical data for standard test compounds and/or elicitation.

When any informative prior is used, there is the possibility of a prior-data conflict. The risk is highest when the prior obtained by the meta-analysis has a large effective sample size (relative to the sample sizes used in the study). If such a mismatch occurs, then this typically reduces confidence in the results. Should this scenario occur in the preclinical setting, then it may be possible to rerun a relatively small study to bring clarity. Alternatively, if in the planning stage one was concerned about the prior-data conflict, then one could use a pre-specified rule for discarding the prior or use a robust mixture prior [7] in future studies. The latter has the effect of dynamically down-weighting the prior where there is a discrepancy between the prior and the data. It is important to stress that any prior, including robust mixture priors, must be specified before running the study, and it must not be changed or down-weighted after seeing the data.

An additional benefit of using the Bayesian approach in all our examples is that there is no need to resort to asymptotic results which would be a concern for small *in vivo* studies. Also, the Bayesian approach allows for informative probability statements about possible effect sizes, which are relevant for decision making to progress compounds to subsequent stages of the discovery process.

When considering the adoption of a Bayesian approach, the issue of implementation and software must be considered. Using a Bayesian analysis will typically necessitate the involvement of a statistician rather than a scientist carrying out the analysis and may involve longer turn-around times. If a specific Bayesian analysis is adopted for a regularly run model, it may be possible to develop a bespoke software application, but the cost-effectiveness of such an application would need to be considered.

All our Bayesian analyses were conducted using OpenBUGS, called from R with R2OpenBUGS. Examples of the OpenBUGS models are given in the Appendices. Frequentist analysis of variance was performed using the lm() function in R.

In summary, we think there is considerable scope for the increased use of Bayesian methodology in preclinical *in vivo* research which could bring benefits in terms of reducing animal numbers by formally incorporating information from historical control animals. Other benefits include a more informative and rigorous analysis of non-standard data, i.e. data that do not comply with the routine assumptions of Normality and homogeneity of variance.

Acknowledgements

We would like to acknowledge Joe Rastrick, Alex Vugler, Katie Donovan, Eric Detrait, and Etienne Hanon for supplying the data used in the examples and for discussions on the utility of the methodologies and Anastasiia Raievska for her statistical contribution to the spiked data example.

Appendix 5.1: OpenBUGS Model for Bayesian Meta-Analysis (Section 5.2)

```
model{
### Likelihood ###
for(i in 1: N){
        Y[i] ~ dnorm(theta[StudyID[i]], tau.within)
}

### Prior for study means ###
for (h in 1:H) {
        theta[h] ~ dnorm(mu, tau.between)
}

### Predictions  ###
theta.star     ~ dnorm(mu, tau.between)
                 # predicted theta for a new study
new.animal     ~ dnorm(theta.star, tau.within)
                 # observation for animal from a new
                 study
tau.within.8     <- tau.within * 8
                 # precision for a study with n=8
new.study.mean.8 ~ dnorm(theta.star, tau.within.8)
                 # prediction for a new study observed
                 # mean with n=8
```

```
### Non-informative priors & hyperpriors  ###
mu              ~ dnorm(0.0, 1.0E-6)
SD.between      ~ dnorm(0.0, 1.0E-4)T(0,)
                  # Half-normal prior for between-study SD
tau.within      ~ dgamma(0.01, 0.01)

### Converting between precision and SD  ###
tau.between     <- pow(SD.between, -2)
SD.within       <- pow(tau.within, -0.5)
}
```

Notes:

1) In the data, Y[] is a vector containing the variable to be analysed, and StudyID[] is a vector indicating the study the data comes from.
2) The "within" and "between" components of the variable names correspond to the "a" and "s" suffices, respectively, in Section 5.2.
3) new.study.mean.8 corresponds to $\overline{Y}^*$ in Section 5.2.

Appendix 5.2: OpenBUGS Model for Outlier Mixture Model (Section 5.3)

```
model {
 ###    Likelihood   ###
 for (j in 1:N) {
           Y[j] ~ dnorm(mu[Treat[j]], taumix[T[j]])
           Outlier[j] ~ dbern(p)
           T[j] <- Outlier[j] + 1
 }

taumix[1] <- tau
taumix[2] <- tau/5

###   Priors  ###
for (k in 1:5){
        mu[k] ~ dnorm(0, 1.E-4)
}
tau ~ dgamma(0.01, 0.01)
p ~ dbeta(1, 19)

###  Derived parameters ###
sigma2 <- 1/tau
```

```
sigma  <- sqrt(sigma2)
for (i in 1:5) {
        Geo[i] <- pow(10, mu[i])
}

for (i in 2:5) {
        Comp[i] <- mu[i] - mu[1]
        Ratio[i] <- pow(10, -Comp[i])
}
}
```

Notes:

1) In the data, Y[] is a vector containing the log-transformed variable to be analysed. Results are back-transformed to the original scale to give geometric means and ratios.

Appendix 5.3: OpenBUGS Model for Censored Normal Model (Section 5.4)

```
model {
###    Likelihood  ###
for (j in 1:N) {
          Y[j] ~ dnorm(t[Treat[j]], tau[Treat[j]])C(,0)
                 # Censored Normal distribution with
                 # mean and precision both depending on
                 # treatment
}

#   Priors and derived parameters
for (m in 1:6) {
        t[m]      ~ dnorm(1, 1.E-4)
                 # Priors for individual treatment means
        tau[m] ~ dgamma(0.01, 0.01)
                 # Priors for individual treatment precisions
        sigma2[m] <- 1/tau[m]
        sigma[m] <- sqrt(sigma2[m])
}

for (k in 2:6) {
        Comp[k] <- t[k] - t[1]
}

}
```

References

[1] Walley, R., Sherington, J., Rastrick, J., Detrait, E., Hanon, E., and Watt, G. (2016). Using Bayesian analysis in repeated preclinical in vivo studies for a more effective use of animals. *Pharmaceutical Statistics*, 15(3): 277–285.

[2] Natanegara, F., Neuenschwander, B., Seaman, J. W., Kinnersley, N., Heilmann, C. R., Ohlssen, D. and Rochester, G. (2014). The current state of Bayesian methods in medical product development: Survey results and recommendations from the DIA Bayesian Scientific Working Group. *Pharmaceutical Statistics*, 13: 3–12. DOI:10.1002/pst.1595.

[3] The National Centre for the Replacement Refinement and Reduction of Animals in Research website; the ARRIVE guidelines section. www.nc3rs.org.uk/arrive-guidelines (last accessed 23rd November 2015).

[4] Neuenschwander, B., Capkun-Niggli, G., Branson, M., and Spiegelhalter, D. J. (2010). Summarizing historical information on controls in clinical trials. *Clinical Trials*, 7: 5–18. DOI:10.1177/1740774509356002.

[5] Evans, M., Hastings, N., and Peacock, B. *Statistical Distributions* (3rd edn). Wiley, New York, 2000.

[6] Gelman, A. (2006). Prior distributions for variance parameters in hierarchical models (comment on article by Browne and Draper). *Bayesian Analysis*, 1: 515–534. DOI:10.1214/06-BA117A.

[7] Schmidli, H., Gsteiger, S., Roychoudhury, S., O'Hagan, A., Spiegelhalter, D., and Neuenschwander, B. (2014). Robust meta-analytic-predictive priors in clinical trials with historical control information. *Biometrics*, 70: 1023–1032. DOI:10.1111/biom.12242.

[8] Woodward, P. (2011). *Bayesian Analysis Made Simple: An Excel GUI for WinBUGS*. Boca Raton, FL, CRC Press

[9] Schmidli, H., Wandel, S., and Neuenschwander, B. (2013). The network meta-analytic predictive approach to non-inferiority trials. *Statistical Methods in Medical Research*, 22: 219–240. DOI:10.1177/0962280211 432512.

[10] Gsteiger, S., Neuenschwander, B., Mercier, F., and Schmidli, H. (2103). Using historical control information for the design and analysis of clinical trials with overdispersed count data. *Statistics in Medicine*, 32; 3609–3622. DOI:10.1002/sim.5851.

6

Planning a Model-Based Bayesian Dose Response Study

Neal Thomas

Pfizer, Inc.

CONTENTS

6.1 Introduction

Statisticians have developed and promoted many methods for designing and analyzing dose response studies. Most of these methods seek to minimize modeling assumptions, for example, pairwise comparisons with hypothesis testing, normal dynamic linear models ([Berry et al., 2002]), and MCP-MOD (Bretz et al. [2005]). In contrast, clinical pharmacologists have long used $E_{\max}$ models to approximate dose response curves. Three recently completed meta-analyses of dose response studies found that with rare exceptions, the $E_{\max}$ models describe dose response well. Use of this model produces much improved designs and analyses. Although simple in its mathematical form, the $E_{\max}$ model is difficult to use with clinical data, which often has low signal-to-noise and limited dosing ranges. In addition to validating the model, the meta-analyses yield informative Bayesian prior distributions that substantially improve our ability to fit and use the $E_{\max}$ model.

The planning of a phase II dose response study using Bayesian methods to quantitatively assess the properties of potential designs and analyses is described here. The planning begins with an empirical basis for the $E_{\max}$ dose response model specification and a prior distribution for the model parameters. The $E_{\max}$ model, which is ubiquitous in pharmacology, will be described. Subsequent assessments of different designs and estimators include performance averaged over the empirically based prior distribution and repeated-sampling assessments conditional on specific sets of parameter values. Model-based Bayesian and maximum likelihood estimators (MLEs) are evaluated. Simple pairwise comparisons of dose group sample means are reported as a standard benchmark. An example based on a recent study is presented. The outcomes and dose levels are masked. The example demonstrates how the results of the recent meta-analyses combined with information specific to a new compound can be applied in an empirically based Bayesian design and analysis of dose response. An R package implementing these methods allows rapid implementation. The Bayesian approach yields substantial improvements in the estimation of dose response curves in many practical settings. The use of Markov Chain Monte Carlo (MCMC) numerical methods also makes it practical to implement dose selection criteria that are very specific to the specialized conditions that arise for many compounds.

Three recent meta-analyses of dose response studies are summarized in Section 6.2. A parametric $E_{\max}$ dose response model is described. Prior distributions for some of the parameters are also specified. The prior distributions were derived as part of the meta-analyses. The example is described in Section 6.3. It covers the design of a dose response trial for a continuous endpoint following a successful proof of concept study comparing a single high dose to placebo. This includes the decision criteria selected by the clinical team. The properties of the dosing designs and the various analyses are in Section 6.4. Computer code producing the primary results is given in Section 6.5.

6.2 Dose Response Meta-Analyses

Three meta-analyses of dose response studies provide an empirical basis for specifying a parametric model for most dose response studies. The meta-analyses summarize the dosing designs, dose group means (continuous endpoints), and proportions (binary endpoints) for the primary efficacy endpoints for more than 150 compounds. The meta-analyses yield predictive prior distributions for some of the parameters in the recommended $E_{\max}$ model. The first meta-analysis (Thomas et al. [2014]) included all dose response studies that demonstrated differentiation from placebo for small-molecule compounds developed at a large sponsor between 1998 and 2009. Patient-level data were available for most compounds. Details of the inclusion and exclusion criteria are given in the publication. Only active compounds were included for reasons of feasibility, but approximately one half of the included compounds were not continued to phase III development for a variety of reasons. The second meta-analysis (Thomas and Roy [2017]) included small-molecule compounds approved by the FDA between 2009 and 2014. The inclusion and exclusion criteria for the studies and compounds were similar to those in the first meta-analysis. It includes compounds from numerous sponsors, but only successful compounds are included, and only summary data are available. The third meta-analysis (Wu et al. [2017]) included biological compounds with published dose response studies, which also tend to have evidence of a drug effect because this is typically a requirement for publication. The meta-analyses include 95 small-molecule compounds and 71 biological compounds. An example of the estimated dose response curves for an extensively studied small-molecule compound is displayed in Figure 6.1.

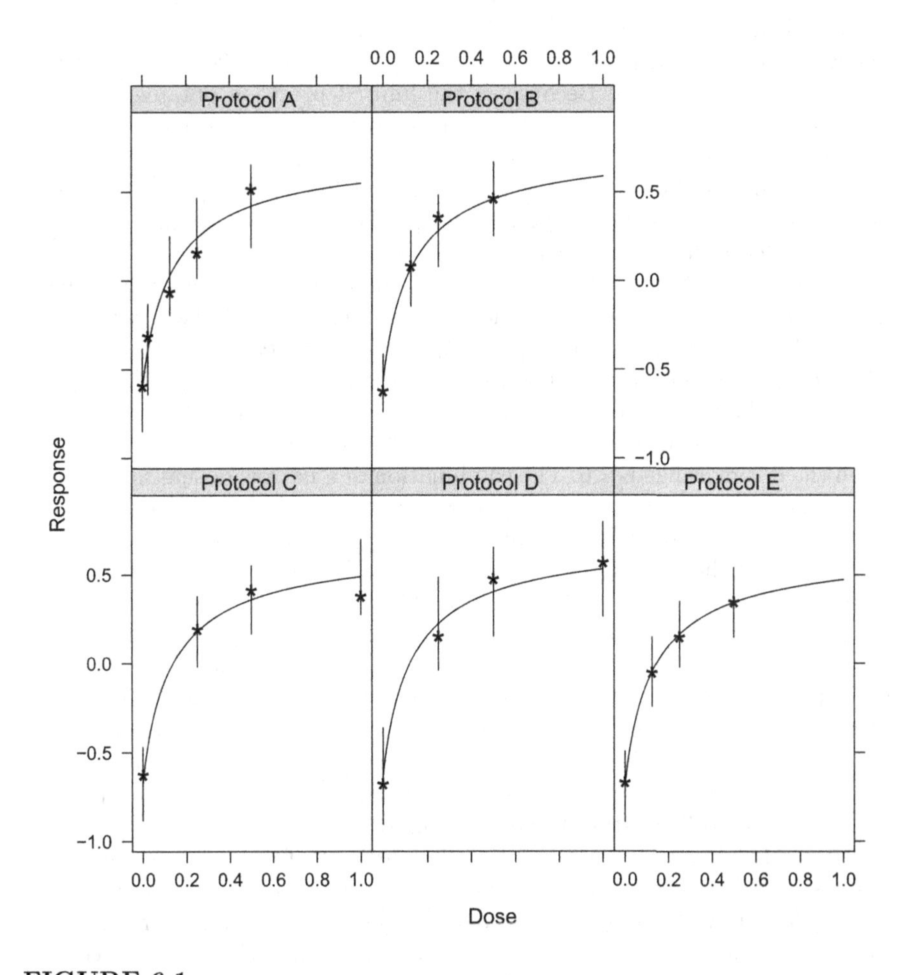

FIGURE 6.1
Dose response curves from five studies for a PDE5 inhibitor. The points are dose group means, the curves are sigmoidal $E_{\max}$ models with a separate placebo response for each study, and the bars are 95% prediction intervals.

The model fit displayed in Figure 6.1 is an $E_{\max}$ model. For a continuous endpoint, which is denoted by Y, the distribution within each dose group is normal with homogeneous residual variance. The mean is given by the sigmoid function,

$$E(Y \mid D) = E_0 + \frac{E_{\max} D^{\lambda}}{ED_{50}^{\lambda} + D^{\lambda}}, \tag{6.1}$$

where D is dose and $E_0, ED_{50}, E_{\max}, \lambda$ are unknown parameters. The residual variance is denoted by σ^2. The mean response changes monotonically from E_0 at $D = 0$ to a maximum of $E_0 + E_{\max}$. The parameter, λ, is called the Hill parameter. The model in Equation (6.1) is called a hyperbolic $E_{\max}$ model when $\lambda = 1$ MacDougall [2006]. The model yields an increasingly steep sigmoidal shape as λ increases. The model can be derived from receptor occupancy models in pharmacology (Iliadis and Macheras [2006], Källén [2007], Goutelle08 et al. [2008]). Because of the limited precision obtained in most clinical dose response studies, other families of models monotonically increasing to an asymptote can yield fits that are practically indistinguishable from the $E_{\max}$ model. The $E_{\max}$ model is preferred for dose response modelling because of its prominence in pharmacology. For binary endpoints, an $E_{\max}$ model is used on the logistic scale.

A high percentage (>90%) of the compounds in the meta-analyses have data that are well-described by the $E_{\max}$ model. Most of the fitted curves are similar to the hyperbolic $E_{\max}$ shape within the observed dosing range with Hill parameters near 1. The ratio of the dose yielding 90% of the effect divided by the dose yielding only 10% of the effect is always 81 for this curve with $\lambda = 1$. Most compounds were studied in dosing ranges <20 fold, and often the dosing range is <6. The combination of a narrow dosing range and a low signal-to-noise ratio yields sample means that change largely due to sampling variability. This often produces unusual patterns of apparent response that obscures the underlying curve. These conditions also make fitting nonlinear models difficult. Non-convergence of iterative maximization algorithms is common, and asymptotic approximations are poor. The data do not dominate inference, and some parameters may not be identified, for example, when all of the studied doses are on the plateau or there are no doses close to the plateau. In this context, it is very important to have informative prior distributions. Because there are numerous dose response studies completed each year, this data can be used to select specific parametric models and provide information on difficult-to-estimate parameters.

Hierarchical modeling of the dose response curves from the numerous compounds included in the three meta-analyses yielded posterior predictive distributions for the ED_{50} and $E_{\max}$ parameters in the sigmoidal $E_{\max}$ model for a future compound. Various forms of the hierarchical model were explored; the results from one of the hierarchical models are reported here and used to form prior distributions in the following sections. It is anticipated that these predictive distributions will be updated as more data becomes available, and the updated distributions will be available in the supporting R package, *clinDR*.

The posterior predictive distribution for the Hill parameter, λ, is a rescaled beta distribution which is restricted to the range $(0, 6)$: $\lambda \sim 6 \times \text{Beta}(3.03, 18.15)$. The distribution is concentrated near 1 with a median value of 0.79. It assigns low probability to steep dose response curves with $\lambda > 3$. The predictive distribution for the ED_{50} was normalized by dividing the ED_{50} by its predicted value at the time the phase II study is planned. This prediction is denoted by P_{50}. Some typical sources of information for this prediction are illustrated in the example in Section 6.3. These early predictions of the ED_{50} are a necessary part of early drug development, as they are required for decisions about what doses to manufacture. If a formal P_{50} is not available, past experience suggests that the midpoint between the two lowest (non-placebo) available doses is a useful proxy for the implied predicted potency. The posterior predictive distribution for the ED_{50}/P_{50} for a new compound is a log-t distribution: $\log(ED_{50}/P_{50}) \sim 0.6t_3 + 0.79$. The λ and ED_{50} parameters estimated from the same data for the same compound are often highly dependent in the posterior distribution, but the posterior distribution of the hierarchical parameters is approximately independent, so the λ and ED_{50} parameters are approximately independent in their posterior predictive distribution.

An additional finding from the meta-analyses, which guides the construction of the prior distribution, is the observation that placebo response often varied between studies of the same compound for unexplained reasons. The treatment differences from different studies, however, appeared consistent after accounting for sampling variability. These findings are consistent with broader findings from the Cochrane reviews showing heterogeneity in treatment difference estimates between pharmaceutical products and placebo is amongst the lowest of various effect estimates (Turner et al. [2012], Rhodes et al. [2015]). The prior distributions and models will include separate independent expected placebo response for different studies. The treatment differences with placebo across studies are assumed to be the same.

6.3 Example

The example is based on a small-molecule compound in phase II development for the treatment of a degenerative chronic disease. The primary endpoint was a continuous variable with well-studied properties that were published in several previous studies. Results are masked here by norming the highest dose considered to 1.0 and the within-dose group standard deviation (SD) to 1.0. The compound was previously evaluated in some other indications. A single-dose proof of concept (POC) study in the current indication was completed before planning for the dose response study began.

Three sources of information were used to predict the ED_{50}. The primary source was data on a mechanistic biomarker that can be measured after a single dose. Data on the biomarker from patients in a different indication, and from healthy volunteers in a phase 1B multiple ascending dose study, were sufficient to estimate an $E_{\max}$ model yielding an $\widehat{\mathrm{ED}}_{50} = 0.1$. The compound also produces a small reduction in systolic blood pressure. Data combined from several early-stage clinical trials were modeled yielding an $\widehat{\mathrm{ED}}_{50} = 0.05$ for this endpoint. In vitro data was also collected for another biomarker that is upstream' from the primary biomarker. Modeling of this data, combined with pharmaco-kinetic data, yielded an $\widehat{\mathrm{ED}}_{50} = 0.025$. The relative consistency of the estimates based on differing data sources on different endpoints is encouraging, but the standard errors (SEs) for the estimates were large, and the ED_{50} on different endpoints can differ. The mechanistic biomarker was judged to be the most predictive of the clinical endpoint, so its $\widehat{\mathrm{ED}}_{50}$ was selected: $P_{50} = 0.1$. The distribution for the ED_{50}/P_{50} then completes the prior distribution specification for the ED_{50}. The distribution of the ED_{50}/P_{50} is the posterior-predictive distribution derived in the meta-analyses and described in Section 6.2.

This example has three estimates of the ED_{50}, which is somewhat unusual, but most compounds reaching phase II of development have evidence of efficacy from animal or other pre-clinical experiments, biomarker data in healthy volunteers, data from related compounds previously studied, etc. to form a projected P_{50}. Assessing uncertainty in the P_{50} is often difficult, and SEs may not be available. The empirical evidence on the accuracy of the P_{50} from meta-analyses of past dose response studies provides a useful measure of uncertainty and serves as a check even when alternative estimates of uncertainty are available, as these are often based on assumption-laden models relating diverse data sources.

Another source of prior information is a 12-week POC study with 150 patients on the 0.5 dose and 50 patients on placebo. Denote the mean effect of the 0.5 dose versus placebo by $E(0.5)$. The primary analysis of the POC study was Bayesian, with an informative prior distribution for the placebo response estimated from hierarchical modelling of several historical studies. The historical placebo prior information was approximately equivalent to 100 concurrent controls. This prior distribution (after rescaling to match the transformed masked data) has a normal distribution: $N(-0.063, 0.136^2)$. A relatively diffuse prior distribution restricted to biologically plausible values was used for the drug effect. The resulting posterior distribution from the POC study was $E(0.5) \sim N(-0.289, 0.127^2)$.

The Bayesian dose response analysis will combine the data from the POC and dose finding studies with a model that assumes a common drug effect across the two studies but allows for potentially different placebo response in each protocol. These assumptions regarding the heterogeneity of results between studies for the same compound are supported by the meta-analyses of dose response studies. The mean placebo responses for the two studies are assigned independent placebo prior distributions, which were constructed for the POC study: $N(-0.063, 0.136^2)$. A diffuse normal prior distribution will be used for the $E_{\max}$ parameter, which only restricts it to biologically plausible values. The prior distribution for the λ parameter is the predictive distribution from the dose response meta-analyses. A diffuse uniform distribution is used for the residual SD, σ, constrained only by biological plausibility. It is centered about the (rescaled) SD estimated from the POC study. The prior distribution for the ED_{50} has already been described. Consistent with the dose response meta-analyses, the model parameters are assumed independent.

For study planning, the numerical evaluation of the posterior distribution of the $E_{\max}$ model parameters that will result from combining the POC data and the planned dose response study data is simplified by approximating the existing POC posterior distribution to create a 'design' prior distribution (Hobbs et al. [2011]) for the dose response study. The data from the POC study are then combined with the dose response data through the prior distribution rather than as part of the likelihood function. The approximation is formed by combining the posterior distribution for $E(0.5)$ with the prior distributions for the placebo response, ED_{50}, and λ parameters. The distributions for these parameters when combined independently determine a multivariate prior distribution. The independence of (E_0, ED_{50}, λ) is consistent with the meta-analyses. There is weak dependence between these parameters and $E(0.5)$ (e.g., a large ED_{50} predicts a lower $E(0.5)$). This dependence is ignored here. It is computationally convenient to replace the $E(0.5)$ parameter by the $E_{\max}$ parameter. This can be accomplished by simulating the $(E_0, ED_{50}, \lambda, E(0.5))$ parameters from their distributions and then solving for the mean function evaluated at the 0.5 dose for the $E_{\max}$ parameter:

$$E_{\max} = E(0.5)(0.5^{\lambda} + ED_{50}^{\lambda})/0.5^{\lambda}. \tag{6.2}$$

A histogram of 10,000 independent draws from the resulting prior distribution for $E_{\max}$ is displayed in Figure 6.2. The prior distribution is approximated by the normal density in the display, which is $N(-0.310,\ 0.203^2)$. As with the $E(0.5)$ parameter, the $E_{\max}$ is assumed to be independent of the other parameters.

The resulting 'design' prior distribution is used when simulating potential data from the dose response study and when computing the Bayesian analyses applied to the simulated data, as described in Section 6.4. A future version of the software for assessing potential designs and analyses will allow the data from an existing study to be incorporated directly into the dose response study planning calculations without approximating the posterior distribution.

The original POC study supported a difference with placebo, but the evidence for an appreciable effect was not strong, so a confirmatory initial test of the null hypothesis is planned. A minimal efficacy target to support phase III development was set at -0.275. A target efficacy was set at -0.488 to support a definitive decision to advance to phase III in the absence of safety findings. The decision criteria stops if the probability of achieving the minimal efficacy is <0.6 for all doses within the studied range (criteria C1). A rapid transition to phase III is planned if the probability of achieving the targeted efficacy is

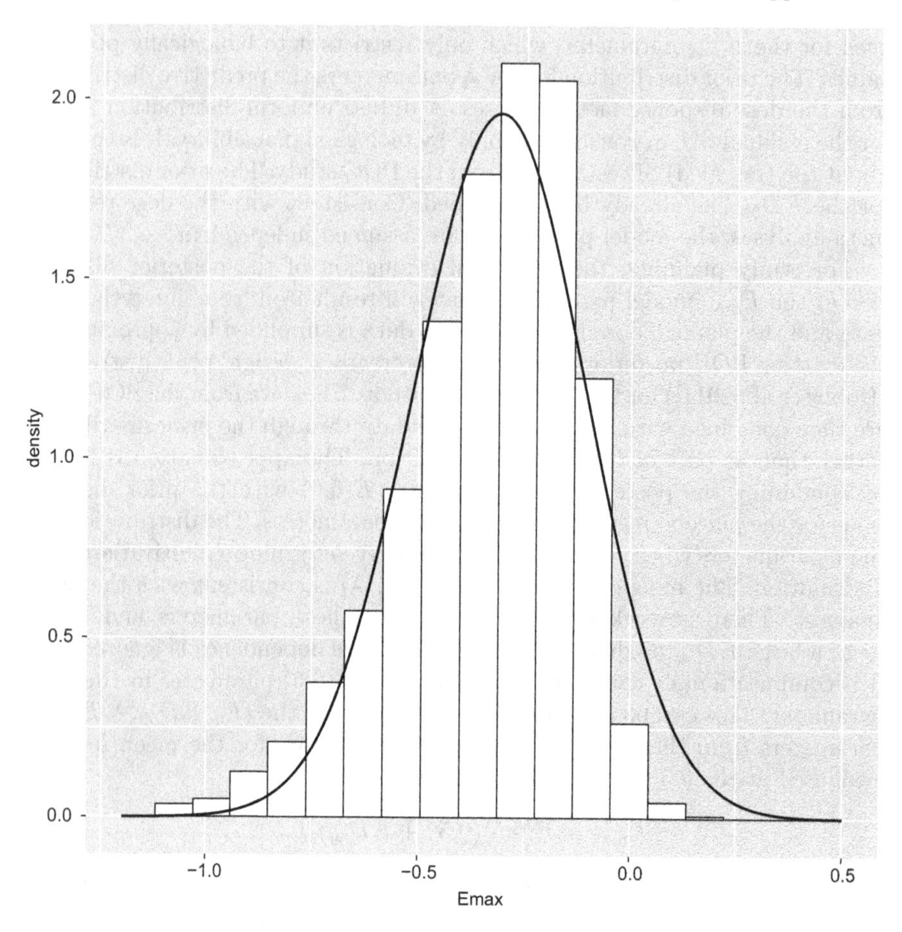

FIGURE 6.2
The prior distribution for the E_{max} parameter and its normal approximation.

$\geq$0.8 for some dose (criteria C2). If more than one dose is identified, a lower dose is preferred, but other endpoints will be considered in final decisions. If neither the minimal criteria to stop is met nor the target criteria to proceed is met, consideration of safety data, other endpoints, and updated commercial assessment will be used to make further development decisions. For regulatory reasons, it was anticipated that substantial evidence establishing dose response would be needed. Proposed criteria to demonstrate dose response specifies that the probability the highest dose has efficacy exceeding the minimal efficacy is $\geq$0.8, while the probability that the lowest dose achieves the minimal efficacy is $\leq$0.5 (criteria C3). This criteria is supplemented by requiring that the sample mean for the lowest dose is not within 0.1 estimated SDs of the sample mean for the highest dose (criteria C4). This criteria is intended to ensure the visual appearance of a dose response.

6.4 Assessing Designs and Analyses

The first design evaluated included all of the available doses: 0, 0.05, 0.10, 0.25, 0.50, and 1. Because the effect estimated in the POC study was low, the maximum feasible sample size of 600 was selected for the first assessment of operating characteristics. Even with the large proposed sample size, the power to achieve the limited objective to differentiate from placebo is moderate, as will be documented. The first design allocates the patients uniformly across the dose groups.

The design and various analyses were evaluated over a factorial collection of model parameters yielding different dose response curves. The different parameter values evaluated are $E_0 = (-0.063, -0.237, 0.111)$, $ED_{50} = (0.1, 0.025, 0.05, 0.4, 1)$, the efficacy at the highest dose $(-0.275$ $-0.382, -0.488)$, and $\lambda = (1, 0.75, 2)$. The $E_{\max}$ model parameters are derived from the other parameter values and the efficacy at the highest dose. Repeated-sampling (1,000) simulated data sets are generated for each fixed combination of the parameters. The first value for each setting corresponds to the center of the prior distribution and is thus most favorable for the Bayesian approach. The remaining parameter values represent different extremes of the prior distributions, some exceeding the $10th$ or $90th$ percentile of the prior distribution. The efficacy values specified for the highest dose are the minimal acceptable efficacy (which is close to the posterior mean for the POC dose), the target efficacy, and the midpoint of these values.

In addition to the Bayesian analysis, the MLE (non-linear least squares) of the $E_{\max}$ model was evaluated. It is common for the MLE computation to fail when there is a limited dosing range or low signal-to-noise. If the sigmoidal $E_{\max}$ estimation fails, then a hyperbolic $E_{\max}$ fit is substituted, and if it also fails, then a model linear in dose, log-linear in dose, or scaled-exponential in dose is selected based on the minimum least squares criteria. Simple pairwise comparison of sample means was also evaluated.

A test of the null hypothesis was performed using the MCP-MOD approach (Bretz et al. [2005]) with contrasts formed from the hyperbolic $E_{\max}$ model with ED_{50} values of $(0.025, 0.05, 0.1, 0.5)$. Power for a one-sided 0.05 test is inadequate at the most likely values of the parameters. When the efficacy at the highest dose matches the minimum target value, and the other parameters are close to their central tendency, the power is only 0.7. The power was not highly sensitive to the shape of the curve, but it was lower for most parameter combinations, and never exceeded 0.78 at the minimum target value. The power exceeded 0.88 when the efficacy at the highest dose was set at the higher target values. Even with 600 patients, the study has adequate power for the limited goal of differentiating from placebo only if the efficacy substantially exceeds the estimate from the POC study.

The square root of the mean square error (RMSE) of the three estimators of treatment differences for each dose versus placebo is displayed in Table 6.1.

TABLE 6.1
RMSE for Bayes and MLE Estimators of Each Dose versus Placebo

	$E_0 = -0.063$, $ED_{50} = 0.1$, $E_{\max} = -0.303$, $\lambda = 1$				
Dose	0.050	0.100	0.250	0.500	1.000
Bayes	0.030	0.044	0.060	0.066	0.070
MLE	0.090	0.111	0.126	0.127	0.128
	$E_0 = -0.237$, $ED_{50} = 0.4$, $E_{\max} = -0.535$, $\lambda = 1$				
Dose	0.050	0.100	0.250	0.500	1.000
Bayes	0.056	0.064	0.065	0.065	0.075
MLE	0.072	0.096	0.114	0.114	0.116
	$E_0 = 0.111$, $ED_{50} = 0.4$, $E_{\max} = -0.683$, $\lambda = 1$				
Dose	0.050	0.100	0.250	0.500	1.000
Bayes	0.024	0.033	0.061	0.099	0.146
MLE	0.072	0.100	0.121	0.116	0.116

$\sigma = 1$. The RMSE for the pairwise comparison is 0.141.

Posterior means are used as the point estimates for the Bayesian method. The table has results for three of the factorial settings of the population model parameters. The upper entry is the setting most favorable to the Bayesian method, with population parameters set at central values in the prior distribution. The middle setting has the E_0 set -1.28 prior SD below the prior mean for E_0 and the $E_{\max}$ parameter -1.1 prior SD below its prior mean. The lower entry has E_0 set 1.28 prior SD above its prior mean and the $E_{\max}$ set -1.84 prior SD below its prior mean. The ED_{50} is four times the P_{50} in both the second and third settings. Except for the highest dose in the third setting, the Bayesian method performed much better than the MLE-based method, which often required modification as noted above. The MLE, in turn, performed better than the pairwise sample mean difference, which has an (unselected) RMSE of 0.141 throughout. Reducing the RMSE of the pairwise comparison to 1.2 (typical MLE RMSE, Table 6.1) would require a sample size multiple increase of 1.38, and reducing it to 0.7 (typical Bayesian RMSE, Table 6.1) would require an increase of 4.1. The large improvements for the Bayes vs ML estimators are due in part to the informative prior distributions for the E_0 and $E_{\max}$ parameters because of the availability of several recent placebo samples and the inclusion of the POC study. The results in Table 6.1 show that the improvements are not highly sensitive to the informative prior distributions requiring only that the population parameters are within the support of the prior distributions. Experience with other settings with less informative prior distributions for E_0 and $E_{\max}$ yielded the same ordering, with substantial but less dramatic improvements.

The coverage of nominal 0.95 confidence/high probability intervals corresponding to the settings in Table 6.1 are given in Table 6.2. Except for the intervals for the high dose in the third setting, the Bayesian intervals included the fixed population values in >0.95 of the simulated samples. Note that the

TABLE 6.2
The Coverage of 95% Intervals for Bayes and MLE Estimators of Each Dose versus Placebo

	$E_0 = -0.063$, $ED_{50} = 0.1$, $E_{\max} = -0.303$, $\lambda = 1$				
Dose	0.050	0.100	0.250	0.500	1.000
Bayes	1.000	0.995	0.988	0.985	0.990
MLE	0.730	0.749	0.768	0.805	0.932
	$E_0 = -0.237$, $ED_{50} = 0.4$, $E_{\max} = -0.535$, $\lambda = 1$				
Dose	0.050	0.100	0.250	0.500	1.000
Bayes	0.989	0.982	0.978	0.986	0.985
MLE	0.739	0.755	0.805	0.853	0.955
	$E_0 = 0.111$, $ED_{50} = 0.4$, $E_{\max} = -0.683$, $\lambda = 1$				
Dose	0.050	0.100	0.250	0.500	1.000
Bayes	1.000	1.000	0.986	0.904	0.761
MLE	0.776	0.796	0.835	0.880	0.956

The $\sigma = 1$.

asymptotic confidence intervals for the MLE were often substantially below the nominal coverage rate. This is common in practical clinical dose response settings. The Bayesian method is biased toward a smaller effect size in the setting most favorable to the drug. Some decision makers may find the bias reducing the drug effect worrisome. Note, however, this favorable condition requires that the POC study result substantially underestimates the effect of the 0.5 dose, and the dose response curve is much steeper than anticipated, which yields a big improvement between the 0.5 and 1.0 doses. The combination of parameters producing bias and low coverage for the Bayesian method is unlikely. There are some other extreme settings in the factorial collection of parameters yielding comparable underestimation of drug effect.

The fixed-population repeated-sampling assessments of operating characteristics summarized in Tables 6.1 and 6.2 do not account for relevance of the settings. A Bayesian assessment of each analysis method (Bayes and non-Bayes) weights the settings based on their likely occurrence by averaging the results over a distribution for the model parameters. The distribution used here is the same prior distribution proposed for the analysis after data collection, but other distributions could be assessed for sensitivity to the choice of prior distribution. The probability of rejecting the null hypothesis obtained by averaging the power results based on the MCP-MOD test is only 0.53. In contrast, the probability is lower, 0.47, if the null test is based on the usual t-test comparing the highest dose to placebo. The RMSE averaged over the prior distribution for each dose versus placebo is in Table 6.3. These results are comparable to the more favorable results in Table 6.1 because these settings receive most of the weight. The coverage probabilities for the Bayesian intervals, which correspond to Table 6.2, are all within simulation error of 0.95 confirming the accuracy of the numerical computations. The results on

TABLE 6.3
RMSE for Bayes and MLE Estimators of Each Dose versus Placebo Averaged over the Prior Distribution

Dose	0.050	0.100	0.250	0.500	1.000
Bayes	0.053	0.060	0.070	0.079	0.089
MLE	0.086	0.101	0.114	0.118	0.130

The RMSE for the pairwise comparison is 0.141

power and RMSE taken together show that the Bayesian methods yield large improvements in the analysis of the dose response, but the probability the compound can produce adequate efficacy is low.

This assessment is refined by assessment of the four decision criteria described in Section 6.3. The probabilities of achieving the criteria are in Table 6.4. The settings are the same as in Table 6.1. Under the most likely setting, there is a high probability of terminating the compound for futility (C1) and essentially no chance of achieving a positive decision to initiate phase III testing (C2). The chance of producing an estimated dose response curve at the end of the study that predicts appreciably different success rates for the low and high doses is less than 0.10 (C3). Criteria (C4) requiring visual distinction between the lowest and highest doses is more often met, but the chance that the sample means from these doses would be indistinguishable still exceeds 0.3. The settings with larger treatment effects yielded more encouraging results, but there is still appreciable chance of terminating the compound, and essentially no chance of strong support to proceed to phase III. The probability of demonstrating a dose response is much improved but still far from assured. The results in Table 6.4 are succinctly summarized by the Bayesian predictive probabilities applied to the four criteria. The Bayesian predictive probabilities for the criteria are $(0.727, 0.01, 0.153, 0.586)$.

Other designs were explored that allocated more patients to the highest dose and placebo. The most efficient design, as measured by the RMSE

TABLE 6.4
The Probability of Achieving Criteria C1–C4 for Different Dose Response Curves

	$E_0 = -0.063$, $ED_{50} = 0.1$, $E_{\max} = -0.303$, $\lambda = 1$			
Criteria	C1	C2	C3	C4
Proportion	0.750	0.000	0.088	0.659
	$E_0 = -0.237$, $ED_{50} = 0.4$, $E_{\max} = -0.535$, $\lambda = 1$			
Criteria	C1	C2	C3	C4
Proportion	0.199	0.004	0.530	0.936
	$E_0 = 0.111$, $ED_{50} = 0.4$, $E_{\max} = -0.683$, $\lambda = 1$			
Criteria	C1	C2	C3	C4
Proportion	0.197	0.004	0.544	0.995

and C1–C4, eliminates the central 0.25 dose. The design also applies a $2:1:1:1:2$ randomization to the remaining doses. The resulting design allocates many more patients (172) to the extreme doses. The resulting RMSE for the $(0, 0.05, 0.1, 0.5, 1)$ doses are $(0.05, 0.056, 0.071, 0.079)$, compared to $(0.053, 0.06, 0.079, 0.089)$ for the original design. The probabilities of achieving the decision criteria under the alternative design are $(0.712, 0.011, 0.177, 0.609)$ compared to $(0.727, 0.01, 0.153, 0.586)$ for the original design. The improvement in the estimation of the dose response curve was not uniform across the entire dosing range (best for highest doses) but exceeded what would be obtained for the original design if its sample size increased by approximately 10%.

Despite the improvement from the more efficient design, the absolute probability that the compound would yield useful efficacy that could be empirically demonstrated remains low. The high costs and long development times required to execute a large phase II study with low probability of success resulted in a decision to terminate development.

6.5 Software for Bayesian Dose Response Analyses

The R package *clinDR* (R Core Team [2015]) was used to create the tables and plots. A typical Bayesian simulation of 1,000 data sets requires <5 min on a laptop computer. The sample code in this section will recreate (within simulation error) the first setting reported in Tables 6.1 and 6.2 and the Bayesian predictive RMSE. The package uses MCMC for the Bayesian computations, which are performed by the R package *Rstan* (R Core Team [2015, 2013]). It uses the R package *DoseFinding* (Bjoern et al. [2016]) to compute the MCP-MOD trend tests. The graphics produced by *clinDR* are implemented using the R package *ggplot2* (Wickam [2009]).

The prior specification is given in listing 6.1. The specification of the fixed-population dose response curve to be simulated is in listing 6.2. The resulting *genfix* object is then input into the simulation code in listing 6.3. The RMSE and coverage probabilities in the tables can be obtained using generic function *summary(outb)*. There are several other generic functions that can be used to summarize the simulation output. For example, to produce a plot of the estimated dose response curve for the first simulated data set, use *plot(outb[1])*. The simulation is changed to report Bayes predictive probabilities in listing 6.4 by replacing the simulation object, *genfix*, with a simulation object, *genran*, which generates the population model parameters from the prior distribution.

There are many optional inputs to the functions that are set at their default values, which are not displayed in the code here. The custom code required to compute criteria C1–C4 has been omitted for brevity.

Listing 6.1
Prior distribution

```
prior<-prior.control(epmu =-0.063,epsd=0.136,
             emaxmu=-0.310,emaxsd=0.203,p50=0.1,
             sigmalow=0.1,sigmaup=2,edDF = 5)
```

Listing 6.2
Simulated data specification

```
dose<-c(0.0,0.05,0.10,0.25,0.50,1.00)
ndose<-rep(100,length(dose))
parm<-c(log(0.1),1,-0.303,-0.063)
mlev<-emaxfun(dose,parm=parm)
genfix<-FixedMean(n=ndose,doselev=dose,meanlev=mlev,
                    resSD=1,parm=parm)
```

Listing 6.3
Simulations

```
### MCP-MOD contrasts
ed50contr<-c(0.0250,0.0500,0.1000,0.5000)
lambdacontr<-rep(1,length(ed50contr))

outb<-emaxsimB(
        nsim=1000, genObj=genfix,prior=prior,modType=4,
        negEmax=TRUE,ed50contr=ed50contr,
        lambdacontr=lambdacontr,
        customCode=NULL,customParms=NULL
        )
summary(outb)
outmle<-emaxsim(nsim=1000,genObj=genfix,modType=4,
          negEmax=TRUE,
          ed50contr=ed50contr,lambdacontr=lambdacontr
          )
summary(outmle)
```

Listing 6.4
Bayesian predictive simulations

```
### bayesian evaluation
genran<-RandEmax(n=ndose, doselev=dose,
           parmEmax=c(-0.310,0.203),
           parmE0=c(-0.063,0.136),
           p50=0.1,
           resSD=1,
           dfSD=200)
outbpred<-emaxsimB(
```

```
        nsim=1000, genObj=genran,prior=prior,modType=4,
        negEmax=TRUE,ed50contr=ed50contr,
        lambdacontr=lambdacontr,
        customCode=NULL,customParms=NULL
        )
summary(outbpred)
```

References

D. Berry, P. Muller, A. Grieve, et al. Adaptive Bayesian designs for doseranging trials. In B. Carlin, A. Carriquiry, C. Gatsonis, et al., editors, *Case Studies in Bayesian Statistics V*. p.99, Springer, New York, 2008.

B. Bjoern, J. Pinheiro, and F. Bretz. *DoseFinding: Planning and Analyzing Dose Finding Experiments*, 2016. https://CRAN.R-project.org/package=DoseFinding. R package version 0.9-14.

F. Bretz, J. Pinheiro, and M. Branson. Combining multiple comparisons and modeling techniques in dose-response studies. *Biometrics*, 61: 738–748, 2005.

S. Goutelle, M. Maurin, F. Rougier, X. Barbaut, L. Bourguignon, and M. Ducher. The hill equation: a review of its capabilities in pharmacological modelling. *Fundamental and Clinical Pharmacology*, 22: 633–648, 2008.

B. Hobbs, B. Carlin, S. Mandrekar, and D. Sargent. Hierarchical commensurate and power prior models for adaptive incorporation of historical information in clinical trials. *Biometrics*, 67: 1047–1056, 2011, doi: 10.1111/j.1541-0420.2011.01564.x.

A. Iliadis and P. Macheras. *Modeling in Biopharmaceutics, and Pharmacodynamics: Homogeneous and Heterogeneous Approaches*. Springer, New York, 2006.

A. Källén. *Computational Pharmacokinetics*. CRC, Boca Raton, FL, 2007.

J. MacDougall. Analysis of dose response studies-E-max model. In N. Ting, editor, *Dose Finding in Drug Development*. Springer, New York, 2006.

R Core Team. *R: A Language and Environment for Statistical Computing*. R Foundation for Statistical Computing, Vienna, Austria, 2015. https://www.R-project.org/.

K. Rhodes, R. Turner, and J. Higgins. Predictive distributions were developed for the extent of heterogeneity in meta-analyses of continuous outcome data. *Journal of Clinical Epidemiology*, 68: 52–60, 2015, doi:10.1016/j.jclinepi.2014.08.012.

Stan Development Team. *Stan: A C++ Library for Probability and Sampling*, 1.3 edition, 2013. http://mc-stan.org/.

Stan Development Team. *Rstan: the r interface to stan, version 2.8.0*, 2015. http://mc-stan.org/rstan.html.

N. Thomas and D. Roy. Analysis of clinical dose response in small molecule drug development: 2009–2014. *Statistics in Biopharmaceutical Research*, 9(2): 137–146, 2017, doi:10.1080/19466315.2016.

N. Thomas, K. Sweeney, and V. Somayaji. Meta-analysis of clinical dose-response in a large drug development portfolio. *Statistics in Biopharmaceutical Research*, 6(4): 302–317, 2014, doi:10.1080/19466315.

R. Turner, J. Davey, S. Thompson, and J. Higgins. Predicting the extent of heterogeneity in meta-analysis, using empirical data from the cochrane database of systematic reviews. *International Journal of Epidemiology*, 41: 818–827, 2012, doi:10.1093/ije/dys041.

H. Wickam. *ggplot2: Elegant Graphics for Data Analysis*. Springer, New York, 2009. ISBN 978-0-387-98140-6, http://had.co.nz/ggplot2/book.

J. Wu, A. Banerjee, B. Jin, S. Menon, S. Martin, and A. Heatherington. Clinical dose-response for a broad set of biological products: A model based meta-analysis. *Statistical Methods in Medical Research*, 2017, doi:10.1177/0962280216684528.

7

Novel Designs for Early Phase Drug-Combination Trials

Ying Yuan and Ruitao Lin

The University of Texas MD Anderson Cancer Center

CONTENTS

Drug-combination therapy provides an effective way to obtain synergistic treatment effects and overcome resistance of monotherapy. In general, drug combinations may involve (1) two or more previously marketed drugs or biologics, (2) two or more new molecular entities, or (3) a mix of previously marketed drugs or biologics and new molecular entities. According to US Food and Drug Administration (FDA) guidelines [1,2], prior to testing a new drug combination in human beings, extensive preclinical studies are required to demonstrate the biological rationale for the combination and to assess the safety of the combination [1]. When such data are not available or indicate safety concerns on the combination, additional toxicology studies are required to address the concerns. Sometimes, the drug-combination trials may involve two or more new investigational drugs that have not been previously studied for any indication; additional considerations are needed for the co-development of the new investigational drugs for use in combination [2]. In drug-combination trials, it is useful to test multiple doses of each drug to identify the optimal dose combination in terms of risks and benefits [2]. As a result, compared to single-agent trials, drug-combination trials have a higher dimension dose searching space, leading to several challenges for designing combination trials, as described in Section 7.1.

Despite the enormous importance and apparent popularity of combination therapies, in terms of the designs used in actual trials of drug combinations, the current status is far from desirable. Riviere et al. [3] reviewed 162 trials published between 2011 and 2013 and found that 88% of the trials used the conventional 3+3 design, which has been widely criticized for its poor operating characteristics, even when used for single-agent trials. They found only one trial that used a new design for combination trials, despite the availability of more than a dozen novel combination trial designs in the (mainly statistical) literature. As we describe herein, the design of drug-combination trials is more complicated and challenging than that for single-agent trials. The goal of this chapter is to clarify some challenges and misconceptions on designing combination trials (Section 7.1) and provide practical guidance and designs to practitioners for conducting drug-combination trials (Sections 7.2 and 7.3).

7.1 Challenges of Designing Combination Trials

Partial order in toxicity A major challenge in designing combination trials is that combinations are only partially ordered according to their toxicity probabilities. Consider a trial combining J doses of agent A, denoted as $A_1 < A_2 < \cdots < A_J$, and K doses of agent B, denoted as $B_1 < B_2 < \cdots < B_K$. Let A_jB_k denote the combination of A_j and B_k, and p_{jk} denote the probability of dose-limiting toxicity (DLT) for A_jB_k. It is typically reasonable to assume that when the dose of one agent (say agent A) is fixed, the toxicity of the combination increases as the dose of the other agent increases (i.e., agent B). In other words, as shown in Figure 7.1, in the dose matrix, the rows and columns are ordered, with the probability of DLT increasing along with the dose. However, in other directions of the dose matrix, e.g., along the diagonals from the upper left corner to the lower right corner, the toxicity order is unknown due to unknown drug–drug interactions. For example, between A_2B_2 and A_1B_3, we do not know which drug is more toxic because the first combination has a higher dose of agent A whereas the second combination has a higher dose of agent B. Thus, we cannot fully rank $J \times K$ combinations from low to high in terms of their DLT rates. This is distinctly different from single-agent trials, for which the dose can be unambiguously ranked assuming that higher dosage yields higher probability of DLT. The implication of such a partial ranking is that conventional single-agent dose-finding designs

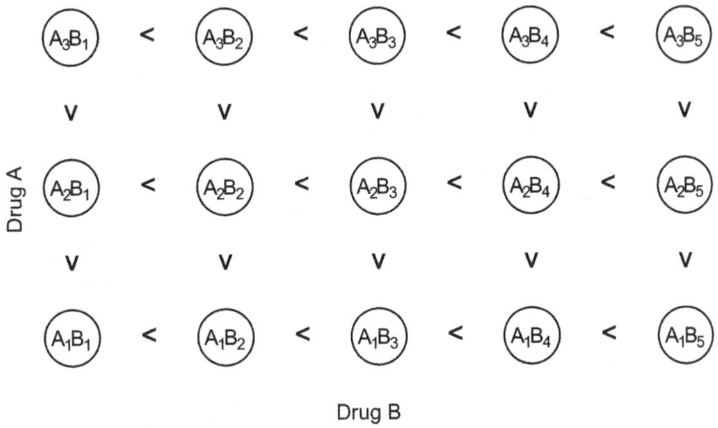

$A_3B_1 < A_3B_2 < A_3B_3 < A_3B_4 < A_3B_5$

$\vee \quad \vee \quad \vee \quad \vee \quad \vee$

Drug A $A_2B_1 < A_2B_2 < A_2B_3 < A_2B_4 < A_2B_5$

$\vee \quad \vee \quad \vee \quad \vee \quad \vee$

$A_1B_1 < A_1B_2 < A_1B_3 < A_1B_4 < A_1B_5$

Drug B

FIGURE 7.1
Partial order in toxicity for drug combinations. $A_{j_1}B_{k_1} < A_{j_2}B_{k_2}$ means that the toxicity probability $p_{j_1k_1}$ is smaller than $p_{j_2k_2}$.

cannot be directly used for finding the maximum tolerated dose (MTD) in drug-combination trials.

MTD contour Another important feature for combination trials is the existence of the MTD contour in the two-dimensional dose space, as shown in Figure 7.2. As a result, multiple MTDs may exist in the $J \times K$ dose matrix. The implication of the MTD contour is that when designing a drug-combination trial, the first and most important question requiring careful consideration is

Are we interested in finding one MTD or multiple MTDs?

As we describe below, the answer to this question determines the choice of different design strategies for drug-combination trials. This important issue, unfortunately, is largely overlooked by existing trial designs.

7.2 Drug-Combination Designs to Find One MTD

7.2.1 Model-Based Designs

Numerous model-based designs have been proposed to find a single MTD for drug combinations. Thall et al. [4] proposed a Bayesian drug-combination dose-finding method based on a six-parameter model, assuming that doses are continuous and can be freely changed during the trial. Yin and Yuan [5,6] proposed Bayesian dose-finding designs based on latent contingency tables [5] and a copula-type model [6]. Braun and Wang [7] developed a dose-finding method based on a Bayesian hierarchical model. Wages, Conaway, and O'Quigley [8]

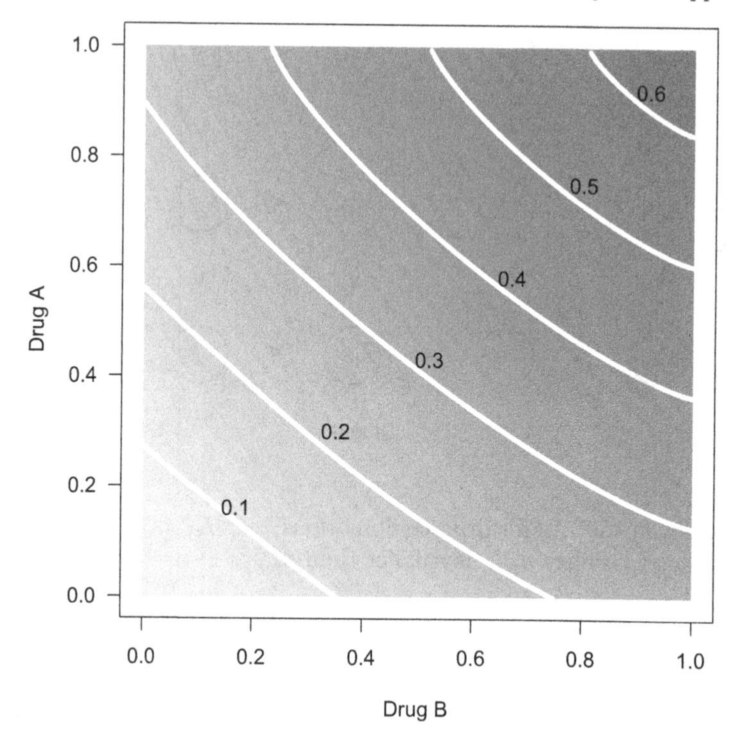

FIGURE 7.2
MTD contour in dose–toxicity surface.

extended the continuous reassessment method (CRM) [9] based on partial ordering of the dose combinations. Braun and Jia [10] generalized the CRM to handle drug-combination trials. Riviere et al. [11] proposed a Bayesian dose-finding design based on the logistic model. Cai, Yuan, and Ji [12] and Riviere et al. [13] proposed Bayesian adaptive designs for drug-combination trials involving molecularly targeted agents. Albeit very different, most of these designs adopt a common dose-finding strategy:

1. Devise a model to describe the dose-toxicity surface.

2. Based on the accumulating data, continuously update the model estimate, and make the decision of dose assignment for the incoming new patient, typically by assigning the new patient to the dose for which the estimated toxicity is closest to the MTD.

3. Stop the trial when the maximum sample size is reached, and select the MTD based on the estimates of toxicity probabilities.

In what follows, we use the copula-type model-based design [6] as an example to illustrate the model-based design. Let p_j be the prespecified toxicity probability corresponding to A_j, $p_1 < p_2 < \cdots < p_J$ and q_k be that of B_k, $q_1 < q_2 < \cdots < q_K$. Typically, before two drugs are combined, they are individually studied in phase I trials. Thus, the values of p_j's and q_k's can be elicited from clinicians, based on the historical data from previous single-agent trials.

The joint toxicity probability of A_jB_k can be modeled using the copula-type regression model [6] as follows

$$p_{jk} = 1 - \{(1 - p_j^{\alpha})^{-\gamma} + (1 - q_k^{\beta})^{-\gamma} - 1\}^{-1/\gamma}, \tag{7.1}$$

where $\alpha, \beta, \gamma > 0$ are unknown model parameters. This model is derived from the Clayton copula function and satisfies natural constraints for drug-combination trials. For example, if the toxicity probabilities of both drugs are zero (i.e., $p_j = q_k = 0$), then the joint toxicity probability $p_{jk} = 0$. If the toxicity probability of either A_j or B_k is one (i.e., $p_j = 1$ or $q_k = 1$), then $p_{jk} = 1$.

Another attractive feature of model (7.1) is that if only one drug is tested, it reduces to the power model used in the CRM. For example, if there is only drug B (i.e., $p_j = 0$), the model degenerates to

$$p_k = q_k^{\beta}, \quad k = 1, \ldots, K.$$

Therefore, the copula-type regression model can be viewed as a generalization of the single-agent CRM to drug-combination trials. This unique feature allows us to easily incorporate prior information of single-agent trials into the combination trials by specifying p_j's and q_k's, which are known as "skeleton" in the CRM [14]. One should not confuse the copula-type regression model (7.1), which involves a univariate toxicity outcome, with the standard copula models that specify bivariate distributions [6]. Figure 7.3 shows the dose-toxicity surface of the copula-type regression model, which provides the coverage of the entire probability domain of (0, 1).

Alternatively, we can also use the logistic regression model for modeling the toxicity probability of A_jB_k [11],

$$\text{logit}(p_{jk}) = \beta_0 + \beta_1 u_j + \beta_2 v_k + \beta_3 u_j v_k, \tag{7.2}$$

where β_0, β_1, β_2, and β_3 are unknown parameters, with $\beta_1 > 0$ and $\beta_2 > 0$ to ensure that the toxicity probability is increasing with the increasing dose level of each agent alone, and $\forall k, \beta_1 + \beta_3 v_k > 0$ and $\forall j, \beta_2 + \beta_3 u_j > 0$ to ensure that the toxicity probability is increasing with the increasing dose levels of both agents together. The u_j and v_k are standardized doses, defined as $u_j = \log\left(\frac{p_j}{1-p_j}\right)$ and $v_k = \log\left(\frac{q_k}{1-q_k}\right)$, where p_j and q_k are the prior estimates of the toxicity probabilities of A_j and B_k, respectively, when they are administered individually as a single agent in previous trials.

Using the prior information to define standardized dose has been widely used in dose-finding trial designs, and it improves the estimation stability and trial performance [15,16].

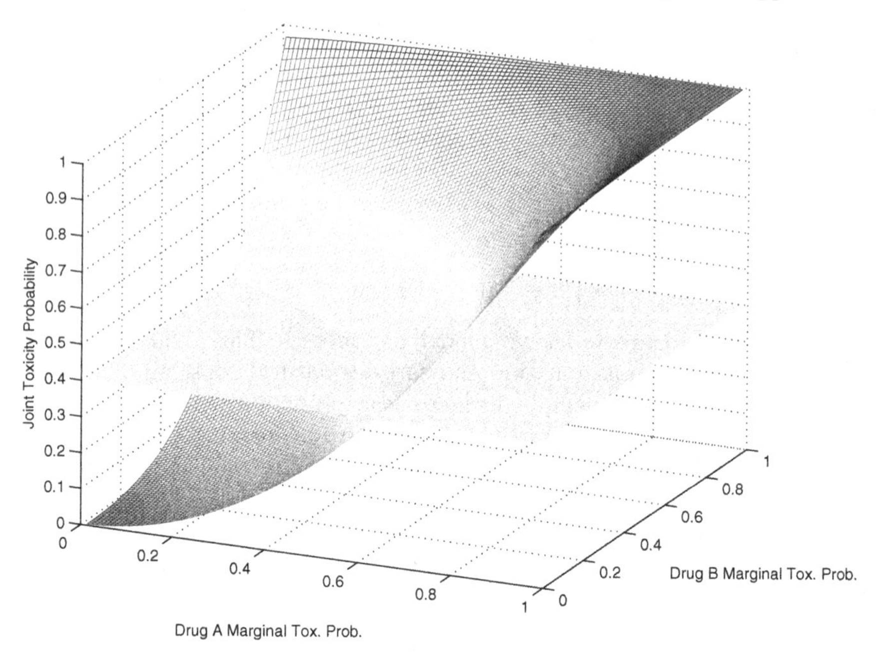

FIGURE 7.3
Dose–toxicity surface of the copula-type regression model.

Denote $\boldsymbol{\theta}$ as the set of unknown parameters of the dose-toxicity model. For example, $\boldsymbol{\theta} = (\alpha, \beta, \gamma)^T$ in the copula-type regression model, and $\boldsymbol{\theta} = (\beta_0, \beta_1, \beta_2, \beta_3)^T$ in the logistic regression model, respectively. Suppose that at a certain stage of the trial, among n_{jk} patients treated at A_jB_k, y_{jk} subjects have experienced DLT. The likelihood given the observed data $\mathcal{D}$ is

$$\mathcal{L}(\boldsymbol{\theta}|\mathcal{D}) \propto \prod_{j=1}^{J}\prod_{k=1}^{K} p_{jk}^{y_{jk}}(1-p_{jk})^{n_{jk}-y_{jk}}.$$

In the Bayesian framework, the joint posterior distribution is given by

$$f(\boldsymbol{\theta}|\mathcal{D}) \propto \mathcal{L}(\boldsymbol{\theta}|\mathcal{D})f(\boldsymbol{\theta}),$$

where $f(\boldsymbol{\theta})$ is the joint prior distribution for $\boldsymbol{\theta}$. For example, $f(\boldsymbol{\theta}) = f(\alpha)f(\beta)f(\gamma)$ in the copula-type regression model, where $f(\alpha)$, $f(\beta)$, and $f(\gamma)$ denote vague gamma prior distributions with mean one and large variances for α, β, and γ, respectively.

Let ϕ denote the target DLT probability for the MTD. The dose-finding algorithm can be described as follows:

1. The first cohort of patients is treated at the lowest dose combination A_1B_1.

2. During the course of the trial, at the current dose combination A_jB_k:
   (i) If $\Pr(p_{jk} < \phi|\mathcal{D}) > c_e$, where c_e is the fixed probability cutoff for dose escalation, the doses move to an adjacent dose combination chosen from $\{A_{j+1}B_k, A_{j+1}B_{k-1}, A_{j-1}B_{k+1}, A_jB_{k+1}\}$, which has a toxicity probability higher than the current doses and closest to π. If the current dose combination is A_JB_K, the doses stay at the same levels.
   (ii) If $\Pr(p_{jk} < \phi|\mathcal{D}) < c_d$, where c_d is the fixed probability cutoff for dose escalation, the doses move to an adjacent dose combination chosen from $\{A_{j-1}B_k, A_{j-1}B_{k+1}, A_{j+1}B_{k-1}, A_jB_{k-1}\}$, which has a toxicity probability lower than the current doses and closest to ϕ_T. If the current dose combination is A_1B_1, the trial is terminated.
   (iii) Otherwise, the next cohort of patients continues to be treated at the current dose combination.
3. Repeat step 2 until the maximum sample size is reached, and select the MTD as the dose whose estimate of toxicity probability is closest to ϕ.

Although the model-based designs perform reasonably well, they are statistically and computationally complicated, leading many practitioners to perceive that decisions of dose allocation arise from a "black box", which limits its application in practice.

7.2.2 Model-Assisted Designs

Model-assisted designs have emerged as an attractive approach for phase I drug-combination trials that combines the simplicity of algorithm-based designs with the superior performance of model-based designs. Model-assisted designs refer to a class of novel designs that use a model for efficient decision making like model-based designs, while their dose escalation and de-escalation rules can be tabulated before the onset of a trial as with algorithm-based designs [17,18]. For drug-combination trials, the model-assisted designs include Bayesian Optimal Interval (BOIN) combination design and keyboard combination design.

7.2.2.1 Bayesian Optimal Interval (BOIN) Combination Design

For ease of exposition, we first describe the BOIN for single-agent trial and then extend it to combination trials. Consider a single-agent trial with J prespecified doses, $d_1 < \cdots < d_J$. Let p_j denote the DLT probability that corresponds to d_j, n_j denote the number of patients who have been assigned to d_j, and y_j denote the number of DLTs observed at d_j, $j = 1, \ldots, J$. Therefore, at a particular point during the trial, the observed data are $\mathcal{D} = \{\mathcal{D}_j, j{=}1, \ldots, J\}$, where $\mathcal{D}_j = (n_j, y_j)$ are the "local" data observed at dose level d_j.

The dose escalation and de-escalation of the BOIN design is straightforward, determined simply by comparing the observed DLT rate at the current dose with a pair of fixed dose escalation and de-escalation boundaries. Specifically, let $\hat{p}_j = y_j/n_j$ denote the observed DLT rate at the current dose and λ_e and λ_d denote the predetermined dose escalation and de-escalation boundaries. Suppose d_j is the current dose level. The BOIN design determines the next dose as follows (see Figure 7.4):

- If $\hat{p}_j \leq \lambda_e$, then escalate the dose to level d_{j+1}.
- If $\hat{p}_j \geq \lambda_d$, then de-escalate the dose to level d_{j-1}.
- Otherwise (i.e., $\lambda_e < \hat{p}_j < \lambda_d$), stay at the current dose level d_j.

The trial continues until the prespecified sample size is exhausted. At that point, the MTD is selected based on isotonic estimates of the p_j that are

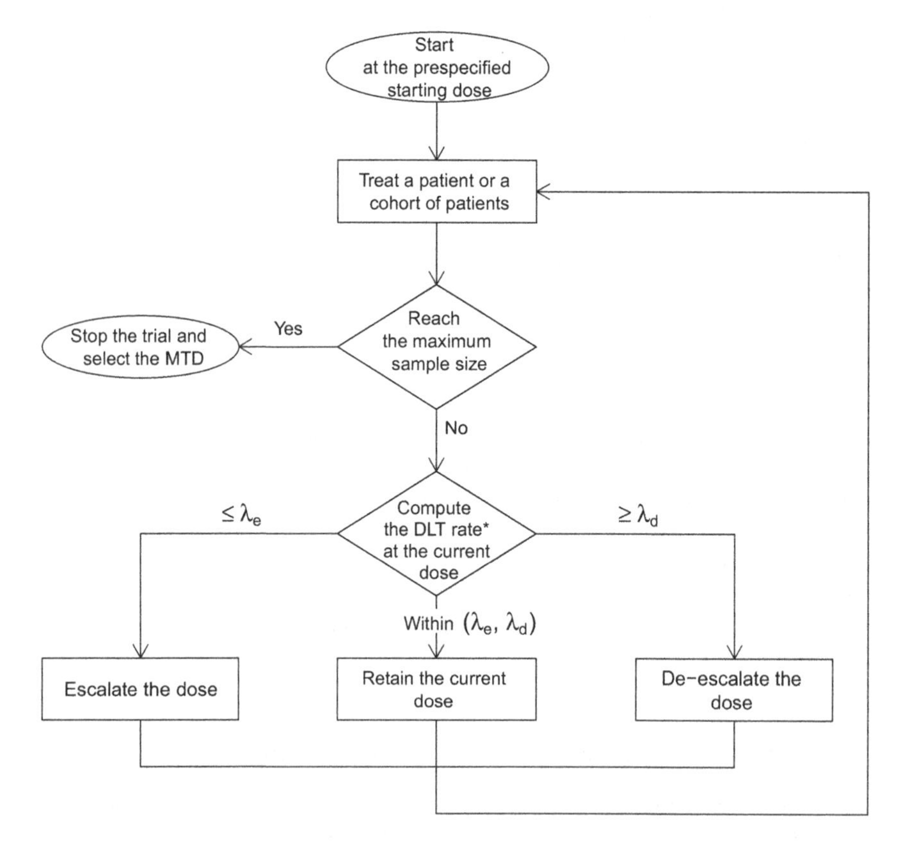

FIGURE 7.4
The flowchart of the BOIN design.

calculated using the pooled adjacent violators algorithm [19]. During the trial conduct, the BOIN design imposes a dose elimination (or overdose control) rule as follows: if $\Pr(p_j > \phi | \mathcal{D}_j) > 0.95$ and $n_j \geq 3$, dose level j and higher are eliminated from the trial, and the trial is terminated if the lowest dose is eliminated, where $\Pr(p_j > \phi | \mathcal{D}_j)$ is evaluated based on the posterior distribution in Equation (7.5).

To determine the dose escalation and de-escalation boundaries (λ_e, λ_d), the BOIN design requires the investigator(s) to specify ϕ_1, which is the highest DLT probability that is deemed to be underdosing such that dose escalation is required, and ϕ_2, which is the lowest DLT probability that is deemed to be overdosing such that dose de-escalation is required. Liu and Yuan [20] provided general guidance to specify ϕ_1 and ϕ_2 and recommended default values of $\phi_1 = 0.6\phi$ and $\phi_2 = 1.4\phi$ for general use. When needed, the values of ϕ_1 and ϕ_2 can be calibrated to achieve a particular requirement of the trial at hand. For example, if more conservative dose escalation is required, setting $\phi_2 = 1.2\phi$ may be appropriate. Given ϕ_1 and ϕ_2, the optimal escalation and de-escalation boundaries (λ_e, λ_d) that minimize the decision error of dose escalation and de-escalation arise as,

$$\lambda_e = \frac{\log\left(\dfrac{1-\phi_1}{1-\phi}\right)}{\log\left\{\dfrac{\phi(1-\phi_1)}{\phi_1(1-\phi)}\right\}}, \qquad \lambda_d = \frac{\log\left(\dfrac{1-\phi}{1-\phi_2}\right)}{\log\left\{\dfrac{\phi_2(1-\phi)}{\phi(1-\phi_2)}\right\}}. \tag{7.3}$$

Table 7.1 provides the dose escalation and de-escalation boundaries (λ_e, λ_d) for commonly used target DLT rate ϕ using the recommended default values $\phi_1 = 0.6\phi$ and $\phi_2 = 1.4\phi$.

For example, given the target DLT rate $\phi = 0.25$, the corresponding escalation boundary $\lambda_e = 0.197$ and the de-escalation boundary $\lambda_d = 0.298$, that is, escalate the dose if the observed DLT rate at the current dose $\hat{p}_j \leq 0.197$ and de-escalate the dose if $\hat{p}_j \geq 0.298$. It has been shown that (λ_e, λ_d) are the boundaries corresponding to the likelihood ratio test and Bayes factors, and thus the resulting BOIN design is optimal with desirable finite-sample and large-sample properties, i.e., long-memory coherence and consistency [20].

Lin and Yin [21] extended the BOIN to combination trials. Suppose A_jB_k is the current dose and $\hat{p}_{jk}$ is the observed DLT rate at A_jB_k. Because the decision rule of dose escalation and de-escalation is determined by the local

TABLE 7.1
The dose escalation and de-escalation boundaries for different target DLT rates

	Target DLT Rate ϕ					
Boundaries	**0.15**	**0.2**	**0.25**	**0.3**	**0.35**	**0.4**
λ_e	0.118	0.157	0.197	0.236	0.276	0.316
λ_d	0.179	0.238	0.298	0.358	0.419	0.479

data $\mathcal{D}_{jk}$ observed at the current dose, the BOIN decision rule described above can be directly used for drug-combination trials. If $\hat{p}_{jk} \leq \lambda_e$, we escalate the dose; if $\hat{p}_{jk} \geq \lambda_d$, we de-escalate the dose; otherwise, we retain the current dose. For combination trials, the only issue is that when we decide to escalate the dose, there are more than one options: we can escalate the dose of A or the dose of B. Similarly, when we decide to de-escalate the dose, we can de-escalate the dose of A or the dose of B. To address this issue, define admissible dose escalation and de-escalation sets $\mathcal{A}_E = \{A_{j+1}B_k, A_jB_{k+1}\}$ and $\mathcal{A}_D = \{A_{j-1}B_k, A_jB_{k-1}\}$. Lin and Yin [21] proposed the following rule: When the BOIN rule says escalation, we escalate to the dose combination that belongs to $\mathcal{A}_E$ and has the highest value of $\Pr(p_{jk} \in (\lambda_e, \lambda_d)|\mathcal{D}_{jk})$; and when the BOIN rule says de-escalation, we escalate to the dose combination that belongs to $\mathcal{A}_D$ and has the highest value of $\Pr(p_{jk} \in (\lambda_e, \lambda_d)|\mathcal{D}_{jk})$. That is, we always move toward the dose that is most likely to be in the acceptable range (λ_e, λ_d). In summary, the BOIN combination trial design is described as follows.

BOIN drug-combination design

(a) Patients in the first cohort are treated at the lowest dose combination A_1B_1 or a prespecified dose combination.

(b) Suppose the current cohort is treated at dose combination A_jB_k, to assign a dose to the next cohort of patients.

- If $\hat{p}_{jk} \leq \lambda_e$, we escalate the dose to the combination that belongs to $\mathcal{A}_E$ and has the largest value of $\Pr\{p_{j'k'} \in (\lambda_e, \lambda_d)|\mathcal{D}_{jk}\}$.
- If $\hat{p}_{jk} \geq \lambda_d$, we de-escalate the dose to the combination that belongs to $\mathcal{A}_D$ and has the largest value of $\Pr\{p_{j'k'} \in (\lambda_e, \lambda_d)|\mathcal{D}_{jk}\}$.
- Otherwise, if $\lambda_e < \hat{p}_{jk} < \lambda_d$, then the dose stays at the same combination A_jB_k.

(c) This process is continued until the maximum sample size is reached or the trial is terminated because of excessive toxicity.

(d) Select the MTD as the dose combination whose isotonic estimate of DLT probability is closest to ϕ.

As the observed DLT rate $\hat{p}_{jk}$ is the most natural and intuitive estimate of p_{jk} that is accessible by non-statisticians, the use of $\hat{p}_{jk}$ to determine the dose escalation and de-escalation makes the BOIN design simpler and more transparent than most existing designs. It is particularly easy for clinicians and regulatory agents to assess the safety of a trial using the BOIN design, thanks to the feature that the BOIN design guarantees de-escalating the dose when

$\hat{p}_{jk} \geq \lambda_d$. For example, given a target DLT rate $\phi = 0.25$, we know *a priori* that a phase I trial using the BOIN design guarantees de-escalating the dose if the observed DLT rate is higher than $\lambda_d = 0.298$ (i.e., the default value). Accordingly, the BOIN design also allows users to easily calibrate the design to satisfy a specific safety requirement mandated by regulatory agents through choosing an appropriate target DLT rate ϕ or ϕ_2. For example, supposing for a phase I trial with a new compound, the regulatory agent mandates that if the observed toxicity rate is higher than 0.25, the dose must be de-escalated. We can easily fulfill that requirement by setting the target DLT rate $\phi = 0.21$, under which the BOIN automatically guarantees de-escalating the dose if the observed toxicity rate $\hat{p}_{jk} > \lambda_d = 0.25$. Such flexibility and transparency renders the BOIN design an important advantage in practice. Lin and Yin [21] showed that despite being simple, the BOIN combination design yields excellent performance comparable to and often superior to more complicated model-based designs.

7.2.2.2 Keyboard Combination Design

Keyboard combination design provides another model-assisted design for finding MTD in drug-combination trials. Unlike BOIN, which makes the decision of dose escalation and de-escalation based on the observed DLT rate $\hat{p}_{jk}$ at the current dose A_jB_k, the keyboard design makes the decision based on the posterior distribution of p_{jk}. Again, for ease of exposition, we introduce the methodology of keyboard design for single-agent trial and then extend it to combination trials.

The keyboard design starts by specifying a target toxicity interval $\mathcal{I}_{\text{target}} = (\phi - \epsilon_1, \phi + \epsilon_2)$ (referred to as the target key), where ϵ_1 and ϵ_2 denote tolerable deviations from ϕ such that any dose with a toxicity probability within that interval can be practically viewed as the MTD. Then, the keyboard design populates this interval toward both sides of the target key, forming a series of equally wide keys that span the range of 0–1 (see Figure 7.5). For example, given $\phi = 0.2$, the target key may be defined as $(0.15, 0.25)$ with $\epsilon_1 = \epsilon_2 = 0.05$. Then, one key of width 0.1 is formed on the left side of the target key, i.e., (0.05, 0.15), and seven keys of width 0.1, i.e., $(0.25, 0.35), \ldots, (0.85, 0.95)$, are formed on the right side of the target key. We denote the resulting intervals/keys as $\mathcal{I}_1, \ldots, \mathcal{I}_K$. In some cases, the DLT probability values at the two ends (e.g., <0.05 or >0.95 in the example) may not be covered by the keys because they are not wide enough to form a key of width 0.1. As explained in Yan et al.[17], ignoring these "residual" DLT probabilities at the two ends does not pose any issue for decision making.

The keyboard design assumes

$$y_j | n_j, p_j \sim \text{Binom}(n_j, p_j) \qquad (7.4)$$
$$p_j \sim \text{Beta}(1,1) \equiv \text{Unif}(0,1),$$

i.e., a beta-binomial model, and thus, the posterior distribution arises as

$$p_j | \mathcal{D}_j \sim \text{Beta}(y_j + 1, n_j - y_j + 1), \text{ for } j = 1, \ldots, J. \qquad (7.5)$$

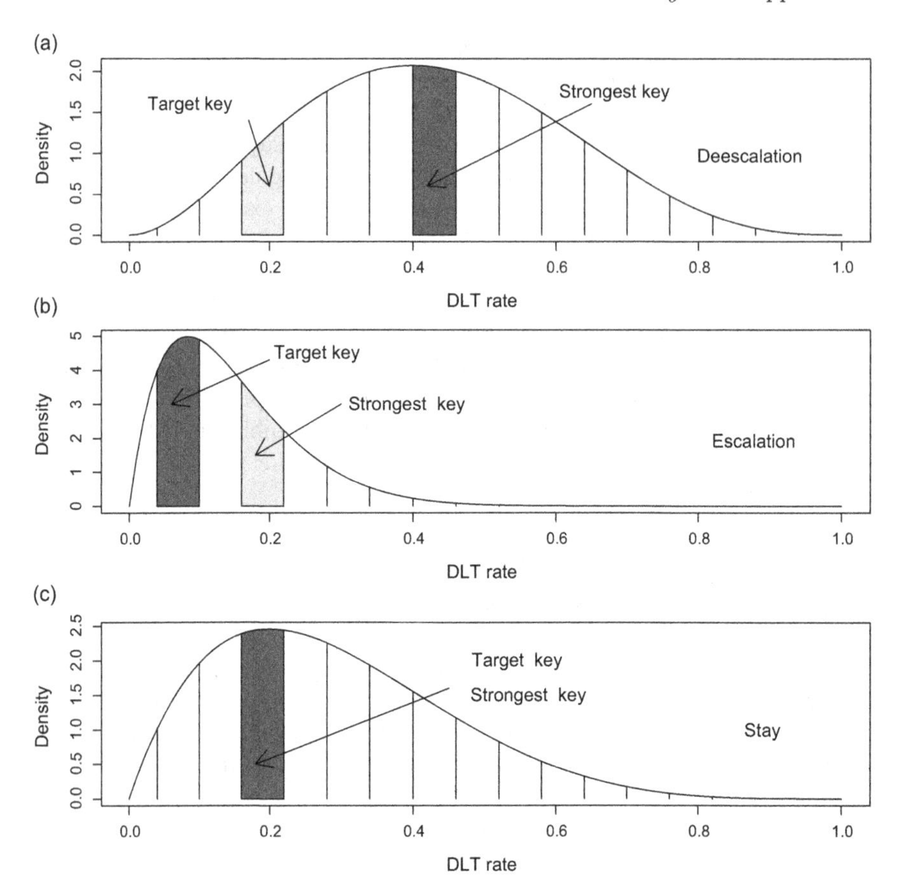

FIGURE 7.5
The dose escalation/de-escalation rule of the keyboard design. If the strongest key is on the right side of the target key, de-escalate the dose (a); if the strongest key is on the left side of the target key, escalate the dose (b); if the strongest key is the target key, retain the current dose (c). The curves in the panels are the posterior distributions of the DLT rate of the current dose.

To make the decision of dose escalation and de-escalation, given the observed data $\mathcal{D}_j = (n_j, y_j)$ at the current dose level j, the Keyboard design identifies the interval $\mathcal{I}_{\max}$ that has the largest posterior probability, i.e.,

$$\mathcal{I}_{\max} = \text{argmax}_{\mathcal{I}_1,\ldots,\mathcal{I}_K}\{\Pr(p_j \in \mathcal{I}_k \mid \mathcal{D}_j);\ k = 1, \ldots, K\},$$

which can easily be evaluated based on p_j's posterior distribution given by Equation (7.5), assuming that p_j follows a beta-binomial model (7.4). $\mathcal{I}_{\max}$ represents the interval that the true value of p_j is most likely located, referred to as the "strongest" key by Yan et al. [17]. Graphically, the strongest key is the one with the largest area under the posterior distribution curve of p_j

(see Figure 7.5). If the strongest key is on the left (or right) side of the target key, that means that the observed data suggest that the current dose is most likely underdosing (or overdosing), and thus dose escalation (or de-escalation) is needed. If the strongest key is the target key, the observed data support that the current dose is most likely to be in the proper dosing interval, and thus it is desirable to retain the current dose for treating the next patient.

Suppose d_j is the current dose level. The keyboard design determines the next dose as follows:

- If the strongest key is on the left side of the target key (denoted as $\mathcal{I}_{\max} \prec \mathcal{I}_{\text{target}}$), then escalate the dose to level d_{j+1}.
- If the strongest key is the target key (denoted as $\mathcal{I}_{\max} \equiv \mathcal{I}_{\text{target}}$), then stay the current dose level d_j.
- If the strongest key is on the right side of the target key (denoted as $\mathcal{I}_{\max} \succ \mathcal{I}_{\text{target}}$), then de-escalate the dose to level d_{j-1}.

The trial continues until the prespecified sample size is exhausted. At that point, the MTD is selected based on isotonic estimates of the p_j that are calculated using the pooled adjacent violators algorithm [19]. As the decision of dose escalation and de-escalation is based only on the local data at the current dose, it is possible that the dose transition oscillates between a safe dose and the next higher dose that is toxic. To avoid that issue, the keyboard design includes a dose exclusion/safety stopping rule: if $\Pr(p_j > \phi \mid \mathcal{D}_j) > 0.95$, dose level j and higher are excluded from the trial. If the lowest dose is excluded, the trial is stopped for safety.

Similar to BOIN, because the decision rule of the keyboard design only depends on the observed data at the current dose, the dose escalation rule above can be directly used in combination trials to guide the dose escalation and de-escalation. Table 7.2 shows the dose escalation and de-escalation rule

TABLE 7.2
Escalation and de-escalation rules for the keyboard design for up to 16 patients treated at the current dose

	Number of Patients Treated at the Current Dose															
	1	**2**	**3**	**4**	**5**	**6**	**7**	**8**	**9**	**10**	**11**	**12**	**13**	**14**	**15**	**16**
	$\phi = 0.2$ with the target key $= (0.17, 0.23)$															
Escalate if $y_j \leq$	0	0	0	0	0	1	1	1	1	1	1	2	2	2	2	2
De-escalate if $y_j \geq$	1	1	1	1	2	2	2	2	3	3	3	3	3	4	4	4
	$\phi = 0.3$ with the target key $= (0.25, 0.35)$															
Escalate if $y_j \leq$	0	0	0	0	1	1	1	1	2	2	2	2	3	3	3	3
De-escalate if $y_j \geq$	1	1	2	2	2	3	3	3	4	4	4	5	5	5	6	6

for targets $\phi = 0.2$ and 0.3. Again, the challenge is that there is more than one option to escalate or de-escalate the dose. Along the similar line as the BOIN combination design, Pan, Lin, and Yuan [22] proposed to use $\Pr(p_{jk} \in \mathcal{I}_{\text{target}}|\mathcal{D}_{jk})$ to choose the appropriate dose for escalation and de-escalation. The keyboard combination design is summarized as follows.

Keyboard drug-combination design

(a) Patients in the first cohort are treated at the lowest dose combination A_1B_1 or a prespecified dose combination.

(b) Suppose the current cohort is treated at dose combination A_jB_k, given the observed data $D_{jk} = (n_{jk}, y_{jk})$, we identify the strongest key $\mathcal{I}_{\max}$ based on the posterior distribution of p_{jk} and assign a dose to the next cohort of patients as follows:

- If $\mathcal{I}_{\max} \prec \mathcal{I}_{\text{target}}$, we escalate the dose to the combination that belongs to $\mathcal{A}_E$ and has the largest value of $\Pr\{p_{j'k'} \in \mathcal{I}_{\text{target}}|\mathcal{D}_{jk}\}$.
- If $\mathcal{I}_{\max} \succ \mathcal{I}_{\text{target}}$, we de-escalate the dose to the combination that belongs to $\mathcal{A}_D$ and has the largest value of $\Pr\{p_{j'k'} \in \mathcal{I}_{\text{target}}|\mathcal{D}_{jk}\}$.
- Otherwise, if $\mathcal{I}_{\max} \equiv \mathcal{I}_{\text{target}}$, then the dose stays at the same combination A_jB_k.

(c) This process is continued until the maximum sample size is reached or the trial is terminated because of excessive toxicity.

(d) Select the MTD as the dose combination whose isotonic estimate of DLT probability is closest to ϕ.

7.3 Drug-Combination Design to Find Multiple MTDs

The primary motivation of combining drugs is to achieve synergistic treatment effects. Because of the existence of the MTD contour and the fact that doses on the MTD contour may have different efficacy due to drug–drug interactions, for many drug-combination trials, it is of intrinsic interest to find multiple MTDs. The efficacy of the MTDs can be evaluated in subsequent phase II trials or simultaneously in phase I–II trials. Given a prespecified $J \times K$ dose matrix, finding the MTD contour is equivalent to finding an MTD, if it exists, in each row of the dose matrix. Without loss of generality, we assume that $J \leq K$. That is, drug B has more dose levels than drug A.

Finding the MTD contour is substantially more challenging than finding a single MTD. This is because, in order to find all MTDs in the dose matrix, we must explore the whole dose matrix using the limited sample size that is characteristic of phase I trials; otherwise, we risk missing some MTDs. Yuan and Yin [23] proposed a general divide-and-conquer strategy for finding multiple MTDs:

Divide the two-dimensional dose-finding problem into a series of simpler one-dimensional dose-finding problems that can be easily conquered by existing single-agent dose-finding methods.

Taking that strategy, Zhang and Yuan [24] proposed the waterfall design to find the MTD contour. In contrast to numerous drug-combination designs for finding a single MTD, a very limited number of designs have been proposed to find the MTD contour. Mander and Sweeting [25] proposed a product of independent beta probabilities escalation (PIPE) design to find the MTD contour based on Bayesian model averaging, without assuming a parametric form on the dose-toxicity curve. Zhang and Yuan [24] compared the waterfall design with the PIPE design and showed the superior performance of the waterfall design; see that article for details.

7.3.1 Waterfall Design

7.3.1.1 Partition Scheme

As illustrated in Figure 7.6, the waterfall design partitions the $J \times K$ dose matrix into J subtrials (or blocks), within which the doses are fully ordered. These subtrials are conducted sequentially from the top of the matrix to the bottom, which is why we refer to the design as the waterfall design. The goal of the design is to find the MTD contour, which is equivalent to finding the MTD, if it exists, in each row of the dose matrix. The waterfall design can be described as follows:

1. Divide the $J \times K$ dose matrix into J subtrials $S_J, \ldots, S_1$, according to the dose level of drug A:

$$\begin{aligned}
S_J &= \{A_1B_1, \ldots, A_JB_1, A_JB_2, \ldots, A_JB_K\}, \\
S_{J-1} &= \{A_{J-1}B_2, \ldots, A_{J-1}B_K\}, \\
S_{J-2} &= \{A_{J-2}B_2, \ldots, A_{J-2}B_K\}, \\
&\vdots \\
S_1 &= \{A_1B_2, \ldots, A_1B_K\}.
\end{aligned}$$

   Note that subtrial S_J also includes lead-in doses $A_1B_1, A_2B_1, \ldots, A_JB_1$ (the first column of the dose matrix) to impose the practical consideration that the trial starts at the lowest dose. Within each subtrial, the doses are fully ordered with monotonically increasing toxicity.

2. Conduct the subtrials sequentially using the BOIN design (or other single-agent dose-finding method) as follows:

   (a) Conduct subtrial S_J, starting from the lowest dose combination A_1B_1, to find the MTD. We call the dose selected by the subtrial the "candidate MTD" to highlight that the dose selected by the individual subtrial may not be the "final" MTD that we will select at the end of the trial. The final MTD selection will be based on the data collected from all the subtrials. The objective of finding the candidate MTD is to determine which subtrial will be conducted next and the corresponding starting dose.

   (b) Assuming that subtrial S_J selects dose $A_{j^*}B_{k^*}$ as the candidate MTD, next, conduct subtrial S_{j^*-1} with the starting dose $A_{j^*-1}B_{k^*+1}$. That is, the next subtrial to be conducted is the one with the dose of drug A that is one level lower than the candidate MTD found in the previous subtrial. After identifying the candidate MTD of subtrial S_{j^*-1}, the same rule is used to determine the next subtrial and its starting dose. See Figure 7.6 for an example.

   (c) Repeat step (b) until subtrial S_1 is completed.

3. Estimate the toxicity probability p_{jk} based on the toxicity data collected from all the subtrials using matrix isotonic regression [26]. For each row of the dose matrix, select the MTD as the dose combination that has the estimate of toxicity probability that is closest to the target toxicity rate ϕ unless all combinations in that row are overly toxic.

In step 2, the reason that subtrial S_{j^*-1} starts with dose $A_{j^*-1}B_{k^*+1}$ rather than the lowest dose in that subtrial (i.e., $A_{j^*-1}B_2$) is that $A_{j^*-1}B_{k^*+1}$ is the lowest dose that is potentially located at the MTD contour. Starting from $A_{j^*-1}B_{k^*+1}$ allows us to quickly reach the MTD. Using Figure 7.6 as an example, the first subtrial S_3 identified the dose A_3B_2 as the MTD, and thus the second subtrial S_2 starts from the dose A_2B_3. It is not desirable to start from the lowest dose A_2B_2 because the partial ordering informs us that A_2B_2 is below the MTD. Starting at the lowest dose in this example will waste patient resources and expose patients to low doses that may be subtherapeutic.

Using the results of each subtrial to inform the design (e.g., the dose range and the starting dose) of subsequent subtrials is a key feature of the waterfall design. Such information borrowing allows the design to explore the two-dimensional dose space efficiently using a limited sample size and decreases the chance of overdosing and underdosing patients.

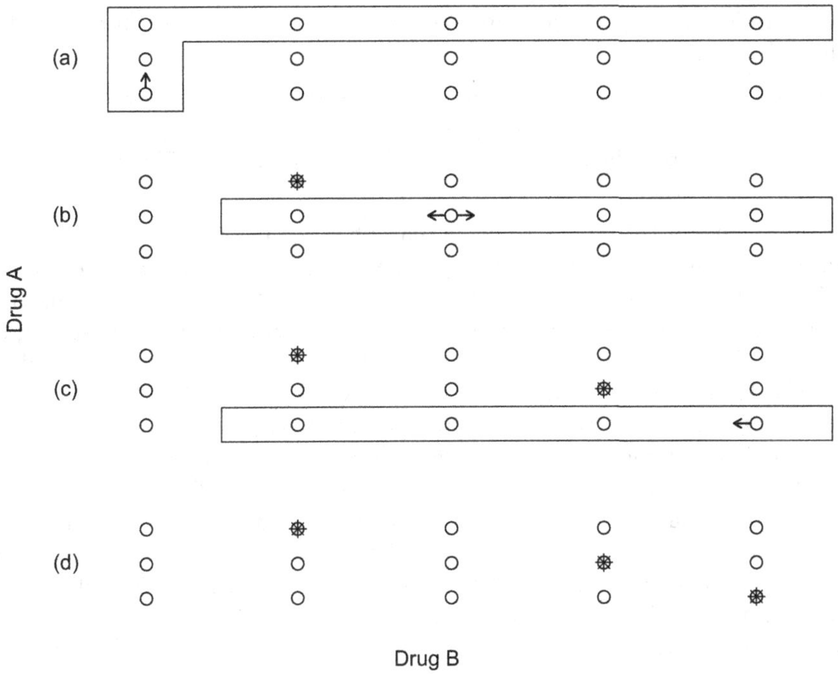

FIGURE 7.6
Illustration of the waterfall design for a 3×5 combination trial. The doses in the rectangle form a subtrial, and the asterisk denotes the candidate MTD. As shown in panel (a), the trial starts by conducting the first subtrial with the starting dose A_1B_1. After the first subtrial identifies A_3B_2 as the candidate MTD, we then conduct the second subtrial with the starting dose A_2B_3, as shown in panel (b). After the second subtrial identifies A_2B_4 as the candidate MTD, we conduct the third subtrial with the starting dose A_1B_5, as shown in panel (c). After all subtrials are completed, we select the MTD contour based on the data from all subtrials, as shown in panel (d).

7.3.1.2 Conducting Subtrials

As the doses in each subtrial are strictly ordered, applying the BOIN design to the subtrial is straightforward. The key issue is to determine when we should end the current subtrial and initiate the next one. One straightforward way is to prespecifiy a maximum sample size for each subtrial. When the maximum sample size is reached, we stop the subtrial, determine the candidate MTD, and initiate the next subtrial. This approach works well for standard phase I one-dimensional dose finding but is not efficient for conducting multiple subtrials. This is because, depending on the distance between the starting dose and the MTD, as well as the shape of the dose-toxicity curve, the subtrials often require different sample sizes to identify the MTD. Based on this consideration, we propose and recommend the following stopping rule for subtrials:

> At any time of the subtrial, if the total number of patients treated at the current dose reaches a certain prespecified number of, say `n.earlystop`, patients, we stop the subtrial and select the candidate MTD and initiate the next subtrial.

The rationale for the stopping rule is that when the patient allocation concentrates at a dose, it indicates that the dose finding might have converged to the MTD, and thus we can stop the trial and claim the MTD. This stopping rule allows the sample size of subtrials to be automatically adjusted according to the difficulty of the dose finding (e.g., the distance between the starting dose and the MTD, and the shape of the dose-toxicity curve). Another attractive feature of the above approach is that it automatically ensures that a certain number of patients are treated at the MTD. Conventionally, we achieve this by adding cohort expansion after identifying the MTD. In practice, we recommend `n.earlystop` > 9 to ensure reasonable operating characteristics. Although the above stopping rule provides an automatic, reasonable way to determine the sample size for a subtrial, in some cases, it is desirable to put a cap on the maximum sample size of subtrials. This can be done by adding an extra stopping rule:

> Stop the subtrial j if its sample size reaches $N_j^{\max}$, where $N_j^{\max}$ is the prespecified maximum sample size for subtrial j.

As a rule of thumb, we recommend $N_j^{\max} = 4\times$(the number of doses in the jth subtrial), for $j = 1, \ldots, J$. This means that given $J \times K$ dose combinations, the maximum total sample size for the trial is $4 \times J \times K$. For example, for a 3×5 combination, as shown in Figure 7.6, the first subtrial contains seven doses, and the second and third subtrials contain four doses each. The recommended sample sizes are 28, 16, and 16 for three subtrials, respectively, resulting a total sample size of 60 patients. This may seem large; however, given that there are a total of 15 doses, 60 patients actually is not a very large sample size. To see this, considering a single-agent trial with 15 doses, the maximum sample size under the $3 + 3$ design is 90 patients. In practice, the recommended sample sizes `n.earlystop` and $N_j^{\max}$ should be further calibrated using simulation until attaining desirable operating characteristics, which can be readily done using R package "BOIN" described later.

An alternative stopping rule is based on the confidence interval (CI) of the estimate of toxicity probabilities. For example, stop the subtrial when the CI of a dose contains the target toxicity probability ϕ and its width is narrower than a certain value. Our numerical study shows that after appropriate calibration, these two approaches have virtually the same performance. This is somewhat expected because the width of the CI is essentially determined by the sample size (i.e., `n.earlystop`). Thus, we recommend the stopping rule that is based on the number of patients treated because it is more transparent, in particular to clinicians, and is easy to implement.

7.4 Software

The software for the BOIN design is available in three forms, including a stand-alone graphical user interface based Windows desktop program which is freely available from MD Anderson Software Download Website `https://biostatistics.mdanderson.org/softwaredownload/Single Software.aspx? Software_Id=99`, Shiny online apps freely available at `http://trialdesign.org`, and R package "BOIN" available from the CRAN. The Keyboard design can be implemented using the Shiny online app freely available at `http://trialdesign.org`. For users who are not familiar with R, Windows desktop program is a good option with intuitive graphical user interface and the function to automatically generate the protocol template for the trial (see Figure 7.7). Here, we provide a brief overview of R package "BOIN". The manual for the package can be found in `https://cran.r-project.org/web/packages/BOIN/index.html`, and a statistical tutorial for using the package to design drug-combination trials can be found in `http://odin.mdacc.tmc.edu/ yyuan/index_code.html`.

- BOIN drug-combination design to find a single MTD
  - `get.boundary(···)`; This function is used to generate escalation and de-escalation boundaries for conducting trials.
  - `next.comb(···)`; This function is used to determine the dose combination for the next cohort of new patients given the currently observed data.
  - `select.mtd.comb(···)`; This function (with argument `MTD.contour= FALSE`) is used to select the MTD at the end of the trial based on isotonically transformed estimates.
  - `get.oc.comb(···)`; This function (with argument `MTD.contour= FALSE`) is used to generate the operating characteristics of the BOIN drug-combination design.
- Waterfall design to find the MTD contour
  - `get.boundary(···)`; This function is used to generate escalation and de-escalation boundaries for conducting subtrials.
  - `next.subtrial(···)`; This function is used to obtain the dose range and the starting dose for the next subtrial when the current subtrial is completed.
  - `select.mtd.comb(···)`; This function (with argument `MTD.contour= TRUE`) is used to select the MTD contour at the end of the trial based on isotonically transformed estimates.
  - `get.oc.comb(···)`; This function (with argument `MTD.contour= TRUE`) is used to generate the operating characteristics of the waterfall design for drug-combination trials.

(a)

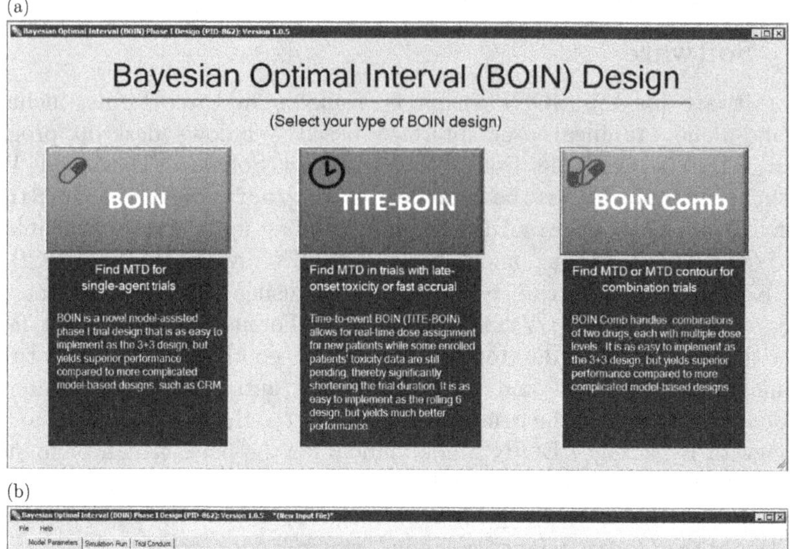

(b)

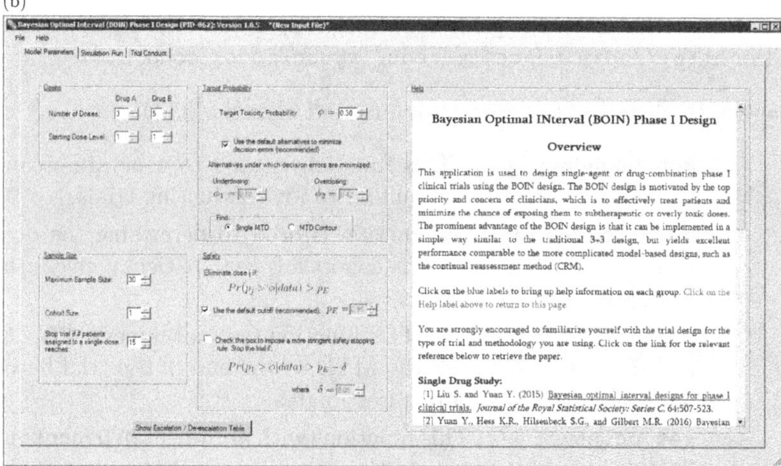

FIGURE 7.7
Windows desktop program for BOIN design. (a) Launch window and (b) BOIN drug-combination design window.

7.5 Trial Examples

Example 7.1: Drug-Combination Trial to Find a Single MTD Using BOIN Drug-Combination Design

Consider the drug-combination trial that combines three doses of gemcitabine (i.e., 600, 800 or 1,000 mg/m^2) and five doses of MK-8776 (i.e., 10, 20, 40, 80, or 112 mg/m^2). The objective is to find an MTD with a target DLT rate of

0.3, among a total of 15 dose combinations. The maximum sample size is 30 patients, treated in cohort size of 3.

To design and conduct this trial, we first ran function

```
R> get.boundary(target=0.3, ncohort=10, cohortsize=3),
```

yielding the dose escalation and de-escalation boundaries as shown in Table 7.3.

Using the BOIN drug-combination design, the trial started by treating the first cohort of three patients at the lowest dose combination (1,1), at which no DLT was observed. The observed data were

$$n = \begin{pmatrix} 3 & 0 & 0 & 0 & 0 \\ 0 & 0 & 0 & 0 & 0 \\ 0 & 0 & 0 & 0 & 0 \end{pmatrix} \quad y = \begin{pmatrix} 0 & 0 & 0 & 0 & 0 \\ 0 & 0 & 0 & 0 & 0 \\ 0 & 0 & 0 & 0 & 0 \end{pmatrix},$$

where n records the number of patients treated at each dose combination, and y records the number of patients who experienced DLT at each dose combination. In matrixes y and n, entry (j, k) records the data associated with combination (j, k). To determine the dose for the second cohort of patients, we called function

```
R> next.comb(target=0.3, npts=n, ntox=y, dose.curr=c(1, 1)),
```

which recommended to escalate the dose to combination (1, 2). Therefore, we treated the second cohort of patients at dose combination (1, 2). In the second cohort, one patient experienced DLT, so the updated data matrices became

$$n = \begin{pmatrix} 3 & 3 & 0 & 0 & 0 \\ 0 & 0 & 0 & 0 & 0 \\ 0 & 0 & 0 & 0 & 0 \end{pmatrix} \quad y = \begin{pmatrix} 0 & 0 & 0 & 0 & 0 \\ 0 & 0 & 0 & 0 & 0 \\ 0 & 0 & 0 & 0 & 0 \end{pmatrix}.$$

To determine the dose for the third cohort of patients, we again called

```
next.comb(target=0.3, npts=n, ntox=y, dose.curr=c(1, 2))
```

TABLE 7.3
Dose escalation and de-escalation rule for the BOIN design

	Number of patients treated									
	3	**6**	**9**	**12**	**15**	**18**	**21**	**24**	**27**	**30**
Escalate if # of DLT <=	0	1	2	2	3	4	4	5	6	7
De-escalate if # of DLT >=	2	3	4	5	6	7	8	9	10	11
Eliminate if # of DLT >=	3	4	5	7	8	9	10	11	12	14

with updated y, n, and dose.curr. The function recommended escalating the dose to (1, 3) for treating the third cohort of patients. We repeated this procedure until the maximum sample size was reached. Figure 7.8 shows the dose assignments for all 30 patients. For example, at dose combination (3, 4) when completing the eighth cohort, there were two DLTs, based on the accumulating toxic information on this dose combination level, the function recommended de-escalating the dose to combination (3, 3). When the trial was completed, the number of patients treated at each dose combination and the corresponding number of patients who experienced toxicity at each dose combination were

$$n = \begin{pmatrix} 3 & 3 & 3 & 0 & 0 \\ 0 & 0 & 3 & 0 & 0 \\ 0 & 0 & 12 & 6 & 0 \end{pmatrix} \qquad y = \begin{pmatrix} 0 & 0 & 0 & 0 & 0 \\ 0 & 0 & 0 & 0 & 0 \\ 0 & 0 & 4 & 4 & 0 \end{pmatrix}.$$

We called function

```
R> select.mtd.comb(target=0.3, npts=n, ntox=y, MTD.contour=FALSE),
```

which recommended dose combination (3, 3) as the MTD.

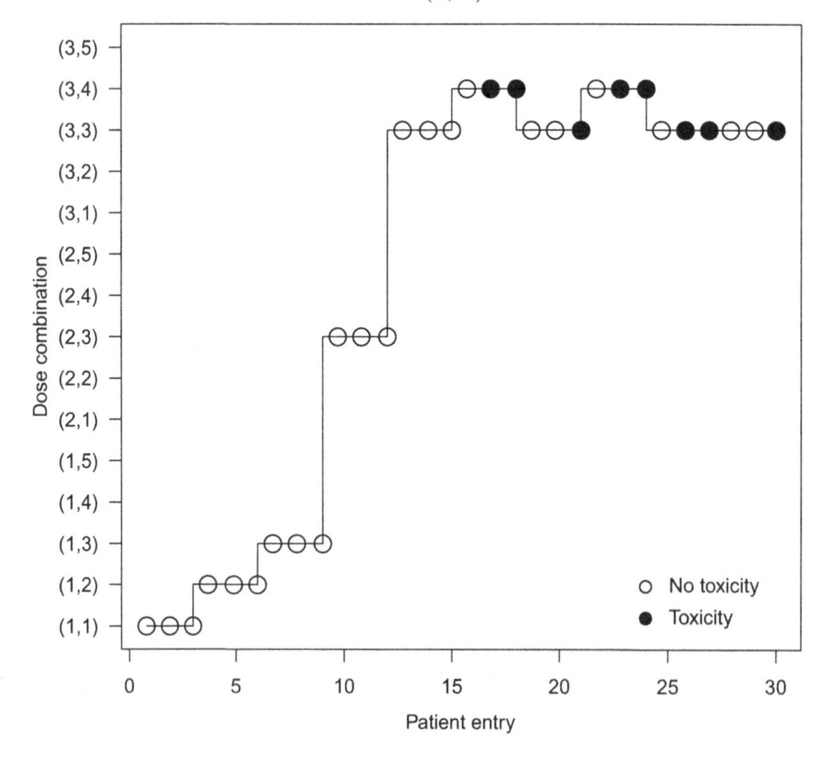

FIGURE 7.8
Illustration of a 3×5 combination trial using the BOIN drug-combination design with a cohort of 3.

Example 7.2: Drug-Combination Trial to Find the MTD Contour Using Waterfall Design

Consider a drug-combination trial similar to Example 7.2, which combines three doses of gemicitabine (i.e., drug A) and five doses of MK-8776 (i.e., drug B). The objective now is to find the MTD contour (multiple MTDs) with a target DLT rate of 0.25. As shown in Figure 7.6, using the waterfall design, the trial started with the first subtrial, which consisted of seven ordered combinations $\{A_1B_1 \rightarrow A_2B_1 \rightarrow A_3B_1 \rightarrow A_3B_2 \rightarrow A_3B_3 \rightarrow A_3B_4 \rightarrow A_3B_5\}$. This subtrial was conducted using the BOIN design in a way similar to those as described in Example 7.1. The starting dose for this subtrial was A_1B_1, and `n.earlystop` was set as 12 such that the subtrial stopped when the number of patients treated at any of the doses reached 12. The first subtrial resulted in the following data:

$$n = \begin{pmatrix} 6 & 0 & 0 & 0 & 0 \\ 6 & 0 & 0 & 0 & 0 \\ 9 & 12 & 0 & 0 & 0 \end{pmatrix} \quad y = \begin{pmatrix} 0 & 0 & 0 & 0 & 0 \\ 1 & 0 & 0 & 0 & 0 \\ 1 & 3 & 0 & 0 & 0 \end{pmatrix}.$$

Based on the data, we called function `next.subtrial(·)` to obtain the doses for the next subtrial.

```
   R> n <- matrix(c(6, 0, 0, 0, 0, 6, 0, 0, 0, 0, 9, 12, 0, 0, 0),
ncol=5, byrow=TRUE)
   R> y <- matrix(c(0, 0, 0, 0, 0, 1, 0, 0, 0, 0, 1, 3, 0, 0, 0),
ncol=5, byrow=TRUE)
   R> next.subtrial(target=0.25, npts=n, ntox=y)

   Next subtrial includes doses: (2, 2), (2, 3), (2, 4), (2, 5)
   The starting dose for this subtrial is: (2, 3)
```

Therefore, we conducted the second subtrial with doses $\{A_2B_2 \rightarrow A_2B_3 \rightarrow A_2B_4 \rightarrow A_2B_5\}$ using the BOIN design with the starting dose A_2B_3. After the second subtrial was completed, the observed data were

$$n = \begin{pmatrix} 6 & 0 & 0 & 0 & 0 \\ 6 & 0 & 3 & 12 & 0 \\ 9 & 12 & 0 & 0 & 0 \end{pmatrix} \quad y = \begin{pmatrix} 0 & 0 & 0 & 0 & 0 \\ 1 & 0 & 0 & 3 & 0 \\ 1 & 3 & 0 & 0 & 0 \end{pmatrix}.$$

We called `next.subtrial(·)` again using the above updated data to obtain the doses for the third subtrial.

```
   R> n <- matrix(c(6, 0, 0, 0, 0, 6, 0, 3, 12, 0, 9, 12, 0, 0, 0),
ncol=5, byrow=TRUE)
   R> y <- matrix(c(0, 0, 0, 0, 0, 1, 0, 0, 3, 0, 1, 3, 0, 0, 0),
ncol=5, byrow=TRUE)
```

```
R> next.subtrial(target=0.25, npts=n, ntox=y)

Next subtrial includes doses: (1, 2), (1, 3), (1, 4), (1, 5)
The starting dose for this subtrial is: (1, 5)
```

The third subtrial included doses $\{A_1B_2 \rightarrow A_1B_3 \rightarrow A_1B_4 \rightarrow A_1B_5\}$, with starting dose A_1B_5. After completing the third subtrial using the BOIN, the trial was completed and resulted in the following final data

$$n = \begin{pmatrix} 6 & 0 & 0 & 6 & 12 \\ 6 & 0 & 3 & 12 & 0 \\ 9 & 12 & 0 & 0 & 0 \end{pmatrix} \qquad y = \begin{pmatrix} 0 & 0 & 0 & 1 & 3 \\ 1 & 0 & 0 & 3 & 0 \\ 1 & 3 & 0 & 0 & 0 \end{pmatrix}.$$

Based on the final data, we ran `select.mtd.comb(·)` to select the MTD contour:

```
R> select.mtd.comb(target=0.25, npts=n, ntox=y,
MTD.contour=TRUE)

The MTD contour includes dose combinations (1, 5) (2, 4) (3, 2)
Isotonic estimates of toxicity probabilities for combinations
are:
                    0.01 NA NA 0.17 0.25
                    0.12 NA 0.12 0.25 NA
                    0.12 0.25 NA NA NA

NOTE: no estimate is provided for the doses at which no
patient was treated.
```

Thus, we selected dose combinations A_1B_5, A_2B_4 and A_3B_2 as the MTD contour, which had the estimated toxicity rate of 0.25.

7.6 Discussion

We have provided practical guidance and strategies to design phase I drug-combination trials. We showed that the choice of design for drug-combination trials critically depends on whether the goal of the trial is to find one MTD or multiple MTDs (i.e., the MTD contour). When the goal of the drug-combination trial is to find a single MTD, the BOIN and keyboard drug-combination designs provide simple, practical, and robust ways to find the MTD. When the goal of the combination trial is to find the MTD contour, the waterfall design is a good choice. The waterfall design is based on rules (or algorithms) and does not need complicated calculations and model fitting, which makes the design more likely to be used in practice by clinical investigators. The BOIN drug-combination and waterfall designs can be easily implemented using freely available R package "BOIN". The keyboard combination design can be implemented using the Shiny online app freely available at `http://trialdesign.org`.

A natural extension of the waterfall design is to seamlessly combine it with a phase II trial, resulting in a seamless phase I–II waterfall design. Specifically, after identifying the MTD contour using the waterfall design, we can seamlessly move the identified MTDs to phase II, where we randomize patients to these MTDs and identify the most efficacious MTD for further confirmative phase III trials. Details on this phase I–II approach and related software are provided by Zhang and Yuan [24] and Yuan and Yin [27].

Bibliography

[1] U. S. Food, Drug Administration/Center for Drug Evaluation, and Research. Nonclinical safety evaluation of drug or biologic combinations. *FDA Guidance for Industry*, 2006.

[2] U. S. Food, Drug Administration/Center for Drug Evaluation, and Research. Codevelopment of two or more new investigational drugs for use in combination. *FDA Guidance for Industry*, 2013.

[3] M.K. Riviere, F. Dubois, and S. Zohar. Competing designs for drug combination in phase I dose-finding clinical trials. *Statistics in Medicine*, 34(1):1–12, 2015.

[4] P.F. Thall, R.E. Millikan, P. Mueller, and S.J. Lee. Dose-finding with two agents in phase I oncology trials. *Biometrics*, 59(3):487–496, 2003.

[5] G. Yin and Y. Yuan. A latent contingency table approach to dose finding for combinations of two agents. *Biometrics*, 65(3):866–875, 2009.

[6] G. Yin and Y. Yuan. Bayesian dose finding in oncology for drug combinations by copula regression. *Journal of the Royal Statistical Society: Series C (Applied Statistics)*, 58(2):211–224, 2009.

[7] T.M. Braun and S.F. Wang. A hierarchical bayesian design for phase I trials of novel combinations of cancer yherapeutic agents. *Biometrics*, 66(3):805–812, 2010.

[8] N.A. Wages, M.R. Conaway, and J. O'Quigley. Continual reassessment method for partial ordering. *Biometrics*, 67(4):1555–1563, 2011.

[9] J. O'Quigley, M. Pepe, and L. Fisher. Continual reassessment method: a practical design for phase i clinical trials in cancer. *Biometrics*, 46:33–48, 1990.

[10] T.M. Braun and N. Jia. A generalized continual reassessment method for two-agent phase I trials. *Statistics in Biopharmaceutical Research*, 5:105–115, 2013.

[11] M.K. Riviere, Y. Yuan, F. Dubois, and S. Zohar. A bayesian dose-finding design for drug combination clinical trials based on the logistic model. *Pharmaceutical Statistics*, 13(4):247–257, 2014.

[12] C.Y. Cai, Y. Yuan, and Y. Ji. A bayesian phase I/II design for oncology clinical trials of combining biological agents. *Journal of the Royal Statistical Society: Series C*, 63:159–173, 2014.

[13] M.K. Riviere, Y. Yuan, F. Dubois, and S. Zohar. A bayesian dose-finding design for clinical trials combining a cytotoxic agent with a molecularly targeted agent. *Journal of the Royal Statistical Society: Series C*, 64:215–229, 2015.

[14] G. Yin and Y. Yuan. Bayesian model averaging continual reassessment method in phase i clinical trials. *Journal of the American Statistical Association*, 104(487):954–968, 2009.

[15] S. Chevret. *Statistical Methods for Dose-Finding Experiments*, volume 24. Wiley Online Library, Hoboken, NJ, 2006.

[16] S. Zohar, M. Resche-Rigon, and S. Chevret. Using the continual reassessment method to estimate the minimum effective dose in phase ii dose-finding studies: a case study. *Clinical Trials*, 10(3):414–421, 2013.

[17] F. Yan, S.J. Mandrekar, and Y. Yuan. Keyboard: a novel bayesian toxicity probability interval design for phase i clinical trials. *Clinical Cancer Research*, 23(15):3994–4003, 2017.

[18] H. Zhou, T.A. Murray, H. Pan, and Y. Yuan. Comparative review of toxicity probability intervals for phase i clinical trials. *Statistics in Medicine*, 37:2208–2222, 2018.

[19] R.E. Barlow, D.J. Bartholomew, J.M. Bremner, and H.D. Brunk. Statistical inference under order restrictions: the theory and application of isotonic regression. *International Statistical Review*, 41(3), 1973.

[20] S.Y. Liu and Y. Yuan. Bayesian optimal interval designs for phase I clinical trials. *Journal of the Royal Statistical Society: Series C (Applied Statistics)*, 64(3):507–523, 2015.

[21] R. Lin and G. Yin. Bayesian optimal interval design for dose finding in drug-combination trials. *Statistical Methods in Medical Research*, 2015, doi:10.1177/0962280215594494.

[22] H. Pan, R. Lin, and Y. Yuan. Statistical properties of the keyboard design with extension to drug-combination trials. Unpublished manuscript.

[23] Y. Yuan and G. Yin. Sequential continual reassessment method for two-dimensional dose finding. *Statistics in Medicine*, 27(27):5664–5678, 2008.

[24] L. Zhang and Y. Yuan. A simple bayesian design to identify the maximum tolerated dose contour for drug combination trials. *Statistics in Medicine*, 2016, doi:10.1002/sim.7095.

[25] A.P. Mander and M.J. Sweeting. A product of independent beta probabilities dose escalation design for dual-agent phase I trials. *Statistics in Medicine*, 34(8):1261–1276, 2015.

[26] B. Gordon, D. Richard, P. Carolyn, and R. Tim. Isotonic regression in two independent variables. *Journal of the Royal Statistical Society: Series C (Applied Statistics)*, 33(3):352–357, 1984.

[27] Y. Yuan and G. Yin. Bayesian phase I/II adaptively randomized oncology trials with combined drugs. *Annals of Applied Statistics*, 5(2A):924–942, 2011.

[28] Y. Chu, H. Pan, and Y. Yuan. Adaptive dose modification for phase i clinical trials. *Statistics in Medicine*, 35(20):3497–3508, 2016.

[29] Y. Yuan, K.R. Hess, S.G. Hilsenbeck, and M.R. Gilbert. Bayesian optimal interval design: a simple and well-performing design for phase i oncology trials. *Clinical Cancer Research*, 22:4291–4301, 2016.

[30] M.R. Conaway, S. Dunbar, and S.D. Peddada. Designs for single- or multiple-agent phase I trials. *Biometrics*, 60(3):661–669, 2004.

[31] J. O'Quigley and L.Z. Shen. Continual reassessment method: a likelihood approach. *Biometrics*, 52(2):673–684, 1996.

[32] K. Wang and A. Ivanova. Two-dimensional dose finding in discrete dose space. *Biometrics*, 61(1):217–222, 2005.

[33] A.I. Daud, M.T. Ashworth, J. Strosberg, J.W. Goldman, D. Mendelson, G. Springett, A.P. Venook, S. Loechner, L.S. Rosen, F. Shanahan, et al. Phase i dose-escalation trial of checkpoint kinase 1 inhibitor mk-8776 as monotherapy and in combination with gemcitabine in patients with advanced solid tumors. *Journal of Clinical Oncology*, 33:1060–1066, 2015.

8

Executing and Reporting Clinical Trial Simulations: Practical Recommendations for Best Practices

Cory R. Heilmann and Fanni Natanegara

Eli Lilly and Company

Maria J. Costa

Novartis Pharma AG

Matilde Sanchez-Kam

SanchezKam, LLC

John W. Seaman, Jr.

Baylor University

CONTENTS

8.1 Introduction

The International Conference on Harmonization (ICH) E9 guidance, "Statistical Principles for Clinical Trials", describes the fundamental principles regarding the design, conduct, analysis, and evaluation of clinical trials. The guidance focuses on late phase or confirmatory trials that are typically the basis for demonstrating the effectiveness and safety of new medical interventions. It covers the following topics: study design and conduct, bias reduction, multiplicity considerations, evaluation of safety and tolerability, and reporting of results. The guidance does not address the use of specific statistical procedures or methods, but rather its emphasis is on statistical principles.

Although the guidance focuses on frequentist statistical methods, it does not discourage the use of other approaches. It states the following: "Because the predominant approaches to the design and analysis of clinical trials have been based on frequentist statistical methods, the guidance largely refers to the use of frequentist methods when discussing hypothesis testing and/or confidence intervals. This should not be taken to imply that other approaches are not appropriate: the use of Bayesian and other approaches may be considered when the reasons for their use are clear and when the resulting conclusions are sufficiently robust".

The Bayesian approach enables incorporation of prior information from various sources into the statistical model. It affords decision-making based on probabilistic statements that are simpler to convey, particularly to a general audience. It has the potential for increasing the precision of parameter estimates of interest and reducing the size of a clinical trial (FDA CDRH, 2010), thus exposing fewer patients to untested medicines or placebo. Substantial advancements in computational algorithms and computer hardware have greatly reduced the time needed for Bayesian simulations and aided in the increased use of Bayesian methods. The cost of these advantages is, typically, a greater investment in planning the trial, analyzing the results, and reporting the conclusions. More time may be required to interpret and report Bayesian results if the readers and audience are not familiar with Bayesian methods.

On February 5, 2010, the Center for Devices and Radiological Health (CDRH) issued a Guidance for the Use of Bayesian Statistics in Medical Device Clinical Trials. Medical devices usually evolve in small steps with newer versions representing refinements or variants of technologies that are already available. The physical mechanism of action and evolutionary development of medical devices lends itself to the use of a Bayesian approach, as device effects are typically local. Such effects are often predictable given prior information on previous generations of the device. The guidance resulted from the substantial experience that CDRH had with Bayesian methodology (FDA CDRH, 2010). Its purpose is to highlight important statistical considerations in Bayesian clinical trials for medical devices. It describes the requirements for the use of Bayesian methods in the design and analysis of medical device clinical trials, benefits and challenges with the Bayesian approach, and comparisons with traditional methods. The possibility of bringing good technology to market quicker using fewer resources is of great appeal to the medical device industry. Given the high attrition rates for new drug applications seen in the pharmaceutical industry (Sacks et al., 2014), among other ethical and scientific considerations, it is clearly desirable to extend this approach to the development of medicines.

The choice of analysis method requires theoretical support and a careful assessment of the underlying assumptions. For frequentist analyses, these are often standard, and the required assumptions are typically well understood. The same principles apply to clinical trials where Bayesian inference is used. Before studying the operating characteristics of Bayesian estimators and hypothesis tests, we must decide on the model (log-linear, logistic, with random effects, without random effects, etc.) and the experimental design. These decisions will affect operating characteristics. As most Bayesian analyses use Markov chain Monte Carlo (MCMC) sampling rather than closed-form solutions, simulations are required to assess operating characteristics of the model(s). In frequentist statistics, bias and type 1 error rates can be well approximated, and parameters are assumed fixed. In Bayesian analyses, prior distributions and the assumption that the model parameters are random variables affect the assessment of the study design. For example, even defining the null space is non-trivial when different placebo effects, variability, and correlation of parameters can all impact the bias and coverage of credible intervals depending on the prior distributions used. It is important to point out that the terms "type 1 error" and "type 2 error" rates do not exist in the Bayesian framework but are referenced in guidance from regulatory agencies such as the FDA (FDA CDRH, 2010). However, Bayesian decision rules are often used for decision making (e.g. proceed to the next phase if the probability of effect is sufficiently high). Given this, we propose the use of the following terms when assessing operating characteristics throughout this chapter:

- False Go (FG) rate: probability of a "Go" decision when you should Stop (under the null space), analogous to false positive rate or type 1 error in the frequentist framework;

- False Stop (FS) rate: probability of a "No Go" decision when you should Continue (under the assumption that a treatment effect exists), analogous to false negative or type 2 error in the frequentist framework;

- True Go (TG) rate: probability of a "Go" decision when you should Continue (under the assumption that a treatment effect exists), analogous to power in the frequentist framework. This is also known as the *probability of study success*;

- True Stop (TS) rate: probability of a "No Go" decision when you should Stop (under the null space), analogous to true negative rate.

In this chapter, we provide recommendations on developing and writing the simulation plan and simulation report. These documents help establish that, prior to finalizing the study, key aspects of the Bayesian analyses of trial results were considered and that the operating characteristics of the design were understood. This chapter provides suggestions for the level of detail that should be included in the protocol and analysis plan. These details typically include brief summaries of the simulations conducted prior to the start of the study and will be read by individuals with less statistical experience. Our recommendations differ for exploratory and confirmatory trials.

8.2 Simulation Plan

In planning a trial or experiment, it is important that statisticians and their team members agree on the objectives of the study design and on which objective will drive sample size considerations. Per ICH E9, the sample size should always be large enough to provide a reliable answer to the study objective(s). Using a frequentist approach, one can specify the type 1 error, power, and population effect size to determine the required sample size. In most Bayesian models, closed-form solutions are not available; therefore, simulations are necessary to understand the impact of various factors on the operating characteristics of the study. There is a rich literature on this subject. Key references are Adcock (1997); Joseph, du Berger, and Belisle (1997); Wang and Gelfand (2002); and Brutti, De Santis, and Gubbiotti (2008). Depending on the phase of drug development, the simulation plan and results may be included as part of the regulatory documents. For a pivotal phase 3 trial, it is critical to have a full simulation plan and report with the level of detail necessary to understand the operating characteristics and replicate the simulations at the time of protocol development.

Study design will typically be an evolving, iterative process with many stakeholders, from statisticians and clinicians to senior review boards and regulatory agencies. Each individual brings his or her own discipline and experience to bear on the design. Thus, the decision to focus on a specific study design and model can be a complex one. Consequently, we recommend creating a comprehensive simulation plan to capture the key decisions regarding the study design(s) and model(s) of interest. This will help making these decisions more transparent and accountable. The simulation plan should provide details about the following:

- Background of the disease state, compound, study;
- Simulation objectives;

- Factors to consider in the simulation;
- Generation of virtual patients/virtual studies;
- Computational details of the simulation.

Table 8.1 provides additional details on the content of the simulation plan based on each of the aforementioned points. Another advantage of creating such a simulation plan is to enable reproduction of the simulation study results, thereby facilitating validation efforts and future learnings. The suggested content of the simulation plan applies to both exploratory and confirmatory trials. This plan can also extend to simulation of a clinical program with multiple consecutive clinical trials.

The simulation factors of statistical model and prior information are explained further in Section 8.3.

In addition to operating characteristics, it is important to understand the impact of the choice of priors and choice of models using operational criteria such as interval estimate performance (interval width, coverage, variability of upper/lower bounds, skewness). The goal is to characterize inherent variability in the simulation study.

Simulating the performance of a study design under various scenarios as suggested earlier can result in hundreds or even thousands of MCMC runs. Clearly not all can be checked for convergence, but we recommend that this be done for a small random subset of such runs, representing the range of design parameters. We refer to each of these as a convergence diagnostic run (CDR). Some diagnostic tools, such as the Brooks–Gelman–Rubin (BGR) statistic, require multiple chains. Thus, unlike a typical simulation MCMC run, a CDR may have more than one chain.

Chain length leading to convergence is one aspect of the compound approximation required in Bayesian inference using iterative simulation methods. That is, the chain must be long enough for two distinct approximations: First, it must be of sufficient length to enable good approximation of the posterior density. Second, beyond that length, the chain must continue far enough to adequately approximate features of that density, such as tail probabilities. Thus, in addition to checking convergence for the chains in a CDR with the usual tools (see, for example, Carlin and Louis 2009, p. 152ff), it should also be determined if the chain length is sufficient to yield a reasonable approximation of the posterior feature of interest. To this end, *for all* simulation runs, the Monte Carlo standard error (MCSE) and the posterior standard deviation (SD) should be assessed for each parameter of interest, as well as variance components. A reasonable rule of thumb is that the chain should extend long enough beyond convergence to render the MCSE no more than 5% of the posterior SD for the parameter. See Cowles (2013, pp. 137–138), Lunn et al. (2012, p. 77ff), and Flegal et al. (2008).

CDR chains should be inspected more closely, for example, by plotting posterior summary values (mean, SD, percentiles) as a function of the length

TABLE 8.1
Content of Simulation Plan

Section	Content
Background	Provide overview of the disease state, compound being studied, stage of clinical development, relevant internal information, competitor's information if any, study design and objectives, and endpoint(s) of interest.
Simulation objective(s)	Provide information on the objectives of the simulation study. Typically, simulations are performed to optimize the operating characteristics of trial designs such as probability of study success and probability of making a correct decision.
Simulation factors	The design parameters that will be varied to compare the base case design versus alternative design(s) and trial executive parameters should be listed here. Examples are: • prior information based on historical data, experts' opinion, etc.; • study designs—number of dose arms, sample size, randomization ratio, adaptive or non-adaptive, parallel, crossover, noninferiority, superiority, etc.; • endpoint and its variability; • decision rules—clinical meaningful difference and probability level required to trigger certain decisions such as Go/No-Go, adding a new dose, starting next phase, etc.; • adaptive elements if appropriate such as timing of interim analysis, randomization/dose allocation, etc.; • statistical models, including longitudinal model, dose response, etc.; • dropout rate—may use historical rate or any other variation; • enrollment rate—low, base, fast enrollment rate.
Virtual patients/ studies	Provide information on how to generate virtual patient/study data, how many patients per treatment arm, what models are used and assumptions on longitudinal profile if warranted. Typically, a null case or placebo-like response is needed for simulations to understand the FG rate for a given design and model. At least one other scenario should be included to understand the TG rate and/or probability of success for a given design and model.
Simulation specifics	Provide number of simulations per virtual patient/study scenario, the starting seed, number of burn-in iterations, number of MCMC iterations, convergence criteria, and checks.

of the post burn-in chain. Once all such summaries have stabilized within prescribed boundaries, accuracy can be said to have been attained. See, for example, Lunn et al. (2012, p. 79). One can choose a stability boundary, say a small $\varepsilon > 0$, and require that the posterior summary of interest be within plus or minus ε. This is illustrated in Figure 7 of Chapter 19.

Finally, summarizing the results from the simulation study across multiple scenarios can be achieved compactly with graphical tools. An example of such graphics is displayed in Figure 11 of Chapter 19.

From a regulatory perspective, the simulation plan will increase transparency and facilitate discussions with reviewers prior to a decision on the final study design. Refer to Chapter 14 for details on suggested information to include in the protocol. This practice is supported by the CDRH Bayesian guidance and Prescription Drug User Fee Act (PDUFA) VI.

8.3 Simulation Report

8.3.1 Introduction

The simulation report summarizes the key results of the simulation plan. The purpose of this report may differ depending on whether the particular trial is exploratory or confirmatory. For a confirmatory trial with frequentist analyses, the type 1 error rate is typically strongly controlled. When planning simulations for a Bayesian analysis, the simulations should clearly demonstrate the control of the FG rate under different scenarios and different assumptions underlying these scenarios. Due to the MCMC sampling variability, to strictly control the FG rate, it is necessary to show that the difference between the estimated FG rate and the nominal FG rate is within the sampling error. For all types of trials, the operating characteristics need to be clearly outlined. If a key decision is intended to be based on the model results (such as "Go if $>$ $x\%$ probability of at least an effect size of y relative to placebo"), then the following are likely of key interest:

- Probability of the compound passing the criteria based on current knowledge;
- Probability of a compound with a good efficacy profile passing the criteria (TG);
- Probability of a compound with a poor efficacy profile passing the criteria (FG).

8.3.2 Background

The assumptions underlying the Bayesian model should be clearly and concisely laid out. Some compounds or classes of compounds have a pharmacokinetic/pharmacodynamic (PK/PD) model which provides effect sizes and variabilities of both effect sizes and individual response based on the dose of drug administered. For other compounds for which there is no such model, it may still be of interest to assume different potential profiles (such as "effect same as placebo", "weak effect", and "strong effect") in order to confirm that the study design can appropriately distinguish FG and TG decisions.

For simulations that are based on a PK/PD model, an appropriate level of explanation is needed. The model used to simulate the data (if it exists in closed form) should be written out. Otherwise, the model should be schematically shown, indicating the inputs and assumptions. The uncertainty of the model should also be explained. A relatively simple PK/PD model may have a population mean effect, θ, a population variance, σ_m^2, and an individual subject response variance, σ_e^2. The population mean, θ, may be assumed fixed using different scenarios such as optimistic ($\theta + x{\times}\sigma_m$) and pessimistic ($\theta - x{\times}\sigma_m$), for some $x > 0$. Alternatively, θ may be treated as a random variable. For the ith simulation, the population mean, θ_i, is generated from a distribution with mean θ and variance σ_m^2. These assumptions and how they were implemented in the simulations should be clearly explained. Where possible, the saved seed used to simulate should be documented so the analyses can be reproduced.

The joint prior distribution required in a Bayesian analysis needs careful specification and detailed justification, whether it is data-driven or derived from expert opinion. Even the use of relatively non-informative priors demands great care. For example, separate, relatively non-informative priors on two parameters can unintentionally induce an informative prior distribution on a function of these two parameters (Seaman et al. 2012). The use of non-informative prior distributions can lead to a lack of convergence of corresponding MCMC sampler. This is especially true for small studies that have highly parameterized hierarchical models. For related issues, see Chapter 19.

Justification should be given for the prior distributions chosen for all parameters. For example, if a diffuse (non-informative) prior distribution is used, such justification may include the lack of historical data or the operating characteristics desired for the analysis. One may appeal to a prior-to-posterior sensitivity analysis to demonstrate that the chosen prior has limited impact on the posterior distribution. For example, suppose the probability of study success is critically dependent on the length of an interval estimate. Suppose further that a uniform prior on the interval $(0, B]$ is chosen for a standard deviation component of the model. Plotting the interval length as a function of B can reveal influential values of that upper bound. If the resulting interval widths "stabilize" beyond a value, say, B^*, then choosing $B > B^*$ is defensible. See Figure 9 in chapter 19 for an illustration.

As with frequentist methods, a data model must be specified in a Bayesian analysis. Assessing posterior sensitivity to modeling assumptions is prudent. This can include assumptions about error distributions, random effects, variance components, missing data, variable selection, etc. Simulations for sensitivity analysis is a broad topic, part of the general subject of Bayesian robustness. For a thorough treatment of Bayesian robustness, see, for example, Gelman et al. (2014, pp. 435–447) or Carlin and Louis (2009, pp. 181–196).

For example, inferences from distribution-free or other robust methods can be compared to those obtained using a model with distributional assumptions. Analyses that accommodate missing data, such as the use of pattern mixture models or multiple imputation methods, make assumptions about the missingness mechanism, such as missing not at random (MNAR). The sensitivity of such methods to variations in the missingness mechanism can be investigated via simulation.

Bayesian network meta-analyses commonly use random effects. A simulation report may consider use of a corresponding fixed-effects model for comparison. Interim analyses are often used to project end-of-study results. For example, an interim analysis might use an integrated two-component (ITP) model (Fu and Manner, 2010) which predicts the effect at study end with only a fraction of patients at the final visit. If this ITP model is the primary model to predict the final endpoint, then other models such as one using a hierarchical correlation structure rather than a parametric model could be used.

Similarly, a parametric, monotonic dose response model such as an E_{max} model may be selected as the primary analysis for a dose-response study. The operating characteristics of this model might be poor if the monotonicity assumptions do not hold or if the shape of the dose-response curve otherwise deviates from the form assumed by the model. For such cases, it is useful to compare the model's operating characteristics to those of a nonparametric dose-response model. This is especially true if the PK/PD dose model, exposure-response model, or both are not well understood.

8.3.3 Recommendation of Simulation Outputs to Share with Non-Statisticians

Non-statistical members of the study team are key stakeholders who should review the simulation plan outputs. These outputs from the simulation plan provide the justification for the trial design and statistical methods used. They need to include the key summaries to be shared with regulatory reviewers so the team can prepare for potential regulatory questions and responses. Additionally, the probability of achieving pre-specified study criteria based on optimistic or pessimistic assumptions and sensitivity of operating characteristics in the "expected case" to the model assumptions should be shared. This is summarized in the following table where the correct decision is to move forward in the optimistic and expected case and to stop in the pessimistic case:

Scenario	Probability of Claiming Success
Pessimistic	FG
Expected	TG
Optimistic	TG

Similarly, the same information can be shared explicitly calling out the FG, FS, TG, and TS probabilities.

Scenario	FS Probability	TG Probability	EC 6 mg	EC 12 mg
Pessimistic	NA	NA	$X\%$	$100 - X\%$
Expected	$X\%$	$100 - X\%$	NA	NA
Optimistic	$X\%$	$100 - X\%$	NA	NA

8.3.4 Recommendation of Simulation Outputs to Share with Regulatory Reviewers

For transparency with regulatory authorities, the full simulation plan and report should be available upon request. As Bayesian analyses may be implemented in exploratory studies, there may not be a need to include the full report in the key trial documents. For example, the team may include some patient cohorts in a phase 1 trial for internal decision-making in order to potentially terminate a poor compound or narrow the range of doses for future studies. However, for confirmatory trials, the key outputs to include are:

- Control of FG rate including definition of null space;
- Power under different scenarios and under the expected scenario (if available);
- Sensitivity checks of the model.

8.4 Protocol, Analysis Plan

The choice between Bayesian and frequentist analyses should be made prior to study initiation. Switching analysis method for primary and key secondary analyses after observing the data is problematic. The FDA CDRH Bayesian guidance requires that Bayesian trials be designed and analyzed prospectively as with any other trial. The use of the Bayesian approach involves extensive preplanning. Decisions should be made at the design stage regarding the prior information, the information to be obtained from the trial, and the mathematical model used to combine the two.

Additionally, the Bayesian approach can involve extensive mathematical modeling of a clinical trial. Modeling choices regarding the probability distributions chosen to reflect the prior information, the relationships between multiple sources of prior information, the influence of covariates on patient outcomes or missing data and sensitivity analyses on the model choices should be discussed with the appropriate regulatory agencies.

The basic tenets of good trial design are the same for both frequentist and Bayesian trials. Components of a comprehensive trial protocol include the objectives of the trial, endpoints to be evaluated, conditions under which the trial will be conducted, population that will be investigated, and planned statistical analysis. In addition, the FDA CDRH Bayesian guidance suggests that the following statistical issues unique to Bayesian trial should be addressed in the clinical trial protocols and statistical analysis plan.

8.4.1 Prior Information

Different choices of prior information or model can produce different decisions. The prior information and model should have been agreed in advance with the FDA. A change in the prior information or model at a later stage of the trial may lead to questions about the scientific validity of the trial results. There should be clinical and statistical justifications regarding the choice of prior information. It is recommended that sensitivity analysis be performed to check the robustness of models to different choices of prior distributions, as discussed earlier.

8.4.2 Criterion for Success

The criterion for success (as it relates to safety and effectiveness) of the study should be provided. For Bayesian trials, a type of decision rule considers that a hypothesis has been demonstrated if its posterior probability is large enough. When there is prior information, the decision rule is based on the posterior distribution adjusted according to the observed data as determined by the clinical trial design and simulations that ensure FG rate control. For example, Bayesian hierarchical models borrow strength and use prior information from agreed upon previous studies with patient-level data. Non-informative priors are usually used at the highest level of the hierarchy. Without prior information, an adaptive trial (e.g. interim analyses, change to sample size or change to randomization scheme) can be implemented and information can be used during the trial to adapt the trial during its course. For more discussion on adaptive trials, refer to Chapter 12.

8.4.3 Justification for the Proposed Sample Size

As with any clinical trial, the minimum sample size affording good safety and efficacy assessment should be specified. The chosen sample size should include a minimum level of information from the current trial to enable verification of model assumptions, appropriateness of prior information used, and amount of information to be borrowed from other studies.

The method of justification depends on the trial design. For a fixed sample size trial, data may be simulated assuming a range of different true parameter values and different sample sizes. For each simulated data set, the posterior distribution of the parameter and credible interval or posterior probability of study claim should be computed and compared with the true parameter value known in the simulation. Repeated simulations can be used to assess the proposed sample size. For an adaptive trial design, the expected sample size is a function of the study design parameters and trial results. The minimum and maximum sample sizes, the number of interim analyses, and the number of patients at each interim analysis should be provided.

Bayesian sample size determination methods can be used for sample size justification. In such methods, instead of specifying a range of parameter values, a "design prior" is used to generate "true" parameter values. A data set is then generated and analyzed using the proposed Bayesian model but with "analysis priors", typically diffuse distributions. This process is repeated a large number of times, and the sample size is chosen to achieve, for example, a specified average interval width or coverage criteria. For an introduction to such methods, see Brutti et al. (2008).

8.4.4 Operating Characteristics

Operating characteristics summarized in the analysis plan should include the probability of satisfying the study claim given various true parameter values and sample sizes for the new trial and probability of an FG and FS.

8.4.5 Prior Probability of the Study Claim

It is recommended that the prior probability of the study claim be evaluated if using an informative prior distribution. It should not be as high as the success criterion for the posterior probability. This is to ensure that the prior information does not dominate the current data and create a situation where unfavorable results from the proposed study get masked by a favorable prior distribution.

8.4.6 Effective Sample Size

The effective sample size in the new trial is given by

$$\text{ESS} = n * v_1/v_2$$

where n is the sample size in new trial, v_1 is the variance of parameter of interest without borrowing (computed using non-informative prior), and v_2 is the variance of the parameter of interest with borrowing (computed using informative prior).

The quantity (ESS $-$ n) is the number of patients borrowed from the previous trial. It quantifies the efficiency gained from using the prior distribution as well as gauging if the prior is too informative (Morita et al., 2008).

8.4.7 Program Code

It is recommended that the program code and data used to conduct the simulations be submitted.

8.5 Case Example

For the purpose of illustration, we will consider the following scenario. A team is planning a phase 2 dose finding study for type 2 diabetes where the primary endpoint is changed from baseline in percent glycosolated hemoglobin A1c (HbA1c). The study includes a placebo and an active comparator (AC), and the study team is interested in assessing the confidence that the experimental compound (EC) leads to greater reduction in HbA1c compared to the AC. Let θ_{EC}, θ_{AC}, and θ_{PBO} be the mean change from baseline for EC, AC, and placebo, respectively. To achieve this, the team wishes to have at least 80% probability of greater HbA1c reduction in the EC compared to the AC or $\Pr(\theta_{\text{EC}} - \theta_{\text{AC}} < 0 \mid \text{data, prior}) \geq 80\%$.

Although this is a confirmatory phase 2 study, the primary objective is to assess how likely the EC is to show HbA1c reductions greater than placebo, so the inference relative to the active comparator is primarily for internal knowledge and decision making. However, the team may consider applying the same Bayesian model to the placebo comparison, in which case the simulations supporting this study could be of regulatory interest as well.

One case to consider is where the true effect of the treatment is equal to placebo, $\theta_{\text{EC}} = \theta_{\text{PBO}}$ for all doses of drug. Here, there may be a reason to show an FG rate of 2.5%, i.e., the study will only conclude that a dose is superior to placebo 2.5% of the time (equivalent to the type 1 error rate). The team may also be interested in having high probability that at least one dose of the experimental compound is better than the active comparator, i.e., they define a successful study as $\Pr(\theta_{\text{EC}} - \theta_{\text{AC}} < 0) \geq 80\%$.

The null case can be defined in many ways. For example, the experimental compound having the same response as the placebo would be a null case. However, given that the Bayesian model assigns prior distributions to all parameters, simply using flat priors centered at the same value is not sufficient to ensure strong control of the FG rate. For example, if the prior placed on one or more variance parameters leads to negative bias in the variance parameters, the simulated FG rate may still be higher than the desired nominal value. In general, it is unlikely that a closed-form solution exists for a moderately complex Bayesian model and as such, simulation under a reasonable set of scenarios would suffice to show that the simulated FG rate is controlled.

To begin documenting this in a simulation plan, it is first necessary to identify the scenarios to include. Here, the PK-generated data would serve as one possible scenario. It is also possible to create an optimistic and a pessimistic scenario which are at the low and high ends of the range of PK values. It may also be necessary to consider a "null" case where the effect is assumed to be equal to placebo. The scenarios may be summarized as follows:

Scenario	Description
Expected scenario	HbA1c change follows PK model
Optimistic scenario	HbA1c change is greater than the median predicted by the virtual patient model
Pessimistic scenario	Virtual patients generated from a scenario where the HbA1c change is less than the median predicted by the virtual patient model
Null scenario	HbA1c change is equal to the placebo change for each dose

The response for each scenario should be explicitly defined, along with any sources of the information supporting these scenarios. A hypothetical scenario (such as the null scenario) can also be considered with explanation that this is based on assessing the operating characteristics in a hypothetical case, not specific data. As there is expected to be sources of data on placebo response, it is recommended to justify these expectations.

Scenario	Placebo (%)	EC 2 mg (%)	EC 6 mg (%)	EC 12 mg (%)	EC 25 mg (%)	AC (%)
Null	−0.1	−0.1	−0.1	−0.1	−0.1	−1.1
Pessimistic	−0.1	−0.2	−0.5	−0.8	−0.9	−1.1
Expected	−0.1	−0.2	−0.7	−1.2	−1.4	−1.1
Optimistic	−0.1	−0.3	−0.9	−1.5	−1.7	−1.1

For simplicity, we will be assuming a 1.0% subject-level standard deviation of response.

As the study is 12–24 weeks in duration and may have multiple endpoints, there will be some dropouts and multiple measurements of the primary endpoint, HbA1c. As such, a dropout pattern should be defined. For this example, we will assume 15% dropout for each arm and 50 randomized patients per arm. For this study, we will assume for simplicity that there is no interim analysis performed to change any aspect of the trial. If a methodology such as the ITP model (Fu and Manner, 2010) is used for modeling the dropout, that should be specified along with any prior distributions related to this model and the dose-response model.

The simulation specifications indicate what is necessary to set up and execute simulations once the trial design and scenarios are decided. For each scenario, we would depict this as follows:

Simulation Factors	
Software	**FACTS^{TM} Version 5.5**
Starting seed	3,500
Burn-in iterations	50,000
Thinning	10
Posterior samples	50,000
Chains	3

The criteria for assessing convergence and the model checks should also be stated as referenced in Chapter 19. For example, the thinning is set to take every 10th value. One check would be to look at the autocorrelation and specify a maximum threshold. For example, it may be stated that the autocorrelation for each parameter be less than 0.5 or that the effective sample size (ESS) must be greater than 10,000. The criteria for convergence (e.g. BGR criteria <1.01) should also be set.

Where decision criterion 1 is defined as $\Pr(\theta_{\text{EC}} - \theta_{\text{PBO}} < 0 \mid \text{data, prior}) \geq 97.5\%$ and decision criterion 2 is defined as $\Pr(\theta_{\text{EC}} - \theta_{\text{AC}} < 0 \mid \text{data, prior}) \geq 80\%$, the probability of achieving the criteria for each scenario can be represented as in the following table. For the null case, the Pr(Criterion 1) and Pr(Criterion 2) are both FG probabilities as the null case is not superior to placebo or the active compactor. Similarly, the pessimistic case is not superior to the active competitor, and Pr(Criterion 2) is an FG probability. All other probabilities are TG probabilities as all scenarios except the null have each dose level showing greater decrease in HbA1c than placebo, and both the expected and optimistic scenarios have at least their highest dose as showing a greater effect on HbA1c than the active comparator.

Scenario	**Pr(Criterion 1) (%)**	**Pr(Criterion 2) (%)**
Null	1	0
Pessimistic	96	14
Expected	99	93
Optimistic	100	99

Criterion 2 provides ample separation with high TG probabilities for the "expected" and "optimistic" scenarios, while also maintaining low FG probabilities for the "null" and "pessimistic" scenarios. Criterion 1 does not adequately prevent FG decisions in the "pessimistic" scenario. The risk tolerance of decision makers could lead to further optimization of Criterion 1 and also the consideration of additional scenarios.

8.6 Conclusions

In this chapter, we have provided the essential elements of the simulation plan and simulation report along with guidance on language to include in the protocol and analysis plan. As PDUFA VI has provided a pathway for complex innovate designs, the recommendations in this chapter can serve as a basis for considerations in the development of such clinical plans. This is an iterative process with teams coming up with scenarios, possible narrowing down scenarios after the initial simulations, and finally coming up with a set of reasonable designs, assumptions, and model parameters to choose from.

We illustrate the critical outputs and presentation of results through the case example. The MCMC checklist in Chapter 19 provides the key outputs to review in order to confirm model convergence and key sensitivity checks.

References

Adcock CJ. (1997), Sample size determination: A review, *The Statistician*, 46, 261–283.

Brutti P, De Santis F, Gubbiotti S. (2008), Robust Bayesian sample size determination in clinical trials, *Statistics in Medicine*, 27, 2290–2306.

Carlin B, Louis T. (2009, *Bayesian Methods for Data Analysis*, 3rd Ed., CRC Press: Boca Raton, FL.

Cowles M. (2013), *Applied Bayesian Statistics*, Springer: New York.

FDA CDRH (2010) Guidance for Industry and FDA Staff: Guidance for the Use of Bayesian Statistics in Medical Device Clinical Trials.

Flegal J, Haran M, Jones G. (2008), Markov chain Monte Carlo: Can we trust the third significant figure? *Statistical Science*, 23(2), 250–260.

Fu H, Manner D. (2010), Bayesian adaptive dose-finding studies with delayed responses, *Journal of Biopharmaceutical Statistics,* 20, 1055–1070.

Gelman A, Carlin J, Stern H, Dunson D, Vehtari A, Rubin D. (2014), *Bayesian Data Analysis*, 3rd Ed., Chapman & Hall: Boca Raton, FL.

Joseph L, du Berger R, Belisle P. (1997), Bayesian and mixed Bayesian/likelihood criteria for sample size determination, *Statistics in Medicine*, 16, 769–781.

Lunn, D, Jackson, C, Best, N, Thomas, A, Spiegelhalter, D. (2012), *The BUGS Book*, CRC Press: Boca Raton, FL.

Morita S, Thall P, Muller P. (2008), Determining the effective sample size of a parametric prior, *Biometrics*, 64, 595–602.

Sacks LV, Shamsuddin HH, Yasinskaya YI, Bouri K, Lanthier ML, Sherman RE. (2014), Scientific and regulatory reasons for delay and denial of FDA approval of initial applications for new drugs, 2000–2012. *JAMA*, 311(4), 378–384. doi:10.1001/jama.2013.282542.

US Food and Drug Administration (FDA). PDUFA reauthorization performance goals and procedures fiscal years 2018 through 2022, https://www.fda.gov/media/99140/download. Accessed May 22, 2019.

Wang F, Gelfand A. (2002), A simulation-based approach to Bayesian sample size determination for performance under a given model and for separating models, *Statistical Science*, 17, 193–208.

9

Reporting of Bayesian Analyses in Clinical Research: Some Recommendations

Melvin Munsaka

AbbVie, Data and Statistical Sciences

Mani Lakshminarayanan

Statistical Consultant

CONTENTS

9.1 Introduction

There is limited literature on Bayesian reporting of clinical trials and observational studies. Lang and Secic (2006) provided the following fundamental elements for Bayesian reporting: report the pre-trial probabilities and specify how they were determined; report the post-trial probabilities and the corresponding intervals; interpret the post-trial probabilities; and report software and statistical methods and models for the Bayesian calculations. Similar advice is provided by other authors in various journals to authors for manuscripts using Bayesian methods to conduct data analyses. A list of seven items that some international experts believe to be most important when reporting a Bayesian analysis for scientific publications has been proposed by Sung et al. (2005). The set of items were as follows: describing the prior distribution, thorough specification, justification, and sensitivity analysis; presenting the analysis in terms of the statistical model and analytic technique; and presenting the results using a measure of central tendency and variance. There have been some good efforts aimed at improving reporting Bayesian analyses including from Anderson et al. (2001), Basis Group (2014), Bitt and He (2017), Hughes (1991, 1993), Ohlssen et al. (2014), Pullenayegum and Thabane (2009), Pullenayegum et al. (2012), Rietbergen et al. (2017), Schoot and Depaoli (2014), and Weeden et al. (2003). In subsequent sections, some key considerations for good practice in Bayesian reporting are discussed beginning with a general review of the advantages of using Bayesian methods. A checklist taking into account several key aspects of Bayesian analysis is proposed that will be useful in ensuring standardization of the reporting of Bayesian analyses, including design, interpretation, and analyses, and documentation particularly in the regulatory setting. Finally, a hypothetical example is used to illustrate the ideas presented in the checklist.

9.2 A Review of the Literature on Bayesian Reporting

9.2.1 Why Use Bayesian Methods?

There are a number of advantages that the Bayesian approach offers in the context of clinical trial data and for that matter in any area of research, see, for example, Wijeysundera et al. (2009), Walley et al. (2015), Efron (1986), Gelman (2008), and Grieve (2016). One key advantage is that evidence regarding a specific problem can be taken into account through priors, e.g., results of observational studies. Bayesian methods enable evidence from a variety of sources, regarding a specific problem, to be taken into account within a concise modeling framework in which one can make direct probability and predictive statements. Elicitation of prior beliefs or evidence forces researchers into examining different sources leading to better inference. The approach enables predictive statements with uncertainty conditioned on current state of knowledge. It also allows one to account for parameter uncertainty through flexibility of incorporation of full uncertainty in all parameters. Bayesian methods lead naturally to a decision theoretic framework and to more natural and informative setting for decisions from the posterior distributions.

By their construction, the posterior distributions of parameters are not confined to usual (normal) representations or based on asymptotic assumptions and give substantially more information than single point estimates. For a discussion of Bayesian methods as they pertain to application in clinical trials, see Price et al. (2014).

Although the potential utility of Bayesian methods is well understood, see for example Brophy and Joseph (1994), the use of subjective prior information will often cast doubt on the use of these methods and potentially destroy any semblance of objectivity. Different prior distributions can be used which can generate varying results. Bayesian methods use appropriate prior clinical information in a formalized manner. At the heart of this are various questions including how this can be explained in clear and succinct ways in the clinical trial setting and how the prior information can be explained in such a way that it is informative and acceptable. It is clear that the manner in which the analysis is performed and the results are presented is very critical in arriving at any kind of understanding of the results, especially given that there is a need in some way to preserve the objective nature of the trial results in the light of the introduction of what may be considered as subjective as a result of introducing the prior information. Additionally, eliciting prior beliefs is a non-trivial exercise, and there are few guidelines to help the Bayesian analyst, see for example Scott et al. (2011) and Wu et al. (2008). Additionally, there is no agreed standard metric of statistical significance such as a p-value in the Bayesian framework. Often, the difficult level of mathematical models (see, for example, Robert (2014)), difficult implementation (only non-common software), and several steps that are needed to obtain results make Bayesian approaches seemingly more daunting and challenging compared to other approaches. Bayesian methods are also computationally complex to implement and time consuming to perform.

9.2.2 Core Concepts in Bayesian Analysis and Computations

The Bayesian approach comes originally from the work of Bayes and Laplace, and much of the modern theory comes from de Finetti in the 1930s. Chapter 1 of this provides extensive details to Bayesian inference and modeling. A brief summary of comparisons between frequentist and Bayesian inferences involves three major aspects, respectively: repeatability of data (thus providing frequency) versus data being observed from the realized sample, underlying remaining constant through the repeatable process versus unknown parameters getting described probabilistically, and finally, parameters getting treated as fixed vs data remaining fixed. Posterior distribution, calculated based on Bayes theorem, is central to Bayesian inference. It is conditional on observation and is based on sample likelihood. Any calculation of posterior does not require averaging over the unobserved values of the data. From a Bayesian perspective, all inferential questions can be answered through a posterior distribution. Once it is calculated, it is easy to compute point and interval estimates of parameters, predict inference of future data, and make any probabilistic evaluation of hypotheses.

There are many real-life problems that may not lend themselves to defining the prior knowledge about the state of nature in a concrete fashion for researchers to specify an a priori probability function; however, to take advantage of a Bayesian approach, researchers in these cases tend to explore the possibility of specifying mathematically tractable priors, which result in conjugate families when combined with likelibood function. Specifying conjugate priors result in closed form expressions for the posterior, and, as a result, one can calculate summaries of the posterior distribution more easily. But, if a problem requires multiple parameters, then calculating the posterior can get mathematically intractable as it would require specifications of multiple priors, and, when combined with likelihood, evaluation of the integrals in the Bayes formula can be mathematically challenging.

Consequently, proper application of Bayesian methods requires approximating high-dimensional integrals (as opposed to frequentist approaches which mainly rely on maximization rather than integration) which posed severe limitations to Bayesian applications for almost 200 years. Around 1990, a computational revolution happened that changed the nature of Bayesian computation and made it easier and tractable. Over the past 25 years, the Bayesian community has 'discovered' and developed an entirely new simulation-based computing method called Markov Chain Monte Carlo (MCMC) that was already in place in physics Metropolis et al. (1953). A Bayesian simulation relies on two principles: Monte Carlo principle and characterization of higher-dimensional joint densities. A Monte Carlo principle relies on the principle that anything one would like to know about a random variable X can be learned by sampling its density $f(x)$ many times. As a result, precision of what we learn about x is limited by the number of samples from $f(x)$ one can generate, store, and summarize.

The second principle leads to the fact that the higher-dimensional joint densities are completely characterized by lower-dimensional conditional densities. This is the basis for MCMC methods. Two of the most popular MCMC procedures are the Gibbs Sampler and the Metropolis–Hastings algorithm. Here is how a Gibbs sampler works in general:

- First, sample values of the unknown parameters from their conditional posterior distributions. This is an iterative process that starts with some initial guesses as to the parameter values.
- A sample is drawn from the conditional distribution of the first parameter given values of all of the others.

- Then, a sample is drawn from the conditional distribution of the second parameter given the value just obtained for the first parameter and the current values of all of the other parameters.
- This process is repeated until values have been drawn for all of the parameters, at which point one cycle of the sampling process, called Gibbs sampling, is completed.

The Gibbs sampling process is repeated until the distributions stabilize. Under broad conditions, this process provides samples from the joint posterior distribution of all of the parameters. Once the process stabilizes, many samples from the joint distribution are obtained, and inferences are based on these samples and statistics describing them, e.g., posterior mean, quantiles of the distributions, etc. Though a Gibbs sampler is a special case of the Metropolis–Hastings algorithm, these two algorithms are treated separately in the literature due to the difference in their sampling mechanism. These topics are explored further elsewhere in this book.

9.2.3 Literature on Reporting of Bayesian Analyses

Although there is a wide variety of books on Bayesian approaches to the analysis of data and numerous references both theory and applied, not much is written about Bayesian reporting of analyses particularly as it pertains to clinical research. A survey of the literature on Bayesian reporting points out to some common themes that may help in making Bayesian analyses more comprehensible to the readers of these analyses, especially those not familiar with the Bayesian approach. These common themes include motivation of the use Bayesian method, likelihood and prior distributions, posterior distribution and inference, sensitivity analysis, assessment of convergence and diagnostics, interpretation of the results, posterior predictive distribution, and software. Each of these aspects of Bayesian reporting is discussed in the next sections.

9.3 Some Recommendations on Bayesian Reporting

9.3.1 Aim of the Study and Motivation in the Use of Bayesian Methods

Just as in any other research setting, it is important that the research questions and objectives of the study are clear. The questions being addressed need to be clarified along with the description of the design and conduct of the trial(s). It may be additionally useful when appropriate to motivate and provide reasoning for use of Bayesian method outlining advantages over use of frequentist method. Additionally, a prospective analysis plan should be provided, and an effort should be made to define various Bayesian terms especially for the non-Bayesian audience.

9.3.2 Likelihood and Prior Distribution

The description of likelihood function should capture the statistical model that is considered optimal for the problem and the data under consideration. In the context of Bayesian communication, the reader should be provided with the appropriate likelihood function, prior beliefs, and threshold values as appropriate so that he or she can make informed decisions on the conclusions drawn.

The choice of prior distribution (and hyperprior if hierarchical modeling is used) and rationale for choice should be clearly provided. There should be a discussion of the types of the various choices of priors (e.g., informative priors, non-informative priors, vague priors). The sources of prior information and timing of the availability of this information or when the prior was constructed must be clarified, i.e., before or during or after data collection process. From an application perspective, it is important that alternative priors for the purpose of sensitivity analysis are explicitly specified. Description of the priors should also include the parameters for the prior distributions and a clear description of the model for the priors. A good rule of thumb is that if the prior distribution is based on belief, then the posterior distribution should be interpreted as an updated statement of belief. The prior use of non-informative and skeptical priors represents a summary of hard evidence in which case the posterior distribution represents an updated summary of hard evidence.

9.3.3 Posterior Distribution and Inference and Posterior Predictive Distribution

The summaries, which can be numerical and/or graphical, of the posterior distribution and overall model parameters and other quantities of interest should be presented and clearly summarized. In most cases, these should include a presentation of posterior credible intervals and a graphical presentation of the posterior distribution. If either a formal or informal loss function has been described, the results should be expressed in these terms. It is also essential that the likelihood can be reconstructed, usually through information given under evidence from study so that subsequent users can establish the results independently as desired. The posterior distribution then represents a summary of beliefs/evidence about the parameters of interest. This is most fully represented graphically but can be further summarized by giving a 95% credibility interval. The robustness of the posterior distribution for different priors' contribution from the study to, say, a meta-analysis should be highlighted. A careful distinction needs to be made between the report as a current summary for action, in which case a synthesis of all relevant sources of evidence is appropriate, and the report as a contributor of information for future action. The report should include all summary statistics of the parameters that have meaningful interpretation.

9.3.4 Sensitivity Analysis

It is important to present a comprehensive discussion of the sensitivity analyses and results. The approaches used to check model fit and to carry out any sensitivity analyses need to be provided. The findings from the sensitivity analysis and implications for study results and any decisions to be made on the basis of these results need to be highlighted. This means that in the presentations of Bayesian analyses, one should report the sensitivity of conclusions to the choice of the prior distribution. Portrayal of sensitivity can be accomplished by including overlaid graphs of the posterior distributions for a variety of reasonable priors or by tabular presentations of credible intervals, posterior means, and medians. The results of any alternative priors and/or addressing the impact of choice of priors on the conclusions should be noted. In doing so, it would help the reader position the new data in the context of existing knowledge.

9.3.5 Assessment of Convergence and Diagnostics

If a simulation was used, the report should provide the simulation details if applicable. The statement about whether any evidence for non-convergence was found should be included. The implementation of sophisticated methods for fitting models, such as Markov Chain Monte Carlo (MCMC) should be reported in sufficient detail for another user to be able to carry out a similar analysis. In particular, MCMC requires diagnostics to indicate that the posterior distribution has been adequately estimated. A discussion of chain convergence must thus be included. Each model parameter estimated should be monitored to ensure that convergence was established for the posterior. That is, the MCMC should be described in sufficient detail especially regarding evidence that the chains converged with plenty of burn in and there are no orphaned chains and not clumpy, i.e., low autocorrelation from sufficient thinning.

9.3.6 Software Tools Used

Finally, there is a wide selection of software available for Bayesian analysis which may employ different MCMC techniques. It is recommended that the software that was used for analysis should be identified. Various software, both commercial and open source, utilize different algorithms such as Gibbs and the Metropolis–Hastings sampling. It is important to identify the specific sampling algorithm being used in an application. Some of the available software and algorithms are described elsewhere in other chapters in this book.

9.3.7 Interpretation of the Results

Just like in the frequentist setting, Bayesian interpretation of the results with respect to central tendency, standard deviation or credible interval of the parameters of interest can be done based on the posterior distribution or posterior predictive probabilities of certain hypothesis statements. Separate elements of the prior distributions and likelihood should be clearly specified and appropriately justified, so that the posterior distribution may be clearly interpreted. The posterior distributions should be clearly summarized. Reporting of the Bayesian analysis should be such that the reader is able to interpret the analysis and have confidence in the approach. The results of the trial should be described clearly as part of the results' discussion and in enough detail so that another reader could carry out alternative analyses if desired. Key results' presentations can include graphs (triplot), posterior density function (or histogram), central tendency (e.g., mean or median), and spread (e.g., 95% credible interval or standard error) for parameters of interest. The credibility of the results will be greatly enhanced by a brief description of model-checks performed, especially as these relate to questionable aspects of the model. In the clinical trial setting, it is important that one uses Bayesian analyses to illustrate the strength of evidence coming from the trial by showing the sensitivity of the conclusions drawn to choice of prior distribution made. This can include a presentation of posterior credible intervals and a graphical presentation of the posterior distribution.

If the aim is to convince the reader of an effect, or otherwise, then a figure showing the post-trial beliefs in a beneficial effect for various degrees of prior skepticism would provide an adequate sensitivity analysis while avoiding the problem of justifying informative prior distributions. Indeed, the strength of evidence may be further supplemented by showing on the same plot, the association between post-trial belief for more substantial effects and degree of prior skepticism. Thus, an interpretation section could, with good guidance in the text and with the advantage of graphical displays, provide an in-depth assessment of a trials strength of evidence used to support the conclusions drawn. Providing graphical summary of both prior and posterior allows the reader to visually assess the relative change of beliefs. In general, the following should always be provided: point and interval estimates, posterior probability of clinical benefit where applicable, classical results (or results under uniform or minimally informative prior), a description of the details of the statistical modeling along with graphical summary of both prior and posterior which will allow the researcher to visually assess the relative change of beliefs.

9.4 Checklist for Bayesian Reporting of Clinical Trials

In this section, a checklist is proposed that can be useful in ensuring standardization of the reporting process. This checklist is important for a number of reasons and potential benefits. First, there is limited consensus regarding the reporting of Bayesian simulations, design, analysis, and interpretation. Second, Bayesian methods offer important advantages in flexibility and complexity relative to frequentist analysis and therefore require certain aspects be reported and discussed. Third, utilization of such a list could improve the scientific rigor and ensure transparency, thereby enabling replication and verification by other investigators and stakeholders. Fourth, the checklist can provide guidance to reviewers and readers to better understand such designs and analyses. Finally, given that the terminology tends to be different between a Bayesian and traditional frequentist analysis, standardized reporting promotes the use of similar terminology and provides clear distinctions whenever a Bayesian design/analysis is reported. The checklist is provided in Table 9.1.

TABLE 9.1
Checklist for Reporting Bayesian Analyses

Key Consideration	Description
Clear research question and study design	• Does the objective of the study address clear and specific research question(s)? • Are the design and conduct of the study clearly described in the study plan?
Motivation for use of Bayesian method	• Has a justification for the use of Bayesian method been outlined in the protocol? • Have the advantages over frequentist method been described in the protocol?
Method of analysis	• Does the analysis plan discuss the likelihood function that captures the complete statistical model? • Justification for choice of prior(s) and details: – Why a flat prior (if used)? – If an informative prior is used, how were the prior parameters chosen (and key references)? • Plan for sensitivity analysis for various priors.
Details of inference to be done	• Description and justification for inference to be carried out and how decisions will be made (e.g., based on credible intervals) • Provide limitations to analysis and inference (if they exist).
Computation methods and software used	• What software to be used • If MCMC is to be used, provide details as to how convergence will be achieved and confirmed.
Simulation, convergence assessment, and diagnostics	• Have you provided simulation details (if applicable) including the type of software, number of chains, convergence diagnostics?

(*Continued*)

TABLE 9.1 (***Continued***)
Checklist for Reporting Bayesian Analyses

Key Consideration	Description
	• Provide details to model summaries and diagnostics. • Present a comprehensiveness of sensitivity analysis.
Possible limitations of the analysis	• Include an honest appraisal of the strengths and possible weaknesses of the analysis. • Provide additional details regarding posterior distribution and corresponding summaries. • Report summary statistics of the parameters that are theoretically meaningful.
Bayesian results and interpretation	• Bayesian interpretation of the results with respect to central tendency, standard deviation, or credible interval of the parameters of interest, based on the posterior distribution or posterior predictive probabilities of certain hypothesis statements • Reporting of the Bayesian analysis should be such that the reader is able to interpret the analysis and have confidence in the approach. • The results of the trial should be described clearly and in enough detail so that another reader could carry out alternative analyses if desired. • In general, a presentation of posterior credible intervals and a graphical presentation of the posterior distribution • The credibility of the results will be enhanced by a brief description of model-checks performed, especially as these relate to questionable aspects of the model. This requires that the posterior distribution be clearly summarized. • In all Bayesian reporting, the separate elements of the prior distributions and likelihood should be clearly specified and appropriately justified, so that the posterior distribution may be clearly interpreted.

(Continued)

TABLE 9.1 (*Continued*)
Checklist for Reporting Bayesian Analyses

Key Consideration	Description
	• If either a formal or informal loss function has been described, the results should be expressed in these terms. • It is also essential that the likelihood can be reconstructed, usually through information given under evidence from study, so that subsequent users can establish the contribution from the study to, say, a meta-analysis. • There should be a careful distinction between the report as a current summary for action, in which case a synthesis of all relevant sources of evidence is appropriate, and the report has a contributor of information for future action. • If the aim is to convince the reader of an effect, or otherwise, then a figure showing the post-trial beliefs in a beneficial effect for various degrees of prior skepticism would provide an adequate sensitivity analysis while avoiding the problem of justifying informative prior distributions. • Thus, an interpretation section could, with good guidance in the text and with the advantage of graphical displays, provide an in-depth assessment of a trial's strength of evidence used to support the conclusions drawn. • Provide classical results (or results under uniform or minimally informative prior)

9.5 A Hypothetical Application Setting

To illustrate how the checklist may be used for the purpose of providing guidance on reporting a Bayesian analysis, we consider a hypothetical setting involving a proof of concept (PoC) study for a new test drug (Test Drug) versus placebo (Placebo) for psoriasis. We first begin by providing a high-level description of the protocol.

9.5.1 Protocol Outline

Title: A Multi-Centre, Randomized, Double-Blind, Placebo-Controlled Proof of Concept Study, to Assess the Efficacy, Safety and Tolerability of Test Drug versus Placebo in Subjects with Moderate to Severe Psoriasis

Entry Criteria: Assume the usual or typical entry criteria for such a study.

Efficacy Endpoint: The primary endpoint for the study will be based on the Psoriasis Area and Severity Index (PASI) with the assessment based on the proportion of subjects with at least a 75% achievement, i.e., a PASI75 response between Baseline and Week 12. A patient's PASI is a measure of overall psoriasis severity and coverage consisting of two major steps:

- Calculating the BSA (Body Surface Area) covered with lesions and
- Assessment of the severity of lesions

The second step in turn consists of assessing lesions' erythema (redness), induration (thickness), and scaling. In the PASI system, the body is divided into four regions: the head, trunk, upper extremities, and lower extremities. Each of these areas is assessed separately for the percentage of the area involved, which translates to a numeric score that ranges from 0 to 100. All calculations are combined into a single score in the range of 0 (no psoriasis on the body) and up to 72 (the most severe case of psoriasis).

A 75% reduction in the PASI scores (PASI75), the current benchmark of the primary endpoints for most clinical trials of psoriasis, is considered as the primary endpoint for this trial.

9.5.1.1 Sample Size Determination

Bayesian sample sizes can be determined based on lengths and coverages of posterior credible intervals.

For example, if the average coverage criterion is used, then it will ensure that the mean coverage of posterior credible intervals of certain length, weighted by the predictive marginal distribution of the data, is at least $1-\alpha$. Similarly, if the sample size is determined based on average length criterion, then it will ensure that the mean length of $100 \times (1-\alpha)\%$ posterior credible intervals weighted by the predictive marginal distribution is at most the size of the length of the interval.

Finally, based on worst outcome criterion, for a suitably chosen subset of the data space, this criterion will ensure that the subset chosen will consist of the most likely 95% of the possible data; then there is 95% assurance that the length of the $100 \times (1-\alpha)\%$ posterior credible interval will be at most the length of the interval.

R Package *SampleSizeProportions* (Joseph et al., 2015) can be used to determine the sample size based on some of the criteria mentioned here. For example, the function **propdiff.acc** returns the required sample sizes to attain the desired average coverage probability level for the posterior credible interval of fixed length for the difference between the two unknown proportions. Suppose the two priors for the Placebo and the treatment are Beta(4, 6) and Beta(2, 3), respectively, where Beta(α, β) is the Beta distribution with parameters α and β. The fixed length of the posterior credible interval for the difference in the PASI75 scores is 0.20. Under this assumption, the function **propdiff.acc** can be used to get a sample size of $n = 145$ per group.

A Bayesian power (assurance) curve can also be investigated as a function of θ, the critical difference that needs to be detected, and the sample size n. Based on this investigation, sample size can be chosen for a given θ.

9.5.1.2 Analysis Approach

For the primary analysis of PASI75, a simple Bayesian model will be assumed; that is, the PASI75 endpoint is assumed to be discrete and distributed as binomial for both groups, i.e., the treatment arm and the Placebo arms. A hierarchical model may also be explored as necessary to account for covariates, see, for example, Spiegelhalter et al. (2004). In the present setting where there are only primary parameters assumed without any hyperparameters, a simple beta-binomial model setup up will be used. The setup will be as follows: let π_1 and π_2 represent the PASI75 responses from the Treatment Group and Placebo Group, respectively, let X_1 and X_2 be the random variables representing the PASI75 responses for the Treatment and Placebo, respectively, and let n_1 and n_2 be the sample sizes for the Treatment and Placebo, respectively. Then, we have that $X_i \sim \text{Bin}(n_i, \pi_i), i = 1, 2$, where $\text{Bin}(n, \pi)$ is the Binomial distribution with parameters n and π. Under the Bayesian setting, it is natural to model each one of the π's as a Beta distribution with parameters α_i and β_i; or simply put, $\pi_i \sim \text{Beta}(\alpha_i, \beta_i),\ i = 1, 2$, which we use for the prior distribution for each π_i. Using Bayes rule and under conjugacy of the Binomial and Beta distribution, it is fairly straightforward to show that the posterior distribution $\pi_i \mid X_i \sim \text{Beta}(\alpha_i + x_i, \beta_i + n_i + x_i),\ i = 1, 2$ (see Appendix B). For analysis purposes, the quantity of interest for decision-making purposes is: $\theta = \pi_1 - \pi_2$. Specifically, we want to obtain the posterior distribution of θ and use this to address various analysis and inference questions. Although some closed form for the posterior distributions for θ are available, they are not mathematically easily tractable, see for example, Pham-Gia et al. (2017), Kawasaki et al. (2013), and Kawasaki and Miyaoka (2010). The posterior distribution of θ will be evaluated using simulation MCMC approach. From this setup, $\theta = \pi_1 - \pi_2 \mid X_1, X_2$ is just be another distribution with updated parameters. θ can also be estimated from the distribution of π_1 and π_2 independently. The posterior distribution for θ can be obtained from samples obtained by subtracting π_2 from π_1.

Analyses of the primary endpoint will be performed on both the Per Protocol and the Full Analysis Set data. The model will be analyzed using Bayesian software.

9.5.1.3 Considerations for the Prior Distribution

As noted above, the prior distribution on the parameter of interest, that is, the response rate based on PASI75 for both Test Drug and Placebo will be a conjugate beta prior. Elicitation of beta parameters will use some of the popular methods (see, for example, Jenkinson, 2005) that may include some of the following:

- Find an average response for Placebo based on a meta-analysis which can be used on available data and use the fact that mean of Beta(α, β) is $\frac{\alpha}{(\alpha+\beta)}$ and the variance is $\sigma^2 = \dfrac{\alpha\beta}{(\alpha+\beta)^2(\alpha+\beta+1)}$. Note that this may not be unique, for example, if the average response is 40%, some of the choices for (α, β) are $(4, 6)$, $(2, 3)$, etc.

- A location and interval method. In this method, the following two quantities will be sought: response rate that is most likely to occur (mode) and a probability estimate that the response rate falls between two values.

The parameters for the conjugate beta prior, once they are determined as described above, will then be combined with the observed binomial data in the computation of posterior distribution (Beta) using Bayes formula. Summary measures from the posterior distribution will then be presented, including: posterior mean, mode, and median. Interval estimates such as 95% credible and the highest posterior density (HPD) intervals will also be presented.

Inadequacies in design can cover a wide variety of clinical study elements: sample size, endpoints, randomization appropriate controls, schedule of assessments, and a variety of other elements appropriate for the correct execution of a complete Phase-II program. Effect of such inadequacies on how well the trial is ongoing in terms of predicting efficacy can be evaluated using adaptive designs. A simple adaptation such as an interim analysis and having a decision criteria for go/no go decisions can significantly help with learning and understanding the current data and, as a result, can help with addressing any knowledge gaps which may need to be addressed appropriately. Summary measures derived from posterior distribution can be used to make decisions during the conduct of a protocol and decide on whether to:

- Stop the trial for futility

- Change/modify study designs (e.g., drop a treatment arm)

- Predict the final outcome.

One measure that will be used to help with interim decision making is the predictive probability of success (POS) which is defined as the probability of achieving a successful (significant) result at a future analysis given the current interim data. POS is related to conditional power, a well-known frequentist measure discussed in the group sequential literature. POS, as a predictive probability, can also be computed as a weighted average of the conditional powers across the current probability that each success rate is the true success rate (i.e., weighted by the posterior). Another way is to look at POS as a predictive power or as an expected power where the expectation is taken across a range of possible effect sizes, whereas a frequentist definition of statistical power is a conditional value under the assumption that a true effect size is known.

Learning and confirming are the two aspects that should set the paradigm during Early Phase-II to Phase-III in any drug development. Depending on the direction dictated by the magnitude of POS calculated at the interim in a Phase-II PoC trial, three actions are possible: accelerating development of compound toward Phase-III, terminating compound development for this indication if there are issues with safety, and re-evaluating development options including development of further studies to understand efficacy. As the POS is calculated for an efficacy endpoint (PASI75), there are three values that are critical for the decision: the minimum value in efficacy that is acceptable, the value considered to be acceptable to proceed to the next stage, and probabilities with these decisions. Besides specification of these acceptable values, other values will also be considered to evaluate the robustness of the results.

Let π_1 and π_2 denote the responder rate of the Test Drug and of the Placebo, respectively. An estimate of the Placebo response will be obtained using retrospective clinical trial data from Placebo controlled studies data (see Appendix A), and the resulting Placebo response will be used to estimate the beta parameters for the beta prior on the Placebo response. For the treatment group, hypothetical response can be assumed based on an expected outcome that the team would like to see from the trial. For example, the team would like to see a fivefold increase or a tenfold increase in the treatment response. Once this decision is made, an appropriate beta prior will be constructed with beta parameters estimated using the same approach as was used for the Placebo response.

We will use the binomial distributions for π_1 and π_2 and the prior distributions and the observed study data to investigate the posterior distribution for $\theta = \pi_1 - \pi_2$. As noted earlier, the posterior distribution of θ does not possess a straightforward and simple closed form solution that is mathematically tractable. We will thus evaluate the posterior distribution of θ using a simulation approach similar to Kawasaki et al. (2013).

9.5.1.4 Bayesian Analysis Using an MCMC Approach

As noted above, under a simple beta-binomial setup under independence, one can use a closed form to find posterior distribution. On the other if one decides to expand the model to include additional covariates, then models such as logistic regression can be used for further evaluation. Under this scenario since there does not exist a simple closed form solution for posterior distribution, the analysis will use MCMC methods. MCMC methods will be used as a means of sampling from the posterior distribution of $\theta = \pi_1 - \pi_2$. MCMC algorithms will enable calculation of the posterior distribution when it is not in a closed form. Three chains will be run, with three diverse sets of initial values. As part of this simulation, 15,000 samples will be generated, and the first 5,000 will be discarded as burn-in samples.

Simulation diagnostics will be used to ensure that the simulation has been adequately performed and also ascertain the validity of inferences drawn. This will include assessment of convergence to ensure that the chain has successfully targeted the stationary distribution, mixing to ensure that the chain moves through in the support of the posterior distribution rapidly, and sampling intensity to ensure that enough samples have been used to adequately describe the posterior distribution. More specifically, the following visual plots will be used in assessing simulation diagnostics:

Trace plots: these plots document the magnitude of the sample drawn at each iteration of the MCMC procedure. Once the chain has identified the stationary distribution of samples, the samples that are drawn will appear to have been randomly sampled from the same region. Trace plots provide an important tool for assessing mixing of a chain.

Autocorrelation plots: these plots document the correlation of samples at each step of the chain with previous estimates of that same variable, lagged by some number of iterations. Ideally, the autocorrelation declines rapidly, so that it is eventually possible to be confident that the samples from the stationary distribution can be thought of as random and not reliant on initial values in the chain.

Density plots: these plots provide a summary of the sampled values that define the stationary distribution of values, which approximates the posterior distribution of interest. Density plots are smoothed histograms of the samples; that is, they show the posterior function that we are trying to explore and use for inference purposes.

In addition to the visual plots discussed above, a Gelman–Rubin diagnostic check is often used to assess model convergence. This diagnostic involves running multiple MCMC procedures, specifying the same model and prior information, from different starting values, and comparing the variance within each chain with the variance between chains. Lack of model convergence is indicated when the variance between chains is larger than the variance within chains. This can result from multimodal problems in which a chain has identified two or more distinct parameter distributions or from a high degree of correlation in the observed data. Although there are no rules for how many different chains to run, researchers often run the Gelman–Rubin diagnostic for three chains. If these results raise any concern over model convergence, the model may be rerun with small changes such as different initial values or an increased number of samples. Any changes to the planned analysis will be fully documented.

The simulation will involve running multiple MCMC algorithms more than once with different starting points to obtain multiple MCMC samples and checking to see if the same results are obtained and conducting checks to test whether the sample means or other sample moments are significantly different across the different MCMC samples and using trace plots to identify potential problems that may affect the MCMC sample. If there is any evidence that the chain has not converged by the end of the first 5,000 samples (see diagnostic details of how to check), the burn-in should be increased until the chains are consistent with convergence. If the MCMC error of any of the nodes of direct interest is greater than 5% of the posterior standard deviation of that node, then, assuming convergence has occurred, the number of samples post burn-in will be increased so that this MCMC error criterion is met. All other nodes monitored should be an equivalent rule but with a 10% margin. The final trace plots, Brooks–Gelman–Rubin diagnostic plot, and autocorrelation plots will be outputs for inclusion in the study report.

9.5.1.5 Criteria and Decision Rules

It is already known that the existing treatment works fairly well and has a clinical meaningful benefit compared to Placebo. In order to justify moving forward with the Test Drug, it should be shown that the new treatment works at least as well as the existing treatment. Analysis of data from the alternative treatment shows that this treatment has a difference compared with Placebo of around 40% points in the PASI75 rates. Thus, the POC criteria need to consider both clinical significance and relevance. Additional improvement percentages in PASI75 will also need to be considered to evaluate the robustness of results.

9.5.1.6 Assessment Details

At each interim analysis, the Bayesian posterior probability $P(\Delta \mid data)$ will be calculated, where $\Delta \geq 0$ is a pre-specified number. Assuming $\Delta = 0$, the following dual criteria similar to Fisch et al. (2015), for significance and relevance, will be used:

- Significance: $P(\Delta \geq 0 \mid data) \geq 90\%$.
- Relevance: $P(\Delta \geq 0.4 \mid data) \geq 50\%$.

In addition, it is also recommended to consider computing posterior probabilities for constructing three decision zones: Go, No Go, and Indeterminate.

For example, if $\theta = 40\%$, then the zones can be defined as follows:

- Go: if $P(\Delta \geq 0.4 \mid data) \geq 80\%$.
- No Go: if $P(\Delta \geq 0.4 \mid data) \leq 20\%$.
- Indeterminate: $20\% < P(\Delta \geq 0.4 \mid data) < 80\%$.

Note that one can also adopt other approaches such as the one discussed in Sverdlov et al. (2015).

9.5.1.7 Decision Rules

A decision will be made on whether to continue to the next phase of development based on the Bayesian posterior probability and the totality of data taking into account both significance and relevance as noted above and considerations for Go, No Go, and Indeterminate.

9.5.1.8 Information for the Prior Distribution of PASI75 for the Placebo

The data for construction of the prior were derived from historical control data prior to the analysis. The data are part of the Quantify Psoriasis clinical database which is a subset of the Quantify RA clinical database. Included in the Quantify RA clinical database are all randomized controlled trials that provide information on clinical safety and efficacy on all biologics as well as newer synthetic disease-modifying antirheumatic drugs (DMARDs) currently approved or in development for rheumatoid arthritis (RA), psoriatic RA, and psoriasis. For the purpose of this study, information on older treatment options methotrexate (MTX and other DMARDs) was included if they were used as active controls. Appendix A shows the selection of the studies leading to the subset of studies selected in the construction of the prior.

9.5.1.9 Information for Prior Distribution of PASI75 for the Test Drug

The data for construction of the prior for the Test Drug will be based on data from Week 12 values of PASI75 from similar drugs from the literature, for example, Checchio et al. (2017).

9.6 An Example of an Implementation of the Checklist Using the Hypothetical Example Protocol

9.6.1 Statistical Analysis Plan

A Bayesian analysis will be used to estimate the posterior distribution of the treatment effect. The primary efficacy analysis will combine the data from studies with prior beliefs based on historical data about the PASI75 Placebo response at Week 12 which will be compared to the active drug. For the active drug, the prior estimate for PAS175 will be obtained from PAS175 values at Week 12 based on data from similar drugs. The analysis will focus on the posterior distribution of $\theta = \pi_1 - \pi_2$. The setup will be as follows: let π_1 and π_2 represent the PASI75 from the Test Drug and Placebo, respectively. Let X_1 and X_2 be the random variables representing the observed PASI75 responses for the Test Drug and Placebo, respectively, and let n_1 and n_2 be the sample sizes for the Test Drug and Placebo, respectively. Then we have that $X_i \sim \text{Bin}(n_i, \pi_i), i = 1, 2$, where Bin is the Binomial distribution with parameters n and π. Under the Bayesian setting, it is natural to model each of the π_i's as a Beta distribution with parameters α_i and β_i or simply $\pi_i \sim \text{Beta}(\alpha_i, \beta_i),\ i = 1, 2$, which we will use for the prior distribution for each π_1. Using Bayes rule and under conjugacy of the Binomial and Beta distribution, it is fairly straightforward to show that the posterior distribution of each one of $\pi_i | X_i = x_i$ is also Beta distribution, specifically, $\pi_i \mid X_i \sim \text{Beta}(\alpha_i + x_i, \beta_i + n_i + x_i),\ i = 1, 2$.

For analysis purposes, the quantity of interest for decision-making purposes is $\theta = \pi_1 - \pi_2$. Specifically, we want to obtain the posterior distribution of $\theta = \pi_1 - \pi_2$ and use this to address various analysis questions, inference, and decision making. Although some closed forms for the posterior distributions for θ are available, they are not mathematically easy to work with or may require complex programming. Thus, the posterior distribution of θ will be evaluated using a simulation MCMC approach. From this setup, we have that $\theta = \pi_1 - \pi_2 \mid X_1, X_2$ is just another distribution with updated parameters. An estimate for θ can be obtained from the posterior distributions of π_1 and π_2 independently by subtracting $\pi_1 - \pi_2$ to obtain posterior samples of θ.

Analyses of the primary endpoint will be performed on both the Per Protocol and the Full Analysis Set data. The model will be fit using Bayesian analysis software. Average responses will be determined based on these groups of studies, and the beta parameters will be generated based on these average responses.

Suggested elicitation methods can also be used, see, for example, Johnson (2011).

- Occasion and interval method (modal response and a known interval that contains the expected response rate) – Location alone – Mean absolute deviation as the spread

The primary efficacy analysis will also be repeated using a diffuse prior distribution for the Placebo response. Treatment response (as a binomial likelihood) will be simulated based on available literature on PASI75 score for various compounds. The inference on the two responses can then use the following explanation.

9.6.2 Research Questions

The Bayesian analysis that will be used is intended to address the question as to whether there is sufficient evidence of efficacy of the drug relative to Placebo to move forward with development based on the difference $\theta = \pi_1 - \pi_2$, taking into account prior knowledge of the Placebo response. The significance and relevance criteria specified earlier will be used for assessment purposes in the assessment of efficacy along with other criteria and assumptions. Specifically, the following questions will be addressed:

- What are the posterior summaries and how do they compare to assumptions?
- Do the posterior probabilities meet pre-specified thresholds?
- Do the model diagnostics support results?
- Are the results sufficiently robust?
- How do the results compare to the frequentist approach?
- Are decisions made sound and supported by sufficient evidence to further the development to next phase?
- Are there any notable observations from the analysis that may suggest some adaptation?

9.6.3 Motivation for Use of Bayesian Method

The main advantage of the Bayesian approach over the frequentist approach is that it allows incorporation of prior knowledge by specifying appropriate prior knowledge via prior probabilities. Bayesian methods are especially useful for statistical inference of complicated models which may present significant difficulties for frequentist methods where decisions need to be made with probability assignments, as in the present study. Results of frequentist hypothesis tests are usually expressed as p-values which would be difficult to use in the decision criteria such as in decision setup above. The Bayesian approach provides a statistical framework that updates estimates of parameters with emerging data. A Bayesian thinks of parameters as random quantities with distributions, as well as the data. The prior information for the parameters (could be vague or come from external sources) is combined with information provided in data to obtain the posterior distribution of the parameters. The statistical inference will be based on the posterior distribution of the parameters. In this specific setting, the Bayesian approach is considered most appropriate as it facilitates a framework to appropriately address questions in the previous section along with appropriate assessment of significance and relevance.

9.6.4 Data Source for and Prior Distribution of the Placebo PASI75 Response

Data to be used in the construction of the prior were derived from historical control data prior the analysis. The data are part of the Quantify Psoriasis clinical database which is a subset of the Quantify RA clinical database. Included in the Quantify RA clinical database are all randomized controlled trials that provide information on clinical safety and efficacy on all biologics as well as newer synthetic DMARDs currently approved or in development for RA, psoriatic RA, and psoriasis. For the purpose of this study, information on older treatment options (MTX and other DMARDs) was included if they were used as active controls. Appendix A shows the selection of the studies leading to the subset of studies selected in the construction of the prior.

Among those studies that were identified in the initial search, additional studies were also not included in the construction of the prior on account of the drug dosing route, study duration, small sample size, different endpoints, drug dosing frequency, timing, imputation method, potential outliers, and different indications. See Appendix A for details.

9.6.5 Prior Estimate of the Prior for Placebo PASI75 Response via Meta-Analysis

To get estimates of α_2 and β_2 for the prior distribution of the Placebo PASI75 response π_2, a random effects meta-analysis (Appendix E) was performed on the 24 qualifying studies out of the 39 studies to get an estimate of π_2 and its 95% confidence limit interval (CI), $[L, U]$, see, for example, Avci (2017). The upper and lower limits of the CI for π_2, L and U, respectively (i.e., $\pi_2 = 0.0363$, $[L, U] = [0.0297, 0.0429]$), were used to obtain an estimate of the mean and variance of π_2 which in turn were used as input to the R code in Appendix B leading to $\alpha_2 = 14.53966$ and $\beta_2 = 386.002$ for the Beta prior density of π_2.

9.6.6 Data Source and Prior Distribution for the Test Drug PASI75 Response

Under the assumption that the Test Drug works well compared to Placebo and is no better or worse than the drugs in the same therapeutic class, a prior estimate of PASI75 response π_1 for the Test Drug, was obtained from PASI75 responses of various drugs in the same class from Checchio et al. (2017) (Table 1, p. 1009). To estimate α_1 and α_2 for the prior Beta distribution of π_1, for the Test Drug, the upper and lower limits of the CIs for PASI75 at Week 12 were summed up across all drugs to get average values for the lower limit L and the upper limit U. These were in turn used to estimate the mean and variance of π_1. Using the R code in Appendix E led to $\alpha_1 = 1.794813$ and $\beta_1 = 1.424987$.

9.6.7 Observed PASI75 Response from the Study at Week 12

For illustration purposes, we assume that once the study had been run, the observed values at Week 12 were as follows: $x_1 = 110, n_1 = 150, x_1 = 6$, and $n_1 = 150$ (can assume different values if necessary). Together with the prior distributions of π_1 and π_2, these were used to obtain the posterior distribution of $\theta = \pi_1 - \pi_2$ via MCMC. Aside and as noted earlier that from these values and the prior estimated for π_1 and π_2, closed forms of the posterior for θ can be used for inference purposes and decision making. For the purpose of illustration of the reporting considerations discussed in this paper, an MCMC will be used to obtain the posterior distribution of θ.

9.6.8 Computation Methods and Software Used

To obtain the posterior distribution of $\theta = \pi_1 - \pi_2$, an MCMC was used along with the observed values of $X_1 = x_1, X_2 = x_2$ and $N_1 = n_1$ and $N_2 = n_2$ from the study at the final analysis using the values of α and β for each one of π_1 and π_2. Since π_1 and π_2 are independent, the posterior estimate of θ was obtained by the difference in the updated posterior estimated for π_1 and π_2. The computations were done using WinBUGS using the code in Appendix D. Code for STAN and JAGS is also provided.

9.6.9 Diagnostics Assessment of the MCMC Simulation

The simulation outputs are provided in Figure 9.1 from WinBUGS, and an assessment of the diagnostics is provided considering various considerations.

The trace plots for each parameter of interest are provided in Figure 9.1. As can be noted from the trace plots, the magnitudes of the samples drawn at each iteration of the MCMC procedure exhibit a fairly stationary distribution of samples and appear to be randomly sampled from the same region suggesting good mixing of a chain.

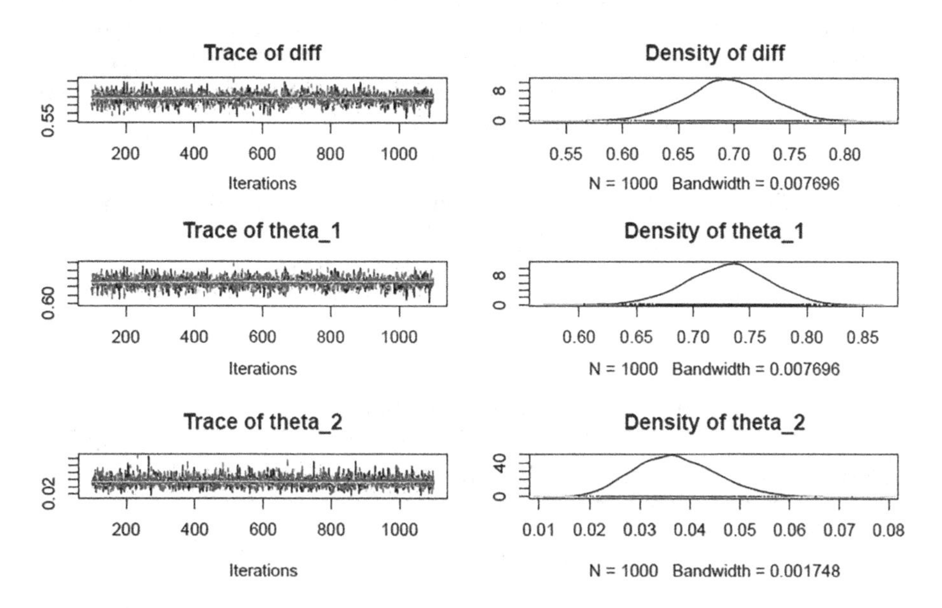

FIGURE 9.1
Convergence, assessment, and diagnostics.

9.6.10 Possible Limitations of the Analysis

Though there are no limitations here except for assessing priors, limitations can occur for complex models that may lead to complexity in computations. There are three areas of limitations that have been well explored in the literature: selection of priors, influence of priors on posterior distributions, and computational complexity. But, over the years, these limitations have been well addressed and dealt with by the practitioners that include use of elicitation methods of prior selection, sensitivity analysis based on different priors and use of powerful software tools, respectively.

As aptly described in the FDA Guidance (2006), use of Bayesian analysis in drug development requires well thought pre-plannig of design, conduct, and analysis of any trial. This is crucial since decisions are made with a Bayesian approach at the design stage regarding priors, the information obtained from the current trial, and the mathematical model (conjugate or otherwise) needed to combine the two. It is important to acknowledge that any change in the prior information during the conduct of the trial may influence the scientific validity of the final trial results. Besides extensive mathematical modeling and computational intensity, use of a Bayesian approach may also require maintaining confidentiality of the data needs (for example, revelation of prior and posterior results), which in turn, may have an impact on ethical considerations. Thus, if adaptive designs are used, then careful planning will be required in order to avoid and minimize any operational bias.

The summary statistics for the posterior distribution of each one of π_1 and π_2 and $\theta = \pi_1 - \pi_2$ are provided in Table 9.2. The mean PAS175 response for Test Drug (π_1) and that of the Placebo (π_2) are 0.7301 and 0.03748, respectively. The average value for $\theta = \pi_1 - \pi_2$ is 0.6926. Both $P(\Delta \geq 0 \mid data) \geq 90\%$ and $P(\Delta \geq 0.4 \mid data)$ are greater than 99.9% showing that both clinical significance and relevance criteria are met.

9.6.11 Describe Posterior Distribution of Parameters and Other Summaries

The density plots for each parameters of interest were provided in Figure 9.1.

TABLE 9.2
Descriptive Summary of the Posterior Distribution

Parameter	Mean	Standard Deviation	2.5% Percentile	Median	97.5% Percentile
π_1	0.7	0.01	0.6	0.7	0.8
π_2	0.01	0.01	0.01	0.01	0.1
θ	0.7	0.01	0.6	0.7	0.8
$P(\theta \geq 0 \mid data)$	99.9	99.9	99.9	99.9	99.9
$P(\theta \geq 0.4 \mid data)$	99.9	99.9	99.9	99.9	99.9

9.6.12 Bayesian Result and Interpretation and Decision Making

Based on the posterior summaries of π_1, the PAS175 for Test Drug, and π_2, the PAS175 for Placebo posterior distribution of $\theta = \pi_1 - \pi_2 = 0.6922$, we have Tables 9.3 and 9.4 for PoC decision making.

Observed PAS122 Value	Prior Distribution for π_1 and π_1	Significance $P(\theta \geq 0 \mid data)$ $> 90\%$	Relevance $P(\theta \geq 0.4 \mid data)$ $> 50\%$	Decision Zone for 40%
$x_1 = 110$	π_1			
$n_1 = 150$	$\alpha_1 = 14.5396$	$99.9 > 90\%$	$99.9 > 90\%$	Go
	$\beta_1 = 386.00$			
$x_2 = 6$	π_1			
$n_2 = 150$	$\alpha_2 = 1.794813$			
	$\beta_2 = 1.424987$			

Decision Zone for 40%

- **Go: if** $P(\Delta \geq 0.4 \mid data) \geq 80\%$.
- **No Go: if** $P(\Delta \geq 0.4 \mid data) \leq 20\%$.
- **Indeterminate:** $20\% < P(\Delta \geq 0.4 \mid data) < 80\%$.

Both significance and relevance criteria were met along with the decision zone for a 40% difference suggesting PoC was met.

9.6.13 Sensitivity Analysis

In order to assess the robustness of the results to the choice of the prior, non-informative Uniform priors (Beta(1, 1)) were also used for both π_1 and π_2. The key summary data for the posterior distributions are provided in Tables 9.3 and 9.4.

TABLE 9.3
Descriptive Summary of the Posterior Distribution

Parameter	Mean	Standard Deviation	2.5% Percentile	Median	97.5% Percentile
π_1	0.7	0.01	0.6	0.7	0.8
π_2	0.01	0.01	0.01	0.01	0.1
θ	0.7	0.01	0.6	0.7	0.8
$P(\theta \geq 0 \mid data)$	99.9	99.9	99.9	99.9	99.9
$P(\theta \geq 0.4 \mid data)$	99.9	99.9	99.9	99.9	99.9

TABLE 9.4
Significance and Relevance

Observed PAS122 Value	Prior Distribution for π_1 and π_1	Significance $P(\theta \geq 0 \mid data)$ $> 90\%$	Relevance $P(\theta \geq 0.4 \mid data)$ $> 50\%$	Decision Zone for 40%
$x_1 = 110$ $n_1 = 150$	π_1 $\alpha_1 = 1$ $\beta_1 = 1$	$99.9 > 90\%$	$99.9 > 90\%$	Go
$x_2 = 6$ $n_2 = 150$	π_1 $\alpha_2 = 1$ $\beta_2 = 1$			

Decision Zone for 40%

- **Go: if** $P(\Delta \geq 0.4 \mid data) \geq 80\%$.
- **No Go: if** $P(\Delta \geq 0.4 \mid data) \leq 20\%$.
- **Indeterminate:** $20\% < P(\Delta \geq 0.4 \mid data) < 80\%$.

From the above sensitivity analysis, it can be seen that the results are reasonably robust with regard to the choice of the prior.

9.6.14 Discussion and Conclusion

In this chapter, it is argued that despite the wide variety of books and numerous journal articles written on Bayesian approaches in the analysis of data, guidance regarding reporting of these analyses is lacking particularly as this pertains to clinical research. The sparse available literature on reporting of Bayesian analyses may preclude comprehensible Bayesian analyses to the reader, especially those not familiar with the approach. A variety of these key considerations for good practice in Bayesian reporting are discussed, and a checklist taking into account several key aspects of Bayesian analysis is proposed that can be useful in ensuring standardization of the reporting of Bayesian analyses. A simple application of the checklist is provided using a hypothetical example. It is hoped that the proposed checklist along with the example provided will help provide some guidance towards standardization of reporting of Bayesian analyses.

9.7 Appendices

9.7.1 Appendix A: Table for Prior Information

Year	Trial Type	Treatment Arms	Randomized	Treated	Dosing Frequency	Route	Indication	Endpoint	Duration	Imputation	Percent Response
2012	Efficacy	1	38	38	0,1,2,q2w	sc	Psoriasis	PASI75	12	Unknown	0
2008	Unknown	3	58	58	q2w	sc	Psoriasis	PASI75	12	NRI	6.8
2002	Efficacy	1	186	186	qw	iv	Psoriasis	PASI75	14	Unknown	3.8
2003	Efficacy	1	NA	168	qw	im	Psoriasis	PASI75	14	Unknown	5
2010	Efficacy	2	18	18	sd	iv	Psoriasis	PASI75	4	Unknown	0
2008	Unknown	1	53	53	q2w	sc	Psoriasis	PASI75	16	LOCF	18.9
2008	Unknown	1	53	53	q2w	sc	Psoriasis	PASI75	16	NRI	18.9
2001	Unknown	1	11	11	0,2,6	iv	Psoriasis	PASI75	10	NRI	18.2
2005	Unknown	1	77	76	0,2,6,q8w	iv	Psoriasis	PASI75	10	NRI	2.6
2007	Unknown	1	208	207	0,2,6	iv	Psoriasis	PASI75	10	NRI	1.9
2001	Efficacy	1	59	59	qw	iv	Psoriasis	PASI75	100	Unknown	10
2003	Unknown	1	NA	55	biw	sc	Psoriasis	PASI75	12	LOCF	1.8
2007	Efficacy	1	4	4	sd	sc	Psoriasis	PASI75	24	Unknown	0
2012	Efficacy	1	27	27	0,2,4,q4w	sc	Psoriasis	PASI75	12	LOCF	7.7
2008	Unknown	1	30	30	qw	sc	Psoriasis	PASI75	12	NRI	3.3
2003	Unknown	1	168	166	biw	sc	Psoriasis	PASI75	12	LOCF	3.6
2006	Unknown	1	52	52		sc	Psoriasis	PASI75	12	NRI	4
2010	Efficacy	4	46	46	q2w	sc	Psoriasis	PASI75	16	NRI	4.3
2012	Efficacy	2	484	484	0,4,8	sc	Psoriasis	PASI75	12	NRI	4.5
2011	Efficacy	1	68	68	0,4,8,biw	sc	Psoriasis	PASI75	12	NRI	7.4
2011	Efficacy	1	72	72	0,4,8,biw	sc	Psoriasis	PASI75	12	NRI	6.9
2011	Efficacy	1	60	60	0,4	sc	Psoriasis	PASI75	12	NRI	5
2008	Unknown	3	255	255	0,4,q12w	sc	Psoriasis	PASI75	12	NRI	3.1
2008	Unknown	3	410	410	0,4,q12w	sc	Psoriasis	PASI75	12	NRI	3.7
2005	Unknown	1	204	193	biw	sc	Psoriasis	PASI75	12	NRI	2.9
2005	Unknown	1	204	193	biw	sc	Psoriasis	PASI75	12	LOCF	3.1
2008	Unknown	1	398	398	q2w	sc	Psoriasis	PASI75	16	NRI	6.5

(Continued)

Year	Trial Type	Treatment Arms	Randomized	Treated	Dosing Frequency	Route	Indication	Endpoint	Duration	Imputation	Percent Response
2004	Unknown	1	51	51	0,2,6	iv	Psoriasis	PASI75	10	NRI LE/ LOCF missing	5.9
2007	Unknown	1	64	67	sd	sc	Psoriasis	PASI75	12	LOCF	1.6
2007	Unknown	1	64	67	sd	sc	Psoriasis	PASI75	12	Unknown	1.1
2006	Unknown	1	309	307	biw	sc	Psoriasis	PASI75	12	LOCF	4.9
2008	Unknown	1	NA	46	qw	sc	Psoriasis	PASI75	12	LOCF	2.2
2012	Efficacy	1	88	88	bid	oral	Plaque psoriasis	PASI75	16	LOCF	5.7
2012	Efficacy	5	22	22	q4w	sc	Plaque psoriasis	PASI75	12	LOCF	9.1
2012	Efficacy	4	50	NA	bid	oral	Plaque psoriasis	PASI75	12	LOCF	2
2012	Efficacy	4	50	NA	bid	oral	Plaque psoriasis	PASI90	12	LOCF	0.77
2012	Efficacy	4	50	NA	bid	oral	Plaque psoriasis	PASI75	12	NRI	2
2012	Efficacy	1	45	45	0,2,6,q8w	iv	Plaque psoriasis	PASI75	10	Unknown	2.2
2013	Efficacy	1	282	282	bid	oral	Psoriasis	PASI75	16	LOCF	5.3

9.7.2 Appendix B: Beta-Binomial Bayesian Setup

- $Y \sim \text{Binomial}(n, \theta)$
- **Prior**

$$\begin{aligned} \theta &\sim \text{Beta}(\alpha, \beta) \\ \Rightarrow & P(\theta) = \frac{\Gamma(\alpha+\beta)}{\Gamma(\alpha)\Gamma(\beta)} \theta^{\alpha-1}(1-\theta)^{\beta-1}, \\ & \alpha, \beta > 0, \quad \theta \in [0,1] \end{aligned}$$

- Also we have

$$E(\theta|\alpha,\beta) = \frac{\alpha}{\alpha+\beta} \text{ and } Var(\theta|\alpha,\beta) = \frac{\alpha\beta}{(\alpha+\beta)^2(\alpha+\beta+1)}$$

Posterior

- From Bayes theorem, i.e., $P(\theta|y) \propto P(y|\theta)P(\theta)$. So

$$\begin{aligned} P(\theta|y) &\propto \binom{n}{y}\theta^y(1-\theta)^{n-y} \times \frac{\Gamma(\alpha+\beta)}{\Gamma(\alpha)\Gamma(\beta)}\theta^{\alpha-1}(1-\theta)^{\beta-1} \\ &= \binom{n}{y}\frac{\Gamma(\alpha+\beta)}{\Gamma(\alpha)\Gamma(\beta)}\theta^y(1-\theta)^{n-y} \times \theta^{\alpha-1}(1-\theta)^{\beta-1} \end{aligned}$$

- Note that

$$\theta^y(1-\theta)^{n-y} \times \theta^{\alpha-1}(1-\theta)^{\beta-1} \quad \text{(called the Kernel)}$$

- The Kernel can be written as

$$\theta^{y+\alpha-1}(1-\theta)^{n-y+\beta-1} \equiv \text{Beta}(y+\alpha, n-y+\beta)$$

- So, the Posterior is

$$\theta|y \sim \text{Beta}(y+\alpha, n-y+\beta)$$

- It follows that

$$\begin{aligned} E(\theta|y) &= \frac{y+\alpha}{n+\alpha+\beta} \\ Var(\theta|y) &= \frac{(y+\alpha)(n-y+\beta)}{(n+\alpha+\beta)^2(n+\alpha+\beta+1)} \end{aligned}$$

We can write this simply as

- **Model**

$$p(y|\theta) \propto \theta^y(1-\theta)^{n-y}$$

- **Prior**

$$p(\theta) \propto \theta^{\alpha-1}(1-\theta)^{\beta-1}$$

- **Posterior**

$$p(\theta|y) \propto \theta^{\alpha+y-1}(1-\theta)^{\beta+n-y-1}$$

9.7.3 Appendix C: Estimating the Parameters of the Beta Distribution

Given the mean and variance of π, the parameters α and β can be estimated from the formula for the mean and variance of the Beta distribution shown in Appendix A. The R code below can be used to solve for α and β.

```
EstBetaParams <- function(mu, var) {
                 {alpha <- ((1 - mu) / var - 1 / mu) * mu ^ 2
                 beta <- alpha * (1 / mu - 1)
                 return(params = list(alpha = alpha, beta = beta))
                 }
```

Also if the modal value of is provided along with the 100% chance that the response rate falls in the interval $[U, L]$, then α and β can also be estimated as follows: suppose the modal response is denoted as m and a 100% chance that the response rate falls between $[U, L]$. Then, the following two equations need to be solved iteratively to get values for (α, β), the beta parameters:

$$\frac{\alpha - 1}{\alpha + \beta - 2} = \hat{m}, \quad \int_L^U \frac{\Gamma(\alpha + \beta)}{\Gamma(\alpha)\Gamma(\beta)} r^{\alpha-1}(1-r)^{\beta-1} dr = \mu.$$

For example, the mode of the distribution of a response is 0.95, and we also assume that the chance that the response is less than 80% is very small. For this special case, the two equations to be solved can be written as

$$\frac{\alpha - 1}{\alpha + \beta - 2} = 0.95 \implies \alpha = 19\beta - 18,$$

$$\int_0^{18} \frac{\theta^{\alpha-1}(1-\theta)^{\beta-1}}{\text{Beta}(\alpha, \beta)} d\theta = 0.00001 \implies \int_0^{18} \frac{\theta^{19\beta-18}(1-\theta)^{\beta-1}}{\text{Beta}(19\beta - 18, \beta)} d\theta - 0.00001 = 0.$$

The R code below can be used in the above setting to solve for α and β.

```
fun <- function(b) {
       answer <- pbeta(0.8, ((19*b)-18),b)-0.0001
       return(answer)
       }
       uniroot(fun,lower=1, upper=100)
          # $root will provide the value of b
       fun1 <- uniroot(fun,lower=1,upper=100)
       a <- fun1$root*19-18
       print(a)
       }
```

9.7.4 Appendix D: Code for Meta-Analysis of Placebo Data

```
library(meta)
m1 <- metaprop(4:1, c(10, 20, 30, 40))
N <- c(38,58,76,55,4,27,30,166,52,46,484,68,72,60,255,410,
       193,193,398,67,67,307,46,22)
Resp <- round(N*(c(0,6.8,2.6,1.8,0,7.7,3.3,3.6,4,4.3,4.5,7.4,
       6.9,5,3.1,3.7,2.9,3.1,6.5, 1.6,1.1,4.9,2.2,9.1))/100)
m3 <- update(m1, sm="PRAW")
forest(m3)
```

9.7.5 Appendix E: Estimate of Alpha and Beta for the Test Drug

```
#Test Drug Estimate
MeanLower <-  c(53.6, 62.2,20.7,34.2,62.6,69.0,71.7,57.0,67.2,
                51.9,17.0,28.2,48.1,7.09,17.9)/100
MeanUpper <-  c(76.8,84.1,44.0,60.6,85.8,89.2,90.7,79.5,87.9,
                76.4,40.0,52.5,72.5,24.1,39.8)/100
estBetaParams <- function(mu, var)
{alpha <- ((1 - mu) / var - 1 / mu) * mu ^ 2
beta <- alpha * (1 / mu - 1)
return(params = list(alpha = alpha, beta = beta))
}
estBetaParams(mean(c(MeanLower,MeanUpper)),
              var(c(MeanLower,MeanUpper)))
```

9.7.6 Appendix F: Code for MCMC Computation for $\theta = \pi_1 - \pi_2$

WinBUGS Code

```
library(R2WinBUGS)
library(coda)

#data
mydata <- list(s_1=110,n_1=150,s_2=6,n_2=150)

# initialization
inits1 <- list(theta_1=0.6,theta_2=0.3)
inits2 <- list(theta_1=0.9,theta_2=0.7)
inits3 <- list(theta_1=0.5,theta_2=0.1)
inits <- list(inits1, inits2, inits3)

#Without using the coda package
```

```
post<-bugs(mydata,inits=inits,model.file=paste
                       (path,"bugsBetaBinomial.txt", sep=""),
          parameters=c("theta_1", "theta_2", "diff", "P",
                       "P_40"), n.chains=3, n.thin=100,
          n.burnin=1000, n.iter=10000, debug=F, codaPkg=F,
          # bugs.directory="C:/Tmp/WinBUGS14")
          bugs.directory="C:/Program Files (x86)/WinBUGS14")

hist(post$sims.list$theta_1,xlab="theta_1", probability = T,
     col="yellow",
     main="Posterior distribution of theta_1",cex.main=.8 )

hist(post$sims.list$theta_2,xlab="theta_2", probability = T,
     col="yellow",
     main="Posterior distribution of theta_2",cex.main=.8 )

hist(post$sims.list$diff,xlab="diff", probability = T,
     col="yellow",
     main="Posterior distribution of diff",cex.main=.8 )

hist(post$sims.list$P, xlab="P", probability = T, col="lightblue",
     main="Posterior distribution of P",cex.main=.8)

hist(post$sims.list$P_40, xlab="P_40", probability = T,
     col="lightblue",
     main="Posterior distribution of P",cex.main=.8)

print(post)
plot(post)
plot(density(post$sims.list$theta_1))
plot(density(post$sims.list$theta_2))
plot(density(post$sims.list$diff))

#Using the coda package

post<-bugs(mydata,inits=inits,model.file=paste(
                    path,"bugsBetaBinomial.txt", sep=""),
          parameters=c("theta_1", "theta_2", "diff", "P",
                    "P_40"), n.chains=3, n.thin=1,
          n.burnin=100, n.iter=1100, debug=F, codaPkg=T,
          # bugs.directory="C:/Tmp/WinBUGS14")
          bugs.directory="C:/Program Files (x86)/WinBUGS14")

coda.post <-read.bugs(post)
```

```
plot(coda.post)
autocorr.plot(coda.post)
model {
s_1 ~ dbin(theta_1,n_1)
s_2 ~ dbin(theta_2,n_2)
theta_1 ~ dbeta(1.794813,1.424987)# informative priors
theta_2 ~ dbeta(14.53966,386.002)
diff <- theta_1-theta_2           # difference in response rates
P <- step(diff)                   # Pred prob that
                                    difference >=0
P_40 <- step(diff-0.40)           # Pred prob that
                                    difference-0.40 >=0
}
```

Can replace above with beta(1,1) for the non-informative prior

R JAGS Code

```
library("runjags")
n_chains <-3

results_jags <- run.jags("model_jags_informative_priors.txt",
                         monitor = c("theta_1", "theta_2",
                         "diff"), n.chain = n_chains,
                         burnin = 1000, sample = 4000,
                         adapt = 1000, method = "rjparallel")
results_jags
plot(results_jags)
summary(results_jags)
coda::effectiveSize(results_jags)
results_mcmc = as.mcmc.list(results_jags)
autocorr.plot(results_mcmc[, -3])
gelman.diag(results_mcmc[, -3])

library(mcmcplots)
denplot(results_mcmc[, -3])
traplot(results_mcmc[, -3])

# Model model_jags_informative_priors.txt file
model {
s_1 ~ dbin(theta_1,n_1) # Model the data
s_2 ~ dbin(theta_2,n_2)
theta_1 ~ dbeta(1.794813,1.424987) # informative priors
theta_2 ~ dbeta(14.53966,386.002)
diff <- theta_1-theta_2 # difference in response rates
P <- step(diff) # Pred prob that difference >=0
```

```
P_40 <- step(diff-0.40) # Pred prob that difference-0.40 >=0
}
# data
data {
list(s_1=110,n_1=150,s_2=6,n_2=150)
}

# initialisation values for 3 chains
inits{
list(theta_1=0.6,theta_2=0.3)
}
inits{
list(theta_1=0.9,theta_2=0.7)
}
inits{
list(theta_1=0.5,theta_2=0.1)
}
```

STAN Code

```
library(rstan)
rstan_options(auto_write = TRUE)
options(mc.cores = parallel::detectCores())
# set-up initial values for current run
n_chains <- 3
init_3 <- list(list(theta_1 = 0.6, theta_2 = 0.3),
         list(theta_1 = 0.9,theta_2 = 0.7), list(theta_1 = 0.5,
         theta_2 = 0.1))
# default for warmup is iter/2, default iter is 2000
results_stan_tp <- stan(file =
                        "model_tp_40_informative_priors.stan",
                        data = list(s_1 = 110, n_1 = 150,
                         s_2 = 6, n_2 = 150),
                        pars = c("theta_1", "theta_2", "diff",
                        "diff_40"),
                        init = init_3, chains = n_chains,
                        iter = 6000, warmup = 2000)
results_stan_tp
library(shinystan)
results_stan_tp_shinystan <- as.shinystan(results_stan_tp)
results_stan_tp_aaa <- launch_shinystan
                                  (results_stan_tp_shinystan)
save(results_stan_tp_aaa, file = "results_stan_tp.RData")
load("results_stan_tp.RData")
launch_shinystan(results_stan_tp_aaa)
# plotting posterior distribution of the two response rates
```

```
post_diff <- As.mcmc.list(results_stan_tp, pars = c("theta_1",
                                                    "theta_2"))
library(coda)
plot(post_diff)
# Model model_tp_40_informative_priors.stan file
model {
s_1 ~ dbin(theta_1,n_1) # Model the data
s_2 ~ dbin(theta_2,n_2)
theta_1 ~ dbeta(1.794813,1.424987) # informative priors
theta_2 ~ dbeta(14.53966,386.002)
diff <- theta_1-theta_2 # difference in response rates
P <- step(diff) # Pred prob that difference >=0
P_40 <- step(diff-0.40) # Pred prob that difference-0.40 >=0
}
# data
data {
list(s_1=110,n_1=150,s_2=6,n_2=150)
}
# initialisation values for 3 chains
inits{
list(theta_1=0.6,theta_2=0.3)
}
inits{
list(theta_1=0.9,theta_2=0.7)
}
inits{
list(theta_1=0.5,theta_2=0.1)
}
```

Note: The same files can be used for sensitivity analyses with priors replaced by non-informative priors of $\alpha = 1$ and $\beta = 1$.

References

Anderson, D. R., W. A. Link, D. H. Johnson, and K. P. Burnham. 2001. Suggestions for presenting the results of data analysis. *J. Wildlife Manage.* **65**: 373–78.

Avci, E. 2017. Using informative prior from meta-analysis in Bayesian approach. *J. Data Sci.* **15**: 575–88.

BaSiS. 2014. The BaSiS Group - Bayesian standards in science. Standards for reporting of Bayesian analysis in the scientific literature. http://lib.stat.cmu.edu/bayesworkshop/2001/BaSis.html.

Bitt, J. A. and Y. He. 2017. Bayesian analysis: A practical approach to interpret clinical trials and create clinical practice guidelines. *Circ. Cardiovasc. Qual. Outcomes* **10**: 1–11.

Brophy, J. M. and L. Joseph. 1994. Placing trials in context using Bayesian analysis: GUSTO revisited by Reverend Bayes. *JAMA* **273**: 871–875.

Checchio, T., S. Ahadieh, P. Gupta, J. Mandema, L. Puig, R. Wolk, H. Valdez, et al. 2017. Quantitative evaluations of time-course and treatment effects of systemic agents for psoriasis: A model-based meta-analysis. *Clin. Pharmacol. Ther.* **102**. https://ascpt.onlinelibrary.wiley.com/doi/epdf/10.1002/cpt.732.

Efron, B. 1986. Why isn't everyone a Bayesian? *Am. Stat.* **40**: 1–5.

FDA. 2006. FDA – Guidance for industry and FDA staff – Guidance for the use of Bayesian statistics in medical device clinical trials. www.fda.gov/downloads/MedicalDevices/%20DeviceRegulationandGuidance/GuidanceDocuments/ucm071121.pdf.

Fisch, R., I. Jones, J. Jones, J. Kerman, G. K. Rosenkranz, and H. Schmidli. 2015. Bayesian design of proof-of-concept trials. *Ther. Innovation Regul Sci.* **49**: 155–62.

Gelman, A. 2008. Objections to Bayesian statistics. *Bayesian Anal.* **3**: 445–50.

Grieve, A. 2016. Idle thoughts of a 'well-calibrated' Bayesian in clinical drug development. *Pharma. Stat.* **15**: 96–108.

Hughes, M.D. 1991. Practical reporting of Bayesian analyses of clinical trials. *Drug Inform. J.* **25**: 381–93.

Hughes, M.D. 1993. Reporting Bayesian analyses of clinical trials. *Stat. Medic.* **12**: 1651–63.

Jenkinson, D. 2005. The elicitation of probabilities: A review of the statistical literature. https://pdfs.semanticscholar.org/e456/433ddb9c1848c31102afa30ca3ea479ce180.pdf.

Johnson, S. R. 2011. Bayesian inference: Statistical gimmick or added value? *J. Rheumatol.* **38**: 794–96.

Joseph, L., R. du Berger, and P. Belisle. 2015. Package SampleSizeProportions - Calculating sample size requirements when estimating the difference between two binomial proportions. https://cran.r-project.org/web/packages/SampleSizeProportions/SampleSizeProportions.pdf.

Kawasaki, Y. and E. Miyaoka. 2010. A posterior density for the difference between two binominal proportions and the highest posterior density credible interval. *J. Japan Stat. Soc.* **40**: 265–75.

Kawasaki, Y., A. Shimokawa, and E. Miyaoka. 2013. Comparison of three calculation methods for a Bayesian inference of $P(\pi_1 > \pi_2)$. *J. Mod. Appl. Stat. Methods.* DOI: 10.22237/jmasm/1383279240, http:// digitalcommons.wayne.edu/jmasm/vol12/iss2/15.

Lang, T. A. and M. Secic. 2006. *How to Report Statistics in Medicine: Annotated Guidelines for Authors.* Philadelphia, PA: American College of Physicians.

Metropolis, N., A. W. Rosenbluth, M. N. Rosenbluth, A. H. Teller, and E. Teller. 1953. Equations of state calculations by fast computing machines. *J. Chem. Phys.* **21**: 1087–92.

Ohlssen, D., K. L. Price, H. A. Xia, H. Hong, J. Kerman, H. Fu, G. Quartey, C. R. Heilmann, H. Ma, and B. P. Carlin. 2014. Guidance on the implementation and reporting of a drug safety Bayesian network meta-analysis. *Pharm. Stat.* **13**: 55–70.

Pham-Gia, T., N. V. Thin, and P. P. Doan. 2017. Inferences on the difference of two proportions: A Bayesian approach. *Open J. Stat.* **7**: 1–15.

Price, K. L., H. A. Xia, M. Lakshminarayanan, D. Madigan, D. Manner, J. Scott, J. D. Stamey, and L. Thompson. 2014. Bayesian methods for design and analysis of safety trials. *Pharm. Stat.* **13**: 13–24.

Pullenayegum, E. M. and L. Thabane. 2009. Teaching Bayesian statistics in a health research methodology program. *J. Stat. Educ.* **17**. ww2.amstat.org/publications/jse/v17n3/pullenayegum.html.

Pullenayegum, E. M., Q. Guo, and R. B. Hopkins. 2012. Developing critical thinking about reporting of Bayesian analyses. *J. Stat. Educ.* **20**: 1–15.

Rietbergen, C., T. P. A. Debray, I. Klugkist, K. J. K. M. Janssen, and K. G. M. Moons. 2017. Reporting of Bayesian analysis in epidemiologic research should become more transparent. *J. Clin. Epidemiol.* **86**: 51–8.

Robert, C. P. 2014. Bayesian computational tools. *Annu. Rev. Stat. Appl.* **1**: 153–57.

Schoot, R. van de and S. Depaoli. 2014. Bayesian analyses: Where to start and what to report. *Eur. Health Psych.* **16**: 75–84.

Scott, J. A., A. L. Hand, and L. S. Sian. 2011. BayesWeb: A user-friendly platform for exploratory Bayesian analysis of safety signals from small clinical trials. *J. Biopharm. Stat.* **21**: 1030–41.

Spiegelhalter, D. J., K. R. Abrams, and J. P. Myles. 2004. *Bayesian Approaches to Clinical Trials and Health-Care Evaluation.* New York: Wiley.

Sung, L., J. Hayden, M. L. Greenberg, G. Koren, B. M. Feldman, and G. A. Tomlinson. 2005. Seven items were identified for inclusion when reporting a Bayesian analysis of a clinical study. *J. Clin. Epidemiol.* **58**: 261–68.

Sverdlov, O., Y. Ryeznik, and S. Wu. 2015. Exact Bayesian inference comparing binomial proportion, with application to proof-of-concept clinical trials. *Ther. Innovation Regul. Sci.* **49**: 163–74.

Walley, R. J., C. L. Smith, J. D. Gale, and P. Woodward. 2015. Advantages of a wholly Bayesian approach to assessing efficacy in early drug development: A case study. *Pharm. Stat.* **14**: 205–15.

Weeden, S., M. Parmar, and L. S. Freedman. 2003. Bayesian reporting of clinical trials. In *Advances in Clinical Trial Biostatistics*, edited by N. L. Geller, 300–5. Boca Raton, FL: CRC Press.

Wijeysundera, D. N., P. C. Austin, J. E. Hux, W. S. Beattie, and A. Laupacis. 2009. Bayesian statistical inference enhances the interpretation of contemporary randomized controlled trials. *J. Clin. Epid.* **62**: 13–21.

Wu, Y., W. J. Shih, and D. F. Moore. 2008. Elicitation of a Beta prior for Bayesian inference in clinical trials. *Biometrical J.* **50**: 212–23.

10

Handling Missing Data in Clinical Trials with Bayesian and Frequentist Approaches

Xin Zhao
Johnson & Johnson

Baoguang Han, John Zhong, and Stacy Lindborg
Biogen

Neal Thomas
Pfizer Inc.

G. Frank Liu
Merck & Co., Inc.

CONTENTS

10.1 Introduction

Missing data is common in clinical trials, but is often inadequately handled which could have an impact on estimation and testing of treatment effects. Based on the National Research Council recommendations (NRC 2010), regulatory agencies are requesting more thorough analyses on missing data from sponsors in recent years. With the advancement of flexible general-purpose Bayesian software packages such as WinBUGS, SAS Proc MCMC, and Stan, it is relatively simple to develop Bayesian methods or multiple imputation approaches to address complex missing data problems, while incorporating the uncertainty. In the meantime, several novel frequentist techniques have also been developed that may provide robust analysis to some challenging missing data problems.

This chapter presents two case studies to demonstrate how to utilize Bayesian and multiple imputation approaches for missing data analysis in clinical trials. We also present a third case study where several alternative frequentist approaches are applied for handling missing data due to death. The case studies are selected to represent different types of endpoints in various therapeutic settings. For Bayesian and multiple imputation approaches, the first example as presented in Section 10.2 is a phase II insomnia study where a continuous primary endpoint is derived from daily patient-reported outcome (PRO) diary data by averaging the daily values at each weekly visit. A graphical approach is used to characterize the complex missing data patterns, and multiple imputation methods are implemented to assess the impact of missing data on the analysis of daily diary data. The second example presented in Section 10.3 is a phase II schizophrenia study with a continuous endpoint, where Bayesian approaches corresponding to a few commonly used frequentist methods on handling missing data are developed. For the frequentist approach, the example presented in Section 10.4 is a phase III pediatric spinal muscular atrophy (SMA) study where a key secondary endpoint is on a functional scale with missing outcome in the presence of death. The missing data due to death is a challenging problem. We present several frequentist approaches that were recently developed to handle such problems. In each case study, we discuss the properties, advantage, and flexibility of various methods on handling missing data.

10.2 Case Study 1: Insomnia Study with Missing Daily PRO Diary Data

10.2.1 Introduction

Bayesian modeling to create multiple imputations for missing diary reports is illustrated for a complex longitudinal study evaluating the effect of a treatment for insomnia. The primary analysis originally proposed for the study averaged daily diary values for each patient into a weekly variable. Following the commonly used approach, missing daily values within a week were ignored if there were at least four daily reports. This approach relies on the assumption of missing completely at random (MCAR, Little and Rubin 2002), and standard errors can still be distorted because the variance of each weekly average is assumed to be the same regardless of the number of diaries contributing to the average. This practice is pervasive in therapeutic areas where daily diary data are collected. We use Bayesian models to create multiple imputations, so each completed data set has exactly seven daily diary values contributing to each weekly patient average. Standard methods are then applied to the completed data sets, and the resulting estimates are combined using the standard multiple imputation approach. The example demonstrates that widely available Bayesian software implemented on desktop computers can now be used eliminate the ad hoc approach to analyzing diary data that became common before adequate computing was available.

10.2.2 Study Description

The case study is a randomized double-blind parallel-group placebo-controlled study of a compound for chronic insomnia. There was a one-week blinded placebo run-in period before randomization. There were five treatment groups: placebo and 15, 30, 45, and 60 mg of the active compound. There were approximately 135 randomized patients per group. It was planned that each patient would receive their assigned treatment for four weeks. Patients called a data collection system each morning and responded to questions about their sleep the previous night.

The primary endpoint, subjective time awake after initial sleep onset (SWASO), was derived by averaging the daily values between weekly clinic visits, as is commonly done with daily diary data (e.g., Merck Sharp and Dohme Corp 2015, Farrar et al. 2001). Most of the "weekly" averages were computed based on collection-time intervals that spanned 4–9 days. The proportion of the weekly averaged endpoints computed with at least one missing daily value ranged from 0.3 to 0.5 across the four weekly visits. The missing data rates for the SWASO averages, as defined in the protocol, at Week 4 for the 0, 15, 30, 45, and 60 mg dose groups are 0.18, 0.25, 0.16, 0.16, and 0.25, respectively. The primary analysis in the statistical analysis plan pre-specified a maximum likelihood mixed linear model (MLLM) for the change from baseline in weekly average SWASO, with site, treatment, visit, baseline SWASO, treatment-by-visit interaction, and baseline-by-visit interaction as fixed predictors.

10.2.3 Multiple Imputation Methods Applied to Daily Values

The analysis in this section assumes the missing data are missing at random (MAR, Little and Rubin 2002). The imputations were performed for the square-root-transformed daily values and then back-transformed before weekly averages were computed. Any negative imputed values were set to 0 before back-transformation. One hundred imputed data sets were generated. The weekly averages corresponding to baseline and four post-randomization visits based on the imputed data sets were computed from exactly seven daily values determined by the planned visit schedule. The primary analysis was applied to each imputed data set.

The imputation model assumes a multivariate normal distribution for the transformed daily SWASO values (including the baseline values) with means determined by a multiple linear regression. The transformed SWASO values are denoted by Y_{ij}, where patients are indexed by $i = 1, \ldots, N$, and study days are indexed by $j = -5, \ldots, 29$. The model includes the following fully observed predictors denoted by X_i: age (continuous), sex, race, and one post-randomization variable, the reason for terminating dosing (planned end of study, adverse event (AE) or death, other). The regression parameters associated with X are denoted by β. The dose group is denoted by T_i, with values (0, 15, 30, 45, 60). A potentially different mean value for each study day for each dose is denoted by δ_j^T. An equi-correlated variance matrix was specified through random (normal) patient-specific terms denoted by $\theta_i, i = 1, \ldots, N$, with mean of 0 and variance ψ^2. The residuals about the daily mean values are denoted by $\in_{ij}$, with variance σ^2:

$$Y_{ij} = X_i'\beta + \delta_j^{T_i} + \theta_i + \in_{ij} \tag{10.1}$$

The model in (10.1) was fit, and the imputations were generated using the general-purpose Bayesian Markov Chain Monte Carlo (MCMC) program STAN (STAN Development Team 2013). A diffuse prior distribution (i.e., diffuse normal prior distributions for fixed effects, diffuse gamma distributions for random effects) was utilized. Results from the MLLM and multivariate normal multiple imputation (MVNMI) analyses are in Table 10.1. The reported results include only comparisons of the lowest and highest doses to placebo at Week 4. The imputation approach yielded estimates of treatment effect that were substantively similar to the MLLM, but the estimates for the 15 mg dose trended toward larger effects. The standard errors from the imputed sets were smaller than those produced by the MLLM.

TABLE 10.1
Estimates and Standard Error (SE) for the Week 4 SWASO Endpoint

	15 mg versus PBO			60 mg versus PBO			Pooled[a]
Method	Est	SE	%Mis Info	Est	SE	%Mis Info	Res SD
MLLM	−2.93	5.97		−26.61	6		44.7
MVNMI	−5.2	5.16	12	−26.17	5.07	9	39.3

[a]Estimated residual standard deviation from a model including both dose groups and placebo (PBO).

Most of the difference is due to the smaller residual standard deviation estimated from the imputed data sets, which is displayed in the final column of Table 10.1. It is not apparent why the imputations yielded a smaller residual standard deviation. The differences in the estimates and standard errors did not change the substantive conclusions of this trial, but in a trial with treatment differences near boundaries for statistical significance, changes of the magnitude observed could yield p-values below and above the boundary.

Plots of observed and imputed daily values for individual patients displayed agreement in location and trend over time. Figure 10.1 displays the observed values, and the first five imputed daily values for three patients treated with the 60 mg dose who have common missing data patterns. Patients "239" and "270" were selected for display because they had a pronounced tendency to repeatedly report rounded times (e.g., $\sqrt{60}$ and $\sqrt{120}$), which the normal-based imputation models cannot accurately reproduce. Aside from this common situation, the imputed values appeared in good visual agreement with the observed values.

The flexibility of the general-purpose software can be used to impute from many alternative models for the mean and variance structures. Several generalizations of the model in 10.1 were evaluated and reported in Thomas et al. (2016). Models with more complex variance–covariance structures that better match the observed data can be implemented by re-fitting model (10.1) with functions of time multiplying the random terms θ_i and $\in_{ij}$.

10.2.4 Conclusions

With current desktop computing and general-purpose statistical software, it is feasible to account for missing daily diaries in aggregated endpoints. The most challenging aspect is the specification of models that can adequately represent the missing data. This problem becomes more difficult when pre-specification of the models is required for confirmatory trials. Imputation model sensitivity can be assessed by fitting several models with flexible mean functions and different variance–covariance structures. Multivariate normal models, however, are unlikely to reproduce some of the features present in subjectively reported diary data.

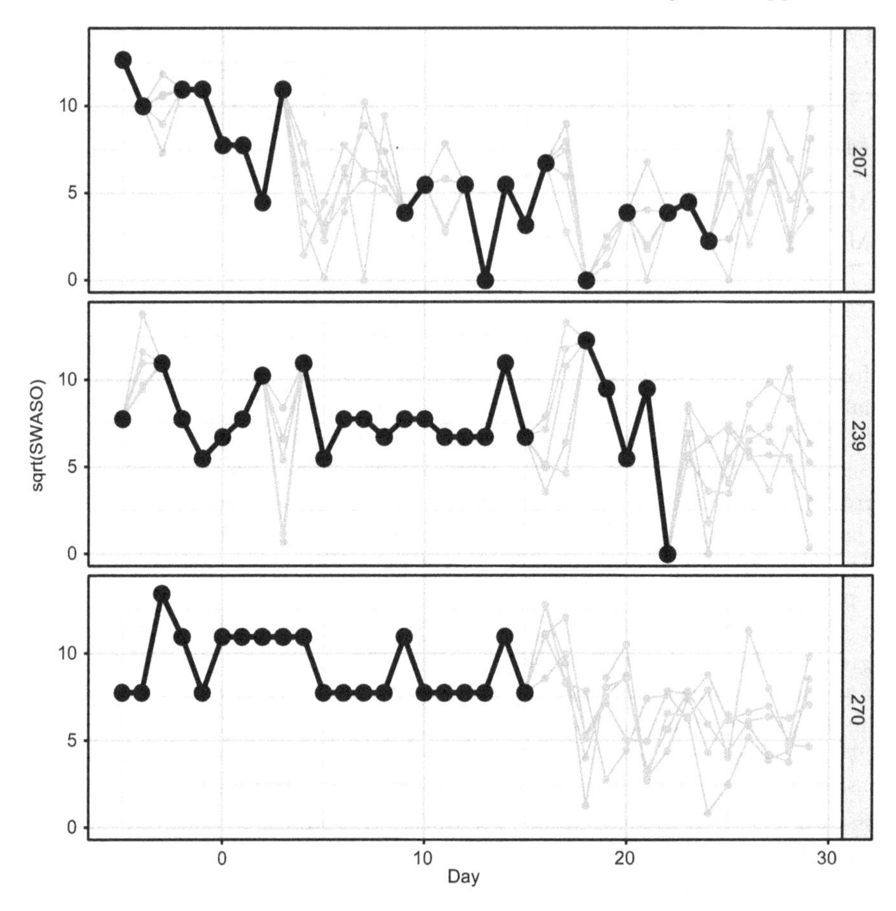

FIGURE 10.1
Longitudinal data for three patients. The observed data are displayed using bold black lines and points. The first five sets of imputed values are displayed using smaller, lighter lines and symbols.

The estimated proportions of missing information computed from the multiple imputation method in Table 10.1 show that recovery of some of the missing information is possible due to correlation between the numerous diary and baseline values. The differences in estimates and standard errors for the treatment effects from the Bayesian multiple imputation method under the MAR assumption were not large enough to change the substantive interpretation of the results. The differences would be large enough, however, to create ambiguity in the results from a trial with smaller treatment effects that achieved borderline statistical significance. Thomas et al. (2016) illustrate how to alter the imputed data sets created under MAR to assess sensitivity to non-ignorable missingness (MNAR) and estimate alternative estimands (National Research Council 2010).

10.3 Case Study 2: Schizophrenia Study with Early Dropouts

10.3.1 Introduction

Bayesian methods provide a natural approach to modeling the uncertainty of missing data in longitudinal trials and have been considered in statistical literature (Daniels and Hogan 2008, Hogan et al. 2014). However, the application of Bayesian methods in real clinical trials is still not common due to the lack of computational software and regulatory considerations. With significant advancement in computation and statistical software, Bayesian method provides a feasible alternative approach for the analysis of longitudinal clinical trials. In this case study, we illustrate Bayesian approaches along with commonly used frequentist methods for handling missing data for a continuous endpoint measured in a real schizophrenia clinical trial. Common sensitivity analysis models including selection model, shared parameter model, and pattern mixture models (PMMs) is implemented using both Bayesian and frequentist methods. This case study demonstrates that Bayesian methods can be used to fit the similar frequentist sensitivity analysis models with some additional flexibility to modify the model specification.

10.3.2 Study Description

The case study data is from a multicenter, randomized, double-blind clinical trial on patients with schizophrenia. The study has three treatment groups: test drug, active control, and placebo with a 2:1:2 ratio. A total of about 200 patients were enrolled from four eastern European countries. The primary endpoint for efficacy is the mean change from baseline in the Positive and Negative Syndrome Scale (PANSS) total score. The PANSS was measured at baseline (end of placebo lead-in period), Day 4, and Weeks 1–4. The primary time point for treatment comparison is at Week 4. In the study, 19%–33% of the patients dropped out from the study prior to Week 4. Majority of the patients dropped out due to the lack of efficacy, especially in the test drug and placebo groups. Graphical summaries of mean profiles by time of dropout are given in Figure 10.2 for active and placebo groups and in Figure 10.3 for test drug and placebo groups. The mean estimates in the plots are based on observed data at each given time point, and a negative change from baseline implies improvement. In general, we can see that for all dropouts, the mean change from baseline values had a worsening trend, which is consistent with majority of dropouts being due to the lack of efficacy.

The primary analysis as specified in the protocol was based on mixed model for repeated measure (MMRM) (see, e.g., Mallinckrodt 2008). The model included factors for baseline PANSS total score, treatment, week, country, and treatment-by-week and baseline-by-week interactions, where week,

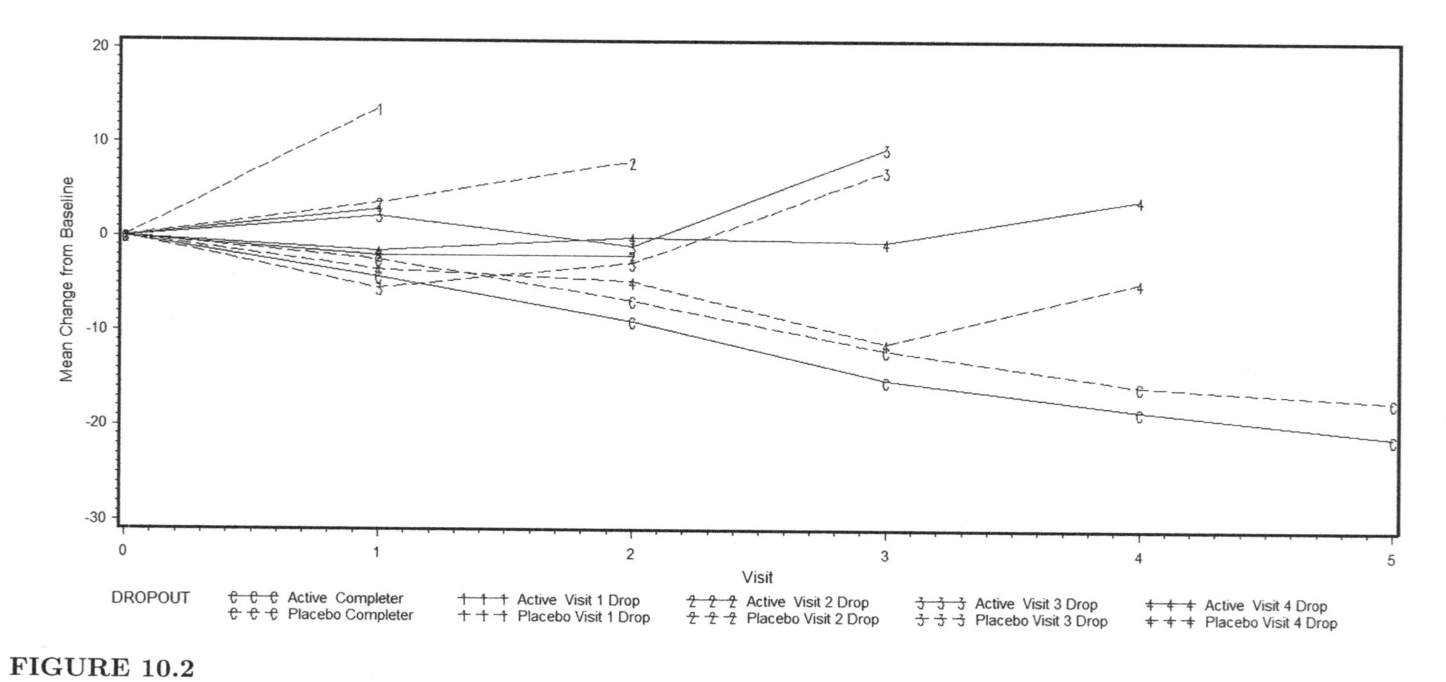

FIGURE 10.2
Visit-wise mean change from baseline for active versus placebo.

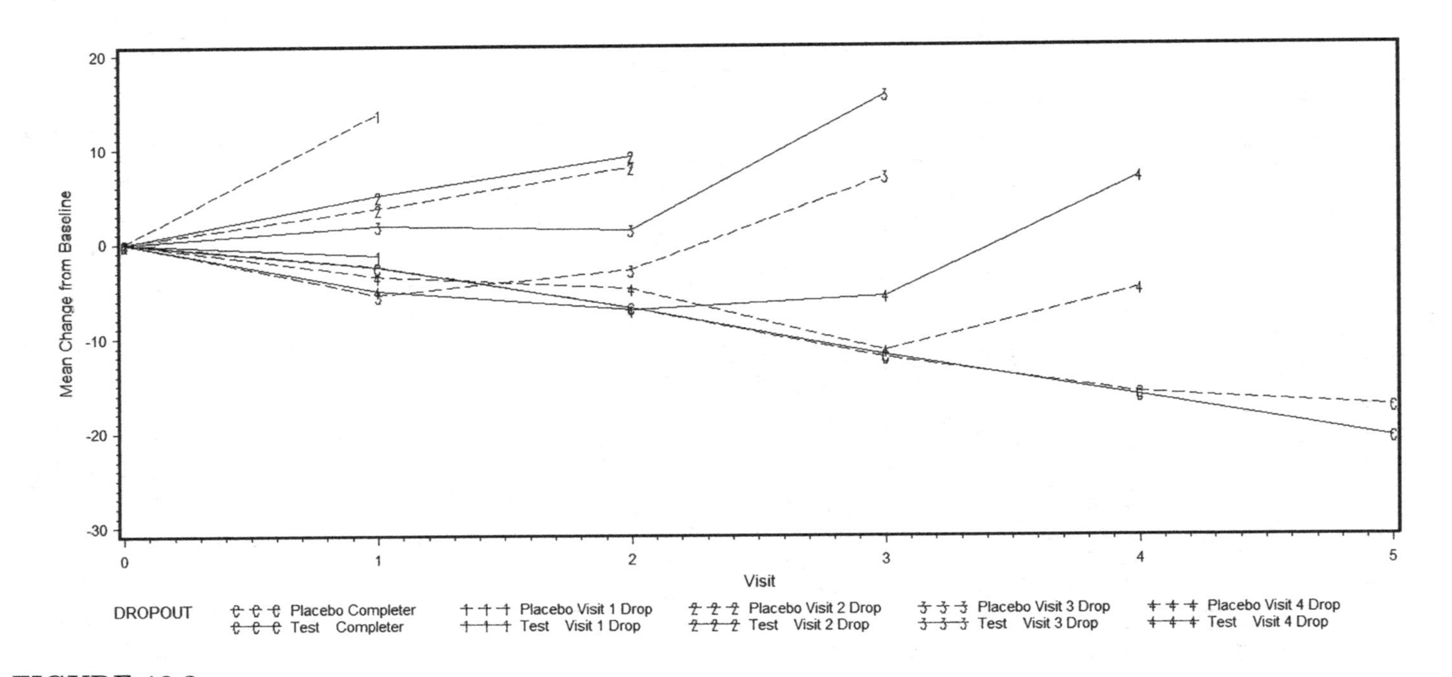

FIGURE 10.3
Visit-wise mean change from baseline for test drug versus placebo.

country, and treatment are treated as categorical variables. An unstructured covariance matrix was used for modeling the intra-subject correlation over the repeated measures. To further simplify the analyses, we removed country from our analysis model as it was not a significant factor in the analysis. The primary analysis results based on MMRM are given in the first column of Table 10.2. The test drug doesn't show the evidence of activity. The point estimate for active control showed some activity, but the effect was not statistically significant either.

10.3.3 Sensitivity Analyses Using Bayesian and Frequentist Methods

One assumption for the MMRM method is that the missing data are missing at random (MAR). This assumption cannot be verified from the observed data. Therefore, sensitivity analysis is suggested for evaluating the robustness of the analysis results against the missing data and MAR assumption.

We first applied a few conventional methods to handle missing data and analyze the case study. The methods are frequentist based and include complete-case analysis, selection models, shared random effect parameter model, and PMM. We then provide Bayesian analogues to the frequentist models including MMRM models under MAR and selection models, shared parameter models, and PMMs under MNAR. In all Bayesian analyses, the missing data are treated as random parameters and will be sampled from the specified joint distribution of the repeated measures. We used SAS Proc MCMC (SAS/Stat 13.2) or STAN for the analyses in which the missing data will be automatically sampled from the specified distribution. Description of the methods and implementation details can be found in Liu et al. (2015). The results for each method are provided in Table 10.2.

The MMRM approach assumes MAR (or MCAR for complete-case analysis). Bayesian approach using vague priors (or non-informative priors) is used to compute the posterior distribution of the parameters in the MMRM model and complete-case analysis. The posterior summaries are shown in the first two columns of Table 10.2. For both models, the results are fairly close to those from the corresponding frequentist methods. This is as expected because the Bayesian models use non-informative priors so that the posterior distribution is similar to the likelihood function from the frequentist method. We also observed that complete-case analysis results are quite different from that of the MMRM analysis with all available data. Especially for the test drug, the completers had larger positive effects compared with that from the MMRM analysis. This implied that the missing data may not be missing completely at random, which is the assumption for complete-case analysis method.

For frequentist selection model, we consider parametric selection model as specified in Diggle and Kenward (1994). For this case study, we used the SAS macro developed by DIA missing data working group (see http://missingdata.lshtm.ac.uk). We performed the analyses separately for

TABLE 10.2
Estimated Mean Difference (SE) Based on Frequentist and Bayesian Methods

Difference	MMRM All Data	MMRM Completers	Selection Model[a]	Shared Parameter[a]	PMM ACMV[b]
Frequentist Methods					
Active versus Placebo	−4.37 (3.75)	−3.22 (3.09)	−7.50 (4.26)	−5.69 (4.36)	−5.99 (3.76)
Test versus Placebo	−0.48 (3.28)	−2.98 (2.76)	−0.46 (3.69)	1.10 (3.71)	−1.97 (3.43)
Bayesian Methods					
Active versus Placebo	−4.33 (3.79)	−3.17 (3.05)	−6.53 (4.33)	−6.43(4.72)	−5.91 (4.21)
Test versus Placebo	−0.48 (3.27)	−2.94 (2.75)	−1.29 (3.74)	0.04(3.86)	−1.91 (3.87)

[a]The selection model and shared parameter model are run separately for active versus placebo, and for test versus placebo, because the available macros only take two treatment groups in the analyses. The Bayesian models are also run separately for consistency.
[b]The imputation is done by treatment group. The frequentist model is fit by SAS PROC MI with 500 imputations.

test drug versus placebo, and for active versus placebo because the macro can only fit data with two treatment groups at a time. The results for fitting these selection models are given in the third column of Table 10.2. Compared to the primary analysis based on MMRM, the selection model result for test versus placebo is similar but that for active versus placebo is somewhat different although the conclusion is similar. This difference might have resulted from the logistic model specification for missing data.

To fit a frequentist-shared parameter model, parametric forms are assumed for the response and dropout models. Here, we consider a quadratic model to allow a potentially non-linear response profile for this case study. The results from fitting these quadratic shared random effects models are provided in the fourth column of Table 10.2. Compared to the primary analysis based on MMRM, the shared parameter model results are somewhat different for both active versus placebo and test versus placebo comparisons, although the conclusions are similar. The difference might have resulted from the quadratic functional form used for the response and shared random effects models. Notice that in the MMRM, a cell-mean model was used for the response, so a different mean parameter is used for each time point and treatment without restriction on the functional form.

PMM approach (Little 1993) has appeal to many scientists as it is transparent in how the observed data is modeled and how the missing data is modeled given the observed data. For this case study, all subjects had observations at Day 4. The dropout pattern is defined as the maximum visit number before a subject dropped out. For example, pattern 1 contains subjects who dropped out at Week 1, pattern 2 contains subjects who dropped out at Week 2, etc., and finally, pattern 5 contains all completers. In the case study analysis, the first two dropout patterns are combined because there were only eight patients who dropped at the second visit. Results using the PMM with the available case missing values (ACMV) restriction comparing frequentist and Bayesian method are provided in the last column of Table 10.2. Both methods provide similar results.

10.3.4 Bayesian Selection Model with Different Specification of Dropout Probability

In the selection models specified in Section 10.3.3, the dropout model has the same parameter over time but different on treatment group. In double-blind clinical trials, we may also expect that the dropout probability may mostly depend on efficacy outcome (observed or unobserved), and the association between dropout and efficacy outcome may be similar across treatment groups (i.e., dropout may not depend on treatment after adjusting for efficacy outcomes as the treatment group is blinded during the study). We developed a Bayesian selection model which allows for different dropout probability patterns to depend on time but not treatment group. The results for estimated means and treatment difference are fairly similar across the methods from conventional MMRM model to different Bayesian selection models. Utilizing the Bayesian approach, we observed that the higher level dependence of missing probability on previous outcomes that are more than one-level away may be ignorable, which is consistent with the fact that the lack of efficacy is the main cause for discontinuation in this trial. Results of the Bayesian model also suggests that the dropout probability is not likely dependent on the missing data in this selection model. This explains why the results are quite similar to that from the MMRM, which is obtained under MAR assumption. Details of the Bayesian selection model specifications and results can be found in Liu et al. (2015).

10.3.5 Conclusions

In this case study, we used a real clinical trial data to explore the applications of Bayesian methods in sensitivity analyses under missing not at random scenarios on continuous repeated measures obtained in a longitudinal trial with missing data. In most of the cases, the results from Bayesian methods with vague (or non-informative) prior were very similar to those from the corresponding frequentist methods. This is helpful for applied statisticians to consider Bayesian approach with non-informative prior as a computational tool to fit complicated statistical models.

For this case study, the conclusions from different methods were consistent, although the estimated treatment effects could be different. In the modified selection model, the parameter estimates were also very similar to that of the original planned mixed model analyses. The analysis indicated that a worsening of response might have led to a higher probability of dropout, which is consistent with the fact that the lack of efficacy was the main cause for discontinuation in this trial.

Most of the frequentist sensitivity analysis methods require special software to implement. For selection models and shared parameter models, we have used the SAS macros developed by DIA missing data working group (available at http://missingdata.lshtm.ac.uk). For PMMs, we applied the new multiple imputation procedure in SAS 9.4. One advantage to use Bayesian approach is that it can be more flexible for users to implement different models, and Markov Chain Monte Carlo methods can be used to obtain the posterior distribution for the parameters of interest.

10.4 Case Study 3: Missing Functional Outcome in Presence of Death

10.4.1 Introduction

This case study is for longitudinally collected functional measures with "missing" (truncated) values due to death. Strictly speaking, the functional measures after death were not simply missing but undefined. As such it may not make sense to estimate the missing data after death. Other types of missing data also occurred in the study. In contrast to the methods presented in other case studies, we focus on statistical methods that avoid the direct estimation of missing values but still provide robust conclusion of efficacy.

10.4.2 Study Description

The considered trial (NCT02193074) was a phase III multicenter, randomized, double-blind, sham-controlled study of nusinersen administered intrathecally to subjects with infantile-onset SMA (Finkel et al. 2017). A total of 122 subjects were randomized 2:1 to receive nusinersen (treatment group) or undergo a sham procedure (control group), respectively. The primary endpoints were the proportion of motor milestone responders and the time to death or permanent ventilation. Based on the positive results from a pre-specified interim analysis of 6-month data, the study was terminated early. Hence, only about 52% subjects had an opportunity to complete the 13-month visit. For the final analysis, the pre-specified efficacy set contained 110 subjects who had the opportunity to be assessed at the Day 183 visit, with 37 subjects from the control group and 73 subjects from the active group. The Children's Hospital of Philadelphia Infant Test of Neuromuscular Disorders (CHOP INTEND) was a key secondary endpoint, and its analysis is the focus of this case study. CHOP INTEND is a validated 16-item and 64-point motor assessment designed specifically to evaluate the motor skills of infants with SMA, a higher score indicating greater motor skill (Glanzman et al. 2010). CHOP INTEND was evaluated twice during the screening period (with mean value of the two measurements serving as the baseline) and on Days 64, 183, 302, and 394.

10.4.3 Missing Data Problems

In this study, three types of missing data had to be addressed. The first type of missing data was due to deaths. In total, there were 29 deaths in the trial with 16 subjects from the control group (43.2%) and 13 subjects from the treatment group (17.8%). The nusinersen treatment group had significantly prolonged overall survival as compared with the control group, with p-value of 0.004 (based on a stratified log rank test). For such a setting, missing data should not be handled as MAR because the distribution of missing data after death, if it exists at all, may differ from that of the observed outcomes. A joint modeling approach has been proposed to handle such problems (Li and Su 2018). However, such models are often sensitive to the parametric model assumptions, and these assumptions are usually hard to justify.

In fact, the CHOP INTEND measure after death was not simply missing but undefined. In the literature, this type of missing is often referred to as "truncated by death" or "censored by death" to distinguish it from other types of missing data where the outcome is merely missing because of inadequate data collection. For such settings, a crude comparison of the outcome, such as CHOP INTEND score, between those who survived in each treatment group may lead to a biased result. Subject survival is a post-treatment event that is related to treatment effect, and a comparison conditional on subject survival will no longer preserve the randomization. This problem receives extensive attention in the literature, and the use of principal stratification to address this issue has been described (Frangakis and Rubin 2002, Egleston et al. 2009, Chiba and VanderWeele 2011). Principal stratification works by comparing treatment effects on the outcome of individuals who would have survived irrespective of which treatment they were given, which is a fair comparison. However, a nontrivial complication to this method is that individuals who would have survived in either treatment group are not identifiable, and many of the statistical and sensitivity analysis techniques developed are difficult to implement in practice (Chiba and VanderWeele 2011).

The second type of missing data in this study was early withdrawal, with one subject from the control group and two subjects from the treatment group. Early withdrawal may be due to the lack of efficacy, the occurrence of adverse events, or other reasons unrelated to treatment. For withdrawal due to reasons unrelated to treatment outcome or lack of efficacy, missing values may be handled as MAR. For all other types of withdrawals, however, missing values are often handled as missing not at random.

The third type of missing data in the study was due to administrative censoring. As the study was terminated early due to meeting pre-specified early stopping criteria, less than 30% of subjects in the efficacy set had the opportunity to complete all the scheduled visits. This type of missing can be considered as MAR.

10.4.4 Methods of Analysis and Results

Four methods are described to address the missing data problems in this study. The first three methods focus on the change of CHOP INTEND total score from baseline, while the fourth method focuses on the rate of change in CHOP INTEND total score.

The first method is the responder analysis, which is usually preferred by regulatory agencies (EMA 2010). Subjects who did not die or withdraw were dichotomized leveraging responder criteria pre-specified in the protocol which was based on whether the change from baseline in CHOP INTEND total score on last available assessment (i.e., at the later of the Day 183, Day 302, or Day 394 visits) was greater than or equal to four points. By using the later outcome on or after Day 183, we allowed the opportunity for the drug to demonstrate clinical meaningful efficacy and avoided imputing missing data due to administrative early study termination. An increase from baseline of four points or more in CHOP INTEND total score is generally considered

TABLE 10.3
Summary of Analysis Based on Responders

	Control	Treatment
Number (%) of evaluable subjects[a]	37(100)	73(100)
Number (%) with change from baseline in total score improvement ≥4 points	1(3)	52(71)
Difference in percentages (treatment − control) and 95% confidence interval[b]		68.53(51.27, 81.99)
p-value (compared to control)[c]		<0.0001

[a]Subjects with opportunity for at least a 6-month (Day 183) assessment.
[b]Exact unconditional confidence interval.
[c]Based on Fisher's exact test.

to be outside of the range of test variability (Finkel et al. 2014, Glanzman et al. 2011). Subjects who died or withdrew from the study were counted as non-responders and were included in the denominator for the calculation of the proportion. As a result, both mortality and early withdrawal were accounted for in the CHOP INTEND responder analysis. Fisher's exact test was used to compare the difference in the proportion of responders between the two groups (control vs. treatment). The p-value and the unconditional confidence interval (CI) for the difference in response rates (Santner and Snell 1980) are provided in Table 10.3.

The second method is called joint rank procedure and is based on a combined assessment of function and survival (CAFS) (Finkelstein and Schoenfeld 1999). Subjects were ranked based on whether they survived, withdrew, or died during the study, with higher ranks assigned for those who survived, medium ranks for those who withdrew, and lower ranks for those who died. Within each category, the subjects were further ranked according to the time to die or withdraw or the change from baseline of CHOP INTEND total score on the last assessment. The subjects were assigned to CAFS score according to their ranks. Treatment groups were compared using an analysis of covariance of subjects' CAFS scores, adjusting for each subject's age at symptom onset, disease duration at baseline, and CHOP INTEND total score at baseline. The results are shown in Table 10.4.

The third method is the trimmed-mean analysis. The trimmed mean is an averaging method that eliminates a partial percentage of the smallest values before evaluating the mean value of the given data (Permutt and Li 2017). Subjects who died or withdrew from the study were considered as having worse outcomes than those who survived and completed the trial and were assigned an arbitrary response score of −999 for purpose of ranking. The response score for subjects who did not die or withdraw was the observed change from baseline in the CHOP INTEND score from the last assessment. Subject response scores in the two groups were then ranked from the largest to the smallest separately. The trimmed mean is the average of the best $1-\alpha$

TABLE 10.4
Summary of Analysis Based on Joint Ranks

CAFS	Control	Treatment
N	37	73
Mean (SD)	−48.4(34.41)	24.5(61.28)
Least squares mean (95% CI)[a]	−47.05(−64.66,−29.44)	23.85(11.39, 36.30)
Least squares difference (95% CI)[a]		70.89(49.13,92.66)
p-Value (compared with control)[a]		<0.0001

[a]Least squares mean, least squares difference, and p-value for treatment comparison based on analysis of covariance of the subject's CAFS scores adjusting for disease duration, age at symptom onset and CHOP INTEND score at baseline.
CAFS, Combined assessment of function and survival; CI, confidence interval; n, number of subjects; SD, standard deviation.

fraction of observations from each group. As all death and withdrawal have to be trimmed off, we chose α as the greater number of proportions of subjects who died or withdrew from the two treatment groups. The p-value associated with the observed difference in the trimmed means was calculated using a reference distribution generated by a permutation test. For each permutation, data were ordered and trimmed equal fractions to trim all death or withdrawal. The results are shown in Table 10.5.

For the final approach, a slope analysis was performed. To account for the outcome for those subjects who died or withdrew from the study, a total score of zero was assigned to the day of death or withdrawal. For each subject in the efficacy set, a slope was calculated by a linear regression of the total score with the actual day of assessment, death, or withdrawal. The slopes were compared between treatment groups using an analysis of covariance adjusting for each subject's disease duration at screening, age at screening, and CHOP INTEND baseline total score. The results are shown in Table 10.6.

TABLE 10.5
Summary of Analysis Based on Trimmed Means

Based on the Top 54.1% of Ranked Subjects in Each Treatment Group	Change from Baseline Control	Change from Baseline Treatment
N	20	39
Mean (SD)	−12.9(8.88)	16.3(5.45)
p-Value (compared with control)[a]		<0.0001

[a]Based on 10,000 permutations.

TABLE 10.6
Rate of Change in Total Score[a]

	Control	Treatment
Number of subjects in efficacy set[a]	37	73
N	37	72
Mean (SD)	−1.14(1.42)	−0.12(0.99)
Least squares mean (95% CI)[b]	−1.17(−1.55, −0.78)	−0.10(−0.38, 0.17)
Least squares difference (95% CI)[b]		1.06(0.59, 1.54)
p-Value (compared with control)[b]		<0.0001

[a]Assigning total score of 0 for subjects who died or withdrew from the study at the time of death or withdrawal.
[b]Least squares means, least squares differences, and p-value for treatment comparison based on analysis of covariance that includes treatment adjusting for each subject's disease duration at screening, age at screening, and CHOP INTEND total score at baseline.

10.4.5 Discussion/Conclusion

In this case study, early dropouts were mainly due to death except for a few cases of voluntary withdrawal (one subject from the sham control and two subjects from the treatment group). For missing data due to voluntary withdrawal, the Bayesian methods or the multiple imputation methods presented in previous sections may be appropriately applied. For dropout due to death, however, the assumption of MAR may not work because the functional scores do not exist for subjects who died. As the functional scores were truncated by death, the exact shape of the functional score after death was unknown. Hence, it would be generally difficult to justify a Bayesian parametric model or a method of multiple imputation. To overcome this problem, all the approaches presented in this case study avoid estimating missing data, which contrasts to other approaches presented in this chapter. Except for the slope analysis, the comparison of CHOP INTEND results was based on the last assessment, thus avoiding estimation of a large proportion (~30%) of missing data on the Day 395 visit due to early termination of the study. In all four methods, subjects who died or withdrew were considered to have unfavorable outcomes, and these subjects contributed to treatment effect comparison in various ways.

Each method has its advantages and disadvantages. The responder analysis offers a straightforward interpretation but suffers from the loss of information by dichotomizing the continuous outcome. The joint rank analysis is a powerful test that assesses both survival and continuous outcome. This analysis, however, is a non-parametric rank analysis, and the magnitude of the difference in CAFS scores between treatment groups cannot be directly compared across trials. Also, separate analyses of the component data for function and survival are still required to understand the specific clinical effects of the study treatment. The trimmed-mean method is a newly proposed method that has not been widely accepted, but it appears to be a fair comparison between treatments. However, the proportion of trimming may be data driven, and the efficiency may also be lost from trimming observed data. Finally, the slope analysis appears to be a reasonably fair comparison, but the results should also be interpreted with caution because the change in CHOP INTEND score may not be linear and the assigned score of zero to the time of withdrawal or death needs justification.

With the four methods, due to the observance of very large treatment effects, all p-values reported are highly significant in this case study, which makes it hard to compare the relative efficiency of the methods. For their application in other studies, the choice of the methods should be determined by the unbiasedness, the statistical efficiency, and interpretability of results. Simulation studies under different scenarios may help assess both the unbiasedness and statistical efficiency. It will also be interesting to investigate how principal stratification method can be applied to the CHOP INTEND data in the trial, and whether this method can offer additional insight to the treatment efficacy.

References

Chiba Y and VanderWeele TJ. A simple sensitivity analysis technique for principal strata effects when the outcome has been truncated due to death. *Am J Epidemiol.* 2011;173:745–51. doi:10.1093/aje/kwq418.

Daniels MJ and Hogan JW (2008). *Missing Data in Longitudinal Studies: Strategies for Bayesian Modeling and Sensitivity Analysis.* Boca Raton, FL: Chapman & Hall/CRC.

Diggle P and Kenward MG. Informative dropout in longitudinal data analysis (with discussion). *Appl Stat.* 1994;43:49–94.

Egleston B, Sharfstein DO, and MacKenzie E. On estimation of the survivor average causal effect in observational studies when important confounders are missing due to death. *Biometrics.* 2009;65:497–504. doi:10.1111/j.1541-0420.2008.01111.x.

EMA (2010). Guideline on Missing Data in Confirmatory Clinical Trials. www.ema.europa.eu/docs/en_GB/document_library/Scientific_guideline/2010/09/WC500096793.pdf.

Farrar J, Young J, LaMoreaux L, Werth J, and Poole M. Clinical importance of changes in chronic pain intensity measured on an 11-point numerical pain rating scale. *Pain.* 2001;94:149–58.

Finkel RS, McDermott MP, Kaufmann P, et al. Observational study of spinal muscular atrophy type I and implications for clinical trials. *Neurology.* 2014;83(9):810–7.

Finkel RS, Mercuri E, Darras BT, et al. Nusinersen versus sham control in infantile-onset spinal muscular atrophy. *N Engl J Med.* 2017;377(18):1723–32. doi:10.1056/NEJMoa1702752.

Finkelstein DM and Schoenfeld DA. Combining mortality and longitudinal measures in clinical trials. *Stat Med.* 1999;18:1341–54.

Frangakis CE and Rubin DB. Principal stratification in causal inference. *Biometrics.* 2002;58:21–9.

Glanzman AM, Mazzone E, Main M, et al. The Children's Hospital of Philadelphia Infant Test of Neuromuscular Disorders (CHOP INTEND): test development and reliability. *Neuromuscul Disord.* 2010;20(3):155–61.

Glanzman AM, McDermott MP, Montes J, et al. Validation of the Children's Hospital of Philadelphia Infant Test of Neuromuscular Disorders (CHOP INTEND). *Pediatr Phys Ther.* 2011;23(4):322–6.

Hogan JW, Daniels MJ and Hu L (2014). Bayesian sensitivity analysis, a book chapter in *Handbook of Missing Data Methodology*, edited by Molenberghs G, Fitzmaurice G, Kenward M, Tsiatis A, and Verbeke G; Chapman and Hall/CRC, Boca Raton, FL.

Li QJ and Su L. Accommodating informative dropout and death: a joint modelling approach for longitudinal and semicompeting risks data. *Appl. Stat.* 2018;67:145–63. DOI: 10.1111/rssc.12210.

Little RJ. Pattern-mixture models for multivariate incomplete data, *J Am Stat Assoc.* 1993;88:125–134.

Little R and Rubin D (2002). Statistical analysis with missing data, Second Edition. New York: Wiley.

Liu, Han B, Zhao X and Lin Q. A comparison of frequentist and Bayesian model based approaches for missing data analysis: Case study with a schizophrenia clinical trial. *Stat Biopharm Res.* 2015;8:1–36.

Mallinckrodt C, Lane P, Schnell D, Peng Y and Mancuso J. Recommendations for the primary analysis of continuous endpoints in longitudinal clinical trials. *Drug Inform J.* 2008;42:303–19.

Merck Sharp & Dohme Corp. Safety and efficacy study of suvorexant in participants with primary insomnia (mk-4305-028). ClinicalTrials.gov. Bethesda (MD): National Library of Medicine (US), 2015. Available from: https://clinicaltrials.gov/ct2/show/NCT01097616; NLMIdentifier:NCT01097616 [Access on 5 March 2015].

National Research Council (2010). *The Prevention and Treatment of Missing Data in Clinical Trials.* Panel on Handling Missing Data in Clinical Trials, Committee on National Statistics, Division of Behavioral and Social Sciences and Education. Washington, DC: The National Academies Press.

Permutt T and Li F. Trimmed means for symptom trials with dropouts. *Pharm Stat.* 2017;16:20–8. doi:10.1002/pst.1768.

Santner TJ and Snell MK. Small-sample confidence intervals for p 1 - p 2 and p 1 / p 2 in 2 × 2 contingency tables. *J Am Stat Assoc.* 1980;75:386–94.

STAN Development Team (2013). Stan: A C++ Library for Probability and Sampling. http://mc-stan.org.

Thomas N, Harel O, and Little R. Analyzing clinical trial outcomes based on incomplete daily diary reports. *Stat Med.* 2016;35:2894–906.

11

Bayesian Probability of Success for Go/No-Go Decision Making

Qi Tang

Digital and Data Sciences, Sanofi US

CONTENTS

To make an informed Go/No-Go decision for an investigational product, all available information should be effectively used to predict the results for ongoing and/or future clinical trials. We will first review literature on the Go/No-Go decision making methodology and case studies, particularly [1], which presented an interesting application of assessing probability of success (POS) in a business acquisition case study involving a key oncology compound in Phase 3 development. Then we focus on Go/No-Go decision making where POS is assessed utilizing all relevant information in a Bayesian way. Bayesian POS provides a quantitative approach to guide decision makers as to how likely a future trial or program is going to be successful. Examples are given to illustrate how to evaluate POS for continuous, binary, and survival endpoints.

11.1 Introduction

The overall probability of a regulatory approval from Phase 1 was about 9.6% according to a recent study report by Biotechnology Innovation Organization (BIO) [2]. This rate is even lower for oncology area where different endpoints are used for early phase and late phase clinical trials. What makes it worse is the increasing cost of clinical trials [3], especially Phase 3 trials, which costs about from US$11.5 million to US$52.9 million on average [4]. To reduce the continuously high failure rate of clinical trials and hence reduce the cost of drug development, a carefully pre-planned quantitative evidence based Go/No-Go decision making exercise, which already has been adopted for Go/No-Do decision making at interim analyses of a clinical trial, needs to be imbedded into the drug development process. The pre-planned quantitative Go/No-Go decision making consists of three steps: evidence collection, modeling and simulation, and determination of decision criteria. Traditionally, only data from the just finished clinical trial were used as evidence for Go/No-Go decision making, and sometimes only summary statistics were used, for example the p-value, which is criticized by the American Statistical Association as been used as the only evidence [5]. Evidence to be collected may not be limited to the just finished clinical trials but also may include clinical data from treatments that share similar mechanism of action and data from other study populations of the same product. For example, [1] describes a case study where clinical trial results of the same disease, colon cancer, and other studies of the investigational agent, bevacizumab, are collected as relevant evidence for predicting the results of an ongoing clinical trial at that time, the National Surgical Adjuvant Breast and Bowel (NSABP) trial, C-08. Although evidence collection is the foundation for scientifically sound Go/No-Go decision making, it is not the focus of this chapter. Throughout this chapter, we focus on modeling and simulation and determination of decision criteria. The modeling and simulation step is about quantification of the strength of the evidence toward a Go decision, which involves synthesis of evidence from all sources and may also need simulation of clinical trials. For example, in [1], a Bayesian method was used to synthesize information and predict clinical trial results based on publically available information about the design and the status of the C-08 trial. The probability of hitting statistical significance at the end of the trial was estimated to be 45%, and the final results of the trial are not significant with p-value 0.15. Since C-08 trial was a key asset of Genetech when Roch was about to acquire Genetech, the estimated probability has helped Roch to determine the price it offered to Genetech's stock. The probability of significant results is often referred to as the probability of success (POS) [6] or assurance [7, 8] in the literature of clinical trials. We will use the POS throughout this chapter, since it becomes popular in recent statistics methodology publication [9, 10, 11], and it is consistent with clinical trial literature beyond statistics [12, 13]. POS can cover a wide range of desirable clinical trial outcomes, such as identifying the target dose, demonstrating comparable efficacies against competitors, achieving clinically meaningful results, where in addition to being significant, the treatment effect is large enough to have clinically meaningful impact to patient's outcomes, etc. Besides predicting the probability of successful outcomes of an ongoing or a new trial, another way to measure the strength of evidence is to make inference about the size of the targeted treatment effect directly. For example, [14] demonstrated a decision making framework based on the probabilities that the treatment effect achieves a lower reference value and a target value, respectively. This framework has been used for all clinical programs in AstraZeneca. The last but not least step of a pre-planned decision making, determining decision criteria, is based on comprehensive clinical trial simulations from the previous step to understand the operating characteristics of a candidate decision criteria. For example, to pre-plan a Go/No-Go rule at the interim analysis of an adaptive clinical trial, extensive simulations under multiple efficacy and/or safety scenarios were performed to understand the pros and cons of each adaptation rule [15, 16, 17, 18, 19]. Comprehensive clinical trial simulations have also made inroads into the Go/No-Go decision making at the end of a clinical trial [20, 14].

11.2 Pre-Planned Go/No-Go Decision Making Framework

To motivate the pre-planned Go/No-Go decision making, let's take a look at a typical post hoc Go/No-Go decision making case study. For example, a long-awaitedPhase 2 study just read out. The hazard ratio of progression free survival is 0.75 with the 95% confidence interval $(0.5, 1.1)$ and two-sided p-value 0.12. The physicians in the team may be very excited because a 25% reduction in hazard is very meaningful in clinical practice. However, the statisticians may not be very excited because the p-value is not significant. What is not very exciting to all of them is that the upper management wanted a comprehensive Go/No-Go decision making exercise to be completed within three weeks because of the competition from other contenders racing to be the first-on-market. Completing such critical task within such short time frame is not easy especially since the collaboration is needed between different functions to predict the likelihood that the product is going to succeed in Phase 3 and its commercial value given predicted Phase 3 study outcomes and product profiles of competitors. Given tight timeline, extensive evidence collection may not be feasible, so only the results of the Phase 2 study are used in the decision making. What is worse is that clinical trial simulations are not conducted because there is not enough time to reach agreement across functions on how the Phase 3 trial shall be designed. In the end, the team delivered the presentation to the upper management claiming that the product has a 70% chance to succeed in Phase 3 without any modeling and simulation support. Why 70%? The team decide that it shouldn't be 90% since the p-value is not significant and the 95% confidence interval covers 1. On the other hand, the team think that it should be higher than 50% given the promising point estimate of the hazard ratio. So, the team takes the average of 90% and 50% and obtains the number to report to the upper management. If the above case study looks familiar to you, it is time to think about how to improve upon the old practice. In contrast to post hoc decision making process, which only starts after early phase clinical trial results are unblinded, the pre-planned decision making process starts before unblinded results are available (Figure 11.1) and ideally as early as possible, which allows sufficient time to collect all available relevant information and prespecify the Go/No-Go decision criterion, and gets updated whenever the competition landscape changes or new evidence emerges. The pre-planned Go/No-Go process reduces the subjectivity from the project team and fixes the issue of limited evidence and inadequate modeling and simulation due to a short timeline to collect comprehensive information. Differences between these two types of decision making frameworks are summarized in Table 11.1.

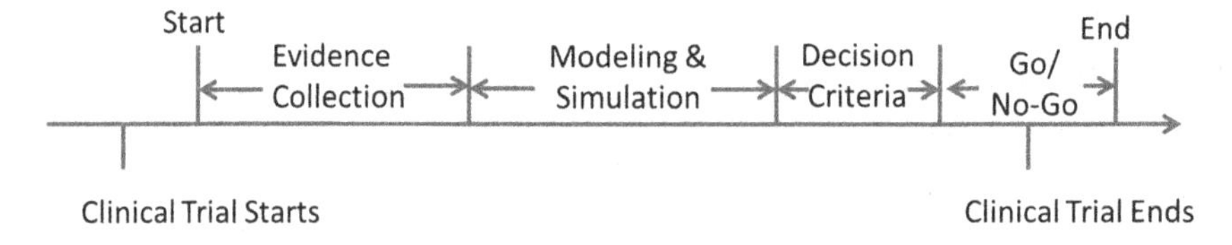

FIGURE 11.1
Pre-planned Go/No-Go decision-making framework.

TABLE 11.1
Comparison between Pre-Planned Go/No-Go Decision Making versus Post Hoc Go/No-Go Decision Making

	Pre-planned	Post Hoc
Starting point	Before unblinding of trial results	After unblinding of trial results
Time for decision making	Sufficient	Insufficient
Evidence collected	Comprehensive	Limited
Integrity	Evidenced based	Mixed with study team's subjectivity
Quality of decisions	High	Low

11.3 Drawback of p-Value and Effect Size Based Go/No-Go Decision Making

In a standard topline results announcement, p-value and effect size are two key statistical summaries that draw most of people's attention and often drives the Go/No-Go decision making. A significant p-value from an early phase study often leads to a Go decision for late stage testing, because a p-value less than 0.05 was often interpreted as the probability that a treatment does not work is 0.05. Another misinterpretation of p-value is that the probability of the effect size smaller than the observed is 0.05, which is actually 0.5 under the use of non-informative flat prior. Recently, American Statistical Association has released an announcement regarding misuse and misinterpretation of p-value for decision making [5] and pointed out that p-value alone is not sufficient for decision making. In clinical development, a great concern regarding the use of p-value for decision making is the reproducibility of clinical trial findings. There were several highly expected experimental treatments that failed to meet primary endpoints in Phase 3, even though they achieved significant results in previous trials in the same or similar patient populations [21, 22, 23], which strongly challenged the way Go/No-Go decisions were made. There are several flaws in p-value or effect size based Go/No-Go decision making.

First, decision making based on p-value less than 0.05 or an observed effect size fails to take into account the uncertainty imbedded in the observed p-value and the effect size. If a new trial is conducted to replicate the results, the new p-value may be smaller or larger than the previously observed one. Second, decision making based on p-value and effect size alone fails to examine the assumption taken as granted that the new trial will have similar study design and study population as the finished trial. A slight change in the inclusion and exclusion criterion for study population or stratification factors in the trial design may have an impact on final results. Thus, a thorough examination of the assumption and bridging of observed clinical trial results and to-be observed trial results is very critical.

11.4 Bayesian POS Based Decision Making

Predicting results of future clinical trials are critical and should be conducted in an appropriate way to take into account uncertainty around the observed data and the concordance between historic and future study populations and study designs. One way to take into account uncertainty is to adopt a Bayesian framework to treat the unknown treatment effect as a random quantity, whose distribution can be modeled by combining observed data and domain knowledge. The uncertainty can be captured in the Bayesian probabilities, such as the probability that a treatment effect is greater than zero. The Bayesian framework also provides flexibilities in predicting results of future clinical trials, such as the probability of a Phase 3 study to yield both significant and clinical meaningful outcomes. The concordance is very challenging to be measured because of the unobserved data from the future trial. However, for Go/No-Go decision making, a spectrum of possible concordance after the weight in from domain experts can be assumed to examine the robustness of a decision, which can be quantified and visualized through a prior for the to-be observed treatment effect. Examples will be given in Sections 11.4.1 to illustrate the Bayesian framework for Go/No-Go decision making.

11.4.1 Decision Making Based on POS

One way of quantifying evidence for Go/No-Go decision making is to evaluate the probability of observing successful clinical trial results in the future, which will be called POS throughout this chapter. A Go decision may be warranted if there is a high POS that an ongoing/future trial will yield statistically significant results and clinically meaningful outcomes. [6] pointed out that POS extended the concept of assurance coined by [8] to any definition of study success going beyond significant p-values. A general formula for POS is given below.

$$\text{POS} = \int_{-\infty}^{\infty} P(\text{Trial or Program Succeeds}|\Theta), P(\Theta|D)d\Theta \tag{11.1}$$

where Θ represents the set of unknown parameters, D indicates the observed historical data and $P(\Theta|D)$ is the posterior density of Θ given historical data. Closed form formulae of POS are available for continuous, binary, and survival types of endpoints under certain assumptions. A simple POS based Go/No-Go decision rule is the following

Go/No-Go Rule Go decision, if POS > threshold; otherwise No-Go.

The details of how to calculate POS will be illustrated in Section 11.4.2, and more complicated Go/No-Go decision rules such as asset value based Go/No-Go decision rules will be discussed in Section11.5.

11.4.2 Calculation of POS

If D comes from only one historical study and the success criterion for a future study is simply achieving statistically significant results, POS becomes the same as assurance, which has convenient closed form formulae, which are tabulated in Table 11.2. For continuous endpoints, $\hat{\sigma}_i$ and σ_i are estimated and unknown true variances for treatment group i, n_i and m_i are the sample sizes for group i in the historical study and the future study, respectively; θ is the unknown true mean difference between groups 1 and 2, and larger θ indicates stronger treatment effect; and $\hat{\theta}_1$ and $\hat{\theta}_2$ denote the estimated

TABLE 11.2
Closed Form Formulae for Probability of Achieving Statistically Significant Results in a Future Trial under Normal Approximations and Empirical Bayes Estimation of Variance

Endpoint	POS	Assumptions & Prior
Continuous	$\Phi\left(\frac{-z_{\alpha/2}\sqrt{\sum_{i=1}^{2}\hat{\sigma}_i^2/m_i}+\hat{\theta}_1}{\sqrt{\sum_{i=1}^{2}\hat{\sigma}_i^2(1/m_i+1/n_i)}}\right)$	$\frac{\hat{\theta}_2-\theta}{\sum_{i=1}^{2}\hat{\sigma}_i^2/m_i}\vert\theta \sim N(0,1)$ & $\theta\vert D \sim N(\hat{\theta}_1, \sum_{i=1}^{2}\hat{\sigma}_i^2/n_i)$
Binary	$\Phi\left(\frac{-z_{\alpha/2}\sqrt{\sum_{i=1}^{2}\hat{p}_i(1-\hat{p}_i)/m_i}+\hat{\theta}_1}{\sqrt{\sum_{i=1}^{2}\hat{p}_i(1-\hat{p}_i)(1/m_i+1/n_i)}}\right)$	$\frac{\hat{\theta}_2-\theta}{\sqrt{\sum_{i=1}^{2}\hat{p}_i(1-\hat{p}_i)/m_i}}\vert\theta \sim N(0,1)$ & $\theta\vert D \sim N(\hat{\theta}_1, \sum_{i=1}^{2}\hat{p}_i(1-\hat{p}_i)/n_i)$
Survival	$\Phi\left(\frac{-z_{\alpha/2}\sqrt{1/(r_1(1-r_1)d_1)}-\hat{\theta}_1}{\sqrt{\sum_{i=1}^{2}1/(r_i(1-r_i)d_i)}}\right)$	$\frac{\hat{\theta}_2-\theta}{r_2(1-r_2)d_2}\vert\theta \sim N(0,1)$ & $\theta\vert D \sim N(\hat{\theta}_1, r_1(1-r_1)d_1)$

mean difference in the historical study and the future study, respectively. For binary endpoints, p_i stands for the event rate in the historical study; θ indicates the unknown true difference of event rates between groups 1 and 2, and larger θ indicates stronger treatment effect; and $\hat{\theta}_1$ and $\hat{\theta}_2$ denote the estimated difference of event rates in the historical study and the future study, respectively. For survival endpoints, d_1 and d_2 denote the total number of events in the historical study and the future study, respectively; r_1 and r_2 indicate the probability of randomizing patients to the treatment group in the historical study and the future study, respectively; θ indicates the unknown true log hazard ratio between groups 1 and 2, and smaller θ indicates stronger treatment effect; and $\hat{\theta}_1$ and $\hat{\theta}_2$ denote the estimated log hazard ratios in the historical study and the future study, respectively.

The normal approximation assumptions hold when the sample size is large enough, and a rule of thumb is that it is greater than 30. Another assumption made for continuous and binary endpoints is that the sum of variances of the two groups in the future study can be estimated by that observed in the historical study, that is,

$$\sum_{i=1}^{2}\sigma_i/m_i = \sum_{i=1}^{2}\hat{\sigma}_i/m_i, \tag{11.2}$$

$$\sum_{i=1}^{2}\text{var}(\hat{q}_i) = \sum_{i=1}^{2}\hat{p}_i(1-\hat{p}_i)/m_i. \tag{11.3}$$

The POS calculated using the formulae in Table 11.2 may be different from the exact calculation using a fully Bayesian approach, which will be described below, but they provide good approximations to the results [7].

When the sample size is small, say less than 30, or the success criterion goes beyond simply statistical significance, or there are multiple historical studies, or more accurate POS calculation is needed, a fully Bayesian approach is desirable with the following simulation steps to compute the POS.

Step 1: Model historical data and obtain a joint posterior distribution of Θ, a set of unknown parameters characterizing the treatment effect and uncertainty of that in the future study, e.g. mean response of each treatment group and/or its standard deviation.

Step 2: Understand the design of the future study.

Step 3: Randomly draw one sample Θ_i from the posterior distribution obtained in Step 1.

Step 4: Simulate N number of clinical trials according to Θ_i and the design of the future study and record whether the success criteria are met using variable S_{ij}. $S_{ij} = 1$ if the success criteria are met for the jth clinical trial.

Step 5: Repeat Step 3 and Step 4 K times.

Step 6: Estimate the POS using the average of S_{ij}, $(KN)^{-1} \sum_{i=1}^{K} \sum_{j=1}^{N} S_{ij}$.

It is helpful for understanding this procedure to relate it with classical power calculation. For power calculation, a set of values of Θ needs to be assumed, and then power can be calculated based on that set of values by repeatedly simulating clinical trials just like Step 4, and then the proportion of successful trials is the power. Thus, the above procedure for POS calculation contains power calculation as a subprocedure and the extra steps to take into account the uncertainty in Θ given historical data, which make it a more appropriate measure of how likely a study is going to succeed than power for Go/No-Go decision making.

11.4.2.1 Continuous Endpoint

To illustrate the above steps for POS calculation of continuous endpoints, consider the following example in the development of an Alzheimer's disease drug. A Phase 2a study was finished with the following results on the primary endpoint, the change of Alzheimer's Disease Assessment Scale-cognitive subscale (ADAS-Cog) 11 (a cognitive test consists of 11 items) total score from baseline to Week 12. The problem is to figure out the POS for the 3, 4, and 5 mg dose groups in a Phase 2b dose-ranging study with four arms and 150 subjects per arm, for which success is defined as one-sided p-value less than 0.05 and placebo adjusted reduction in ADAS-Cog 11 greater than 1. The efficacy of 3 mg dose group can be predicted based on the summary results in Table 11.3, while those of 4 and 5 mg have to rely on predictions from a pharmacokinetic-pharmacodynamic (PKPD) model. Let's first calculate the POS for the 3 mg dose group using an empirical Bayes approach. Let μ_0 and

TABLE 11.3
Summary Results of a Phase 2 Clinical Trial of Patients with Alzheimer's Disease

Treatment	n	LS Mean	SE	Treatment Effect Against Control	One-Sided p-Value
Placebo	65	−0.58	0.64		
1 mg	65	−1.05	0.62	−0.47	0.30
3 mg	65	−1.96	0.63	−1.38	0.06

μ_3 be the true unknown mean responses of the placebo and the 3 mg dose group, respectively. Let σ_0 and σ_3 be the true unknown standard deviations of the placebo and the 3 mg dose group, respectively. Since the ADAS-Cog 11 change score from baseline usually follows normal distribution, the model used for modeling the observed data in the Phase 2a study is the following:

$$\bar{y}_i \; N(\mu_i, \sigma_i^2/65), \text{for } i = 0, 3, \tag{11.4}$$

where $\bar{y}_i$ is the LS Mean in Table 11.3 and under the framework of empirical Bayes, $\sigma_i^2/65$ can be estimated by the observed standard error in Table 11.3 and then assumed to be known. Although there are historical placebo data in Alzheimer's disease, to simplify the example, we take a flat prior on both μ_0 and μ_3 to obtain posterior distributions of μ_i.

$$\mu_i \; N(\bar{y}_i, \sigma_i^2/65), \text{for } i = 0, 3. \tag{11.5}$$

The above posterior distributions serve as priors for predicting results of the next clinical trial. Step 1 of POS calculation is finished. Step 2, the design of the future study, is a parallelly randomized double-blind dose-ranging study with 150 subjects per arm. Steps 3–6 can be implemented in R using the codes below.

```
pval=NULL #vector for storing p-values from simulations
suc=NULL # vector for storing simulated clinical trial result
n.mcmc=10000 # number of simulated trials
n=150 # number of subjects per arm
set.seed(1234) # fix the random seed for reproducible results
mu0=rnorm(n.mcmc,-0.58,0.64)# posterior samples for placebo
mu3= rnorm(n.mcmc,-1.96,0.63)# posterior samples for 3 mg
for (i in 1:n.mcmc) # simulate n.mcmc number of trials
{
 y0.new=rnorm(n,mu0[i],5.24)# simulated placebo responses
 y3.new=rnorm(n,mu3[i],5.24)# simulated treatment responses
 pval[i]=t.test(y3.new, y0.new, alternative = "less")$p.value;
 suc[i]=pval[i]<0.05 & (mean(y0.new)-mean(y3.new))>1
}
POS=mean(suc)# estimated POS
```

The estimated POS for the 3 mg dose group is 63.5%. Similar steps can be followed for POS calculation for 4 and 5 mg dose groups except Step 1, where the posterior will be from a PKPD model. The above POS calculation is for each dose. The POS of the whole study may also be of interest, which can be defined as the dose with the best efficacy meeting the previous success criteria. Then, the R codes above can be easily modified to incorporate this scenario.

11.4.2.2 Binary Endpoint

The discussion above is about continuous endpoints. In the following, we demonstrate how to calculate POS of a future trial with binary endpoints. To estimate POS, we start with Steps 1 and 2. Assume that a fixed design of $n = 65$ per arm is planned for a future study. Assume that the success criterion will be that the observed relative risk reduction is more than 30% and trial results are statistically significant from Fisher's exact test under one-sided type I error of 0.025. In Step 3, to obtain the prior for a new study, use the binomial model in Equation 11.6:

$$y_0 \sim \mathrm{Binom}(\mathrm{p}_0, 26), \mathrm{y}_1 \sim \mathrm{Binom}(\mathrm{p}_1, 26). \tag{11.6}$$

Let $Beta(1, 1)$ be the prior of p_0 and p_1 before the historical trial was conducted. The posterior distributions of p_0 and p_1 are $Beta(1 + y_0, 1 + 26 - y_0)$ and $Beta(1 + y_1, 1 + 26 - y_1)$, respectively. These two distributions can serve as priors of p_0 and p_1 for POS calculation of a new study. Thus, Step 3 is completed. In Step 4, the following R code can be used to generate 10,000 samples of p_0 and p_1 assuming they are independent.

```
n.mcmc=10000; set.seed(1234);
pval=NULL; suc=NULL;
y0=15; y1=9; n=65;
p0=rbeta(n.mcmc, 1+y0, 1+26-y0);
p1=rbeta(n.mcmc, 1+y1, 1+26-y1);
#simulates a new trial based on the ith pair of p_0 and p_1.
y0.new=rbinom(1,n,p0[i]);
y1.new=rbinom(1,n,p1[i]);
#check the success criterion
pval[i]=fisher.test(matrix(c(y0.new,n-y0.new,
y1.new,n-y1.new),nrow=2), alternative = "greater")$p.value;
suc[i]=pval[i]<0.05 & (y0.new - y1.new)/y0.new>0.3;
#estimate the POS
POS=mean(suc)
```

For the above example, the estimated POS is 61%.

11.4.2.3 Survival Endpoint

So far, we have only focused on how to evaluate POS for future randomized studies. However, there is also a critical need for evaluating POS of ongoing studies and POS of open label studies, especially in oncology. Such POS will assist Go/No-Go decisions for expanding the target patient population by either adding more cohorts to the existing study or starting a new clinical trial and also for acquisition or in-licensing. For example, PD-1 and PD-L1 inhibitors work by boosting human's immune response against cancer cells. Thus, if it demonstrated high POS for skin cancer in the middle of a trial, additional cohort of lung cancer patients may be recruited to accelerate the development for treating lung cancer. Also in immunology, IL-4 inhibitors work in a pathway of multiple immune system related disease, and thus a new trial may be started in chronic obstructive pulmonary disease (COPD) patient population if there is high POS for the ongoing trial in asthma to succeed.

For POS of ongoing trials, a procedure below similar to the Steps 1–6 for POS of a future trial can be applied.

Step 1' Jointly model historical data together with unblinded data if any from ongoing clinical trial and obtain a joint posterior distribution of Θ, a set of unknown parameters characterizing the treatment effect and uncertainty of that for patients to be enrolled or already enrolled but endpoints not matured yet in the ongoing study.

Step 2' Understand the design of the ongoing study.

Step 3' Randomly draw one sample Θ_i from the posterior distribution obtained in Step 1'.

Step 4' For each sampled Θ_i, simulate N set of responses of patients to be enrolled and patients whose endpoints are not matured yet in the ongoing trial. The simulated "completed" trials need to match the design and interim results of the ongoing trial. Record whether the success criteria are met at the end of the trial using variable S_{ij}. $S_{ij} = 1$ if the success criteria are met for the jth clinical trial.

Step 5' Repeat Step 3' and Step 4' K times.

Step 6' Estimate the POS using the average of S_{ij}, $(KN)^{-1} \sum_{i=1}^{K} \sum_{j=1}^{N} S_{ij}$.

If the sponsor of an ongoing clinical trial conducts an unblinded interim analysis and would like to make a Go/No-Go decision, unblinded subject level information may provide more accurate POS estimation than the above POS calculated based on blinded information. Below is an example of such a situation with survival endpoint.

For demonstration purposes, consider the case where all subjects have been randomized. Otherwise, enrollment speed needs to be taken into account. Start with Step 1' for calculating POS. In Step 3', to predict events for ongoing subjects, the hazard rates, λ_1 and λ_2, for the control group and treatment group, respectively, need to be modeled. A simple approach is to assume the time-to-event (TTE) follows an exponential distribution in Equation 11.7 [1]. The log hazard ratio, θ, which was estimated by log-rank statistics, follows a normal distribution in Equation 11.7.

$$\text{TTE}_{\text{ij}} \sim \lambda_j e^{-\lambda_j \text{TTE}_{\text{ij}}} \text{ for } i = 1, \ldots, 10 \text{ and } j = 1, 2 \tag{11.7}$$

Then, the likelihood can be written as

$$\lambda_1^{d_1} e^{-\lambda_1 \text{EXP}_1} \cdot \lambda_2^{d_2} e^{-\lambda_2 \text{EXP}_2}. \tag{11.8}$$

In Equation 11.8, EXP_j is the total exposure of subjects in the jth group, and d_j is the number of events in the jth group. The unblinded interim data are tabulated in Table 11.4.

By summing up the exposure time of subjects within each group in Table 11.4, we obtain $\text{EXP}_1 = 61$ and $\text{EXP}_2 = 48$. Similarly, we have $d_1 = 5$ and $d_2 = 2$. Since there is no prior knowledge before this trial, assume a non-informative gamma prior, Gamma$(0.001, 0.001)$, on λ_j. The posterior distribution of λ_j is Gamma$(0.001 + d_j, 0.001 + \text{EXP}_j)$. These posterior distributions serve as prior distributions for the ongoing subjects, and Step 1' is completed. For Step 2', for the purpose of illustration, we consider an event driven ongoing study with 1:1 randomization and total number of events to be 15. The success criterion is that the hazard ratio is significant from the log-rank test, and the estimated hazard ratio is less than 0.75. In Step 3', draw samples from the posterior distributions of λ_1 and λ_2. In Step 4', we consider the

TABLE 11.4
Event Time and Event Status of 20 Subjects Randomized in an Ongoing Trial

Subject ID	Treatment Group	Exposure Time	Event	Ongoing
1	Treatment	3	Yes	No
2	Control	4	No	Yes
3	Treatment	8	No	Yes
4	Control	4	No	Yes
5	Treatment	9	No	Yes
6	Control	1	Yes	No
7	Treatment	1	No	No
8	Control	12	No	Yes
9	Treatment	5	Yes	No
10	Control	8	Yes	No
11	Treatment	5	No	No
12	Control	9	No	Yes
13	Treatment	2	No	Yes
14	Control	1	Yes	No
15	Treatment	5	No	Yes
16	Control	9	Yes	No
17	Treatment	5	No	Yes
18	Control	3	No	Yes
19	Treatment	5	No	No
20	Control	10	Yes	No

simpler case where all patients have been randomized, and we only need to predict the event time for ongoing subjects. The expected number of total events is 15 for this ongoing trial. Since $d_1 + d_2 = 7$ events were already observed with 10 subjects ongoing, it is possible to reach 15 events. Because the study will be finished after only eight more events, simulated events after the first eight events are censored. Next, predict the event times for ongoing subjects. Because of the memoryless property of exponential distributions, the TTE conditional on existing exposure time follows the same exponential distribution as the unconditional one. Thus, the TTE conditional on exposure, TTE_{cij}, is predicted for ongoing subjects according to Equation 11.7 and ith sample of λ_1 and λ_2 and unconditional time-to-event is calculated by summing up TTE_{cij} and EXP_{ij}.

```
set.seed(1234); n.mcmc=10^4; pval=NULL; suc=NULL; library
("survival");
lambda1=rgamma(n.mcmc,5.001, rate=58.001);
lambda2=rgamma(n.mcmc,2.001, rate=45.001);
#read the subject level interim data into a data frame
dat=read.csv("survival␣data.csv");
dat_comp=dat;
flag_cont=dat$Ongoing=="Yes" & dat$Group=="Control";
flag_trt= dat$Ongoing=="Yes" & dat$Group=="Treatment";
t2= dat_comp[flag_trt, "Exposure.Time"];
t1= dat_comp[flag_cont, "Exposure.Time"];
n2= sum(flag_trt); n1= sum(flag_cont);
t1=rexp(n1, lambda1[i])+ t1;
t2=rexp(n2, lambda2[i])+t2;
rk= rank(c(t2,t1));
e2=ifelse(rk[1:n2]<=8, "Yes","No");
e1=ifelse(rk[-(1:n2)]<=8, "Yes","No");
Tmax=max(c(t2,t1)[which(rk<=8)])
if (sum(rk[1:n2]>8)>0)
{
t2[rk[1:n2]>8]=rep(Tmax, sum(rk[1:n2]>8))
}
if (sum(rk[-(1:n2)]>8)>0)
{
t1[rk[-(1:n2)]>8]=rep(Tmax, sum(rk[-(1:n2)]>8))
}
dat_comp[flag_trt, "Exposure.Time"]=t2
dat_comp[flag_cont, "Exposure.Time"]=t1
dat_comp[flag_trt, "Event"]=e2
dat_comp[flag_cont, "Event"]=e1
#check the success criterion using the R code below.
out=survdiff(Surv(dat_comp$Exposure.Time,
```

```
as.numeric(dat_comp$Event)-1)~ dat_comp$Group);
pval[i]= 1- pchisq(out$chisq, 1);
r=out$obs/out$exp;
hr=r[2]/r[1]
suc[i]=pval[i]<0.05 & hr<0.75
# use the proportion of successful trials to estimate
 the POS.
POS=mean(suc)
```

For this example, the estimated POS is 30.0%.

Although unblinded interim analysis provides subject level data for more accurate POS estimation, a penalty on type I error control has to be paid. If such penalty is not desirable or unblinded information is not available, POS can still be calculated using all publically accessible information. Such an example was described in [1]. The background is that in 2010, Hoffman-LaRoche was considering acquisition of Genetech and a registrational trial of bevacizumab (brand name Avastin) in patients with resected Stages II and III carcinoma of the colon. The price that Hoffman-LaRoche would offer for the rest of Genetech stocks that it did not own hinged on the probability that this trial turns out to be successful. To evaluate the POS, for Step 1', data from other colon cancer trials and trials of bevacizumab in other diseases were pooled together by the authors since there are no unblinded results announced publically. Based on historical data, consistent hazard reductions of 10%–15% were observed, and there was little evidence suggesting the reduction can be 30%. A "custom" prior was obtained and discretized for the purpose of interpretation to the clinical team. However, the details of statistical models were not disclosed in [1]. A handful of other priors were also constructed to compare with the "custom" prior and visualized in Figure 11.2, where the y-axis denotes the probability or in other words the relative prior weight of each potential treatment effect in terms of percentage reduction of hazard.

For Step 2', to understand the study design, the study protocol was used to figure out the primary endpoint, primary analysis method, the sample size, and the stopping rules at the interim analysis and the timing of interim analyses. Details can be found in [1]. For Steps 3'–6', because of the discretization of the prior, a shortcut as below was used to calculate POS.

Step 3" Pick a distinct value of Θ, Θ_i, from all of its possible values under the discretized prior distribution of Θ.

Step 4" Same as Step 4'.

Step 5" Repeat Step 3" and Step 4" for all of the possible values of Θ_i under the discretized prior distribution, $i = 1, ..., L$.

Step 6" Estimate the POS using the weighted average of S_{ij}, $(LN)^{-1} \sum_{i=1}^{L} \sum_{j=1}^{N} w_i S_{ij}$, where w_i is the probability of Θ_i in the prior distribution of Θ.

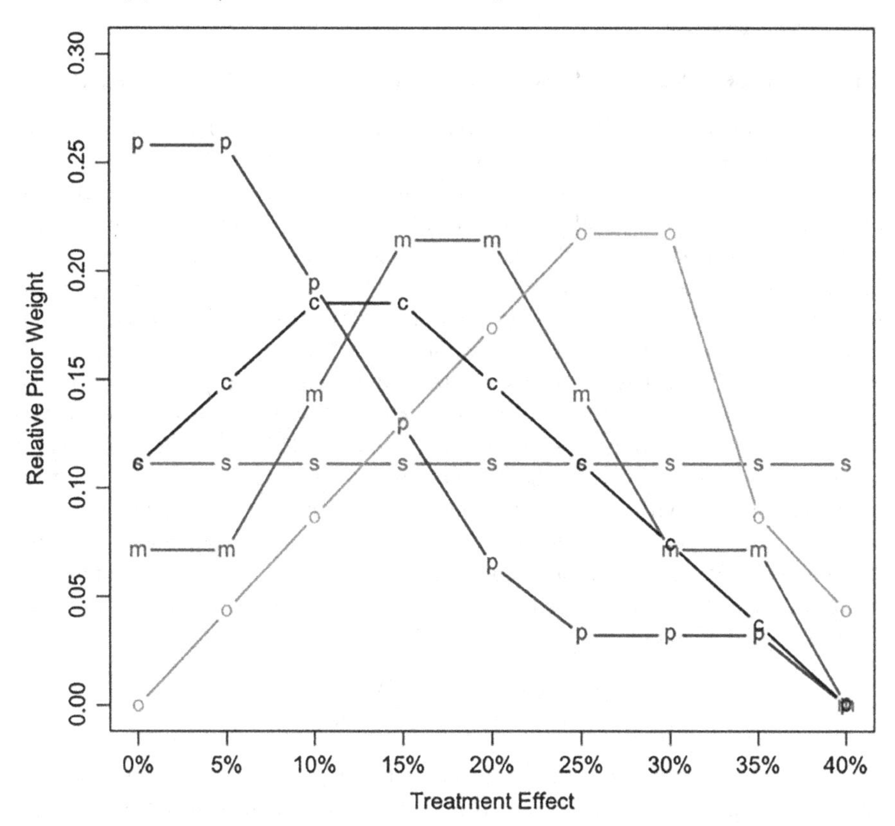

FIGURE 11.2
Priors for POS evaluation from Ref. [1]. The treatment effect is the percentage reduction of hazard due to bevacizumab. Abbreviations: s, simple flat prior; p, pessimistic prior; o, optimistic prior; m, moderate prior; and c, custom prior.

In this example, for Step 3", clinical trials were simulated under each one of the possible percentages of hazard reduction, 0%,...,40%. Then for Step 4" and Step 5", clinical trials were simulated under each of the treatment effects, and only a subset of trials which match the status of the ongoing trial were kept; that is, the trial has passed the first three interim analyses. Lastly, for Step 6", weighted averages were calculated, and the estimated POS under the choice of prior, "custom prior", is 47.5%. The actual study was finished without stopping at any interim analyses. The final estimate treatment effect is 11% reduction of hazard, which is in line with the "custom prior" and failed to achieve statistical significance (p-value = 0.15), which is much more consistent with the estimated POS (47.5%) than the original planned study power (90%).

11.5 Asset Value Based Go/No-Go Decision Making

Go/No-Go decision making often requires prediction of potential financial return from an investigational product besides calculation of POS, especially when such decision making may affect other products in a portfolio of investment. The financial return can be simply calculated as revenue minus cost, which consists of development cost and commercialization cost. Since financial return needs to be calculated for each year after regulatory approval until patent expiration, an aggregated sum of returns is needed for Go/No-Go decision making. Thus, net present value (NPV) is proposed as a measure of the present value of total returns during the life time of a product. It is the difference between the present value of all the cash inflows (revenues) and the present value of cash outflows (costs) over a period of time [24]. It is a method of calculating the profitability of a project or investment. The NPV is the tool of choice for most Go/No-Go for two reasons. First, NPV considers the time value of money, translating future cash flows into today's currency. Second, it provides a concrete number that managers can use to easily compare an initial outlay of cash against the present value of the return. The formula for calculating NPV is below.

$$\text{NPV} = \sum_{n=1}^{T} \frac{\text{FV}}{(1 + \text{DiscountRate})^n},$$

where n is the year whose cash flow is being discounted and FV is the projected cash flow, which is the cash inflow minus cash outflow, for each year.

11.5.1 Combine POS with NPV to Calculate Expected NPV

The prediction of NPV relies on the assumption that a product can demonstrate a certain profile of efficacy and safety advantages over competitors in registrational clinical trials. Based on limited knowledge of the product's clinical performance, the probability of achieving such a profile in a clinical trial may not be 100%, and there is a certain chance of failing to demonstrate favorable profile in a clinical trial. Thus, expected NPV (eNPV), which is the probability weighted average of the NPVs under all the possible development outcomes, is a more reliable measure of expected return and has been used in clinical trial or clinical program optimization. eNPV can be calculated using the formula below:

$$\text{eNPV} = \sum_{i=1}^{K} \text{NPV}_i \cdot \text{Prob}(\text{Outcome}_i), \tag{11.9}$$

where NPV_i is the NPV under the ith study outcome, $i = 1, \ldots, K$, and $\text{Prob}(\text{Outcome}_i)$ is the probability of observing the ith outcome conditional on all the available information about the product.

TABLE 11.5
Probability of Success and NPVs of Two Study Design Options

Study Option	Study Outcome	Probability (%)	NPV
80% Powered study	Success	70	\$3 billion
	Failure	30	\$ −0.2 billion
90% Powered study	Success	73	\$2.5 billion
	Failure	27	\$ −0.25 billion

To illustrate the utility of eNPV, consider an oncology example, where a Phase 2 study has been finished and there is an option of conducting a 80% powered Phase 3 study versus the option of conducting a 90% powered Phase 3 study. Due to the challenge of enrollment, a 90% powered study will push the product launch date 6 months later and thus miss the opportunity of being the first-to-market, which leads to smaller market share and lower NPV. The POS and NPV of the two options are summarized in Table 11.5.

For each study in Table 11.5, there are two possible outcomes, success or failure, where success indicates regulatory approval and achievement of a certain benefit risk profile. If the Phase 3 trial is successful, the NPVs are \$3 and \$2.5 billion for 80% powered study and 90% powered study, respectively; otherwise, the NPVs are \$ −0.2 and \$ −0.25 billion for the two studies. According to Equation 11.9, the eNPVs are $70\% \cdot 3 - 30\% \cdot 0.2 = \$2.04$ billion and $73\% \cdot 2.5 - 27\% \cdot 0.25 = \$1.75$ billion for the two studies. Thus, based on eNPV, an 80% powered study shall be recommended. Of course, sensitivity analyses on POS and NPV calculation shall be conducted to understand the robustness of the conclusion against miscalculation in POS or NPV.

More complicated examples using eNPV for clinical program optimization can be found in [25], where eNPV was used as one of the decision criteria to choose the optimal clinical development plan among three clinical development plans. For each plan, the POSs are considered together with the NPVs for all possible outcomes. Plan A is a development plan which utilizes limited Phase 2 studies before starting two large Phase 3 studies. Plan B utilizes a powered imaging study in Phase 2 prior to Phase 3 to reduce the risk of Phase 3 failure. Plan C uses an adaptive Phase 2b/3 study to reduce risk and speed development. Plan C has the highest eNPV and is the chosen plan. Please refer to [25] for details.

11.6 Conclusions and Discussions

Traditionally, the planning of Go/No-Go decision did not happen until clinical trial results are unblinded which leaves the study team limited time to quantify the evidence supporting a Go/No-Go decision based on totality of data. Thus, a prospective approach is proposed to replace the traditional retrospective method for Go/No-Go decision making. The proposed approach can be applied at the end of a Phase 2 study, at an interim analysis of an ongoing study, and in a due diligence project to support Go/No-Go decision making. In the new approach, Bayesian POS is a critical factor for decision making, which was illustrated using four case studies of a future or an ongoing clinical trial for three types of endpoints: continuous, binary, and survival.

Another critical factor for Go/No-Go decision making is NPV, which measures the cumulative net cash flow through the whole life cycle of a product in current value of money. However, evaluation of NPV is based on an assumption of a certain benefit–risk profile of a product, which is uncertain. Thus, combining POS with NPV is desirable to provide a more accurate measure of expected return, which is called eNPV. eNPV can be instrumental in Go/No-Go decision making, and the robustness of eNPV needs to be studied before a firm decision is made, since a slight change in any of the assumptions used in POS or NPV calculation may alter eNPV a lot.

References

[1] Kristine R Broglio, David N Stivers, and Donald A Berry. Predicting clinical trial results based on announcements of interim analyses. *Trials*, 15(1):73, 2014.

[2] David W Thomas, Justin Burns, J Audette, A Carrol, C Dow-Hygelund, and M Hay. *Clinical Development Success Rates 2006–2015*. San Diego, CA: Biomedtracker/Washington, DC: BIO/Bend, Amplion, 2016.

[3] Roger Collier. Rapidly rising clinical trial costs worry researchers, 2009.

[4] Aylin Sertkaya, Hui-Hsing Wong, Amber Jessup, and Trinidad Beleche. Key cost drivers of pharmaceutical clinical trials in the United States. *Clinical Trials*, 13(2):117–126, 2016.

[5] Ronald L Wasserstein and Nicole A Lazar. The ASA's statement on p-values: Context, process, and purpose, 2016.

[6] Christy Chuang-Stein. Sample size and the probability of a successful trial. *Pharmaceutical Statistics*, 5(4):305–309, 2006.

[7] Ding-Geng Chen and Shuyen Ho. From statistical power to statistical assurance: It's time for a paradigm change in clinical trial design. *Communications in Statistics-Simulation and Computation*, 46(10):7957–7971, 2017.

[8] Anthony O'Hagan, John W Stevens, and Michael J Campbell. Assurance in clinical trial design. *Pharmaceutical Statistics*, 4(3):187–201, 2005.

[9] Christy Chuang-Stein, Simon Kirby, Jonathan French, Ken Kowalski, Scott Marshall, Mike K Smith, Paul Bycott, and Mohan Beltangady. A quantitative approach for making go/no-go decisions in drug development. *Drug Information Journal*, 45(2):187–202, 2011.

[10] Tony Sabin, James Matcham, Sarah Bray, Andrew Copas, and Mahesh KB Parmar. A quantitative process for enhancing end of phase 2 decisions. *Statistics in Biopharmaceutical Research*, 6(1):67–77, 2014.

[11] Joseph G Ibrahim, Ming-Hui Chen, Mani Lakshminarayanan, Guanghan F Liu, and Joseph F Heyse. Bayesian probability of success for clinical trials using historical data. *Statistics in Medicine*, 34(2):249–264, 2015.

[12] Michael Hay, David W Thomas, John L Craighead, Celia Economides, and Jesse Rosenthal. Clinical development success rates for investigational drugs. *Nature Biotechnology*, 32(1):40–51, 2014.

[13] Joseph A DiMasi, Henry G Grabowski, and Ronald W Hansen. Innovation in the pharmaceutical industry: New estimates of R&D costs. *Journal of Health Economics*, 47:20–33, 2016.

[14] Paul Frewer, Pat Mitchell, Claire Watkins, and James Matcham. Decision-making in early clinical drug development. *Pharmaceutical Statistics*, 15(3):255–263, 2016.

[15] Shein-Chung Chow and Mark Chang. Adaptive design methods in clinical trials—A review. *Orphanet Journal of Rare Diseases*, 3(1):11, 2008.

[16] Xian Zhou, Suyu Liu, Edward S Kim, Roy S Herbst, and Jiun-Kae Jack Lee. Bayesian adaptive design for targeted therapy development in lung Cancer—A step toward personalized medicine. *Clinical Trials*, 5(3):181–193, 2008.

[17] Frank Bretz, Franz Koenig, Werner Brannath, Ekkehard Glimm, and Martin Posch. Adaptive designs for confirmatory clinical trials. *Statistics in Medicine*, 28(8):1181–1217, 2009.

[18] FDA Draft Guidance. Adaptive design clinical trials for drugs and biologics. *Biotechnology Law Report*, 29(2):173, 2010.

[19] Jürgen Hummel, Song Wang, and John Kirkpatrick. Using simulation to optimize adaptive trial designs: Applications in learning and confirmatory phase trials. *Clinical Investigation*, 5(4):401–413, 2015.

[20] Yiyi Chen, Zunqiu Chen, and Motomi Mori. A new statistical decision rule for single-arm phase ii oncology trials. *Statistical Methods in Medical Research*, 25(1):118–132, 2016.

[21] Pam Belluck. Eli Lilly's experimental Alzheimer's drug fails in large trial. www.nytimes.com/2016/11/23/health/eli-lillys-experimental-alzheimers-drug-failed-in-large-trial.html. Accessed: 2018-01-02.

[22] Pam Belluck. Bristol Myers: Opdivo failed to meet endpoint in key lung-cancer study. www.wsj.com/articles/bristol-myers-opdivo-failed-to-meet-endpoint-in-key-lung-cancer-study-1470400926. Accessed: 2018-01-02.

[23] U.S. Food and Drug Administration. Twenty two case studies where phase 2 and phase 3 trials had divergent results.

[24] Carl-Fredrik Burman and Stephen Senn. Examples of option values in drug development. *Pharmaceutical Statistics*, 2(2):113–125, 2003.

[25] Steven A Julious and David J Swank. Moving statistics beyond the individual clinical trial: Applying decision science to optimize a clinical development plan. *Pharmaceutical Statistics*, 4(1):37–46, 2005.

12

Simulation for Bayesian Adaptive Designs—Step-by-Step Guide for Developing the Necessary R Code

J. Kyle Wathen

Janssen R&D

CONTENTS

12.1 Introduction

Simulating a clinical trial can be a complex process; however, the knowledge gained during the simulation process can greatly improve the likelihood of success for the entire program. Simulations are typically done to access various statistical properties, but additional information can also be obtained. To determine the required details for simulation, a team is often forced to think more critically about all aspects of the trial. With the advances in computing speed and availability of large grid computing, it is now possible to include more complex settings and evaluate more realistic situations. The use of clinical trial simulation (CTS) is greatly increased over the last 10 years. Many pharmaceutical companies have published examples involving simulations [1–7]. Through simulation, the team can often reduce uncertainty and likelihood of costly errors, and increase the understanding of potential risks before the trial ever enrolls a patient.

In many research institutions and pharmaceutical companies, statisticians are typically focused on clinical development involving patients with a particular disease such as breast cancer, cardiovascular disease, or infectious disease. This approach is useful because it allows the statistician to gain a deeper more complete understanding of the specifics of a disease that may be needed to design a more informative clinical trial. However, due to the amount of trial-related efforts (from initiation to reporting) required from the statistician, this typically leaves very little time for considering simulation or adaptive aspects because both of these also require a large time and effort commitment as well as specialized knowledge such as for developing simulation software if none is available. Typically, a trial development team consists of multiple experts from clinical, regulatory, logistics, marketing, and statisticians. Including a statistician that is an expert in clinical trial design and simulation on the trial development team can simplify consideration of non-standard approaches and result in ways to improve the design and increase likelihood of success. The simulation expert will work very closely with the trial statistician and will require input from all members of the trial development team to achieve a simulation that reflects what is realistic and closely matches how the trial will be conducted in practice.

In this chapter, we develop a series of R [8] (version 3.4.0) programs to simulate a Bayesian adaptive design. While other programming languages, like C++, may be more computationally efficient, R is a commonly used programming language among statisticians. Several factors contribute to the popularity of R: it (1) is available as freeware, (2) can be extended by users and packages shared with others, (3) is easy to learn and begin development when compared to other languages like C++, and (4) is versatile with numerous free packages developed by the R users. To make this concept concrete and ease the learning curve for beginners, we begin with a simple fixed sample Bayesian example. Next, we proceed to a more complex design with multiple interim analysis. Finally, we conclude with a Bayesian adaptive design that incorporates a binary outcome that incorporates Bayesian adaptive randomization with the stopping rules and randomization probabilities that are continually updated. The goal of this chapter is to provide users with an example which begins with a simple design and builds upon the design to gain the skills necessary to simulate many adaptive designs. The goal of this chapter is not to promote any one methodology or design approach. As such, each of the complete designs is available in a GitHub repository located at https://github.com/kwathen/IntroBayesianSimulation. In the repository, there are three examples, and each example is a standalone R Studio [9] project. There are many topics in the context of Bayesian simulation and software development such as prior elicitation, Markov chain Monte Carlo

(MCMC), code testing, validation, and generic functions that are beyond the scope of this chapter, but the reader is encouraged to learn about each of these topics prior to designing more complex simulations.

The remainder of this chapter is organized as follows:

- Section 12.2 introduces some basic simulation terminology.
- Section 12.3 describes the Bayesian adaptive design we would like to simulate.
- Section 12.4 provides a few programming best practices.
- Section 12.5 begins the development of a simulation program from scratch in R where we lay out the basic functions needed for the simulation.
- Section 12.6 improves the example started in Section 12.5 and adds the adaptive randomization feature.
- Section 12.7 extends the previous examples to include a futility rule and improvement of the adaptive randomization feature.

This chapter is intended to develop a set of skills and go through the thought process of how to simulate a Bayesian adaptive design. The reader may refer to other sources for best practices [10].

12.2 Common Simulation Terminology

For ease of explanation, we introduce simulation terminology that is used throughout this chapter. We explain concepts in the context of a two-arm trial comparing the standard of care (S) to an experimental (E) treatment. We also use the simplifying assumption that a positive change on the primary outcome is desirable. The concepts and approaches of simulation discussed here can easily be adapted to fit more complex settings. We begin with a simple approach and extend the design.

There are numerous commercial and freeware packages available for CTS. Some more frequently used commercial packages are EAST® [11] and ADDPLAN® [12]. In addition, some freeware packages can be downloaded from The University of Texas M.D. Anderson Cancer Center Quantitative Research Computing software download [13]. However, it is not uncommon that a customized simulation software must be developed to address the adaptive elements, logistical considerations, or other decision points that may have to be considered in any adaptive clinical trial.

Regardless of what software is used, we use the term "**Virtual Trial Simulator (VTS)**" to reference the complete collection of components necessary to simulate and understand the clinical trial design. Conceptually, the

VTS consists of several components that each perform a specific task. The VTS consists of a virtual trial, simulation model(s), analysis model(s), and decision rules, and one typically supplies scenarios, or hypothetical situations, they want to simulate.

We refer to the trial being simulated as the ***virtual trial***. The virtual trial consists of the analysis model and decision rules. The goal of the simulation is to utilize software to make the virtual trial design match what will be done when the trial is conducted. Using this approach, the virtual trial is simulated repeatedly by enrolling ***virtual patient*** or computer-generated patients, which allows the clinical trial team to obtain the average behavior or ***Operating Characteristics (OCs)***. The OCs consist of frequently required things such as false-positive, power, and average sample size. Commercial software will often provide a wide variety of additional information that is used in understanding the performance of different trial designs. A major advantage of custom software is the ease of obtaining any summary information that could be used for decision making, such as average number of patient events at a given point in the trial or the probability of making various decisions under many cases.

A subtle, but important, aspect of simulating a clinical trial is to understand the difference between the analysis model and the simulation model. Statisticians and clinical teams are familiar with the analysis model, because the ***analysis model*** is used during interim analysis and final analysis to analyze the data obtained from the trial and are typically documented in the statistical analysis plan for the trial. However, the simulation model is not part of a typical design because no simulation is done. To simulate the trial, the ***simulation model*** provides the details about how the trial will be simulated. The details include specifics about how the virtual patients will be generated and how the logistical aspects of the trial will be simulated. For virtual patients, the simulation model includes details about how the primary outcome and other patient outcomes and characteristics, if needed, will be generated during the simulation. It is important to note that the analysis model and simulation model do not need to be the same. For example, the simulation model may include a patient covariate, such as age, that impacts the primary patient outcome but the patient's age in not part of the analysis model. If the analysis model and the simulation model do not match, then the simulation will highlight how sensitive the analysis model and decision making is to departures from what is expected. For the trial, the simulation model describes how aspects of the trial will be simulated. For example, the simulation model provides details about how the patient arrival times in the virtual trial are generated. In practice, it is common for the patient accrual to increase over the first several months of the trial and the details on how this is incorporated in the simulation would should be part of the simulation model.

12.3 Desired Bayesian Adaptive Design

In this section, we describe the desired Bayesian design in detail. The subsequent sections begin with a simple design and walk through the steps to obtain a working R program for simulating the design described in the chapter. Statisticians are often trained and educated on how to simulate data from various distributions and commonly focus on fixed designs that allow for the simulation of sufficient statistics rather than patient-level data. Simulating a trial at the sufficient statistic level is much easier, and less flexible, than simulating a trial at the patient level. The trial we are developing in this example is a randomized, two-arm trial where the primary outcome is binary and is observed, on average, 2 months after treatment. Before each patient enrolls, we want to utilize all available data to compute the probability that one treatment is better than the other and this probability is used to adapt the randomization probabilities and checking stopping rules. If it is likely that one treatment is superior, then we want to stop the trial and select the superior treatment. We would like to have a minimum of 20 patients enrolled in the study before we stop for futility or success and before we begin any adaptive features. Based on historical data, the recruitment will ramp up over the first 6 months of the study. From the simulation, we would like to compute the OCs such as the probability of selecting each arm, probability of selecting an arm before the last patient is enrolled, the average length of the study, and the average number of patients on each treatment.

While this design may seem simple on the surface, there are many subtle details that the simulation should account for and care must be taken when developing the necessary software. For example, a patient's outcome is known once we simulate a value, say from a Bernoulli distribution, but in the real trial, it will be 2 months between treatment and evaluation of response, on average, and this fact should not be ignored when simulating the trial. In addition, in an adaptive trial that is repeatedly looking/analyzing the data, the simulation must account for the recruitment process. More specifically, in this example, one must simulate the enrollment time of the patient to know when their binary outcome is observed. If recruitment is fast, then less data will be observed and thus available for making decisions. For example, suppose the maximum number of patients enrolled was 50 and 25 patients are enrolled each month, then by the time the first patient's outcome is observed at about 2 month, the trial has likely reached maximum enrollment and no adaptions can be made. This is in contrast to a case where two patients are recruited each moth and thus only four patients are expected to enroll between the time a patient enrolls and their outcome is available to be included in the analysis. There are many subtleties that should be accounted for, and some are addressed in subsequent sections.

For ease of expiation, we denote the two treatments by S, for standard of care, and E for experimental. We assume a maximum of $N = 200$ patients will be enrolled with the goal to compare treatments S and E based on the binary responses with probability θ_S and θ_E, for S and E, respectively. We assume θ_S and θ_E follow independent Beta(a_j, b_j) priors for $j = S$ or E. It is well known that, after observing x_j response out of n_j patients treatment with $j = S$ or E, the resulting posterior is a Beta($a_j + x_j$, $b_j + n_j - x_j$). We calculate the posterior probability the E is better than S, $p_{E>S} = \Pr(\theta_E > \theta_S \mid \text{data})$, and thus, $p_{S>E} = 1 - p_{E>S}$. In the trial, $p_{E>S}$ and $p_{S>E}$ are utilized to make decisions and adapt randomization. It is important to note that the data that goes into the calculations in the simulation would only be the data that is available at the time the calculation was computed.

12.4 Programming Best Practices

Before we begin with the example of developing simple simulation, we introduce the conventions that are utilized in this chapter. These are simple practices and guidelines we have utilized in developing simulations that are extremely complex and developed over months or even years in some cases. All examples are developed using a project in RStudio version 1.4.423 [9]. While this is not a requirement, RStudio provides many nice features that make it easier to develop projects and reduce the risk of bugs or errors. All code is developed using the following conventions.

12.4.1 Programming Conventions

1. Use descriptive names for functions and variables.
2. Variable and function names should be camel case (first letter of important words are in capital, no spaces); e.g., RunAnalysis and MakeDecision are better than run_analysis or run.ana.
3. Use the following prefixes to help others understand what you are doing with a variable

    a. Prefix integer variable with an n then camel case; e.g., nQtyOfReps would be an integer variable for the quantity of replications, nQtyOfPats = quantity of patients.
    b. Prefix double of float variables with d, e.g., dMean would be a double/float variable for mean.
    c. Prefix vectors with v, e.g., vMeans would be a vector of means.
    d. Prefix matrix with m, e.g., mVarCov would be a matrix for the variance–covariance.
    e. Prefix list with an l, e.g., lData would be list of data.

f. Prefix a class variable with a c, e.g., cAnalysis = structure(list(), class= "TTest").

4. Avoid using . in function and variable names. This recommendation is based on how R name highlighting is done and variable names with a . do not properly highlight.
5. Try to make functions short with a specific task. A general rule of thumb is 1–2 screens of code for a function, and any more code you should consider creating more functions. This approach helps with testing.
6. When given the option between a really efficient, very difficult to follow approach and a less efficient easily understood option, ALWAYS take the less efficient option that is easy to follow and less likely to have bugs. If you are trying to improve the speed of your simulations, you can always re-write to the more efficient and would be able to keep both and compare results for testing purposes.
7. The idea is to make the simulation source code read like a book rather than making it unnecessarily long and complicated to follow.
8. For larger projects, use multiple R files that each contains functions with a specific task. For example, you may want to have a file called SimulatePatientOutcome.R and include various functions for simulating patient outcomes.

12.5 Make a Plan, Start Simple—Example 1

As the desired design increases in complexity so does the amount of subtle details that must be accounted for and included in the simulation. Before considering the details of a simulation, we begin by considering things that are required, regardless of the complexity of the design. For example, rather than begin with the complex Bayesian adaptive design, we could consider a simple design with two treatments Standard (S) and Experimental (E), binary data that is observed at 1 month after enrollment, equal randomization for the entire trial, no early superiority or futility checking, and at the end of the study, if $p_{E>S} > 0.9$, then the trial will conclude E is superior to S, and if and $p_{S>E} > 0.9$, then the trial will conclude S is superior to E. This simplified design would eliminate several aspects that are more difficult to implement. In addition, for the simplified design, we assume a set number of patients are accrued each month.

Getting started can often be a daunting task. Much like writing a paper, it is a good idea to start with a draft outline. Start by listing tasks that need to be developed in order to simulate the simplified design. The following list is an example of the tasks to begin with.

1. Patient variables
   a. Arrival or enrollment time
   b. Treatment
   c. Primary outcome (binary in this case)
   d. Time the outcome is observed, 1 month after arrival in this simplified example.
2. Simulate arrival times.
3. Compute the calculation of interest, namely $p_{E>S}$ and $p_{S>E}$.
4. Decision at the end of the study.

Once the brainstorming is complete, it is time to start development. In the package from Github.com, there are three examples. Example 1 is one potential solution to the simplified trial described in this section. Many people new to CTS develop a simulation program in a single .R file. While this approach is sufficient for very small projects, it can quickly become very difficult document to follow, test, and check for accuracy. A much better approach is to start with an outline above and decide which pieces could be created in separate file. For example, in the current example, we begin with a simple randomizer that does equal randomization, whereas in the complex design, we have an adaptive randomizer that incorporates all available patient data. Since we are planning on modifying the randomizer, creating a Randomizer.r file would be a great approach. By doing this, one can find all code relevant for how patients are randomized in a single file. This also creates a central place to keep all randomizers that are utilized. By creating separate files for each important task, the project becomes much more manageable and simplifies development of a test project.

Example 12.1: Project Overview

Main.r—This file contains the main R code to set up the inputs, simulates a single trial, creates a for-loop to simulate many virtual trials, and produces simple summaries. This is the main file in the project and is responsible for sourcing the remaining files in the project.

Randomizer.r—This file contains the function(s) to randomize patients. For this example, this file will only contain one function but more will be added in the subsequent examples.

SimulatePatientOutcome.r—This file contains functions used to generate the patient outcome based on the treatment they receive.

AnalysisMethods.r—The file that contains functions that pertain to calculations that are based on patient data

SimulateTrial.r—The file that contains the SimulateSingleTrial (discussion below) function that will simulate 1 virtual trial.

Functions.r—Miscellaneous functions such as simulating the patient enrollment times and a function to make decisions based on $p_{E>S}$ and $p_{S>E}$. If one wanted to have more complex decision-making functions or accrual patterns (such as a ramp-up), then additional R files could be added so that the code that performs the similar tasks is in the same file.

This simplified trial could be simulated in a few steps with less functions; however, the R code in Example 12.1 provides a starting template to simulate the more complex design. The function SimualteSingleTrial, located in SimulateTrial.R, is critical to understand. To develop this function, it is very useful to start with pseudo, or fake, code outlining the steps in simulating a single virtual trial. For example, one may write the following description followed by pseudo code for how to simulate a single trial:

Verbal Description

1. Simulate the arrival times for all patients.
2. Randomize each patient to a treatment.
3. Use the treatment to simulate patient outcomes.
4. Analyze the data to compute $p_{E>S}$ and $p_{S>E}$ utilizing all patient data.
5. Make a decision based on $p_{E>S}$ and $p_{S>E}$.

Pseudo Code 1.0

```
SimulateArrivalTimes( )
For each patient
{
        RadomizePatient( )
        SimulatePatientOutcome( )
}
RunAnalysis( )
MakeDecision( )
```

If you understand the pseudo code above, there is a clear connection between it and the R code developed (see Figure 12.1). The pseudo code above corresponds to the red squares in Figure 12.1. In the R function listed in Figure 12.1, lines 12–17 are doing a few preliminary "book keeping" variable initializations. The for-loop starting on line 25 is of particular interest because this loop is going through and randomizing each patient and simulating each patient's outcome based on the treatment they receive. In this example, it would be more efficient to do a vectorized approach like the following:

Vectorized Version

```
nMaxQtyOfPats  <- 200
vRespRate      <- c( 0.2, 0.6)
vTreat         <- rbinom( nMaxQtyOfPats, 1, 0.5 ) + 1
vOut           <- rbinom( nMaxQtyOfPats,1,
                  vRespRate[ vTreat ])
```

```
5  SimulateSingleTrial <- function( nMaxQtyOfPats, dQtyPatsPerMonth,
6                                   dPriorAS, dPriorBS,
7                                   dPriorAE, dPriorBE,
8                                   dPU,
9                                   dTrueRespRateS,
10                                  dTrueRespRateE
11                                  )
12 {
13     #Setup the variables needed in this function
14     vPatOutcome <- rep( NA, nMaxQtyOfPats )       # Vector that contians the patients outcome
15     vTreat      <- rep( NA, nMaxQtyOfPats )       # Vector that contains the patients treatment S = 0, E = 1
16     vQtyPats    <- rep( 0, 2 )                    # Vector to keep track of the number of patients on S and E
17
18     #Simulate arrival times and times the outcomes are observed
19     vStartTime  <- SimulateArrivalTimes(dQtyPatsPerMonth, nMaxQtyOfPats )
20     vObsTime    <- vStartTime + 1  # Note: In this example we observe the outcome 1 month after they enroll (or are treated)
21
22
23     #For loop to randomize and simulate the patient outcomes.
24     #Note: In this example this could be done easier, but this is a building block for when we want ot update randomization before each patient
25     for( i in 1:nMaxQtyOfPats )
26     {
27         vTreat[ i ]         <- GetTreatment( 0.5 )
28         vPatOutcome[ i ]    <- SimulatePatientOutcome( vTreat[ i ], dTrueRespRateS, dTrueRespRateE )
29
30         vQtyPats[ vTreat[ i ] + 1 ] <- vQtyPats[ vTreat[ i ] + 1 ] + 1
31     }
32
33     dCurrentTime    <- vObsTime[ nMaxQtyOfPats ] + 0.00001  #Adding 0.0001 to make sure all patient outcomes are observed
34     dProbSGrtE      <- RunAnalysis( dCurrentTime, vPatOutcome, vTreat, vObsTime, dPriorAS,dPriorBS, dPriorAE, dPriorBE )
35     dProbEGrtS      <- 1.0 - dProbSGrtE
36
37     nDecision       <- MakeDecision( dPU, dProbSGrtE, dProbEGrtS )
38
39     #Build the return list - in a large scale simulation you may not want to return the patient data
40     lRet <- list( nDecision = nDecision, dProbEGrtS = dProbEGrtS, vQtyPats = vQtyPats,
41                   vPatOutcome = vPatOutcome, vTreat = vTreat, vStartTime = vStartTime, vObsTime = vObsTime )
42     return( lRet )
43
44 }
```

FIGURE 12.1
SimulateSingleTrial taken from SimulateTrial.r squares indicate where the R code is that accomplishes each of the tasks in Pseudo Code 1.0

The vectorized version would be more efficient because it does not use for-loops and avoids repeated calls to various functions. However, this version is less readable, especially the last line as it always raises the question: What response rate is used for each patient?

The "for-loop" approach in Figure 12.1 offers several advantages: (1) readability, (2) easily extended to allow for adaptive randomization, and (3) simulation of patient outcome could easily be extended to account for patient covariates.

Developing complex simulations is often a delicate balance between readability and easy of extensions versus efficiency. On larger scale complex projects, it is advised to start small and add on as is done in this chapter. If efficiency becomes an issue and extensions are not needed, less readable more efficient version can be implemented and utilize both version for testing purposes.

12.6 Extending the Simple Simulation—Example 2

In this section, we add some very important aspects to the simulation program developed in the previous section. In particular, we go through the steps and requirements to add the adaptive randomization feature and update the section that accounts for the fact those patients' outcomes are observed, on average, 2 months after treatment. We also add in the input variable to allow the user to easily input the minimum number of patients enrolled before adaptive randomization begins.

Due to the way the project has been organized into files with functions for specific tasks, the required changes to the R code should be more straightforward than trying to scroll through one long R file that does the entire thing. One could use the following thought process in planning the changes:

1. Addition of an input parameter for the minimum number of patients. This impacts main.r (where we define the input), and this variable needs to be sent to the SimulateSingleTrial( ) function (located in SimulateTrial.R).
2. Patient outcomes are observed 2 months after treatment, on average. This impacts the SimulateSingleTrial( ) function, and we need to create a function to simulate the time; see SimulatePatientOutcome.r for the new function SimulateOutcomeObservedTime( ).
3. Create a new randomizer that will perform the adaptive randomization, GetTreatmentAdaptiveRandomization() function found in Randomizer.r.
   a. Since this approach will compute $p_{E>S}$ and $p_{S>E}$ utilizing all patient data before each patient is enrolled, we need to modify the RunAnalysis() function to only use available data.
   b. It is useful to track the randomization probabilities for each patient.

While the additions may seem minor, for the above thought process, the additions require creation of two new functions and modifications to the following files: (1) Main.r, (2) SimulateTrial.r, (3) SimulatePatientOutcome.r, and (4) Randomizer.r. After all modifications have been implemented to accomplish the changes, it is a good practice to run a small simulation and check that everything appears to be running correctly. An even better goal is to develop test functions to verify that the code performs as desired as the project is being built; however, this concept is beyond the scope of this chapter. Therefore, we settle for some basic print statements that will allow us to check specific facts. For example, since patient outcomes take approximately 2 months to observe and approximately 7.5 patients are accrued each month, we should expect about 15 (2 months * 7.5 patients/month) fewer outcomes than the number of patients enrolled when a new patient is enrolled. Example print statements have been added to SimulateSingleTrial() and RunAnalysis( ), and as expected, there are typically 15 fewer patient outcomes in the analysis than there are enrolled in the virtual study.

After the changes have been added, we have the major components of the desired trial implemented, only lacking a few aspects such as early stopping for futility or superiority. It is very helpful to simulate several example trials and plot various values of interest. Since we are altering the randomization ratio before each patient based on the current data, it would be useful to

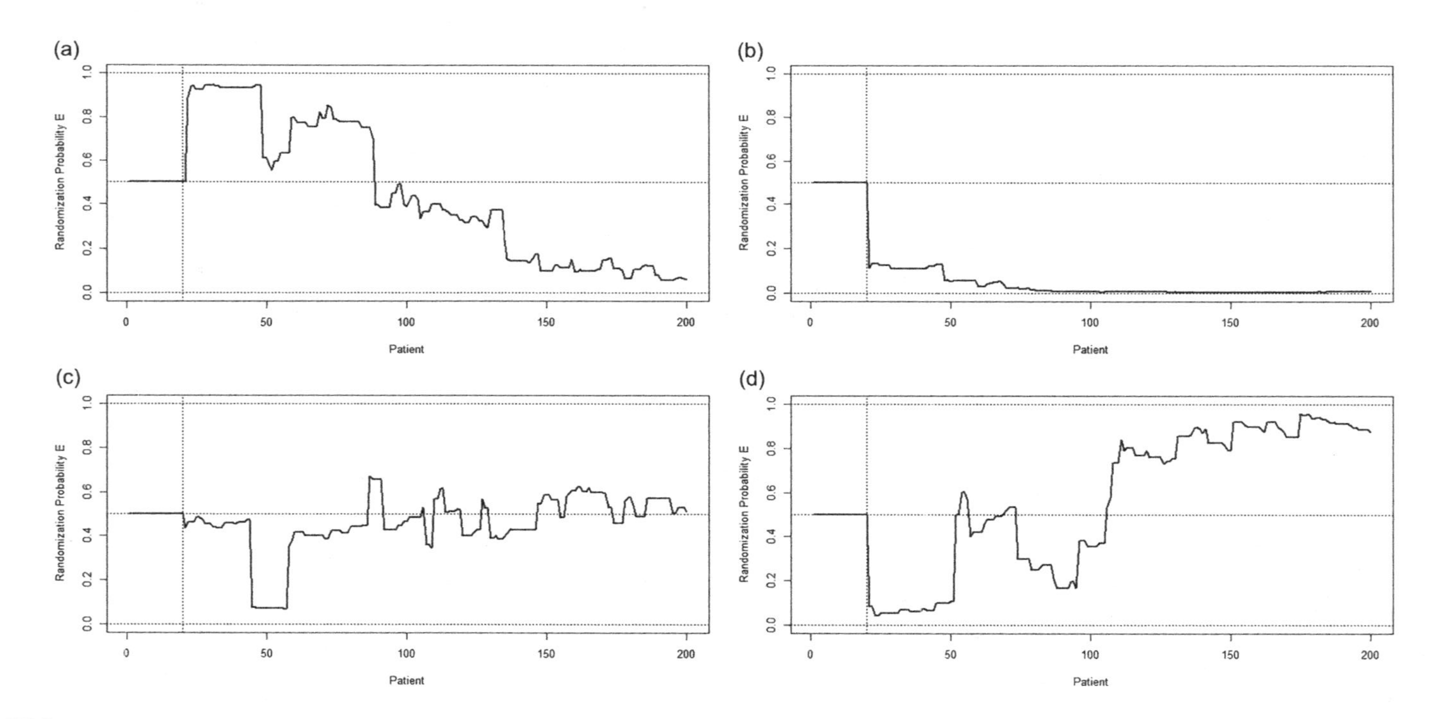

FIGURE 12.2
Randomization probability of E for each patient in four virtual trials.

plot the randomization probabilities. Figure 12.2 illustrates the randomization probabilities of four virtual trials where the true response rate for both S and E is 0.2. The dotted horizontal line represents equal randomization, and the vertical dotted line is at 20 patients, the point when the randomization probabilities may be altered. It is interesting to note how much the randomization probability can fluctuate. Examples like the trial displayed in Figure 12.2c are very important because they demonstrate a behavior where the randomization gets "stuck" very early assigning all patients to S. It is a good idea to simulate many, many virtual trials at this point gain insight into the adaptive method that goes beyond the usual averages that can often hide potential problems. If you conducted a trial and the results were like the one shown in Figure 12.2c, it would likely be unsatisfactory due to a small number of patients assigned to the experimental treatment.

Based on the simulated virtual trials and some preliminary simulation results at this stage, the desired design could be modified. In particular, instead of randomizing to E with probability $p_{E>S}$, the final design employs a function of $p_{E>S}$ for randomization as described in Wathen and Thall [14] details provided in the next section. To understand the value of this change, one could simulate a scenario using $p_{E>S}$ as the randomization, simulate the same scenario utilizing a function of $p_{E>S}$, $f(p_{E>S})$, and compare the probability of ending up with a trial similar to Figure 12.2c.

12.7 Extending to the Desired Design—Example 3

In this section, we extend the two previous examples to include design elements to achieve the desired adaptive design. In particular, the remaining trial design features that need to be implemented in this section include a stopping rule that is checked prior to every patient enrolled, and the enrollment rate increases over time and alters the randomization to utilize a function of $p_{E>S}$ and $p_{S>E}$. Each improvement adds a layer of complexity to the R code that is typical in simulating adaptive clinical trials. The changes will impact many things, such as what is reported for each virtual trial (e.g., sample size), difficulty checking the code for accuracy, and of great importance, the time required to simulate many virtual trials.

The combination of altering the randomization probabilities and possibly of stopping the trial before each patient enrolls is a difficult concept for many to grasp when they initially start designing adaptive trials and developing custom code for the adaptive design. In a fixed sample design, or limited number of analysis, one typically simulates the patient data in blocks and combines the blocks of simulated patient data at later stages in the simulation. In contrast, when adaptions are made frequently and the adaptions impact how the remainder of the trial is conducted, it is much more difficult to simulate data upfront, and one must keep track of what data would be available at a given analysis. This concept is especially important in the case of patient outcomes that are not observed immediately, such as the delayed binary outcome in this example or time-to-event data. The basic outline of this adaptive trial simulation may look like the following:

Simplified loop for simulating an adaptive clinical trial

Step 1: Simulate the arrival time of the next patient.

Step 2: Check the stopping rules to see if the trial should be stopped.

Step 3: If the trial is not stopped, update randomization probabilities if needed, and randomize the patient.

Step 4: Simulate the patient outcome as a function of the treatment; this also includes the time the outcome is observed.

Step 5: Repeat steps 1–4.

To help make the example loop more concrete, we assume it takes 2 months for the patient outcome to be observed and the loop would look like the following abbreviating Step X by SX:

Loop 1:

S1: Patient 1—Simulated arrival time is $A_1 = 0$ months.
S2: No data is available; thus, the trial is not sopped.
S3: Patient is randomized to S, denoted by $\text{Trt}_1 = 0$.
S4: Outcome for patient 1 is simulated as a non-response, denoted by $X_1 = 0$, and is obverted at time $XT_1 = A_1 + 2$ months $= 2$.

Loop 2 (second time through the loop):

S1: Patient 2—Simulated arrival time is $A_2 = 0.5$ months.
S2: Check the stopping rule. Current time in the trial is equal to when the patient enrolls: Current Time CT $= 0.5$. Since the outcome for patient 1 would not be observed until CT $\geq XT_1$, no data is available.
S3: Patient is randomized to E, denoted by $\text{Trt}_2 = 1$.
S4: Outcome for patient 2 is simulated as a response, denoted by $X_2 = 1$ at time $XT_2 = A_2 + 2 = 2.5$.

Special Note: Before each patient enrolls, a new dataset is created and used for analysis and randomization. This is an important point because as more virtual patients are enrolled, it would be a major error to include outcomes that have been simulated but not observed. It is best to develop a function to create the dataset for analysis, so it can be tested for accuracy.

Loop 3: At this point in trial 2, virtual patients are enrolled.

S1: Patient 3—Simulated arrival time is $A_3 = 2.1$ months.
S2: Check the stopping rule. Current time in the trial is equal to when the patient enrolls: Current Time CT $= 2.1$. Only data that has $XT_i \leq$ CT is available. Thus, the data for S is 0 responses, 1 non-response, and no data available for E.

S3: Patient is randomized to S, denoted by $\text{Trt}_3 = 0$.

S4: Outcome for patient 3 is simulated as a response, denoted by $X_3 = 1$ at time $XT_3 = A_3 + 2 = 4.1$.

The above example demonstrates the importance of "book" keeping in simulating adaptive trials to keep track of what data is available when an analysis is conducted. Failing to implement the book keeping for making datasets is a typical mistake made by people when they begin developing custom adaptive designs. This concept is important to keep in mind when extending the code for the new features.

After considering the three extensions for this example, it is best to make a list of the code that is needed and how it will impact the current project. The thought process would look like the following:

1. Addition of early stopping rule for futility/superiority. To accomplish this task, a new file called StoppingRules.R has been added to Example 3. The MakeDecision function is moved to this file as well since the CheckStoppingRule( ) function will utilize the MakeDecision( ) function.
2. Addition of ramp-up in accrual. Modify the SimulateArrivalTimes ( ) function to utilize a vector of expected number of patients accrued each month. Since the anticipated number of patients accrued each month is expected to increase during the early stages of the trial, a new function called SimulateAMonthOfAccrualTimes() is added that generates 1 month of patients given a specified recruitment rate.
3. Modification of randomization probabilities to allow for adaptive randomization probabilities that are based on a function of $p_{E>S}$. In addition, we would like to bound the randomization to an interval, for example, [0.1, 0.9] to prevent long streaks of one treatment assignment. These changes require new input parameters in the Main.r. Consequently, this modification impacts the SimualteSingleTrial() and GetTreatmentAdaptiveRandomization( ) functions found in SimulateTrial.R and Randomizer.r, respectively.

In particular, the randomization probability ρ_E to E is defined as

$$\rho_E = \frac{p^r_{E>S}}{p^r_{E>S} + p^r_{S>E}}$$

In addition, we restrict the randomization to either arm by a minimum randomization probability, denoted by dMinRandProb. Therefore, we restrict the randomization probabilities to the interval [dMinRandProb, 1 – dMinRandProb]. The restriction is done to limit the long periods of time where the same treatment is given (see Wathen and Thall [12] for more information about this topic).

These changes listed above impact several files and adds an additional file. One thing to keep in mind when implementing changes and extending the code base, it is good practice to consider more readable code, rather than combining several things into one line of code. This is of particular importance when developing in R. For example, in order to restrict the randomization to the interval `[ dMinRandProb, 1 - dMinRandProb ]`, assuming `dRandProbS` = randomization probability for S, a single line could be written as follows:

```
dRandProbS <- min( 1- dMinRandProb,
max( dMinRandProb, dRandProbS))
```

However, this approach is less readable than was done in the code (Randomizer.r) lines 47–48 as follows:

```
dRandProbS <- max( dMinRandProb, dRandProbS)
dRandProbS <- min( 1-dMinRandProb, dRandProbS)
```

The second approach is implemented to make it easier for users to read and understand. When a task takes more than few lines of code, it can be beneficial to create a function to perform the task rather than writing extensive comments. For example, the above two lines of code could be replaced by

```
dRandProbS <- ConstrainRandomization( dRandProbS, dMinRandProb )
```

where

```
ConstrainRandomization <- function( dRandProbS, dMinRandProb )
{
        dRandProbS <- max(dMinRandProb, dRandProbS)
        dRandProbS <- min(1-dMinRandProb, dRandProbS)
        return( dRandProbS )
}
```

The approach is much easier to read and follow, and is significantly easier to test and verify it performs as desired.

12.8 Possible Extensions and Improvements

Example 3 provides a complete solution to simulate one scenario at a time of a Bayesian outcome adaptive randomization design with early stopping and continual updates to the randomization rate. All CTSs require more than one scenario to be simulated. In addition, it is often important to run through "What if" questions, such as "What if there is a hidden patient covariate that interacts with treatment?"

In order to facilitate the simulation of more than one scenario, it is often best to develop a function to simulate one scenario. Using the final code for Example 3, one could refactor the for-loop on lines 57–70 of Main.r and create a SimulateScenario function. This function would require the same input parameters as SimulateSingleTrial and one additional parameter that would specify the number of replications to simulate the scenario. This function would make it much easier for simulating many scenarios and provide full understanding of how the design would perform in practice.

Another useful extension would be to create an additional function for SimulatePatientOutcome that would simulate the patients from a heterogeneous population to help understand what impacts a hidden covariate would have on the design. As a simple example, one could simulate a binary variable for age (age <50 year or age ≥ 50) in the SimulatePatientOutcome function that, when equal to 1, would cause the patients to be more likely to respond to E. This type of simulation is very useful to help understand the risks of a design, but is often not conducted.

12.9 Conclusions

In this chapter, we have developed a simulation package in R that demonstrates many important features and skills that are needed to simulate more complex adaptive clinical trials. We began with a simple fixed sample size design and provided two extensions to accomplish the desired Bayesian outcome adaptively randomized design where randomization probabilities were updated prior to each patient enrolling. In the ever-changing world of clinical trials, simulations can provide a clear picture of how various clinical trial designs, Bayesian or frequentist, will perform in realistic scenarios, not just limits based on theory that may not apply. Utilizing the skills developed in this chapter, a statistician may alternate between communication with team member and developing simulation code to educate the team on potential gains or pitfalls in proposed designs.

Appendix: Quick Guide to Examples 1–3

Example 1	Simulate a fixed sample design
	Introduction to file layout in the project
	Project overview by file
Example 2	Extend the fixed design
	Add a minimum number of patients
	Add time between treatment and evaluation of outcome
	Add adaptive randomization function
	Update randomization probabilities
Example 3	Extend Example 2
	Add early stopping rule
	Add ramp-up in accrual
	Alter randomization as a function of $p_{E>S}$ and $p_{S>E}$

References

[1] Bhatt DL, Stone GW, Mahaffey KW, et al. Effect of platelet inhibition with cangrelor during PCI on ischemic events. *New England Journal of Medicine.* 2013; 368: 1303–1313.

[2] Donohue JF, Fogarty C, Lotvall J, et al. Once-daily bronchodilators for chronic obstructive pulmonary disease: Indacaterol versus tiotropium. *American Journal of Respiratory and Critical Care Medicine.* 2010; 182: 155–162.

[3] Kitazawa Y, KP2035 Study Group. Phase III double-blind study of latanoprost/timolol combination (KP2035) in patients with primary open-angle glaucoma or ocular hypertension. *Japanese Journal of Clinical Ophthalmology.* 2009; 63(5): 807–815. (in Japanese). Abstract in English: http://medicalfinder.jp/ejournal/1410102731-e.html.

[4] Satoh J, Yagihashi S, Baba M, Suzuki M, Arakawa A, Yoshiyama T, Shoji S. Efficacy and safety of pregabalin for treating neuropathic pain associated with diabetic peripheral neuropathy: A 14 week, randomized, double-blind, placebo-controlled trial. *Diabetic Medicine.* 2010; 28(1): 109–116.

[5] Heritier S, Lo SN, Morgan CC. An adaptive confirmatory trial with interim treatment selection: Practical experiences and unbalanced randomization. *Statistics in Medicine.* 2011; 30(13): 1541–1554.

[6] RP103-03: ClinicalTrials.gov Identifier: NCT01000961 http://clinical trials.gov/ct2/show/NCT01000961

[7] MacArthur RD, Hawkins T, Brown SJ, LaMarca A, Chaturvedi P, Ernst J. ADVENT trial: Crofelemer for the treatment of secretory diarrhea in HIV+ individuals. *Poster presented at the Conference on Retroviruses and Opportunistic Infections*, March 5–8, 2012, Seattle, WA.

[8] R Core Team (2013). R: A language and environment for statistical computing. R Foundation for Statistical Computing, Vienna, Austria. www.R-project.org.

[9] RStudio Team (2016). RStudio: Integrated Development for R. RStudio, Inc., Boston, MA. www.rstudio.com.

[10] Mayer, C, Wathen JK. Best practices in clinical trial simulations for adaptive study designs. BOOK TITLE; 2018.

[11] EAST®, www.cytel.com/software/east.

[12] ADDPLAN® Software, www.iconplc.com/innovation/addplan.

[13] M.D. Anderson Cancer Center Quantitative Research Computing, http://biostatistics.mdanderson.org/SoftwareDownload/.

[14] Wathen, JK, Thall, PF. A simulation study of outcome adaptive randomization in multi-arm clinical trials. *Clinical Trials*. 2017; 14(5): 432–440.

13

Power Priors for Sample Size Determination in the Process Validation Life Cycle

Paul Faya
Eli Lilly and Company

John W. Seaman, Jr. and James D. Stamey
Baylor University

CONTENTS

13.1 Introduction

The life cycle of a drug product, after pharmaceutical development, progresses through three stages: technology transfer, manufacturing for commercial distribution, and, finally, product discontinuation. Before commercial manufacturing, pharmaceutical firms are required to validate their production process.

This is a regulatory requirement per Parts 210 and 211 of the Code of Federal Regulations (CFR) Title 21. Its purpose is to establish scientific evidence that the production process can consistently deliver a quality product (FDA, 2011). The European Medicines Agency (2014) makes similar recommendations in its guideline for process validation. This requires analysis of validation batches, the number of which should be based on the manufacturer's prior knowledge of the process. The assurance of drug quality, safety, and efficacy is vitally dependent on this process validation and is a difficult challenge faced by pharmaceutical manufacturers.

In this chapter, we offer an approach to this problem based on Bayesian sample size determination (SSD) and assurance. Bayesian methods are ideal for incorporating prior information about the manufacturer's experience and are well-suited for implementation in the process validation life cycle. Indeed, the life cycle shares similarities with clinical trials for medical devices, where prior information from evolutionary product development has facilitated adoption of Bayesian methods. The use of Bayesian methods in device trials is the subject of an FDA guidance on the use of Bayesian statistics (FDA, 2010).

Evidence of process consistency has historically relied on the production of three consecutive validation batches. That this may not provide adequate proof of process consistency is widely acknowledged by industry practitioners and regulators. The FDA's 2011 revised guidance for industry places process validation in line with a three-stage life cycle approach. Stage I (Process Design) utilizes development and scale-up activities to establish the manufacturing process. Reproducible commercial manufacturing is the focus of Stage II (Process Qualification), in which the process design is challenged under "process performance qualification" (PPQ). Ongoing assurance that the process remains in a state of control is established in Stage III (Continued Process Verification). The revised guidance recommends that manufacturers rely on scientific rationale and a thorough understanding of the process to justify the decision for the number of validation batches. Furthermore, as the guidance notes, "the number of samples should be adequate to provide sufficient statistical confidence of quality both within a batch and between batches".

The FDA requirement that the number of validation batches for Stage II be based on prior process knowledge has proven particularly challenging. Pharmaceutical firms are expected to make a scientific proposal justifying the number of validation batches to be used. To assist manufacturers in this determination, the International Society for Pharmaceutical Engineering published a discussion paper in 2012 titled "Stage 2 Process Validation: Determining and Justifying the Number of Performance Qualification Batches". That paper acknowledges the difficulty of the task and suggests methods for determining the number of batches needed for PPQ. In addition, to date there have been at least four research papers published on this subject.

Wiles (2013) proposes a frequentist approach to continual recalculation of a process capability index after each new PPQ batch, proceeding until a lower confidence bound on the index exceeds a minimum capability criterion. This method does not prescribe a requisite sample size or a level of assurance that Stage II manufacturing will yield a "positive outcome".

A confidence interval on a product's critical quality attribute can be constructed based on previously collected product data and historical evidence of batch-to-batch variability of comparable products. Pazhayattil et al. (2016) determine the minimum number of batches by comparing this frequentist interval estimate to specification limits.

A Bayesian approach proposed by Yang (2013) considers the issue of SSD via the application of the classical conjugate beta-binomial structure. Each PPQ batch passes or fails its acceptance criteria. This beta-binomial approach does not model between-batch variation. Furthermore, Yang does not consider the issue of prior effective sample size despite using a data augmented prior.

LeBlond and Mockus (2014) use a Bayesian one-way normal random effects model to compute the posterior probability of passing the United States Pharmacopoeia (USP) compendial standard <905> (USP, 2011). Sampling from the joint posterior distribution of the parameters of interest in the model, the authors construct a tri-linear interpolation table to determine the posterior probability of passing the compendial standard. This is used, together with operating characteristic (OC) curves, to determine the number of batches for Stage II.

There is a rich literature on Bayesian SSD methods as applied in the drug development process (see Eaton et al., 2013; Muirhead and Şoaita, 2013; Chuang-Stein and Yang, 2010; Brutti et al., 2008; Whitehead et al., 2008; Patel and Ankolekar, 2007; Chuang-Stein, 2006; Wang et al., 2006; O'Hagan et al., 2005; Spiegelhalter et al., 2004; O'Hagan and Stevens, 2001; Lee and Zelen, 2000; Joseph et al., 1997; Spiegelhalter and Freedman, 1986). Standard classical designs rely on the conditional power function for SSD (see Armitage et al., 2008). However, the conditional power approach suffers from the problem of "local optimality" since it depends critically on the chosen design values for the parameter and the uncertainty of such values is not accounted for, as noted by Brutti et al. (2008). A Bayesian SSD approach uses probability distributions to model uncertainty on the design values and affords generation of predicted "future data" (Rubin and Stern, 1998) that can be used to evaluate the probability of rejecting the null hypothesis.

In what follows, we extend the Bayesian SSD approach and the closely related concept of Bayesian assurance (O'Hagan and Stevens, 2001) to the process validation life cycle. Specifically, we use the two-prior approach (Wang and Gelfand, 2002) to determine Stage II sample sizes. As with most Bayesian procedures, the approach is simulation-based, making extensive use of Markov chain Monte Carlo (MCMC) methods.

The chapter is organized as follows. In Section 13.2, we present the statistical model and detail the application of Bayesian SSD in Stage II. An example using potency uniformity as the critical quality attribute of interest is given in Section 13.3. We briefly consider an assurance-based plan for making the transition from Stage I to Stage III of the process validation life cycle in Section 13.4. Concluding remarks are given in Section 13.5.

13.2 Statistical Method

13.2.1 Hierarchical Model

For our development, we focus on the potency or "content uniformity" attribute. Potency refers to the content of drug substance, or active pharmaceutical ingredient (API), present in a tested dosage unit. It is expressed as a percentage of the drug's label claim (LC) and is tested to ensure the consistency of dosage units in a batch. For example, if the LC of a drug is 2.0% active, then an assay result of 1.9% implies a potency value of 95%. The API content in each unit is expected to be within a narrow range around the LC. It is subject to specification limits at the time of manufacture and throughout the product's shelf life.

Critically, the FDA guidance recommends consideration of both within-batch and between-batch quality in the determination of Stage II sample sizes. To accommodate this requirement, we model drug potency with a normal hierarchical structure, with unknown means and variance components, that is, a one-way normal random effects model. Both frequentist and Bayesian methods use this traditional multi-level model in settings where data is available from multiple groups (in our case, batches) and the between-group variation is of interest (see Gelman and Hill, 2006). This structure allows for the "borrowing of strength" across all of the potency values (Gelman et al., 2013).

Let $\mathbf{Y} \equiv [y_{ij}]$ be an $n_0 \times k_0$ matrix, where y_{ij} denotes the potency of the jth unit in the ith batch in Stage I, $i = 1, \ldots, n_0$ and $j = 1, \ldots, k_0$. We assume that $y_{ij} \sim \mathrm{N}(\mu_i, \sigma_W^2)$. For our hierarchical model, we need priors on the batch means, μ_i, and the within-batch variance, σ_W^2.

It is reasonable to assume that the manufacturing conditions are the same across batches. This includes the facility, utilities, equipment, operating parameters, processing limits, components, and raw materials. Therefore, it is reasonable to assume that batches are exchangeable so that $\mu_i \sim \mathrm{N}(\mu_B, \sigma_B^2)$, $i = 1, \ldots, n_0$. For the within-batch variance, we assign an inverse-gamma prior, denoted by $\mathrm{IG}(\lambda, \phi)$, with density

$$\frac{\phi^\lambda}{\Gamma(\lambda)}(1/\sigma_W^2)^{\lambda+1} e^{-\phi/\sigma_W^2}, \; 0 < \sigma_W^2 < \infty, \; \lambda, \phi > 0.$$

We follow Christensen et al. (2011) in our choice of an inverse-gamma prior for the variance parameter. They suggest a straightforward way to elicit the inverse-gamma shape parameters from an expert. The model is summarized as follows:

$$y_{ij}|\mu_i, \sigma_W^2 \sim \mathrm{N}(\mu_i, \sigma_W^2), \tag{13.1}$$

$$\mu_i|\mu_B, \sigma_B^2 \sim \mathrm{N}(\mu_B, \sigma_B^2), \tag{13.2}$$

and

$$\sigma_W^2 \sim \mathrm{IG}(\lambda, \phi). \tag{13.3}$$

We denote the corresponding densities for Equations (13.1)–(13.3) by $f(y_{ij}|\mu_i, \sigma_W^2)$, $\pi(\mu_i|\mu_B, \sigma_B^2)$, and $\pi(\sigma_W^2)$, respectively. (We assume that the choice of normal parameters yields negligible probability mass below zero since potencies are, of course, positive.)

We give μ_B and σ_B^2 independent normal and inverse-gamma priors, respectively:

$$\begin{aligned}\pi(\mu_B, \sigma_B^2) &= \pi(\mu_B)\pi(\sigma_B^2) \\ &= \mathrm{N}(\mu_0, \sigma_0^2) \times \mathrm{IG}(\alpha, \beta).\end{aligned}$$

Then, the joint posterior for $(\mu_B, \sigma_B^2, \sigma_W^2)$ is

$$\begin{aligned}\pi(\mu_B, \sigma_B^2, \sigma_W^2|\mathbf{Y}) &\propto l(\mathbf{Y}|\mu_B, \sigma_B^2, \sigma_W^2)\pi(\mu_B, \sigma_B^2|\mu_0, \sigma_0^2, \alpha, \beta)\pi(\sigma_W^2|\lambda, \phi) \\ &= \left[\prod_{i=1}^{n_0}\prod_{j=1}^{k_0} f(y_{ij}|\mu_i, \sigma_W^2)\right]\left[\prod_{i=1}^{n_0}\pi(\mu_i|\mu_B, \sigma_B^2)\right] \\ &\quad \times \pi(\mu_B)\pi(\sigma_B^2)\pi(\sigma_W^2),\end{aligned} \tag{13.4}$$

where $l(\cdot)$ is the likelihood function. In the next section, we consider prior elicitation for a Bayesian SSD approach.

13.2.2 Sample Size—The Two Priors

Bayesian sample size procedures often use what Brutti et al. (2008) call a "two priors" approach, including an informative *design prior* and a relatively non-informative *analysis prior*. In our approach, the design prior represents what the FDA guidance describes as "the entire compilation of knowledge and information gained from the design stage". In Section 13.2.2.1, we detail a method for constructing design priors. We use the analysis prior in fitting our model once the Stage II data are obtained (Wang and Gelfand, 2002) and, in contrast to conventional Bayesian SSD, we propose using the design prior as the analysis prior, albeit "discounted" in the form of a power prior (Ibrahim and Chen, 2000). The discounting level can be determined based on a firm-level validation policy related to an acceptable level of "borrowing" from Stage I to Stage II. We describe these analysis priors in Section 13.2.2.2.

We use these design and analysis priors in exploring the probability of a successful qualification at different number of batches. The method is necessarily based on MCMC methods, as the posteriors are not of closed form. We consider computational details in Section 13.3. For each value, the probability of success is computed until a desired level is achieved. The method proceeds as follows. Here we take $\boldsymbol{\theta} = (\mu_B, \sigma_B, \sigma_W) \in \Theta$.

1. Construct an initial informative prior, $\pi_0(\boldsymbol{\theta})$. This can be based on expert knowledge and historical understanding of the product and process being validated. Alternatively, if the manufacturer prefers a primarily "data-driven" analysis, then $\pi_0(\boldsymbol{\theta})$ can be relatively non-informative.
2. Compute the posterior, $\pi_D(\boldsymbol{\theta}|\mathbf{Y}_D)$, where $\mathbf{Y}_D$ represents potency data gained from Stage I:
$$\pi_D(\boldsymbol{\theta}|\mathbf{Y}_D) \equiv \frac{l(\boldsymbol{\theta}|\mathbf{Y}_D)\pi_0(\boldsymbol{\theta})}{\int_\Theta l(\boldsymbol{\theta}|\mathbf{Y}_D)\pi_0(\boldsymbol{\theta})d\boldsymbol{\theta}}.$$
This posterior is our design prior.
3. Make a random draw from $\pi_D(\boldsymbol{\theta}|\mathbf{Y}_D)$, denoted $\boldsymbol{\theta}_* = (\mu_B^*, \sigma_B^*, \sigma_W^*)$.
4. Generate k potency values from each of n batches using $\boldsymbol{\theta}_*$. Denote this simulated sample by $\mathbf{Y}_*$; it represents a set of batch potency results from a Stage II study.
5. Combine the data, $\mathbf{Y}_*$, with the analysis prior, $\pi_A(\boldsymbol{\theta})$: The posterior distribution is
$$\pi_A(\boldsymbol{\theta}|\mathbf{Y}_*) \equiv \frac{l(\boldsymbol{\theta}|\mathbf{Y}_*)\pi_A(\boldsymbol{\theta})}{\int_\Theta l(\boldsymbol{\theta}|\mathbf{Y}_*)\pi_A(\boldsymbol{\theta})d\boldsymbol{\theta}}.$$
6. Let $\tilde{y}$ be the "next" predicted potency value from our process. Draw a sample of size $k \times n = n_P$ from the posterior predictive distribution,
$$p(\tilde{y}|\mathbf{Y}_*) \equiv \int_\Theta l(\tilde{y}|\boldsymbol{\theta})\pi_A(\boldsymbol{\theta}|\mathbf{Y}_*)d\boldsymbol{\theta}.$$
Denote this generated vector as $\tilde{\mathbf{y}}_{n_P}$.
7. Evaluate $\tilde{\mathbf{y}}_{n_P}$ with a chosen test statistic used to assess product quality.
8. Repeat steps 3–7 a large number of times to evaluate the probability of success based on the posterior predictive distribution of the test statistic.

In the following sections, we describe the process for construction of the design prior, $\pi_D(\boldsymbol{\theta})$, and the analysis prior, $\pi_A(\boldsymbol{\theta})$.

13.2.2.1 The Design Prior

Decisions about sample size are ultimately the responsibility of the manufacturer, so, at the design stage, it is sensible for the company to use all of its prior information, as noted by O'Hagan and Stevens (2001). Incorporating such information into an informative design prior, $\pi_D(\boldsymbol{\theta})$, is perhaps the most challenging step in the Bayesian SSD. Using this prior in Stage II entails combining data from batches filled during Stage I together with expert knowledge of the process being validated, reflected in $\pi_0(\boldsymbol{\theta})$. This approach thereby permits the data, expert opinion, or a combination of both elements to influence the design prior, depending on the validation scenario. Alternatively, a more traditional design prior approach can be taken by using only historical data and excluding expert opinion. By doing so, $\pi_D(\boldsymbol{\theta}|\mathbf{Y}_D)$ is driven by the data, and $\pi_0(\boldsymbol{\theta})$ is taken to be a relatively non-informative prior. This can be done by modifying the hierarchical model in Section 13.2.1 and assigning diffuse priors for σ_W, σ_B, and μ_B:

$$\sigma_W \sim \mathrm{U}(0, A), \tag{13.5}$$

$$\sigma_B \sim \mathrm{U}(0, B), \tag{13.6}$$

and

$$\mu_B \sim \mathrm{N}(0, 1000).$$

Motivation for the use of uniform priors on standard deviation components is discussed in Gelman (2006). Choices for A and B should be made large enough to render the medians and 95% credible interval widths for σ_W and σ_B stable, as we illustrate in Section 13.3.1. Thus, the joint posterior for $\boldsymbol{\theta}$ defined in (13.4) becomes

$$\pi(\boldsymbol{\theta}|\mathbf{Y}_*) \propto \left[\prod_{i=1}^{n}\prod_{j=1}^{k} f(y_{ij}|\mu_i, \sigma_W^2)\right]\left[\prod_{i=1}^{n} \pi(\mu_i|\mu_B, \sigma_B^2)\right] \pi(\mu_B)\pi(\sigma_B^2|B)\pi(\sigma_W^2|A).$$

Note that not all data generated during Stage I should be used in the elicitation of the design prior. This is because, at this stage, data are derived from experimental, demonstration, clinical, or R&D batches. Data from these batches and the expected Stage II results may differ greatly depending on the extent to which critical process parameters (CPP) have been adjusted during Stage I. Consequently, only the data from Stage I batches manufactured under expected commercial CPP settings should be used in the construction of the design prior. As the FDA guidance indicates, Stage II will "confirm the process design", thereby ensuring that the process "performs as expected".

13.2.2.2 The Analysis Prior

If we use a relatively non-informative analysis prior, as is typical in Bayesian SSD analyses, the simulated Stage II data in our process will drive the sample size conclusion. This practice allows for "convergence" between Bayesian and frequentist inference (O'Hagan and Stevens, 2001). Instead, we recommend a relatively informative analysis prior, borrowing knowledge gained from Stage I. To this end, the design prior can be converted into a power prior (Ibrahim and Chen, 2000), with the power parameter's value chosen so as to bound the assurance borrowed from Stage I in a way dictated by guidelines in the master validation plan. The power parameter, $a_0 \in [0, 1]$, serves as a relative precision parameter for the Stage I data. Its value dictates the heaviness of the tails of the design prior for the components of $\boldsymbol{\theta}$. Smaller values of a_0 yield heavier tails, thereby determining the influence of the Stage I data on the analysis prior. Given our data from Stage I, $\mathbf{Y}_D$, our analysis prior is defined for $a_0 \in [0, 1]$ as

$$\pi_A(\boldsymbol{\theta}|\mathbf{Y}_D, a_0) \equiv \frac{[l(\boldsymbol{\theta}|\mathbf{Y}_D)]^{a_0}\pi_0(\boldsymbol{\theta})}{\int_\Theta [l(\boldsymbol{\theta}|\mathbf{Y}_D)]^{a_0}\pi_0(\boldsymbol{\theta})d\boldsymbol{\theta}}. \tag{13.7}$$

The choice of a_0 is critical. Placing a prior on a_0, while recommended in the paper by Ibrahim and Chen (2000), is now known to be problematic (see, for example, Neuenschwander et al., 2009 and Duan et al., 2006a,b). We recommend choosing a value for a_0 consistent with an overall validation policy. That policy should establish a limit on the acceptable amount of assurance to be borrowed from Stage I for use in the Stage II analysis, thereby constraining the value of a_0. We consider this issue further in the next section.

13.2.3 Bayesian Assurance

O'Hagan and Stevens (2001) introduce the concept of Bayesian assurance, similar to frequentist power, in the context of clinical trial design. It is the unconditional probability that a trial will lead to a specific outcome. Specifically, assurance is the expected power where the expectation is taken with respect to the prior probability distribution of $\boldsymbol{\theta}$ (O'Hagan et al., 2005). For our purposes, we can think of assurance as the probability, denoted by γ, that we will obtain a successful result in our Stage II PPQ study. Thus, the sample size is chosen sufficiently large as to yield a probability of at least γ of obtaining a successful result.

In Stage II, the assurance of success must be based on a statistic or "metric" related to critical quality attributes. Achieving a target for this metric is the goal, specified as a minimum level of assurance. The Bayesian SSD approach affords generation of predicted Stage II values for quality attributes, thereby enabling computation of assurance *a priori* for our chosen test statistic.

The choice of the minimum level of assurance for Stage II is problematic. The FDA guidance does not prescribe target levels but notes that "success at this stage signals an important milestone in the product lifecycle. A manufacturer must successfully complete PPQ before commencing commercial distribution of the drug product". Thus, the choice of the assurance level for Stage II should be made judicially, based on a risk analysis related to the quality attributes under examination. Assurance rarely reaches 100%, regardless of sample size. A preliminary analysis of the largest achievable assurance may indicate if the design prior is "good enough" to continue the study. We illustrate this in Section 13.3.

The overall validation strategy should indicate the amount of assurance to "borrow" or "leverage" from Stage I. Such borrowing of information is reasonable, especially since the design prior has been constructed using data that is representative of the commercial process. Furthermore, the FDA guidance suggests that for Stage II studies, "data from laboratory and pilot studies can provide additional assurance that the commercial manufacturing process performs as expected". Hence, from a regulatory perspective, such borrowing is acceptable. A manufacturer may determine, for example, that no more than 50% assurance may be borrowed from Stage I. Compliance with this policy can be achieved by calibrating the power parameter, a_0 in Equation (13.7). We illustrate this calibration in Section 13.3.

In order to limit the number of Stage II batches required to achieve a target assurance level, a "top-off" policy using Stage III data is a reasonable option. Increased monitoring and data acquisition in Stage III could provide sufficient data to achieve the desired assurance. We consider the Stage II to III transition strategy in Section 13.4.

13.3 An Example

In this section, we illustrate the use of our method for the validation of a drug product where potency is the critical quality attribute of interest. Suppose that the product specification for potency of the active ingredient in the drug product is 90%–110% of LC. Thus, our lower specification limit (*LSL*) is 90%, while the upper specification limit (*USL*) is 110%.

To evaluate potency results, we use the process capability index as the test statistic. Process capability measures the ability of a process to produce product within specification (Bothe, 1997). The index is defined as

$$C_{pk} = \min\left(\frac{USL - \bar{y}}{3s}, \frac{\bar{y} - LSL}{3s}\right),$$

where $\bar{y}$ and s denote the sample mean and sample standard deviation, respectively, for all of the potency values observed during Stage II qualification. In our example, the PPQ protocol establishes an acceptance criterion of $C_{pk} > 1.15$. Hence, the assurance of success is defined as $\gamma = \Pr(C_{pk} > 1.15)$.

The software package rjags in the R environment was used to perform the simulations presented throughout this example.

13.3.1 Design and Analysis Priors

Suppose we have data from four hypothetical "demonstration" and "R&D" batches produced during Stage I of process validation. Suppose further that, for each batch, ten assays were performed for potency. These results are summarized in Table 13.1. We take the potency values in Table 13.1 as having been generated from the final commercial process and formulation.

Here, we construct the design prior without expert opinion, using only Stage I data. Software tools to assist in construction of informative priors for

TABLE 13.1
Unit (j) Potencies for Stage I Demo Batches (i)

	$j=1$	2	3	4	5	6	7	8	9	10
$i=1$	99.4	102.0	102.5	96.7	99.1	102.8	98.3	98.1	103.1	96.6
2	100.4	100.5	99.5	99.2	101.0	100.2	98.7	97.6	102.7	99.3
3	99.9	97.1	102.3	100.6	97.0	98.8	101.7	99.1	102.4	98.8
4	100.6	96.0	97.1	98.4	100.2	97.5	99.9	101.6	99.7	101.8

$\pi_0(\boldsymbol{\theta})$ using expert opinion are available. ParameterSolver is such a tool and is freely available from the MD Anderson Cancer Center's Division of Biostatistics website. For example, if expert opinion suggests that $\sigma_W^2 \approx 4.5\% \pm 1.5\%$ and $\sigma_B^2 \approx 2\% \pm 1\%$ with 95% probability, then the software can be used to select suitable inverse-gamma priors, in this case $\sigma_W^2 \sim$ IG(32,134) and $\sigma_B^2 \sim$ IG(13,21). For our illustration, assuming no such expert advice, we calibrate the uniform priors for σ_W^2 and σ_B^2 by selecting values for the upper limits A and B in Equations (13.5) and (13.6) per Section 13.2.2.2. See Figure 13.1a. There we have graphed the posterior median and 95% credible intervals for σ_W for increasing values of A. At approximately $A = 4$, the median and the credible interval width stabilize. Thus, $A = 5$ is a conservative choice for the uniform prior on σ_W^2 at the design stage. Similarly, we set $B = 10$ for the uniform prior on σ_B^2.

The joint design prior structure is thus

$$
\begin{aligned}
y_{ij}|\mu_i, \sigma_W^2 &\sim \mathrm{N}(\mu_i, \sigma_W^2),\\
\mu_i|\mu_B, \sigma_B^2 &\sim \mathrm{N}(\mu_B, \sigma_B^2),\\
\sigma_W^2 &\sim \mathrm{U}(0, 5), \qquad (13.8)\\
\mu_B &\sim \mathrm{N}(0, 1000), \qquad (13.9)
\end{aligned}
$$

and

$$\sigma_B^2 \sim \mathrm{U}(0, 10). \qquad (13.10)$$

The posterior, $\pi_D(\boldsymbol{\theta})$, is computed using MCMC methods implemented in rjags. A single chain was used, with 50,000 burn-in iterations, followed by 100,000 iterations for inference and a thinning factor of 2. Convergence was verified via conventional diagnostic tools—see, for example, Gelman et al. (2013). The corresponding marginal posterior densities for μ_B, σ_B^2, and σ_W^2 are shown in Figure 13.2.

To construct the analysis prior, we suppose that the manufacturer has established a validation policy that, at most, 50% assurance may be borrowed from Stage I into the Stage II study. Assurance for this example has been defined as $\gamma = \Pr(C_{pk} > 1.15)$. Hence, we select a_0 so that $\Pr(C_{pk} > 1.15|a_0, \mathbf{Y}_D) = 0.5$. The assurance at the design stage is calculated using the posterior predictive distribution,

$$p_D(\tilde{y}|\mathbf{Y}_D, a_0) \equiv \int_\Theta l(\tilde{y}|\boldsymbol{\theta})\pi_D(\boldsymbol{\theta}|\mathbf{Y}_D, a_0)d\boldsymbol{\theta}.$$

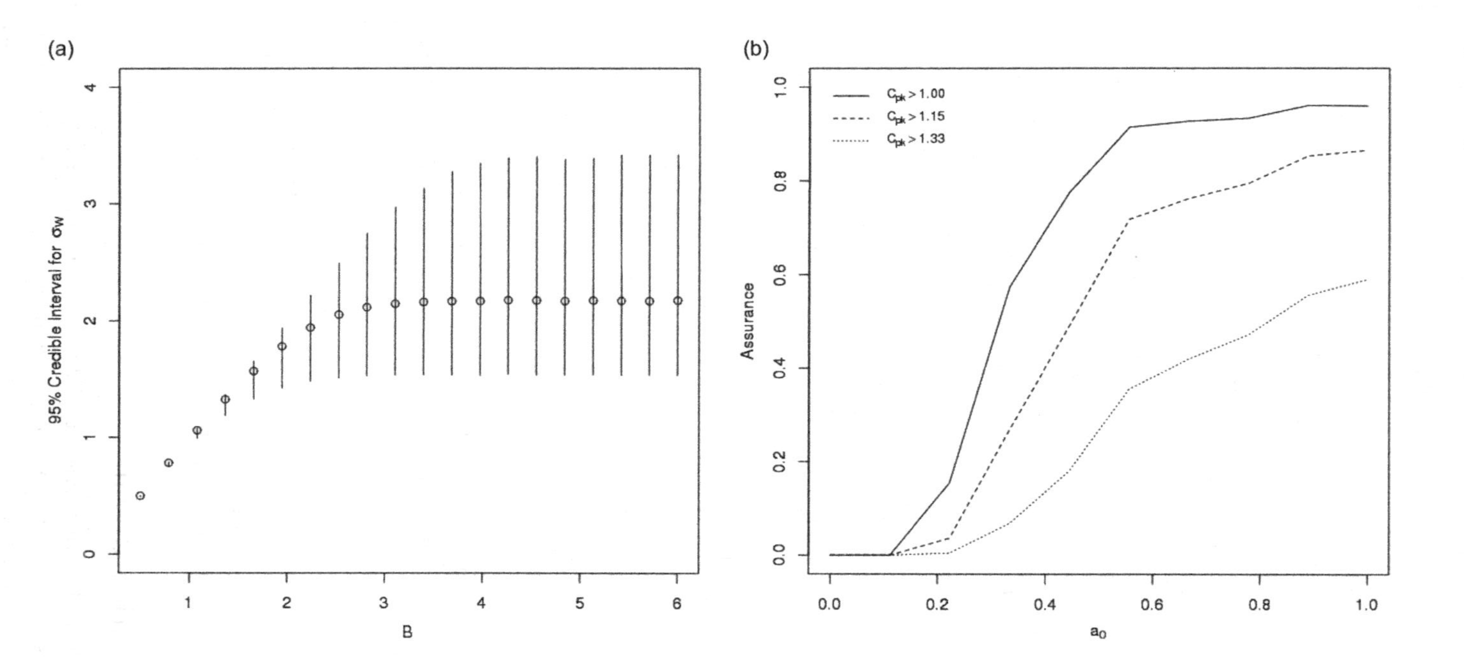

FIGURE 13.1
(a) Median values and credible interval widths for σ_W as functions of A. (b) Design-level assurance levels for increasing values of a_0 across three acceptance criteria for C_{pk}.

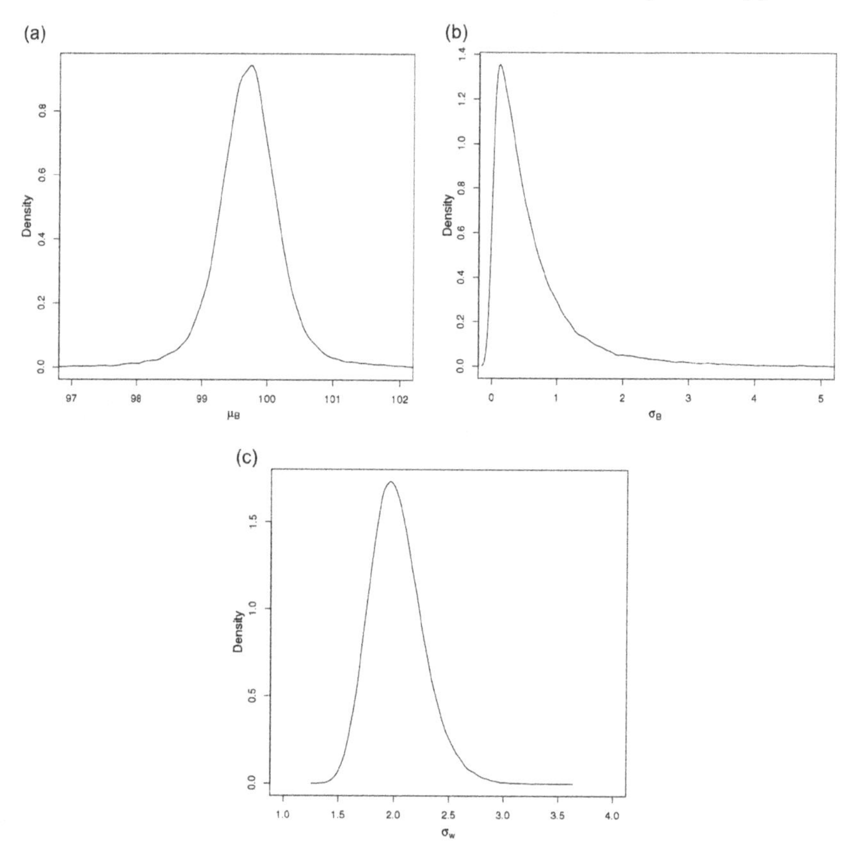

FIGURE 13.2
Design prior densities for μ_B, σ_B^2, and σ_W^2.

Numerous draws from $p_D(\tilde{y}|\mathbf{Y}_D, a_0)$ can be used to calculate C_{pk} values and the assurance value computed as $\gamma = \sum_i^N I\{C_{pk(i)} \geq 1.15\}/N$, where I is the indicator function, N denotes the number of draws, and $C_{pk(i)}$ is the ith calculated process capability value. In Figure 13.1b, we plot the assurance, γ, against increasing values of a_0. We see that $\gamma = 0.5$ at approximately $a_0 = 0.4$. It is also convenient and useful for decision makers to compare assurance across different acceptance criteria for C_{pk} when varying a_0. Figure 13.1b shows that for $C_{pk} \geq 1.00, 1.15$, and 1.33, the maximum assurance levels achievable at the design stage are approximately 95%, 85%, and 60%, respectively.

13.3.2 Simulation Results

The simulation described in Section 13.2.2 was carried out as follows. We kept k fixed at ten assays per batch and considered several values of n. We calculated a total of 1,000 posterior predictive values of C_{pk} for each n. Thus, for each n, we obtained an approximation of the distribution of $C_{pk}(\tilde{\mathbf{y}}_{n_P})$. Assurance, $\gamma = \Pr(C_{pk} > 1.15)$, was then approximated by observing the proportion of simulated C_{pk} values exceeding 1.15. The resulting assurance curve as a function of the number of PQ batches is depicted in Figure 13.3. It is based on five simulations at 1, 2, 3, 4, and 5 batches. Sensitivity of approximate assurance values with respect to the choice of the C_{pk} limit was also investigated. To this end, we include in Figure 13.3 additional assurance curves for $\Pr(C_{pk} > 1.00)$ and $\Pr(C_{pk} > 1.33)$. Decision makers can use such a graph to weigh the cost of running PQ batches against the predicted assurance of "success" for the Stage II study. Suppose, for example, our goal is $\Pr(C_{pk} > 1.15) \geq 75\%$. Then, we select four batches for the study. Note that the assurance curve starts to plateau as the number of batches increases. This is because the data begins to dominate the posterior and we reach an approximate maximum assurance level based on the elicited design prior.

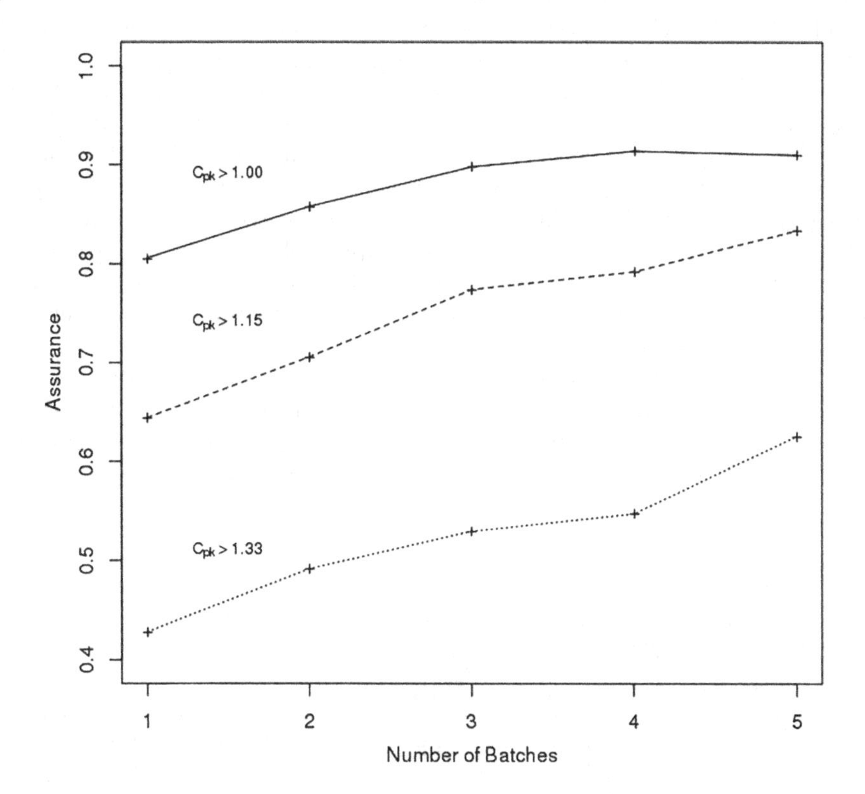

FIGURE 13.3
Assurance curves for different lower limits on C_{pk} values across number of batches for Stage II.

13.4 Transition to Stage III Monitoring

In the example above, a sample size of four batches is selected for Stage II qualification. The type of drug and manufacturing process will determine the practicality of this choice for a Stage II study. Practical or not, the manufacturer may consider an assurance "top-off" strategy, relying upon a period of enhanced monitoring during Stage III meant to supplement knowledge gained during Stage II. In this section, we present an example of a transition process from Stage II to Stage III, utilizing the idea of Bayesian assurance. We build upon the example presented in Section 13.3, but assume a decision by the firm to manufacture only two Stage II batches, and top-off the target assurance level in Stage III.

Manufacturers often divide Stage III of the process validation life cycle into two sub-stages: Stage III-A, intended for enhanced monitoring, and Stage III-B for routine monitoring. Stage III-A is usually performed under an established protocol with pre-defined acceptance criteria. A structured approach to transitioning from Stage II to Stage III could be designed into the process validation approach by specifying minimum assurance levels for Stage II and Stage III-A. For example, the validation plan could require a minimum of 70% assurance to be achieved for Stage II and 85% assurance for Stage III-A. These assurance "hurdles" can vary greatly depending on process knowledge and risk understanding. Such a structured approach could reduce the number of batches required for Stage II qualification, thereby allowing earlier commercial production under an enhanced period of Stage III-A monitoring. Table 13.2 summarizes this structured transition approach, starting from the selection of the power prior parameter, a_0, yielding 50% assurance at Stage I, to routine monitoring in Stage III. Here, 70% and 85% assurance levels are achieved after two and five batches, respectively.

This structured transition approach affords increased flexibility in determining the number of batches required for Stage II by using information across the process validation life cycle. In addition to being consistent with existing practice, this method also provides a clear assurance-based, hands-off transition from Stage I to Stage III of the process validation life cycle. Of course, the

TABLE 13.2
Assurance-Based Transition Plan from Stage I to III

Stage	Required Assurance	Planned Batches	Activity
I	50%	($a_0 = 0.4$)	Process design
II	70%	2	Qualification
III-A	85%	5	Enhanced monitoring
III-B	N/A	N/A	Routine monitoring

choice and justification of the assurance "hurdles" in this structured transition will be a challenge for the manufacturer.

13.5 Discussion

Choosing the number of Stage II batches in the process validation life cycle requires prior knowledge and data, as well as understanding the within- and between-batch variability of the manufacturing process. In this chapter, we have shown that a hierarchical Bayesian model is ideally suited to address this SSD problem. Our method utilizes prior knowledge and data to generate simulated predicted data for Stage II qualification. Assurance of success in Stage II qualification can then be evaluated using a process capability metric. A major advantage of the approach presented here is that the simulated data from the posterior predictive distribution may be evaluated using any metric preferred by the user. This provides great flexibility for the decision maker. Furthermore, it is often much easier for validation and quality decision makers to think in terms of assurance of success rather than traditional Type I and Type II error probabilities.

Prior elicitation is clearly the most challenging aspect of the Bayesian assurance method we propose. In our example, we restricted the informativeness of the design prior, basing it on Stage I data only, and excluded the influence of expert opinion. We did so to present an example that is primarily data-driven and more likely to be accepted by regulatory bodies. Construction of the design prior was assumed to be aided by the availability of development data generated from the proposed commercial process and formulation. Where such prior data and/or knowledge is limited or not available, the design prior elicitation process will be more challenging and may require more sophisticated methods. Furthermore, as the design prior becomes less informative (more "diffuse"), resulting sample sizes may become impractically large for some manufacturers. In such cases, robust risk analyses and a strong Stage II-to-Stage III transition plan will be critical factors in justifying a reasonable number of batches for PPQ. A possible solution is an extended Stage III-A that reduces the sample size burden for Stage II.

Finally, the example presented in Section 13.3 focused on drug potency uniformity as the critical quality attribute driving the SSD decision. Other attributes related to the physical, chemical, biological, or microbiological property or characteristic of the drug may also need to be considered in the sample size decision. In that broader context, the Bayesian model we propose could be extended to a multivariate scenario while still following the same hierarchical, two-stage prior process. This is an area of future research to be explored.

13.6 Appendix

13.6.1 JAGS Model

13.6.1.1 Design Prior/Analysis Prior

The jags model for the design prior is provided below and uses the standard "zeros trick" defined in the BUGS Manual (www.openbugs.net/Manuals/Tricks.html) as well as in *The BUGS Book: A Practical Introduction to Bayesian Analysis* by Lunn et al. (2012). The trick allows for sampling from a distribution that is not included in the list of standard distributions of the software package. In our case, the non-standard distribution is the normal likelihood with a power prior parameter, a. For the design prior, $a = 1$. For the analysis prior, a is adjusted accordingly.

```
model{
  #Likelihood
  R<-100000
  for (i in 1:N) {
  for (j in 1:K) {
  #zeros[i,j]<-0
  zeros[i,j]~dpois(zeros.mean[i,j])
  zeros.mean[i,j]<- -log(L[i,j])+R
  L[i,j]<-pow((tau.W)/(2*3.14159),a/2)
  *exp(-(a*tau.W/2)*(Y[i,j]-mu[i])*(Y[i,j]-mu[i]))
  }
  mu[i]~dnorm(mu.B,tau.B)
  }
  tau.B<-1/pow(sigma.B,2)
  tau.W<-1/pow(sigma.W,2)

  #Priors
  sigma.W~dunif(0,A)
  mu.B ~ dnorm(C,tau.D)
  sigma.B~dunif(0,B)

  }
```

13.6.1.2 Analysis Prior

The jags model for the analysis prior is provided below. Note that the parameter values for the prior distributions of μ_B, σ_W, and σ_B are fitted from the posterior distributions of the analysis prior model defined in Section 13.6.1.1.

```
model
{
  #Likelihood
  for (i in 1:N) {
  for (j in 1:K) {
  Y[i,j]~dnorm(mu[i],tau.W)
  }
  mu[i]~dnorm(mu.B,tau.B)
  }

  #Priors
  tau.W<-1/pow(sigma.W,2)
  tau.B<-1/pow(sigma.B,2)
  sigma.W~dgamma(lambda,phi)
  sigma.B~dgamma(alpha,beta)
  mu.B ~ dnorm(C,tau.D)
  tau.D<-1/pow(sigma.D,2)

}
```

References

Armitage, P., Berry, G., and Matthews, J. N. (2008). *Statistical Methods in Medical Research.* John Wiley & Sons: New York.

Bothe, D. R. (1997). *Measuring Process Capability: Techniques and Calculations for Quality and Manufacturing Engineers.* McGraw-Hill: New York.

Brutti, P., De Santis, F., and Gubbiotti, S. (2008). Robust Bayesian sample size determination in clinical trials. *Statistics in Medicine*, 27(13):2290–2306.

Christensen, R., Johnson, W., Branscum, A., and Hanson, T. E. (2011). *Bayesian Ideas and Data Analysis: An Introduction for Scientists and Statisticians.* CRC Press: Boca Raton, FL.

Chuang-Stein, C. (2006). Sample size and the probability of a successful trial. *Pharmaceutical Statistics*, 5(4):305–309.

Chuang-Stein, C. and Yang, R. (2010). A revisit of sample size decisions in confirmatory trials. *Statistics in Biopharmaceutical Research*, 2(2):239–248.

Duan, Y., Smith, E. P., and Ye, K. (2006a). Using power priors to improve the binomial test of water quality. *Journal of Agricultural, Biological, and Environmental Statistics*, 11(2):151–168.

Duan, Y., Ye, K., and Smith, E. P. (2006b). Evaluating water quality using power priors to incorporate historical information. *Environmetrics*, 17(1):95–106.

Eaton, M. L., Muirhead, R. J., and Soaita, A. I. (2013). On the limiting behavior of the "probability of claiming superiority" in a Bayesian context. *Bayesian Analysis*, 8(1):221–232.

European Medicines Agency (2014). Guideline on process validation for finished products—Information and data to be provided in regulatory submissions.

Food and Drug Administration (2010). Guidance for the use of Bayesian statistics in medical device clinical trials.

Food and Drug Administration (2011). Guidance for industry: Process validation: General principles and practices.

Gelman, A. (2006). Prior distributions for variance parameters in hierarchical models (comment on article by Browne and Draper). *Bayesian Analysis*, 1(3):515–534.

Gelman, A. and Hill, J. (2006). *Data Analysis Using Regression and Multilevel/Hierarchical Models.* Cambridge University Press: Cambridge.

Gelman, A., Carlin, J. B., Stern, H. S., Dunson, D. B., Vehtari, A., and Rubin, D. B. (2013). *Bayesian Data Analysis.* CRC press: Boca Raton, FL.

Ibrahim, J. G. and Chen, M.-H. (2000). Power prior distributions for regression models. *Statistical Science*, 15:46–60.

Joseph, L., Du Berger, R., and Belisle, P. (1997). Bayesian and mixed Bayesian/likelihood criteria for sample size determination. *Statistics in Medicine*, 16(7):769–781.

LeBlond, D. and Mockus, L. (2014). The posterior probability of passing a compendial standard, part 1: Uniformity of dosage units. *Statistics in Biopharmaceutical Research*, 6(3):270–286.

Lee, S. J. and Zelen, M. (2000). Clinical trials and sample size considerations: Another perspective. *Statistical Science*, 15(2):95–110.

Lunn, D., Jackson, C., Best, N., Thomas, A., and Spiegelhalter, D. (2012). *The BUGS Book: A practical Introduction to Bayesian Analysis.* CRC press: Boca Raton, FL.

Muirhead, R. J. and Şoaita, A. I. (2013). On an approach to Bayesian sample sizing in clinical trials. In *Advances in Modern Statistical Theory and Applications: A Festschrift in honor of Morris L. Eaton* (eds. Galin Jones, Xiaotong Shen), pp. 126–137. Institute of Mathematical Statistics: Beachwood, OH.

Neuenschwander, B., Branson, M., and Spiegelhalter, D. J. (2009). A note on the power prior. *Statistics in Medicine*, 28(28):3562–3566.

O'Hagan, A. and Stevens, J. W. (2001). Bayesian assessment of sample size for clinical trials of cost-effectiveness. *Medical Decision Making*, 21(3):219–230.

O'Hagan, A., Stevens, J. W., and Campbell, M. J. (2005). Assurance in clinical trial design. *Pharmaceutical Statistics*, 4(3):187–201.

Patel, N. R. and Ankolekar, S. (2007). A Bayesian approach for incorporating economic factors in sample size design for clinical trials of individual drugs and portfolios of drugs. *Statistics in Medicine*, 26(27):4976–4988.

Pazhayattil, A., Alsmeyer, D., Chen, S., Hye, M., Ingram, M., and Sanghvi, P. (2016). Stage 2 process performance qualification (PPQ): A scientific approach to determine the number of PPQ batches. *AAPS PharmSciTech*, 17(4):829–833.

Rubin, D. B. and Stern, H. S. (1998). Sample size determination using posterior predictive distributions. *Sankhyā: The Indian Journal of Statistics, Series B*, 60:161–175.

Spiegelhalter, D. and Freedman, L. (1986). A predictive approach to selecting the size of a clinical trial, based on subjective clinical opinion. *Statistics in Medicine*, 5(1):1–13.

Spiegelhalter, D. J., Abrams, K. R., and Myles, J. P. (2004). *Bayesian Approaches to Clinical Trials and Health-Care Evaluation*, vol. 13. John Wiley & Sons: New York.

United States Pharmacopoeia (2011). USP <905> uniformity of dosage units.

www.usp.org/sites/default/files/usp_pdf/EN/USPNF/2011-02-25905UNIFORMITYOFDOSAGEUNITS.pdf.

Wang, F. and Gelfand, A. E. (2002). A simulation-based approach to Bayesian sample size determination for performance under a given model and for separating models. *Statistical Science*, 17: 193–208.

Wang, S.-J., Hung, H., and O'Neill, R. T. (2006). Adapting the sample size planning of a phase III trial based on phase II data. *Pharmaceutical Statistics*, 5(2):85–97.

Whitehead, J., Valdés-Márquez, E., Johnson, P., and Graham, G. (2008). Bayesian sample size for exploratory clinical trials incorporating historical data. *Statistics in Medicine*, 27(13):2307–2327.

Wiles, F. (2013). Risk-based methodology for validation of pharmaceutical batch processes. *PDA Journal of Pharmaceutical Science and Technology*, 67(4):387–398.

Yang, H. (2013). How Many Batches Are Needed for Process Validation under the New FDA Guidance? *PDA Journal of Pharmaceutical Science and Technology*, 67(1):53–62.

14

Bayesian Approaches in the Regulation of Medical Products

Telba Irony and Lei Huang
CBER

CONTENTS

14.1 Introduction

The use of Bayesian designs for clinical trials in the regulatory setting has been advocated because of promising advantages that could bring faster access of safe and effective medical treatments to patients who need them [1–3]. Since a practical path to using Bayesian methods has been provided by modern developments in computing power and Markov chain Monte Carlo (MCMC) algorithms, Bayesian trial designs have been increasingly earning enthusiasts [1]. Although medical product development has not been a space with extensive applications of Bayesian approaches, we have seen more and more examples emerging recently. One of the main obstacles for using Bayesian methods in the regulatory setting is the resistance against the perceived added subjectivity inherent to the use of prior distributions. Furthermore, the regulatory practice of controlling type I error rate has been another hurdle hindering the growth of Bayesian methods in medical product development. Since external prior information may assign different probabilities to different points in the parameter space, including the null space, the definition of type I error, a frequentist concept, is ambiguous within the Bayesian framework. Consequently, there is considerable disagreement on how to control type I error rate in a Bayesian trial design that borrows strength from external sources. As one can imagine, the imposition of an extra frequentist condition mismatching the Bayesian paradigm has oddly stumbled the advancement of Bayesian methods in this field. Rather, a decision analysis approach to assess benefit-risk of medical products would be more suitable for using Bayesian methods. Nevertheless, in the absence of a major paradigm shift, control of type I error rate remains a standard approach in regulatory applications. This may in part explain why we have seen more use of Bayesian adaptive designs with non-informative priors than Bayesian designs that incorporate outside information via informative priors, as the type I error rate inflation incurred by interim analyses is generally mild as compared to the inflation generated by priors. Despite that, there has been a considerable increase in the Bayesian approach in the regulatory setting because of the possibility to assess type I error rates using simulations. Therefore, we are optimistic that Bayesian methods will bloom in medical product development, and even more examples will come into sight. In this chapter, we present examples of Bayesian clinical trial designs and discuss them from a regulatory perspective.

The chapter focuses on the use of prior information to increase the power of clinical trials, the use of Bayesian adaptive designs, and the use of simulations to assess the operating characteristics of the Bayesian designs. Subjectivity can be a primary concern for using prior information, and we discuss how to handle subjectivity associated with the use of information outside the clinical trial. We also discuss Bayesian adaptive designs, including their advantages. Finally, we note that it has been customary to carry out simulations to assess and to control the operating characteristics of Bayesian designs, such as type I error rate, and to optimize key design features. The need for simulating trials at the design stage, and the benefits and hurdles of conducting such simulations are also discussed in this chapter.

14.2 The Use of Prior Information

A distinct advantage of the Bayesian approach lies in its ability to formally combine prior information with current data to augment the current trial, often resulting in increased precision. The use of prior information through Bayes theorem is straightforward and mathematically sound [1]. Since no ad-hoc procedures are needed to make use of the totality of the data, transparency and consistency are preserved. When prior information is available, it can be used to increase the power of a clinical trial, or equivalently reduce the sample size and the length of the trial. The Bayesian approach is particularly advantageous in the rare disease space where increasing power by using any available information is crucial.

Prior information can be obtained from several sources. For example, data derived from clinical trials of a medical product conducted outside the United States can be a source of prior information for new clinical trials conducted within the United States. In addition, data from clinical trials of similar or earlier generations of a medical product, if legally available, can also be used as prior information.

Data from adult trials can be used as prior information to power clinical trials of medical products to be indicated for pediatric use. For many medical products approved in the United States, development and evaluation is typically done initially for adult populations and then later evaluated for use in children. While there are a host of factors to consider when extrapolating available clinical data from adults to support pediatric indications, prior adult data can be used to power a pediatric clinical trial. Using extrapolated prior information from adults to children would enable industry to conduct trials with a smaller pediatric sample size to support a pediatric indication [4].

Clinical data registries, when available, may also play a role in providing prior information. Clinical data registries typically contain information on health status of patients sharing a common medical condition and the health care they received. There are many types of clinical data registries, including those that focus on a disease or condition (e.g., cystic fibrosis), and a procedure (e.g., coronary artery bypass grafting surgery), or track the performance of a device (e.g., artificial joint). Since data registries are observational, and data quality is typically not as high as in a formal clinical trial, the amount of information that can be borrowed (i.e., the confidence in the prior information) is generally lower. In other words, it is expected that a larger discount of the registry information will be applied.

It is worthwhile to mention that information on the natural history of a disease can form a prior distribution to augment the information provided by a small inactive control group. It may be desirable to include a minimal number of subjects in an inactive control group, especially when the disease is life-threatening or irreversibly debilitating. Prior information obtained from natural history studies conducted from registries or other sources may be used to reduce, or even eliminate, the number of subjects needed for the inactive control group [5,6]. In a paper published in 1976, Pocock [7] provides an approach to combining the control group in a randomized study with historical controls where historical data were incorporated into a prior for the parameters of the control group, resulting in a more efficient use of subjects in clinical studies.

The following examples illustrate how prior information can be effectively used to power clinical trials.

Example 14.1

This example is taken from Irony and Pennello [8]. The T-Scan 2000 electrical impedance imager is indicated for use as an adjunct to mammography to provide physicians with additional information to guide a recommendation regarding further breast examination for patients with equivocal mammograms. This device had been studied in a variety of trials. However, the demonstration of benefit was inconclusive due to various issues. In a multicenter trial conducted in the United States, the T-Scan 2000 was not targeted to suspicious areas based on mammography results, as intended to be used clinical practice, and in two Israeli sites, the device was not used as an adjunct to mammography. While the device was used as intended in an Italian study, the sample size in that trial was too small to draw firm conclusions. To resolve the impasse, a hierarchical multinomial logistic model was proposed to allow the Italian study to borrow considerable strength from the US and Israeli studies. The model would work well if the differences in sensitivity and specificity among studies were not too large. By using the Bayesian hierarchical model, the company demonstrated that targeted use of T-Scan 2000 in conjunction with mammography was superior in both sensitivity and specificity to the use of mammography alone, with a 98% probability that using T-Scan 2000 would result in fewer total biopsies. Based on this evidence, the T-Scan 2000 electrical impedance imager was granted approval for marketing in 1999. More information can be found in Ref. [9].

Example 14.2

This example, taken from Gsteiger et al. [5], provides an illustration of how historical information can be borrowed to power a control group in a clinical trial. Multiple sclerosis (MS) is an unpredictable and often disabling disease of the central nervous system that disrupts the flow of information within the brain, and between the brain and body. Lesions detected on MRI scans of the brain are used to evaluate MS. The number of lesions is typically considered as an over-dispersed count variable. To design a placebo-controlled clinical trial with relapsing-remitting MS, historical placebo control groups from nine studies with the same indication and primary endpoint were available (six trials from the literature and three in-house studies), rendering a total of 1,926 subjects (1,432 from external trials and 504 from in-house trials). To account for over-dispersion, a Negative Binomial-Normal hierarchical model was assumed for the number of lesions. Relatively non-informative priors were used for the hyperparameters—an exponential distribution with mean 3 for the over-dispersion parameter κ, a Normal (0, 100^2) for the population mean η, and a half-normal (scale $=$ 1) distribution for the population standard deviation τ. The posterior distribution of the parameters to be used in the new trial as prior information was obtained via meta-analysis of the historical control data. The effective sample size of the historical control group was approximately $n = 45$. Sensitivity analyses were conducted by using different prior distributions for τ, the parameter representing the heterogeneity among the trials that provided the historical controls. Different priors for τ yielded fairly similar posterior distributions demonstrating low sensitivity to the distribution of τ.

Example 14.3

Hemophilia is a congenital bleeding disorder occurring predominantly in males, characterized by a deficiency of either coagulation factor VIII (FVIII—hemophilia A) or coagulation factor IX (FIX—hemophilia B). Xyntha is a recombinant antihemophilic coagulation factor indicated for the control and prevention of bleeding episodes, and perioperative management of adults and children with hemophilia A. The Food and Drug Administration (FDA) review and approval document for Xyntha are publicly available at Ref. [10].

The standard of care for moderate to severe hemophilia A is routine prophylaxis with or on-demand administration of coagulation factor VIII (FVIII). This treatment tends to be very effective except when patients develop an alloantibody (i.e., inhibitor) against the exogenous FVIII. Inhibitors bind FVIII and prevent the hemostatic action, requiring alternative treatment with higher costs and potentially increased risk. Therefore, it is crucial that a low inhibitor-formation rate for a proposed hemophilia A treatment be demonstrated. In the pivotal study of Xyntha, 2 out of 89 patients (2.2%) developed an inhibitor. The Clopper–Pearson 95% two-sided exact confidence interval (CI) for inhibitor rate is given by (0.3%, 7.9%), where the upper limit of the CI is above the FDA's proposed limit of 6.8%. A Bayesian approach was proposed to estimate the inhibitor-formation rate for Xyntha, where two historical studies conducted by the same sponsor were considered as sources of prior data. In the first study, 1 of the 113 patients who received the predecessor treatment developed an inhibitor, and in the second study, 3 of the 110 patients who received Xyntha prior to a manufacturing change developed an inhibitor. The posterior distribution was a Beta(5, 220) based on these two historical studies if a non-informative Beta(1, 1) prior was used. This Beta distribution was discounted 50% to a Beta(2.5, 110) to account for partial exchangeability of the historical data with new data from the proposed clinical trial. Assuming a Beta(2.5, 110) distribution as the prior, the posterior distribution of the inhibitor-formation rate was given by a Beta(4.5,197) distribution after observing two inhibitions in 89 subjects. An estimated upper limit of a 95% credible interval was calculated to be 4.17%, below the 4.4% acceptable limit proposed by the sponsor. Based on these results, Xyntha was approved by FDA in 2008. Alternatively, Lee and Roth [11] proposed a relatively non-informative Beta(0.3, 3.9) prior, which has a median of approximately 2% and is in line with data from large population surveys of inhibitor incidence, in a Bayesian analysis of inhibitor-formation rate in a different setting.

Example 14.4

Hepatitis B is a liver infection caused by the hepatitis B virus (HBV). For some people, hepatitis B can become a long-term, chronic infection leading to serious health issues, including cirrhosis and liver cancer. The best way to prevent hepatitis B is vaccination. Currently, two hepatitis B vaccines (Engerix-B®

and Recombivax HB®) and a few combination vaccines that include hepatitis B antigen have been approved by the FDA for preventing hepatitis B. Both Engerix-B and Recombivax-B are approved to be administered on 0-, 1-, and 6-month schedules. Because some subjects do not complete the 3-dose series [12], which results in compromised protection against HBV, there is a need for a more potent vaccine requiring fewer doses, which could improve compliance and offer better protection against HBV. A 2-dose vaccine for hepatitis B (Heplisav-B) was studied, and the results were submitted to the FDA for consideration. The information for this example was obtained from the FDA advisory committee meeting materials for Heplisav, which is publicly available at Ref. [13].

Non-inferiority of immunogenicity was established in two pivotal studies (HBV-10 and HBV-16). However, in a third large Phase 3, randomized safety study (HBV-23), there was an imbalance of acute myocardial infarction, with 14 events occurring in the Heplisav-B group versus one event in the Engerix-B group, with an estimated relative risk of 6.97. Such an imbalance was not observed in the HBV-10 or HBV-16 pivotal studies. To scrutinize the safety signal, FDA conducted several Bayesian analyses to calculate the posterior probabilities of relative risk exceeding 1, 2, 3, and 5. The extreme cases in which full-borrowing (i.e., pooling HBV-10, HBV-16, and HBV-23 studies) and no-borrowing (i.e., only including HBV-23 data) were presented to FDA's Vaccines and Related Biological Products Advisory Committee, as other potential borrowing scenarios would fall between the two extremes. A non-informative Beta distribution was used as the prior distribution for event rates of acute myocardial infarction for Heplisav-B and Engerix-B in Study HBV-23 in the no-borrowing scenario. In the full-borrowing case, the same non-informative Beta prior was assigned to Heplisav-B and Engerix-B groups in Studies HBV-10 and HBV-16. The resulting posterior distributions, Beta distributions due to conjugacy, were used as the priors for Heplisav-B and Engerix-B groups in Study HBV-23.

14.2.1 Prior Information to Enhance Development and Use of Medical Products

Incorporating prior information using a Bayesian approach may be applied for other purposes, in addition to increasing study power. In the following examples, we illustrate how prior information can be effectively applied to enhance development and clinical use of medical products.

Example 14.5

As mentioned in Example 14.3, the standard of care for hemophilia is regular or on-demand administration of FVIII or FIX concentrate to minimize bleeding episodes. Currently, the dosing regimen of FVIII or FIX for a given patient relies solely on empirical information, patient weight, and clinical experience. However, some have suggested that such an approach to dosing may not be optimal given there are wide differences in patient metabolism rates, resulting in different drug half-lives among patients. An alternative approach would be to use patient-specific pharmacokinetic (PK) data to inform decision-making on dosing regimen. Bjorkman et al. [14,15] showed that a prior distribution for PK parameters can be obtained from a population PK model based on available activity-time profiles from four studies. For individual patients, the distribution of PK parameters is updated based on limited individual-specific PK information (as few as two to three PK samples) combined with the population prior [14,15]. The posterior probabilities of individual PK parameters are then used in simulations to help physicians determine appropriate dosing regimens for a given patient.

Example 14.6

Phase I studies in oncology are usually designed to find out the maximum tolerated dose (MTD), which is the maximum dose associated with acceptable dose-limiting toxicity (DLT). In a typical dose-finding study, a set of candidate doses are pre-determined. Over the course of the study, groups of subjects are recruited sequentially such that the first group is exposed to the lowest dose. The dose to be used in the next group may be escalated or de-escalated from current dose level depending on the number of subjects experiencing DLT in the current group. This process continues until certain criteria are met and the MTD is identified. Often the MTD is determined via the classical 3 + 3 design [16] despite criticisms levied against this design [17]. The 3 + 3 design is a rule-based dose-finding method, where groups of 3 subjects are exposed to a higher dose level, a lower dose level, or the same dose level depending on the toxicity experienced by previous groups according to pre-defined rules. As an alternative, model-based dose escalation methods have gained popularity in recent years due to their flexibility related to variable group sizes and their superiority in targeting the MTD.

An example using a Bayesian Logistic Regression Model (BLRM) is described in Neuenschwander et al. [18]. The logistic model is fitted for the probability of observing DLT for each dose d, denoted by p_{d}, and Normal priors are assigned to the regression parameters. Once the DLT results for a group are available, the posterior distribution of the regression parameters is updated, and the posterior mean of p_{d} is calculated. Typically, but not always, the posterior mean of p_{d} increases as the dose level increases. The dose for which the posterior mean of p_{d} is within the targeted DLT range is selected for the next group. For example, if the target is to find out the dose for which p_{d} is no more than 25% DLT, the dose for which the posterior mean of p_{d} is almost 25% is selected for the next group. The prior for the regression parameters in the logistic model may be estimated from historical information if available. For example, a random-effects meta-analysis may be carried out to evaluate between-trial variability [11], which in turn can suggest a strong or a weak prior for the regression parameters. The same method can be used in determining the MTD for a combination therapy with priors derived from historical data collected in monotherapies. A case study of nilotinib in combination with imatinib in adult patients with imatinib-resistant gastrointestinal stromal tumors was presented in Bailey et al. [19].

14.2.2 Considerations When Using Prior Information

The use of prior information in the regulatory setting is usually subject to extra scrutiny and demands strong clinical and statistical justification since the degree of exchangeability between prior and current data is subjective. Comprehensive discussions with clinicians and expert reviewers should be involved in the process. Non-informative priors may elude the subjectivity issue and are less controversial in the regulatory setting. However, using an informative prior distribution is of great interest when available, and it is the only mathematically sound way to account for the totality of the data and increase statistical power.

The following measures may minimize concerns regarding the use of informative priors and increase their acceptability in the regulatory setting.

The first measure is to minimize selection bias when choosing sources of prior information. While the FDA recommends that as many sources of "good" prior information as possible be identified, the assessment of "goodness" is subjective. In general, studies with similar protocols (endpoints, target population) and time frame for data collection may be considered appropriate for constructing the prior [1]. However, it is always of concern that sponsors may selectively choose studies with the most favorable results when identifying prior information for Bayesian analyses. To guard against selection bias, both favorable and non-favorable prior information should be included, especially when the effective sample size of the prior distribution is large. A complete list of available studies and comprehensive review of currently available results are necessary when constructing a valid prior. Ideally, the criteria for inclusion of available studies as prior information should be determined before examining results of those studies to minimize selection bias.

The second measure is to discount the prior information to compensate for the assumption of exchangeability and to allow for "partial exchangeability". In Example 14.3, the prior for inhibitor rate was discounted by 50%. While several methods for discounting prior information have been developed [20,21], there is currently little agreement on which method is most appropriate and the topic still invites discussions. Although statistical tools such as meta-analytic-predictive priors described in Example 14.6 are available, these methods rely on the quality and quantity of historical information selected and still involve subjectivity such as the choice of hyperparameters, for example. Typically, information collected from prior studies is heavily discounted unless justified otherwise by a good reason. A prior distribution may be too informative, and the effective sample size of the prior may be used as a measure of how informative is the prior. If the prior is too informative, that is, the effective sample size is too large, remedies include decreasing the effective sample size of the prior, increasing the stringency of the success criterion, increasing the sample size of the pivotal trial, or using Bayesian hierarchical models with parameters that decrease the prior strength. The choice of hyperparameters used in the hierarchical model may bring in additional subjectivity. A larger sample size for the pivotal study automatically discounts the prior information and is often justified by the need to gather safety information. For example, in pivotal vaccine studies, the safety database needs to include many subjects to ensure the safety risks are minimal given that vaccines are normally administered to a large, healthy population.

A third way to minimize concerns about using informative prior distributions is to increase the stringency of trial success criterion, which could be translated into either increasing the threshold for the posterior probability of trial success or decreasing the significance level to which the type I error rate needs to be controlled. Although the Bayesian approach relies on evaluation of posterior distributions and not on hypotheses testing and controlling type I error rates, some regulators believe that control of type I error rate at a pre-specified significance level, usually higher than the traditional values, constitutes good practice [1]. This control is done through simulations at the trial design stage and is explained in more detail in Section 14.4. If the traditional 0.05 (two-sided) or 0.025 (one-sided) significance level cannot be relaxed, the practice of controlling type I error rate at the traditional level is equivalent to completely discounting the prior information.

It is important to note that currently, there is no unified approach for defining an acceptable prior. The acceptability of the prior is determined on a case-by-case basis, based on clinical and statistical inputs obtained at the study design stage. Moreover, acceptability of the prior can change over the course of the clinical trial if the state of the science changes. Therefore, choice of a relatively conservative prior is almost always recommended.

Appropriate selection of a prior distribution requires close collaboration between statisticians and clinicians during the study design stage. Statisticians have a critical role to play in guiding clinicians through the decisions involved in selecting a prior in a way that is both clear and comprehensible. It is crucial that statisticians communicate with clinicians, so all understand the impacts of their choices. For example, in the Bayesian Logistic Regression dose-finding model described in Example 14.6, a meta-analysis may provide insights into selection of the appropriate prior. However, simulated trial scenarios and their operating characteristics, such as error rates, should always be presented to clinicians who may provide iterative feedback on whether the dose escalation is too aggressive or too conservative in the simulated scenarios in order to fine-tune the prior distribution until agreement is finally reached.

Readers are encouraged to explore Chapters 2–4 for additional topics with regard to building priors.

14.3 Bayesian Adaptive Designs

An adaptive design is defined as a clinical study design that allows for prospectively planned modifications based on accumulating study data without undermining the study's integrity and validity [22]. Adaptive designs have been increasingly adopted because they usually make trials more efficient.

Often, accurate and precise estimates for important parameters needed for designing a study, such as the variance or the difference to be detected, may not be available, and when the actual parameter values are substantially different from their design-stage estimates, a non-adaptive design based on these estimates may not be appropriate. Often, sample size is too small or too large, and power is under- or over-estimated. An adaptive design, in contrast, provides an opportunity to adjust the study design based on observed interim data, thereby reducing the uncertainty about the parameters' estimates. Possible study modifications include, but are not limited to inclusion/exclusion criteria, randomization procedure, treatment regimens (e.g., dose level, schedule, duration), sample size, and primary and secondary endpoints.

Adaptive designs are natural to the Bayesian approach due to its adherence to the likelihood principle. A Bayesian adaptive design differs from a frequentist's adaptive design because adaptive decisions can be based on Bayesian quantities (posterior probabilities or predictive probabilities) and use Bayesian decision criteria in the final analysis. In addition, a purely Bayesian approach follows the likelihood principle and allows for continuous design adaptation as the trial proceeds, without corrections for multiple looks. However, the FDA continues to recommend that Bayesian adaptive trials be adaptive by design. That means they should be planned in advance and that their operating characteristics (type I error rate, power, and expected sample size) should be assessed in a variety of scenarios. In addition, it is important to establish firewalls in order to minimize operational biases and maintain trial integrity [1]. An example of the use of Bayesian adaptive designs in vaccine development is described in Example 14.7.

A common design is to adapt the sample size. During the interim analysis, the sample size of a study may be reduced according to pre-specified rules if sufficient evidence of efficacy is observed at the interim look or when the observed variance is smaller than the variance estimated at the design stage. In this case, the trial becomes smaller and regulatory decisions may be reached faster, particularly if there is no need for additional safety data. Conversely, the sample size may be increased if the interim analysis reveals that the initial assumptions were not accurate, rendering an underpowered study. In both cases, the predictive probability of trial success conditional on observed interim data is typically calculated and used for decision-making. Thus, the sample size is determined and optimized during the trial, and it is neither too small nor too large. This type of design prevents inconclusive trials, assuring that the questions initially posed are answered with an optimal amount of information.

Bayesian adaptive designs are crucial when historical information is borrowed; that is, an informative prior is used in the analyses. As described earlier in the chapter, one of the challenges in using prior information to construct a prior distribution is determining how much the prior data should be discounted. It is often difficult to answer this question conclusively at the design stage when there is little or no information on how similar the prior studies are to the planned study. An adaptive design would increase the sample size of the current study when it is not similar to the prior studies, which implicitly discounts the prior studies and reduces their importance when compared to the current study. In addition, if hierarchical models are used to combine prior and current studies, the homogeneity parameter can be adaptively estimated. Simply put, if the accumulated data in the current clinical trial shows a different pattern than what was found in prior studies, the historical information may be discounted via adjustments in the homogeneity parameter, requiring a study with a larger sample size to maintain appropriate power. Conversely, if the historical information resembles the pattern of data in the current clinical trial, the final sample size to maintain appropriate power will be smaller.

When using the Bayesian approach, techniques such as statistical modeling of the primary endpoint based on a surrogate endpoint or early follow-up observations may be adopted in conjunction with an adaptive design. This may be advantageous when delayed responses, such as survival times or response status at 24 months post-treatment, are used as the primary endpoint, but subjects are also observed at 12 months. The number of subjects with 24-month outcome data may be limited at the time of an interim analysis, but the number of subjects with 12-month outcome data will be higher. An advantage of the Bayesian approach is that a statistical model embedded in the adaptive design can estimate final outcomes for those subjects who have not yet reached the 24-month follow-up based on their 12-month follow-up results or other surrogates, making use of all available information at the interim analysis. If the correlation between 12- and 24-month results is high, the model is a good predictor, the sample size will be smaller, and the trial can be terminated early. The use of intermediate or surrogate endpoints intrinsically increases the information included in the interim analysis, thereby resulting in an earlier decision (early top for efficacy or futility). A real example of how modeling can be applied in Bayesian adaptive designs is provided in Example 14.8.

Adaptive randomization, more commonly used in exploratory rather than confirmatory studies, is a design in which the probability of patient assignment to a given treatment group is adjusted based on comparative analyses of the accumulated outcomes of patients previously enrolled [22]. Typically, the probability that a new subject will be assigned to a treatment depends on the posterior probability that the treatment is effective, given the results obtained thus far. The higher the posterior probability, the higher the probability of assignment. Adaptive randomization, when implemented correctly, can increase the likelihood that more subjects will be assigned to the most effective treatments, which is ethically appealing. Examples are given in Berry et al. [2].

Example 14.7 For this example, we consider a respiratory disease infection of particular concern for infants and older adults, for which there are no effective treatments. The need for a vaccine as a cost-effective approach to prevent the respiratory disease infection has been unmet. A Phase III clinical trial was designed to establish efficacy of a candidate vaccine. The trial adopted a Bayesian adaptive design where up to three interim analyses and one final analysis were planned, depending on subject accrual rate. Subjects were initially randomized into treatment and placebo at a 1:1 ratio, and subsequently randomized at a k:1 ratio if no safety signal was detected within a given time, where $k > 1$ to enable faster recruitment in the treatment arm. The number of events (number of subjects who undergo the specific respiratory disease during study follow-up) in the treatment and placebo arms was assumed to follow binomial distributions, and a non-informative Beta distribution was assigned as the prior for the parameter in the binomial distributions. The vaccine efficacy (VE) is typically evaluated as $1 - R$, where R is a ratio of risks, incidence rates, or hazards of disease in the vaccinated group relative to the control group. VE is a number that is always less than or equal to 1, with 1 being the case in which the vaccine is 100% effective for all vaccinees. A positive VE (i.e., $0<\text{VE}\leq 1$) indicates that the vaccine is protective, while a negative VE indicates that the vaccine is harmful. Because it is expected that the vaccine will be administered to a broad range of people, the candidate vaccine needs to demonstrate super efficacy (i.e., $\text{VE} > C$, where C is a positive number between 0 and 1) to be licensed. In this example, the posterior probability of the VE was approximated via the Monte Carlo method (simulation), and a Pocock rejection boundary [23] was determined to claim efficacy. The study would claim success if the posterior probability of VE being no less than a clinically meaningful margin, say $X\%$, were above a pre-specified threshold, say C_0, at any efficacy interim analysis or at the final analysis. Futility analyses were more frequent and conducted in a similar fashion based on the posterior probability of VE, with much less stringent choices of X and C_0 to be in line with the futility nature. A minimum sample size (until which no efficacy interim analyses would be performed) was determined based on the need to evaluate safety, and a maximum sample size was determined to yield sufficient power, based on simulations. Extensive simulations were performed to demonstrate that the type I error rate was well controlled at 0.025 under various assumptions.

Example 14.8
Bayesian adaptive designs where statistical models are used to predict the primary endpoint based on early follow-up times or surrogates have been widely used to assess effectiveness of orthopedic devices, especially fusion devices to

treat lumbar and cervical degenerative disc disease (e.g., see InFUSE^{TM} Bone Graft/LT-CAGE^{TM} Lumbar Tapered Fusion Device [24]; the advisory panel meeting transcript can be found in Ref. [25]). The primary endpoint for these trials is the success rate of a composite endpoint that includes fusion, pain, and function at 24 months after implantation. Subjects are usually evaluated on the composite at 3, 6, 12, and 24 months. The design includes interim analyses for sample size determination, for success, and for futility. The decisions are based on the posterior probability that the success rate of the treatment device is not inferior to the control device. These posterior probabilities are calculated using the patients who have 24-month follow-up, but also the ones that only have 12-, 6-, and 3-month follow-up results. A mathematical model embedded in the likelihood function predicts 24-month follow-up results for those patients who have not yet reached the 24-month follow-up visit. If early follow-up results are good predictors of 24-month results, the adaptive design is advantageous and the trial stops early.

Bayesian adaptive designs are highly valuable because the trials are revised as data accrues in order to answer the questions that will inform decisions, minimizing the chance of inconclusive trials. In general, if the treatment is better than predicted at the design stage, the trial stops early. On the other hand, if the treatment is slightly worse than predicted at the design stage, but still clinically meaningful, the trial will enroll more subjects than initially predicted but will still lead to a conclusion. If the sample variability is higher than initially assumed, the trial will also enroll more subjects instead of stopping with inconclusive results.

If the correlation between the results obtained at early follow-up times and late follow-up times is high, the adaptive design will generate substantial savings, and the trial will stop early provided there is no need to collect additional data to assess safety.

Despite their appeal, Bayesian adaptive designs may come at a price. First, a substantial amount of work is involved at the planning stage. Current thinking advises that the operating characteristics of the trial, such as expected sample size, type I error rate, and power, be carefully examined, even though some of these concepts may not be legitimately defined within the Bayesian framework. For example, the notion of type I error rate is unusual in the Bayesian approach, as it usually does not make sense to condition on the null hypothesis when performing a Bayesian analysis. While it may seem odd to evaluate Bayesian designs using frequentist operating characteristics, the regulatory practice of controlling type I error rates for confirmatory studies still applies to Bayesian adaptive designs. Unlike for most frequentist methods, operating characteristics in Bayesian adaptive designs are analytically intractable and need to be calculated via simulations. Hence, simulations play a critical role in study design to ensure that the operating characteristics are acceptable. More details on simulations are presented in the following section. In addition, clinical trials employing Bayesian adaptive designs require statisticians with specific expertise to carry out the interim and final analyses due to the complexity of the design. Although this may not pose a problem for product sponsors, it can be a challenge for others, such as the data monitoring committees (DMCs). In a double-blind adaptive study, the DMC will normally conduct unblinded interim efficacy analyses, interpret the results, and inform the sponsor of their recommendation; thus, it is crucial that independent statisticians in the DMC are familiar with relevant knowledge and perform the efficacy analyses as intended. Finally, there is less availability of software to carry out simulations and computations of Bayesian adaptive designs as compared to traditional adaptive designs, not to mention that Bayesian-specific expertise is needed for running these simulations and computations. Readers are encouraged to refer to Chapters 15 and 16 for additional information on currently available software for Bayesian analysis.

14.4 Simulations of Clinical Trials

A full Bayesian approach in which the posterior probability of effectiveness is the sole criterion for decision-making has been considered in Ref. [26]. Nevertheless, presently, the design of clinical trials in the regulatory setting has been evaluated through its operating characteristics which include type I error rate, power, and average sample size. For most Bayesian designs, these operating characteristics cannot be assessed through closed formulas, requiring simulations of large numbers of trials. The idea is to simulate the proposed trial under several scenarios in which the treatment is effective as well as under scenarios where the treatment not effective, and then evaluate the proportion of simulated trials for which the success criteria are met. These proportions provide an estimate of type II error rates (or power) and type I error rates, as well as the average sample size for a Bayesian adaptive design.

The process involves dividing the parameter space into regions for which the medical product should be approved (to estimate power) and regions for which the product should not be approved (to estimate type I error rate). Dividing the parameter space can be a complex task, especially when multiple parameters are considered as primary (or secondary) endpoints and when nuisance parameters (such as variance or correlation between surrogate and primary endpoints) are relevant to the trial design.

In some cases, a bracketing strategy can be used to evaluate the robustness of the design against potential deviations if a lower limit and an upper limit of potential deviation can be identified, representing two limiting scenarios. If the operating characteristics for both limiting scenarios are acceptable, it is expected that the actual trial will fall between the two cases and should be acceptable as well.

Due to the lack of closed-form solutions in Bayesian adaptive designs, it is difficult to quantitatively assess the impact of different design features on the operating characteristics. For example, in a vaccine study, the value of the posterior probability of VE being greater than a clinically important margin is often used as a criterion for study success. The study will be successful if the posterior probability is larger than a pre-specified value (e.g., 0.97). The choice of 0.97 instead of 0.96 or 0.98 is substantiated through simulations, in order to maintain the type I error rate below the ceiling of 0.05.

Bayesian adaptive trial design features include stopping rules for success and futility, number and timing of interim analyses, prior probabilities, hierarchical model parameters, prior discount factors, the predictive model, minimum and maximum sample sizes, randomization ratio, accrual rate, dose/treatment selection, number of centers, and use of covariates. Simulations conducted at the design stage can be used to assess the operating characteristics for different combinations of these features and to select the optimal combination for a given situation.

For example, simulations may show that the type I error rate of a trial design is too large for regulatory purposes. Remedies for excessive type I error rates include but are not limited to increasing the stringency of trial success criteria, reducing the number of interim analyses, increasing the number of subjects before the first interim analysis, discounting the prior information, and increasing the maximum sample size [1].

In addition to study design optimization, there are other advantages of clinical trial simulations. Simulations can bridge the communication barriers across disciplines because they mimic what could happen in the real world, enabling clinicians to understand the processes and results under various scenarios. In addition, simulations provide the ability to "look into the future" to avoid anticipated regrets. Simulations may reveal scenarios that may prove to be problematic before the actual clinical trial is implemented, allowing investigators to correct the study design before the study starts. Simulations also increase trial predictability and help sponsors prepare and budget for different scenarios and surprises. In short, after an extensive and sound simulation study, the actual trial is not the first trial, but it is a familiar trial because the possible scenarios have already been analyzed during the simulation stage.

Simulations are a great tool to evaluate Bayesian designs, but their review poses unique challenges to regulatory agencies because an extensive number of simulations are expected to be carried out at the design stage to cover numerous possible and plausible scenarios. Regulatory reviewers not only need to verify that the simulation results are appropriately generated but may also need to perform additional simulations to determine whether additional scenarios are deemed necessary. The assumptions of nuisance parameters used to generate simulated scenarios need to be evaluated, and sensitivity analyses may be necessary. The accuracy and precision of the estimators are contingent on the number of iterations. To obtain reliable estimates with acceptable accuracy and precision, tens of thousands of iterations may be needed, followed by the examination of convergence and stability of the estimates to ensure that the number of iterations is large enough. Fortunately, there is an increasing number of commercial simulation packages available, such as EAST and FACTS. Nevertheless, R is the primary software used by statisticians to run simulations due to its versatility to deal with complex study designs. However, R is known for its slow speed when running loops and may not be practical when performing complicated simulations with a large number of repetitions. In these situations, techniques such as parallel computing using a powerful scientific server may be needed to run the simulations. Other software such as C++ and Python can be used when faster processing speed is needed, but the use of these packages may represent a steep learning curve for regulatory reviewers.

In summary, simulation studies generate a large body of information that must be reviewed by the regulatory agencies. To expedite the review process, prompt communication with regulators is highly recommended, and clear documentation provided by sponsors can be very helpful. Submission of program codes and any data used to conduct the simulations is recommended by FDA [1].

Example 14.7 (Continued)
Here, we revisit the respiratory disease example to describe the simulations that were conducted by the sponsor to ensure the operating characteristics of the Bayesian design were robust and acceptable under various deviations from the assumptions. Different ranges of values were assigned for nuisance parameters, so that a broad range of possible scenarios were explored. Since information about the background event rate (i.e., the event rate in the placebo arm) in the literature was inconsistent, a wide range of background rates were simulated to cover numerous possible scenarios. In addition, because the accrual rate could impact the number of interim analyses, different accrual patterns, with constant and varying accrual rates, were simulated to estimate the operating characteristics of the design. Given that the randomization ratio was expected to change from 1:1 to k:1 resulting in the vaccine arm ending up with k times the number of subjects as the placebo arm, simulations were conducted to account for scenarios in which the vaccine arm had more or fewer subjects than the ideal scenario.

For this example, the posterior probability of VE being at least 30% was calculated using Monte Carlo methods, with multiple values generated from Beta distributions to approximate the posterior probability with adequate precision. As described earlier, using software such as the R program to complete these simulations is impractical given that at least 42 scenarios are needed to simulate seven different background rates and six assumed values of VE. Indeed, using a laptop equipped with Intel i5-4300U CPU, it took 16 min to simulate the 42 scenarios once; a total of 2,567 h would be needed to simulate the trial 10,000 times. Fortunately, these same simulations can be completed within hours using parallel computing in a high-performance computing environment at the FDA.

Example 14.8 (Continued)
In cases such as Example 14.8 where Bayesian adaptive designs use statistical models to predict the primary endpoint based on early follow-up times or surrogates, the scenarios to be simulated assess different relationships (i.e., transition probabilities) among the various follow-up points (e.g., 3, 6, 12, 24 months). For example, the following probabilities table could be used as an initial transition probability matrix to generate the first simulated scenario:

	P(Success)	P(Failure)
P(3 month)	0.40	0.60
P(6 month|3-month failure)	0.25	0.75
P(6 month|3-month success)	0.90	0.10
P(12 month|6-month failure)	0.15	0.85
P(12 month|6-month success)	0.90	0.10
P(24 month|12-month failure)	0.15	0.85
P(24 month|12-month success)	0.90	0.10

To span multiple simulation scenarios, several plausible transition matrices can be generated based on input from clinicians. For instance, it is unlikely that a successful treatment result would be observed at an early follow-up point but not at later follow-up. The simulation scenarios also include extreme cases in which there is (1) perfect correlation among follow-up results and (2) complete independence among follow-up results. Such extreme scenarios are helpful in delineating the range of probabilities. In addition, because the trial included interim analyses for sample size determination, futility, and success, various plausible accrual rates and success rates for treatment and control groups can be simulated.

For advice on how to conduct computer simulations more efficiently, readers are encouraged to refer to Chapter 8 of this book.

14.5 Conclusions

The use of Bayesian approaches to clinical trials for regulation of medical products has advanced slowly, mainly due to difficulties in dealing with the subjectivity inherent to the use of prior information coupled with a need to control type I error rate at a fixed significance level, most often 0.05 (two-sided) or 0.025 (one-sided), independently of the context in which the treatment is evaluated. The uptake of Bayesian clinical trials has been higher for medical devices. Notwithstanding, we have been currently observing an uptrend in adopting Bayesian trial designs in the pharmaceutical space.

In the regulatory setting, Bayesian adaptive designs with non-informative priors have been more prevalent, possibly because the inflation of type I error rate incurred by interim analyses is small. Their main advantage is thepossibility of embedding models in the likelihood function, which cannot be done with frequentist adaptive designs. These models can predict final results from data observed at interim looks, increasing the power of the trial and enabling earlier decisions. The use of informative priors has been more limited, even in the case of medical device clinical trials, because all prior information needs to be discounted to control the type I error rate at traditional levels.

Due to the lack of closed mathematical forms, simulations have been used to evaluate the operating characteristics of Bayesian trial designs in the regulatory setting. As a by-product, the same simulations can be used to optimize the design of Bayesian clinical trials and to "look into the future" to avoid "anticipated regret". When the actual trial is conducted, most problems have been already predicted and fixed.

To sum up, Bayesian designs have earned their place in pharmaceutical development, and it is encouraging to increasingly see applications taking advantages of the flexibilities provided by Bayesian methods. Yet, there are many areas waiting to be explored. For example, a full Bayesian decision analytical approach may bring even more impact in the regulatory setting. Regulators are eager to see more Bayesian applications because innovative trial designs can generate enough evidence with smaller sample sizes, result in more ethical trials, and ultimately accelerate the development of safe and effective treatments for the patients who need them. The FDA announced a Complex Innovative Trial Design Pilot Program to support the goal of facilitating and advancing the use of complex adaptive, Bayesian, and other novel clinical trial design [27]. Thirty million Americans (roughly 10% of the population) are affected by 7,000 known rare diseases, and treatments are available for only 500 of those diseases. Patients suffering from the other 6,500 rare diseases are longing for treatments, and the trials for such treatments need to extract enough evidence from small samples. Bayesian approaches may provide the solution. The FDA has recently conducted a public workshop to discuss the use of complex innovative designs in clinical trials of drug and biological products according to the Prescription Drug User Fee Act VI (PDUFA VI), the 21st Century Cures Act, and part of the FDA Reauthorization Act (FDARA) that were enacted by Congress [28]. With the recent calls for using real-world evidence, decision analyses, benefit-risk assessments, patient preferences, and advanced analytics, there is no doubt that the opportunities for practitioners who favor Bayesian methods will abound.

References

[1] FDA Center for Devices and Radiological Health, and Center of Biologics Evaluation and Research. 2010. Guidance for the use of Bayesian statistics in medical device clinical trials. https://www.fda.gov/regulatory-information/search-fda-guidance-documents/guidance-use-bayesian-statistics-medical-device-clinical-trials (Accessed July 3, 2019).

[2] Berry, S. M., B. P. Carlin, J. J. Lee, and P. Muller. 2011. *Bayesian Adaptive Methods for Clinical Trials.* Boca Raton, FL: CRC Press.

[3] Lee, J. J. and C. T. Chu. 2012. Bayesian clinical trials in action. *Statistics in Medicine* 31(25): 2955–72. doi: 10.1002/sim.5404.

[4] FDA Center for Devices and Radiological Health, and Center of Biologics Evaluation and Research. 2016. Leveraging existing clinical data for extrapolation to pediatric uses of medical devices. www.fda.gov/ucm/groups/fdagov-public/@fdagov-meddev-gen/documents/document/ucm444591.pdf (Accessed April 12, 2018).

[5] Gsteiger, S., B. Neuenschwander, F. Mercier, and H. Schmidli. 2013. Using historical control information for the design and analysis of clinical trials with overdispersed count data. *Statistics in Medicine* 32: 3609–22.

[6] Berry, D. A. 2004. Bayesian statistics and the efficiency and ethics of clinical trials. *Statistical Science* 19: 175–87.

[7] Pocock, S. J. 1976. The combination of randomized and historical controls in clinical trials. *Journal of Chronic Disease* 29: 175–88.

[8] Irony, T. Z., and G. A. Pennello. 2001. Choosing an appropriate prior for Bayesian medical device trials in the regulatory setting. In *American Statistical Association 2001 Proceedings of the Biopharmaceutical Section.* Alexandria, VA: American Statistical Association.

[9] Summary of safety and effectiveness data. 1999. www.accessdata.fda.gov/cdrh_docs/pdf/P970033B.pdf (Accessed April 12, 2018).

[10] Xyntha. 2014. www.fda.gov/biologicsbloodvaccines/bloodbloodproducts/approvedproducts/licensedproductsblas/fractionatedplasmaproducts/ucm089069.htm (Accessed April 12, 2018).

[11] Lee, M. L., and D. A. Roth. 2005. A Bayesian approach to the assessment of inhibitor risk in studies of factor VIII concentrates. *Haemophilia* 11: 5–12.

[12] Nelson, J. C., R. C. Bittner, L. Bounds, S. Zhao, J. Baggs, J. G. Donahue, S. J. Hambidge, S. J. Jacobsen, N. P. Klein, A. L. Naleway, K.

M. Zangwill, and L. A. Jackson. 2009. Compliance with multiple-dose vaccine schedules among older children, adolescents, and adults: Results from a vaccine safety datalink study. *American Journal of Public Health* 99: S389–97.

[13] July 28, 2017: Vaccine and Related Biological Product Advisory Committee Meeting Presentations. www.fda.gov/AdvisoryCommittees/CommitteesMeetingMaterials/BloodVaccinesandOtherBiologics/ VaccinesandRelatedBiologicalProductsAdvisoryCommittee/ucm570984.htm (Accessed April 12, 2018).

[14] Bjorkman, S., and P. Collins. 2013. Measurement of factor VIII pharmacokinetics in routine clinical practice. *Journal of Thrombosis and Haemostasis* 11: 180–82.

[15] Bjorkman, S., A. Folkesson, and S. Jonsson. 2009. Pharmacokinetics and dose requirements for factor VIII over the age range 3–74 years. *European Journal of Clinical Pharmacology* 65(10): 989–98.

[16] Storer, B. E. 1989. Design and analysis of Phase I clinical trials. *Biometrics* 45: 925–37.

[17] Reiner, E., X. Paoletti, and J. O'Quigley. 1999. Operating characteristics of the standard phase I clinical trial design. *Computational Statistics & Data Analysis* 30: 303–15.

[18] Beuenschwander, B., M. Branson, and T. Gsponer. 2008. Critical aspects of the Bayesian approach to phase I cancer trials. *Statistics in Medicine* 27: 2420–39.

[19] Bailey, S., B. Neuenschwander, G. Laird, and M. Branson. 2009. A Bayesian case study in oncology Phase I combination dose-finding using Logistic regression with covariates. *Journal of Biopharmaceutical Statistics* 19: 469–84.

[20] Ibrahim, J. G. and M. H. Chen. 2000. Power prior distributions for regression models. *Statistical Science* 15: 46–60.

[21] Hobbs, B. P., D. J. Sargent, and B. P. Carlin. 2012. Commensurate priors for incorporating historical information in clinical trials using general and generalized linear models. *Bayesian Analysis* 7: 639–74.

[22] FDA Center for Device and Radiological Health, Center for Biologics Evaluation and Research. 2016. Adaptive design for Medical Device Clinical Studies. www.fda.gov/downloads/medicaldevices/deviceregulationandguidance/guidancedocuments/ucm446729.pdf (Accessed April 12, 2018).

[23] Pocock, S. J. 1977. Group sequential methods in the design and analysis of clinical trials. *Biometrika* 64: 191–99.

[24] Summary of safety and effectiveness data. 2002. www.accessdata.fda.gov/cdrh_docs/pdf/P000058b.pdf (Accessed April 12, 2018).

[25] Orthopedics and rehabilitation devices advisory panel public meeting transcript. January 10, 2002. http://www.fda.gov/ohrms/dockets/ac/02/transcripts/3828t1.htm (Accessed July 3, 2019).

[26] Proschan, M. A., L. Dodd, and D. Price. 2016. Statistical considerations for a trial of Ebola virus disease therapeutics. *Clinical Trials* 13(1): 39–48.

[27] Complex Innovative Trial Design Pilot Program. 2018. https://www.fda.gov/drugs/development-resources/complex-innovative-trial-designs-pilot-program (Accessed July 3, 2019).

[28] Promoting the Use of Complex Innovative Designs in Clinical Trials. 2018. https://www.fda.gov/drugs/news-events-human-drugs/promoting-use-complex-innovative-designs-clinical-trials (Accessed July 3, 2019).

15

Computational Tools

David Kahle
Baylor University

Michael Sonksen
Eli Lilly & Company

CONTENTS

15.1 Introduction

While computational methods play a central role in modern statistics in general, perhaps nowhere are they more important than in Bayesian statistics. In addition to the study design, data manipulation, modification, and model fitting common to all statistical endeavors, the Bayesian statistician has the added burden of working through Bayes' theorem, a notoriously difficult task only made possible through heavy-duty computational methods invented and implemented in the last few decades.

In this chapter, we turn our attention to practical tools used by Bayesians working in pharmaceutical drug development. We begin in Section 15.2 with a look into the *why* of statistical software for Bayesian statistics: what makes Bayesian statistics uniquely challenging and how computer enabled Monte Carlo methods provide an avenue to inference. In Section 15.3, we present an overview of statistical software commonly used in the area: the statistical computing platform `R` and the Bayesian software `BUGS` and `JAGS`. In Section 15.4, we present a series of concrete implementations of common Bayesian models in cookbook fashion. We then turn to computational aspects of prior elicitation in Section 15.5. We conclude our overview of computational tools in Section 15.6 where we outline the Bayesian approach sample size determination.

15.2 The Monte Carlo Nature of Modern Bayesian Computation

Bayes' theorem is

$$p(\theta|y) = \frac{l(\theta|y)p(\theta)}{\int l(\theta|y)p(\theta)d\theta}, \tag{15.1}$$

where $p(\theta)$ is the prior distribution, $l(\theta|y)$ is the likelihood function, and $p(\theta|y)$ is the posterior distribution. From a conceptual perspective, the posterior distribution $p(\theta|y)$ is the endgame: it contains all of the combined information from the data and the prior, and it is theoretically sufficient for the determination of any action. However, in practice, the posterior is typically only of *penultimate* interest. Practical questions are almost never answered by the posterior itself; rather, the posterior is summarized in a way that provides the ultimate object of interest: a quantity, a plot, etc. In fact, most objects of interest can be represented as expectations with respect to the posterior. For example, in point estimation, the ultimate object of interest is the estimate $\hat{\theta}$, often taken to be the mean of the posterior $\mathbb{E}_{p(\theta|y)}[\theta]$. In interval estimation, we want an interval $[l, u]$ or a region $\mathcal{C}$ that has some pre-specified probability of containing the true value of the parameter θ. Common examples of this in Bayesian statistics are the highest posterior density (HPD) region, which is the smallest set $\mathcal{C}$ to which the posterior assigns the given probability, and, in the one-dimensional case, the equal-tailed posterior interval, which simply uses percentiles to provide an interval, e.g. the 2.5th and 97.5th percentiles for a 95% interval. Both these can be described with expectations.

The primary challenge to the practical application of Bayesian methods is often said to be the intractability of the posterior distribution $p(\theta|y)$. The problem stems from the inability to integrate the function $\tilde{p}(\theta|y) = l(\theta|y)p(\theta)$, the un-normalized posterior, and closely related functions. If we were we able to integrate these functions with ease, we could compute both the posterior and its summaries, since both assume this form. However, in most models of interest, two aspects of $\tilde{p}(\theta|y)$ present a challenge. First, most models have many parameters so that θ is a vector in several dimensions, and thus the integral is a multivariate integral, sometimes over a non-rectangular region. Second, the functional form of $\tilde{p}(\theta|y)$ is almost always very complicated, involving not only algebraic and simple transcendental functions but also special functions. Such integrals rarely admit closed-form antiderivatives. Thus, the naive approach to Bayesian inference suggested by Bayes' theorem – compute the integral in the denominator to obtain the posterior and proceed with inference – cannot be considered a realistic general strategy for Bayesian inference of the kind encountered by the statistician in pharmaceutical drug development.

In light of this, the practitioner may be faced with the question: how might we proceed at all? In some cases, the special structure of the problem can be exploited to enable the determination of the posterior; this is the case with conjugate priors. In other cases, numerical integration can be used; however, numerical integration tends to not scale well into many dimensions. Analytical approximations to the posterior can provide efficient solutions and are routinely used in practice, but they often come with an unknown cost of accuracy and so are generally not considered best practice. Instead, the vast majority of modern Bayesian practice focuses on Monte Carlo strategies for inference.

The term "Monte Carlo" generally refers to mathematical or statistical methods that rely on random numbers to solve problems of interest. While there are many uses of Monte Carlo methods in Bayesian statistics, in this exposition, we focus on the primary use: computing summaries of the posterior distribution that can be represented either directly as expectations respective of the posterior or can be computed algorithmically from such expectations. Perhaps surprisingly, Bayesians using Monte Carlo methods often never compute the posterior distribution $p(\theta|y)$ itself. Instead, they swap the hard problem of computing complicated integrals for the somewhat easier problem of generating random variates from the posterior distribution. The motivation is fairly straightforward: classical statistics is replete with methods to draw inferences about a population given a random sample from it; if we can draw such a sample from the "population" that is the posterior distribution $p(\theta|y)$, we can use those methods to draw inference. Remarkably, in many cases, knowing only the un-normalized version of the posterior $\tilde{p}(\theta|y) = l(\theta|y)p(\theta)$ is sufficient for generating such random samples. Markov chain Monte Carlo (MCMC) algorithms are a very general class of algorithms that are typically used for this process. Gibbs sampling is a particularly popular variant of MCMC that underlies most statistical software used in Bayesian data analysis.

We end this introductory section with a basic yet illustrative demonstration of Monte Carlo methods that showcases the basic ethos of how they are used in Bayesian statistics. Suppose we are interested in the sensitivity θ of a new rapid influenza diagnostic test (RIDT). Further, suppose that of $n = 30$ randomly sampled individuals confirmed to have the flu, $y = 25$ test positive with the new RIDT. How would a Bayesian go about using the data to understand θ? First, a prior distribution is demanded. Since θ is a probability, the beta family of probability distributions is a common choice. Suppose a domain expert specifies a Beta(3, 2) prior so that $p(\theta) \propto \theta^{(3-1)}(1-\theta)^{2-1}$; see Section 15.5 for how this might be obtained. Assuming the subjects have the same likelihood of testing positive and are exchangeable given θ, the likelihood of a particular θ is $l(\theta|y) = \theta^{25}(1-\theta)^5$. Thus, the un-normalized posterior is $\tilde{p}(\theta|y) = l(\theta|y)p(\theta) = \theta^{27}(1-\theta)^6$. Theoretically, the posterior itself is easily recognized from $\tilde{p}(\theta|y)$; it is the density of the Beta(28, 7) distribution, and this fact can be used to compute expectations of interest. But this is a rarity in Bayesian statistics: typically, the posterior distribution of $p(\theta|y)$ cannot be recognized from $\tilde{p}(\theta|y)$ as coming from some nice family. In these cases, Monte Carlo methods are the standard tools of practice.

In order to compute expectations of the form $\mathbb{E}_{p(\theta|y)}[g(\theta)]$ given only $\tilde{p}(\theta|y)$, it suffices to be able to sample from the distribution $p(\theta|y)$, and to do that, it often suffices to know the un-normalized $\tilde{p}(\theta|y) = l(\theta|y)p(\theta)$. In the current example, we may want to know the Bayes estimate for θ, the posterior mean $\mathbb{E}_{p(\theta|y)}[\theta]$. Using a result known as the Fundamental Theorem of Simulation [1], if we can sample points (θ, U) from the bivariate distribution $(\theta, U) \sim \text{Unif}(\mathcal{G})$ with $\mathcal{G} = \{(\theta, u) : 0 \leq \theta \leq 1 \text{ and } 0 \leq u \leq \tilde{p}(\theta|y)\}$, we can sample points from $p(\theta|y)$ by marginalizing out U, i.e. only looking at the θ (first) coordinates of the pairs of sampled values (θ_k, u_k). In other words, we can use the un-normalized $\tilde{p}(\theta|y)$ to generate pseudo-observations from $p(\theta|y)$ via marginalization. In this case, sampling pairs $(\theta, U) \sim \text{Unif}(\mathcal{G})$ can be accomplished easily with rejection sampling: sample uniformly on the rectangle $[0,1] \times [0, M] \supseteq \mathcal{G}$, where M is the maximum value $\tilde{p}(\theta|y)$ takes on the interval $[0, 1]$, discard points that are not in $\mathcal{G}$, and then discard the u values. Figure 15.1 illustrates 1,000 such draws. The desired estimate can then be approximated as $\mathbb{E}_{p(\theta|y)}[\theta] \approx \bar{\theta}_{1,000}$, the average of the θ values of the 1,000 draws. For the points depicted, an estimate of $\bar{\theta} = 0.8004$ was observed. The true value is $28/(28+7) = 0.80$, so the approximation is very good. In fact, we

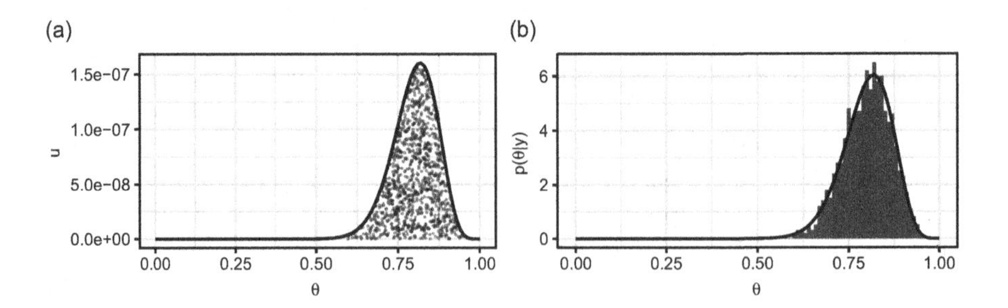

FIGURE 15.1
(a) 1,000 draws from $(\theta, U) \sim \text{Unif}(\mathcal{G})$ using $\tilde{p}(\theta|y)$ and rejection sampling and (b) a histogram of the θ values overlayed with the known true density $p(\theta|y)$.

can use the same sample to determine credible intervals for θ. A 95% equal-tailed interval is given by the 2.5 and 97.5th percentiles of the observations. Of our 1,000 observations, this is found to be $[0.6535, 0.9150]$, which is quite close to the true theoretical values $[0.6547, 0.9130]$, obtained from the known beta posterior. An HPD region can be constructed using a numerical procedure that determines the point c for which the proportion of points falling in $\mathcal{C}_c = \{\theta : p(\theta|y) \geq c\}$ is $1 - \alpha$, which here is an interval but in general need not be. Note that $\log \tilde{p}(\theta|y)$ can be used instead of $p(\theta|y)$ in this calculation to aid in numerical stability. In this case, the 95% HPD interval was found to be $\mathcal{C}_c = [l_c, u_c] = [0.6715, 0.9206]$. In all cases, our estimates can be made arbitrarily accurate by drawing more observations from the distribution; this is a general feature of Monte Carlo methods, whose primary limitation is computational power.

```
Warning: Removed 2 rows containing missing values (geom_bar).
```

15.3 Statistical Software

In this section, we present a brief yet self-contained introduction to software commonly used by Bayesian statisticians in pharmaceutical drug development. As the field is large, any introduction of this kind requires careful selection of the software discussed. Thus, while we might have used `SAS` as a general computing platform, for example, we have chosen to use the software `R`. This is for several reasons. First, `R` is free and open source, so it is available to all users. Second, it is modular, with over 10,000 registered add-on packages created by a vibrant community of developers. Third, `R` is cross-platform: it currently runs on all major operating systems and most computing server platforms, including Windows, MacOS, and several Linux distributions including Debian, Ubuntu, Red Hat/Fedora, and SUSE. Fourth, most Bayesian software programs are designed to work with and through `R` through packages, so `R` can connect to the computational engines of `JAGS`, `STAN`, `Greta`, and `OpenBUGS`, for example. Last, `R` is *popular*; it is used in virtually every domain of applied statistics from academic departments to governmental labs to industry leaders in virtually every sector.

15.3.1 An Overview of Statistical Software Used in Bayesian Statistics

As previously noted, the current programming lingua franca of the statistics world is the programming language `R`. Originally designed by Ross Ihaka and Robert Gentleman, `R` is the modern version of the now antiquated language `S`, a statistical programming language created at Bell Labs in the 1970s [2]. As it is open-source and GNU-compliant free software, it is sometimes referred to as

GNU-S. It was first released in 1993 and has been continuously maintained by a core team of developers ever since. The main R webpage is www.r-project.org.

As downloaded from the above link, R is composed of essentially two parts: an interpreter/computational engine and a graphical user interface (GUI), whose presentation varies somewhat among the supported operating systems. As typically used in practice, the GUI itself is also composed of two parts: a console window and a scripting window; this is displayed in Figure 15.2 as it appears on MacOS. The console window is simply a command line interface where users can type lines of code to be evaluated by the R interpreter. The scripting window is where most users write their code before sending it to the console through a combination of keystrokes, called "sourcing" code or code chunks. This workflow has a few advantages. For one, users typically iterate through chunks of code several times before being content with the end result, interactively modifying it, sourcing it, and reinterpreting the output. This process is significantly faster in a scripting window than in the console, which usually has limited or inconvenient access to the code the user has previously run. More important, however, is the ability to save one's work. The primary advantage of using a scripting window is that it can be saved and either revisited in the future or disseminated to others. When saving an R script file, the .R file extension is customary; otherwise, it is simply a plain-text file.

From the vantage point of industry-grade software development, the GUI distributed with R is very crude. While it does on some platforms provide

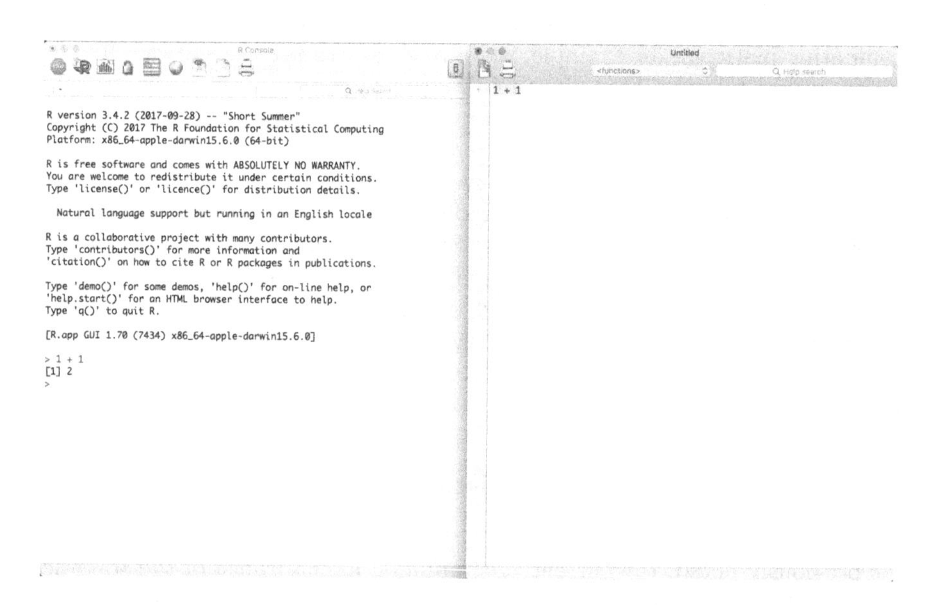

FIGURE 15.2
The R GUI on MacOS. The console is shown on the left and the scripting window on the right.

syntax highlighting, which makes looking through code significantly easier, it is otherwise essentially a very basic text editor like Notepad on a Windows machine or TextEdit on a Mac. Integrated development environments (IDEs) are applications, typically GUI-based, that facilitate writing code in one or more languages. For example, software developers on Windows often use Microsoft Visual Studio, which is Microsoft's flagship IDE; MacOS and iOS developers usually use Xcode, Apple's flagship IDE. Both these applications provide support for many languages such as `C`, `C++`, `Bash`, and others. In fact, as of 2017, Visual Studio can act as an IDE for `R` as well. IDEs typically provide users with the following helpful features: syntax highlighting, code completion, snippets (shorthand expressions that are expanded by the IDE into longer chunks of code used for commonly used constructs), debugging, source code version control, project management, and more. Projects are collections of code and related files that pertain to the same general undertaking: a data analysis, a scholarly article, a clinical trial, etc.

One of the most popular IDEs for coding in `R` is RStudio, created by Rstudio, Inc. Like `R`, RStudio is a cross-platform application, having both desktop and server solutions. The server solution is often used in pharmaceutical drug development (and elsewhere) and allows for RStudio and `R` sessions to be remotely hosted and accessed through one's browser, which allows for a pooling of computational resources (memory, processing, etc.) and their management by a large IT infrastructure. It also allows for things like collaborative coding – users simultaneously editing the same file. Both desktop and server versions are freely available under the AGPL v3 license, and premium versions are available with added features under RStudio's licensing agreement. You can find RStudio for download at www.rstudio.com. The RStudio IDE is displayed in Figure 15.3.

RStudio is an IDE designed for `R`. It has all of the IDE features previously listed as well as many more. However, while RStudio is designed for `R`, it is also designed for tasks generally encountered by `R` users. This includes support for coding in languages commonly used with `R` such as `C++`, shell scripting, and `Python`; `R` package development; document preparation; and dissemination of things produced with `R`. Along these lines, in addition to the tidyverse suite of packages described in Section 15.3.2.4, Rstudio, Inc. offers two technologies created through RStudio: `RMarkdown` and Shiny.

`RMarkdown` (http://rmarkdown.rstudio.com) is an `R`-friendly variant of `Markdown`, a very simple plain-text markup language used to quickly create high-quality documents in a variety of formats. `RMarkdown` files are text files with the file extension `.Rmd` that RStudio can convert into a wide variety of different kinds of files: Microsoft Word files, PDFs, webpages (HTML files), and more. The beauty of `RMarkdown` is that it is very simple and allows users to create all kinds of documents that intersperse ordinary text (like this paragraph) with `R` code and its output in one place. Consequently, `RMarkdown` is very commonly used to create write-ups and presentations for analyses done with `R`.

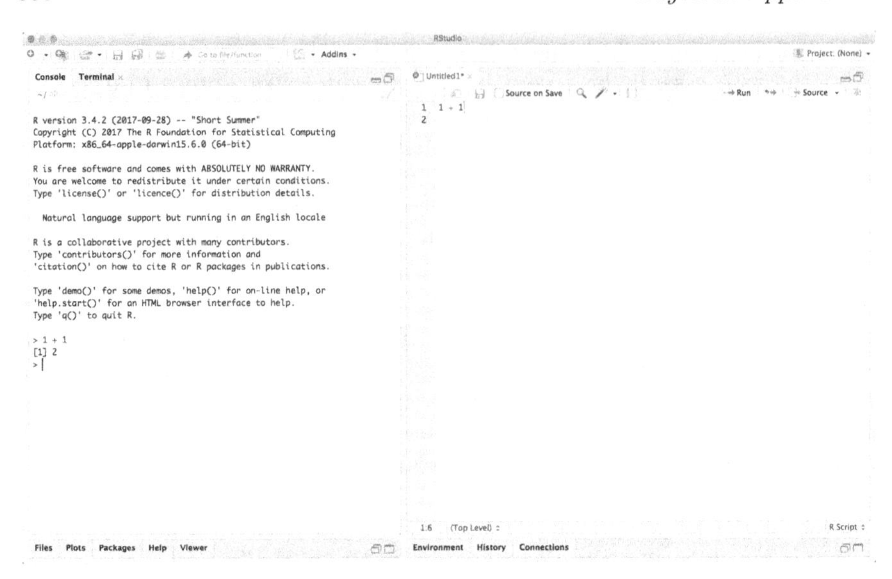

FIGURE 15.3
The RStudio IDE. Note that two panes are minimized at the bottom of the GUI.

Shiny (http://shiny.rstudio.com) is a technology created by Rstudio, Inc. that allows users to easily create interactive webpages called Shiny Apps that are connected to server-side `R` sessions. Created via `R` scripts in RStudio, Shiny Apps are used for all sorts of purposes: disseminating work that users can manipulate (e.g. change graphical or model parameters), creating applications that guide experimental design, or creating computational tools for non-statisticians to use to accomplish a given task. Shiny Apps can be hosted either locally through localhost (i.e. on the user's own computer) or remotely and can be setup to require login credentials. If they are public facing, they are accessible through any web browser on any platform, including those on smart devices. To make publishing Shiny Apps easy, Rstudio, Inc. offers a web hosting service called shinyapps.io (www.shinyapps.io) that enables users to disseminate Shiny Apps directly from RStudio in mere clicks. Like other Rstudio, Inc. products, basic versions of the service are free, and paid tiers offer more features such as more simultaneous users, computational power, etc.

15.3.2 An Introduction to `R`

Written itself mostly in `C`, `R` shares a syntax and an ethos similar to that language, albeit significantly less verbose. Nevertheless, the language is still very distinct and thus demands an exposition on the basic components to get you up and running. For a more thorough treatment, we recommend Hadley Wickham's *Advanced* `R`, available online at https://adv-r.hadley.nz or in paperback from online retailers [3]. http://rseek.org provides a nice Google-based search engine for `R` topics.

15.3.2.1 `R` as Calculator

Basic arithmetic functionality. At a basic level, `R` is a powerful calculator built to do statistical computations. As a consequence, it has all the basic arithmetic functions you might expect: `2 + 3`, `2 - 3`, `2 * 3`, `2 / 3`, and `2^3`. The output format for all these looks analogous to this:

```
2 / 3

[1] 0.6666667
```

You can store variables with either `<-` or `=`, but the two are not the same and the former is more commonly used, e.g. `x <- 2`. To make assignment print the assigned values, you can parenthesize the expression:

```
(x <- 2)

[1] 2
```

`R` has many built-in implementations of special functions such as the exponential `exp()` and logarithm `log()`; trigonometric, e.g. `sin()`; combinatorial, e.g. `factorial()` or `choose()`; and others such as the gamma `gamma()` and beta `beta()` functions. For example:

```
gamma(10)

[1] 362880
```

Two special characters in `R` are `#` and `?`. `#` is the commenting character in the `R` language, so whatever follows it will not be evaluated by the `R` interpreter. This is useful for making simple notes to document what a chunk of code does. `?`, when preceding a function or object, opens the documentation for the referent thing. For example, `?sin` opens the documentation for the function `sin()`.

Distributions and random number generation. In addition to features common to scientific calculators of its caliber, `R` is particularly suited to statistical applications. As a consequence, it has a suite of functions used to deal with common families of probability distributions. Every supported family has four functions that follow a common syntax: `r`/`d`/`p`/`q` followed by the abbreviated name of the distribution, e.g. `rnorm()`, `dnorm()`, `pnorm()`, and `qnorm()`. The `r____()` line of functions generates random values from the distribution; the `d____()` is the probability density/mass function (PDF/PMF); the `p____()` is the cumulative distribution function (CDF); and the `q____()` is the quantile function (inverse CDF). For example, `rnorm(10, mean = 0, sd =1)` generates four values from the standard normal distribution:

```
set.seed(1)
rnorm(4, mean = 0, sd = 1)

[1] -0.6264538  0.1836433 -0.8356286  1.5952808
```

Common discrete distributions are the binomial `rbinom()`, Poisson `rpois()`, geometric `rgeom()`, negative binomial `rnbinom()`, and hypergeometric `rhyper()`. Common continuous distributions are the normal `rnorm()`, chi-squared `rchisq()`, t `rt()`, F `rf()`, exponential `rexp()`, gamma `rgamma()`, beta `rbeta()`, and uniform `runif()`. See the help menu, e.g. `?rbinom`, to understand their parameterizations. The multivariate normal distribution is in the `MASS` package accessible by `MASS::mvrnorm()`, and the Wishart is `rWishart()`. Draws from the discrete uniform distribution can be generated with `sample()`, which performs simple random sampling with or without replacement:

```
sample(5, 1) # 1 observation from the disc. unif. on 1, 2, 3, 4, 5

[1] 4

sample(5, 10, replace = TRUE) # 10 such observations

 [1] 1 2 1 4 2 4 3 4 5 2
```

Statistical summaries. R also has many built-in functions that compute common statistical summaries of data. For example, here's how you compute the mean and median of a set of numbers:

```
simulated_data <- sample(5, 100, replace = TRUE)
mean(simulated_data)

[1] 3.04

median(simulated_data)

[1] 3
```

You can also compute the frequency and relative frequency distributions and then check them:

```
(freq <- table(simulated_data))

simulated_data
 1  2  3  4  5
15 23 21 25 16
```

```
n <- length(simulated_data)
freq / n

simulated_data
   1    2    3    4    5
0.15 0.23 0.21 0.25 0.16

sum(freq / n)

[1] 1
```

s^2, the sample variance, and s, the sample standard deviation, are included:

```
var(simulated_data)

[1] 1.735758

sd(simulated_data)

[1] 1.317482
```

The range `range()`, interquartile range `IQR()`, and median absolute deviation `mad()` are implemented as well. With multivariate datasets whose rows are observations and columns are measures, `colMeans()` computes the multivariate sample mean, `cov()` computes the sample covariance matrix, and `cor()` computes the sample correlation matrix:

```
# generate 100 observations from standard bivariate normal
simulated_multivariate_data <- MASS::mvrnorm(100, mu = c(0, 0),
                                              Sigma = diag(2))

# look at first three observations of simulated data
head(simulated_multivariate_data, n = 3)

           [,1]        [,2]
[1,] -0.3178857  0.35872890
[2,]  0.4888056 -0.01104548
[3,] -2.6586580 -0.94064916

# compute summaries
colMeans(simulated_multivariate_data)

[1] -0.08066956 -0.05763152

cov(simulated_multivariate_data)

           [,1]       [,2]
[1,] 1.05858615 0.05222152
[2,] 0.05222152 0.76897631
```

```
cor(simulated_multivariate_data)

           [,1]       [,2]
[1,] 1.00000000 0.05788019
[2,] 0.05788019 1.00000000
```

Arbitrary percentiles can be computed with `quantile()`. For example, we can combine `rnorm()` with `quantile()` to compute Monte Carlo estimates of the familiar quantities $\pm z_{.025} = \pm 1.96$:

```
quantile(rnorm(1e5, mean = 0, sd = 1), probs = c(.025, .975))

     2.5%     97.5%
-1.962917  1.945382
```

Linear algebra. Linear algebra forms the core of modern scientific computing, so R provides an extensive collection of linear algebra functionality. Vectors are created with the `c()` function:

```
(b <- c(1, 3))

[1] 1 3
```

Vectors are neither row nor column vectors but can act as either depending on the context. Matrices can be created with `matrix()`, which takes in a vector and dimensions and then populates the elements of the matrix by column:

```
(A <- matrix(c(1, 2, 3, 4), nrow = 2, ncol = 2))

     [,1] [,2]
[1,]    1    3
[2,]    2    4
```

Solving the matrix equation $\mathbf{Ax} = \mathbf{b}$ can be done with the `solve()` function:

```
(x <- solve(A, b))

[1]  2.5 -0.5
```

When used with only $\mathbf{A}$, `solve()` computes the inverse of a matrix $\mathbf{A}^{-1}$:

```
solve(A)

     [,1] [,2]
[1,]   -2  1.5
[2,]    1 -0.5
```

All arithmetic operations are vectorized in R, so `x + b` gives what you would expect, the element-wise sum; subtraction, multiplication, division, exponentiation, and so forth all work element-wise in the same way. Consequently, `*` performs element-wise multiplication of matrices (the Hadamard product), not matrix multiplication. Matrix multiplication is performed using `%*%`, for example:

```
A %*% x

     [,1]
[1,]    1
[2,]    3
```

Matrix transposition is performed with `t()`.

Matrix decompositions play an important role in statistics, so R has highly efficient implementations of many. Eigenvalues and eigenvectors can be computed with `eigen()`:

```
eigen(A)

eigen() decomposition
$values
[1]  5.3722813 -0.3722813

$vectors
           [,1]       [,2]
[1,] -0.5657675 -0.9093767
[2,] -0.8245648  0.4159736
```

The Cholesky decomposition is available with `chol()`, the singular value decomposition with `svd()`, and the QR decomposition with `qr()`. Other decompositions are available through add-on packages such as `Matrix` [4].

15.3.2.2 Data Structures in R

As a useful oversimplification, R can be thought of as having four basic data structures that can be categorized by element type (homogeneous/heterogeneous) and dimension (1d, 2d). The 1d data structures are called vectors and come in two forms: atomic vectors and lists. Atomic vectors are linear arrays whose elements are of the same fundamental data type: logical (Boolean), integer, double/numeric (floating point decimal numbers), and character. Lists are linear arrays whose elements can be of differing types. The 2d data structures are either called matrices when their elements are all of the same type or data frames, the difference being that the columns of data frames can be of differing types.

Constructing vectors depends naturally on what kind of vector you would like to create. As we have seen in the context of linear algebra, the `c()` function is the general concatenation function used to create vectors. For example, an atomic vector of doubles can be created like this:

```
c(1.2, 2.3, 3.4)

[1] 1.2 2.3 3.4
```

R's colon operator : is often used to create sequences of integers:

```
1:10

 [1]  1  2  3  4  5  6  7  8  9 10
```

Other sequences are created with the seq() function:

```
seq(0, 1, by = .1)

 [1] 0.0 0.1 0.2 0.3 0.4 0.5 0.6 0.7 0.8 0.9 1.0
```

You can conveniently create equispaced sequences of a given length with a syntax such as seq(0, 1, length.out = 11). You can assign names to atomic vectors when creating them with c():

```
c("a" = 1, "b" = 2, "c" = 3)

a b c
1 2 3
```

The same syntax works using unquoted versions of the names, i.e. c(a = 1, b = 2, c = 3), and in fact this form is more common in practice. We use the quoted version for clarity. When names are supplied with vectors, the resulting structure is referred to as a named vector.

To create a list, you use the list() function:

```
list(1.2, 2.3, 3.4)

[[1]]
[1] 1.2

[[2]]
[1] 2.3

[[3]]
[1] 3.4
```

Notice the difference in how R prints atomic vectors and lists; it reflects the more complex nature of the list. Not only can lists contain elements of different types, but its elements can also have different lengths:

```
list(rnorm(3), c("a", "b", "c", "d"))

[[1]]
[1]  0.2038511 -1.1037830 -0.3944997

[[2]]
[1] "a" "b" "c" "d"
```

In fact, the elements of a list can be just about anything in the R language, including other lists. Just as we can create named atomic vectors, we can create lists. The syntax is entirely analogous, what is important to notice is how R prints these objects to the screen, namely with the names of the elements preceded by $. This recognition will come in handy shortly.

```
list("x" = rnorm(3), "y" = c("a", "b", "c", "d"))

$x
[1]  2.7175667 -0.5167343 -1.0793515

$y
[1] "a" "b" "c" "d"
```

Besides the humble numeric vector, named lists are perhaps the most important data structure in R. They are the return value of most statistical functions. For example, the output of `lm()`, R's function for linear regression, is a named list.

Data frames are the primary data structure in R used to represent spreadsheet shaped datasets. They are created with `data.frame()`:

```
data.frame("x" = c("Male", "Female", "Male"),

                    "weight" = c(192, 141, 177))

       x weight
1   Male    192
2 Female    141
3   Male    177
```

Data frames are named lists whose elements are vectors of the same length. Consequently, they are often confused with matrices. The distinction between matrices and data frames is that the latter can have measures of differing types, which is appropriate for a real dataset.

The most important function in determining the structure of an object in R is the function `str()`. For example, the `cars` dataset comes with R and contains 50 observations of the distance traveled by cars after applying the brakes at given speeds (see `?cars` for details) [5, 6]. When applied to `cars`, `str()` tells us that it is a data frame with 50 rows and 2 columns:

```
str(cars)

'data.frame': 50 obs. of  2 variables:
 $ speed: num  4 4 7 7 8 9 10 10 10 11 ...
 $ dist : num  2 10 4 22 16 10 18 26 34 17 ...
```

`str()` can be used on any object in `R`, be it a vector, list, data frame, or even a more complex object such as a function or environment. Once the structure of an `R` object is known, it is easy to extract pieces of it using one of three operators: `[`, `[[`, and `$`.

`[` is `R`'s basic subsetting operator. To pull out specific elements of a vector, you can use integers or, if named, characters:

```
x <- c("a" = 1, "b" = 2, "c" = 3)
x[2]

b
2

x[c(1,2)]

a b
1 2

x["b"]

b
2
```

Negative integers remove elements, so `x[-3]` is equivalent to `x[c(1,2)]` in the above output. `[` behaves the same way on lists, but the result is always a list, which many people find strange at first. Specifically, if you have a list `y` of length three, `y[2]` returns a list of length one whose contents are the second element of `y`, not the contents themselves. To do that, you use `[[`:

```
y <- list("a" = 1:2, "b" = c("c", "d"))
y[2]

$b
[1] "c" "d"

y[[2]]

[1] "c" "d"
```

Lists can be subset by their names, too. This can be done with either `[[` or `$`:

```
y[["b"]]

[1] "c" "d"

y$b

[1] "c" "d"
```

This latter form is the standard way to access variables in a data frame, which is a kind of list. You can inspect the top rows of a dataset with `head()`:

```
head(cars, 5)

  speed dist
1     4    2
2     4   10
3     7    4
4     7   22
5     8   16
```

And you can extract columns using `$`:

```
cars$speed

 [1]  4  4  7  7  8  9 10 10 10 11 11 12 12 12 12 13 13 13 13
     14 14 14 14
[24] 15 15 15 16 16 17 17 17 18 18 18 18 19 19 19 20 20 20 20 20
     22 23 24
[47] 24 24 24 25
```

The output of `lm()` is typical and illustrative of data structures and subsetting in practice. If we want to fit the linear regression `dist` $= \beta_0 + \beta_1 \times$`speed` and store it into a variable named `mod`, we type:

```
(mod <- lm(dist ~ speed, data = cars))


Call:
lm(formula = dist ~ speed, data = cars)

Coefficients:
(Intercept)        speed
    -17.579        3.932
```

To look at the output, we use `str()` with some additional arguments to make the output easier to read:

```
str(mod, 1, give.attr = FALSE)

List of 12
 $ coefficients : Named num [1:2] -17.58 3.93
 $ residuals    : Named num [1:50] 3.85 11.85 -5.95 12.05 2.12 ...
 $ effects      : Named num [1:50] -303.914 145.552 -8.115 9.885
                  0.194 ...
 $ rank         : int 2
 $ fitted.values: Named num [1:50] -1.85 -1.85 9.95 9.95 13.88 ...
 $ assign       : int [1:2] 0 1
 $ qr           :List of 5
 $ df.residual  : int 48
 $ xlevels      : Named list()
 $ call         : language lm(formula = dist ~ speed, data = cars)
 $ terms        :Classes 'terms', 'formula'  language dist ~ speed
 $ model        :'data.frame': 50 obs. of  2 variables:
```

`mod` is a named list. To extract elements of the Regression, we simply use `$` like this:

```
mod$coefficients

(Intercept)       speed
 -17.579095    3.932409
```

Last, we note that while the return value of most statistical functions in `R` is a named list, it does not always print to the screen as one, which can be misleading. For example, if we just ran the model (or alternatively rerun the model with `lm(dist ~ speed, data = cars)`), we get:

```
mod


Call:
lm(formula = dist ~ speed, data = cars)

Coefficients:
(Intercept)        speed
    -17.579        3.932
```

The reason this does not appear to be a named list is that `R` uses a special function to print things output by the `lm()` function. When a user types in a variable name by itself, as in the above example, `R` silently runs `print()` on it. The function `print()` behaves differently depending on what is being printed; it is a *generic method* in `R`'s S3 object oriented system. The details of object oriented systems and their methods is beyond the scope of this chapter;

what is important to take away is that `str()` is still able to tell us about the structure of any object in R, and once that is known, most can be dissected using the `[`, `[[`, and `$` operators.

15.3.2.3 Functions in R

Functions are the verbs of the R programming language: it is functions that *do* things. Functions are created with `function()`, and they are executed, or "called", using a prefix syntax as in mathematics (e.g. $f(x)$).

```
f <- function(x) {
  2*x
}
f(3)

[1] 6
```

Functions can also be called referring to their arguments directly, e.g. `f(x = 3)`. Functions of many variables are created by simply enumerating the arguments in `function()`, for example `function(x, y)`. They are defaulted by setting arguments with = when the function is defined:

```
g <- function(x = 3) {
  2*x
}
g()

[1] 6
```

It is very common to want to apply a function to a data structure in a specific way. For example, if you have a vector `x`, you may want to apply a function `f` to every element of `x`. While this kind of thing can be accomplished with a `for()` loop, a clearer and more common way to accomplish the task is to use one of R's many `_apply()` type functions like `lapply()`, which accepts a vector and a function and returns the list whose elements are the function applied to each element of the vector:

```
lapply(1:2, f)

[[1]]
[1] 2

[[2]]
[1] 4
```

`sapply()` does the same thing as `lapply()` but attempts to simplify the result from a list to an atomic vector or matrix, depending on the context:

```
sapply(1:2, f)

[1] 2 4
```

When you want to apply a function across the rows or columns of a matrix, you use the `apply()` function. It accepts a matrix, a function, and an indicator to communicate which index to apply the function across (1 for rows, 2 for columns, etc.):

```
(m <- matrix(1:4, nrow = 2))

     [,1] [,2]
[1,]    1    3
[2,]    2    4

apply(m, 1, mean)

[1] 2 3
```

In some cases, there are special functions that perform the exact operation you are evaluating with `apply()`, for example `rowMeans()` here, but those are typically only for basic descriptive statistics.

Lastly, a handy function to know about is `replicate()`. `replicate()` allows a user to evaluate a chunk of code several times, saving the result into a vector (with `simplify= FALSE`). The reason why this is so helpful is that it allows a user to write code that does one iteration of a simulation and then wrap it to essentially say "repeat this" an arbitrary number of times. This allows us to very quickly generate Monte Carlo estimates of unknown quantities. For example, statistical theory provides the basic result that variance of the mean of a random sample of data is equal to that of the population diminished by a factor of the square root of the sample size. That fact can be easily ascertained without theory for individual populations using `replicate()`. For example, if the population is the standard normal distribution and the sample size is $n = 100$, the variance of the sample mean would be $1/10$. To do this via simulation, we can use `replicate()` to generate, say, $N = 10{,}000$ datasets of 100 observations each, `sapply()` to compute the sample mean of each, and then to compute the variance of the means:

```
datasets <- replicate(1e4, rnorm(100, mean = 0, sd = 1),
                      simplify = FALSE)
var( sapply(datasets, mean) )

[1] 0.009727879
```

This result is accurate to a precision referred to as Monte Carlo error depending on the number of Monte Carlo replications N used.

15.3.2.4 Packages and the Tidyverse

As noted in the introduction, the ability to easily share R code with others is one of its greatest advantages. The most common way to share code is bundled in packages, compressed directories of files containing code and documentation. In modern practice, the sharing of packages mainly occurs through two major avenues: the Comprehensive R Archive Network (CRAN, https://cran.r-project.org), which is the more traditional and formal avenue, and GitHub (https://github.com), a free social coding website.

CRAN is the most common way to obtain packages. You can download and install packages with `install.packages()`. In cases where the desired package depends on other packages, R is typically smart enough to get those, too.

```
install.packages("rjags")

trying URL 'https://cran.rstudio.com/bin/macosx/el-capitan/
          contrib/3.4/rjags_4-6.tgz'
Content type 'application/x-gzip' length 278683 bytes (272 KB)
==================================================
downloaded 272 KB

The downloaded binary packages are in
/var/folders/r3/126_d6t55f5d32tplbg5mk1d0c48s9/T//
RtmpDluQOW/downloaded_packages
```

You only need to install a package once until you need to update it, which you do just like installing. To use the code in a package, you need to load it. Packages are loaded with the `library()` function:

```
library("rjags")

Loading required package: coda
Linked to JAGS 4.3.0
Loaded modules: basemod,bugs
```

Once a package is loaded, you have access to all the functions and datasets that the package developer made available. You can see what these are with the `ls()` function:

```
ls("package:rjags")

 [1] "adapt"          "coda.samples"  "dic.samples"     "diffdic"
 [5] "jags.model"     "jags.samples"  "list.factories"  "list.modules"
 [9] "list.samplers"  "load.module"   "parallel.seeds"  "read.bugsdata"
[13] "read.data"      "read.jagsdata" "set.factory"     "unload.module"
```

If you use `ls()` with no arguments (nothing inside the parentheses), `R` tells you what variables are defined in your current workspace.

GitHub is a popular social coding website used by software developers in hundreds of languages. The website has a free basic service and paid tiers for more advanced offerings. It is very common for professional `R` package developers to post their public packages online, both pre and post release. What they post is the raw source code of the package and, oftentimes, an overview exposition. The GitHub page of the package `mpoly` (https://github.com/dkahle/mpoly) provides a simple and typical example [7]. A more elaborate example is the GitHub page for `rstan` (https://github.com/stan-dev/rstan) [8]. Although we won't discuss it in detail here, `rstan` is an important package in the Bayesian statistics community, including applications in pharmaceutical drug development. `rstan` provides a set of tools to interface `R` with the Bayesian software `Stan`. To install a package directly from GitHub, you can use the `devtools` package, available from CRAN [9]. For example:

```
devtools::install_github("dkahle/mpoly")
```

There are two things to notice in the above code. First, we use the double colon operator `::` to call the function `install_github()` directly without loading the `devtools` package. This is generally not a good idea, but it is fine in this case. Second, the GitHub username of the package developer is needed; here that is `dkahle`. If you don't know the username of the developer, you can usually find it with a quick Google search, e.g. "github mpoly".

One more note on GitHub is helpful before proceeding. Newcomers to the area often confuse GitHub with `Git`. `Git` is a Linux command line tool used for version control, i.e. keeping track of all the changes that a programmer makes to a particular collection of code, which it calls a *repository*. GitHub is a website that leverages `Git` to enable remote hosting of repositories, which may be public or private, and encourage collaborative coding. Many companies and other organizations contract with GitHub to foster internal development of software products; this is common in the pharmaceutical industry.

While on the subject of packages, it is helpful to introduce a special ecosystem of packages called *the tidyverse*. At its simplest, the tidyverse (https://github.com/tidyverse) is a collection of a few dozen `R` packages created by RStudio to facilitate a wide range of applications in `R`, ranging from loading data into `R` to data manipulation to graphing. For example, the `ggplot2` package, `R`'s most popular plotting system, is a tidyverse package [10]. `dplyr` and `tidyr` help manipulate datasets [11, 12]. You can install all the tidyverse packages by installing the `tidyverse` package, which simply facilitates loading tidyverse packages [13]. The tidyverse provides new versions of many base-`R` functions that are consistently implemented.

The tidyverse is more than merely a collection of packages that help with various `R` tasks. It is increasingly way of thinking about doing data analysis in `R`, ranging from syntactical flows that encourage good coding practice to the data structures and functions that get the job done. Along these lines, one particularly useful feature of the tidyverse is the pipe operator `%>%`. The pipe operator comes in the `magrittr` package, but it is also included in all tidyverse packages [14]. A simple function, the pipe operator allows you to call a function `f(x, y)` with the seemingly more complex expression `x %>% f(y)`. While small, this shift allows you to write code that you can read from left to write and top to bottom, like English. It is very common to see code that looks like this:

```
dataset %>%
  transformation_one(arg1, arg2) %>%
  transformation_two(arg1) %>%
  model(model ~ specification, data = .)
```

This kind of coding allows a user to take a dataset (or other object) and perform several manipulations on it before doing some ultimate task, such as modeling or graphing. This kind of coding style is very effective while interactively working in `R`.

15.3.3 An Introduction to JAGS

In Section 15.2, we presented an overview of Monte Carlo methods in Bayesian statistics. In practice, sampling from posterior distributions is typically not done with `R` but rather a probabilistic programming language (PPL) that is optimized for the kinds of computations performed, namely MCMC algorithms. In this section, we provide a basic working introduction to one of these PPLs, `JAGS`, which stands for "Just Another Gibbs Sampler", referring to the MCMC variant that predominates in the sampling it performs [15].[1] `JAGS` was written by Martyn Plummer in `C++` as a Bayesian engine accepting model specifications in a language very similar to the older PPL `BUGS`, which first had a Windows command line interface, then was distributed with a Windows-based GUI as `WinBUGS`, which was still later distributed as `OpenBUGS` [16, 17]. `JAGS` runs on all major operating systems and was designed to work with `R`. To make these more accessible to `R` users, `R` packages have been created that make interfaces to each, `R2WinBUGS`, `R2OpenBUGS`, and `rjags` [18, 19].

The basic idea behind MCMC algorithms is to construct a Markov chain whose stationary distribution is the posterior distribution of interest. The goal is to construct a growing sequence of random numbers (or vectors/arrays), each depending on the previous numbers only through the last, which after a time can be considered to be observations sampled from a target distribution, which in Bayesian applications is the posterior distribution. Two of the more popular and easy to implement MCMC variants are the Gibbs sampler and Metropolis-Hastings algorithm. See [20] or [21] for great explanations of these algorithms.

[1] `JAGS` also relies heavily on the slice sampler, its "workhorse" sampler.

To use JAGS, you must first download and install it from https://sourceforge.net/projects/mcmc-jags/. While there, pick up a copy of the user manual. The user manual is the authoritative resource for all things JAGS; it is both accurate and well-written. To use JAGS in R, you also need rjags. From there, using JAGS in R is a four-step process:

1. Specify the model using the JAGS syntax.
2. Aggregate the data and all knowns into a named list.
3. Compile the model with the data and initialize the sampler.
4. Sample from the compiled model.

We illustrate each of these for the following hypothetical dataset $y_1, \ldots, y_{20}$ assumed to have arisen from a $\text{Bin}(35, \theta)$ distribution:

```
set.seed(2)
n <- 35
(y <- rbinom(n, size = 1, prob = .20))

 [1] 0 0 0 0 1 1 0 1 0 0 0 0 0 0 0 1 1 0 0 0 0 0 1 0 0 0 0 0
     1 0 0 0 1 1 0
```

The first step to using JAGS in R is to specify the model. We do this by creating a plain-text file containing the JAGS code. This can be achieved in either of two ways. First, you can open any basic text editor, write the model file, and save it to disk as, say, binom_model.txt. Better, you can write it in R with the cat() function using the file = argument:

```
cat(file = "binom_model.txt", "
model {
  # set the likelihood
  for (i in 1:n) { y[i] ~ dbin(theta, 1) }

  # set the prior
  theta ~ dbeta(1, 1)
}
")
```

This has the advantage of being inside the R file where the rest of the analysis is being performed. Note that here we use a uniform prior on θ. Obviously, JAGS code looks very much like R code. For example, the for() loop syntax is identical; we have included it here inline to conserve space; it is typically placed across several lines as in R. The JAGS model language has many similarities to the R language. The characters <- and = are used for assignment as in R to express deterministic relations, and ~ is used to specify the stochastic relations, e.g. the distribution of random variables.

The second step is to aggregate all the data into a named list. Note that "data" here includes not only the observed quantities, but also all fixed quantities, for example those known by design.

```
binom_data <- list("y" = y, "n" = n)
```

In cases where the data are not stored in global variables, you can specify them directly in the `list()` call.

The third step is to compile the model with the data and initialize the sampler. This is achieved with `jags.model()`.

```
binom_model <- jags.model(file = "binom_model.txt",
                          data = binom_data)

Compiling model graph
   Resolving undeclared variables
   Allocating nodes
Graph information:
   Observed stochastic nodes: 35
   Unobserved stochastic nodes: 1
   Total graph size: 38

Initializing model
```

You can also specify the number of chains (`n.chains`, defaulted to 1), their initial values (`inits`, randomly generated if not specified), and number of preliminary steps to run to tune the MCMC sampler (`n.adapt`, defaulted to 1,000). Note that the tuning iterations are not retained. The result is a `JAGS` model object of class `jags`; it represents a directed acyclic graph that is ready to be sampled from.

The fourth step is to actually do the sampling. This can either be done with `jags.samples()` or `coda.samples()`, but we prefer `coda.samples()` because the return type is a bit nicer:

```
post_draws <- coda.samples(model = binom_model,
                           variable.names = "theta",
                           n.iter = 1e4)
str(post_draws)

List of 1
 $ : 'mcmc' num [1:10000, 1] 0.189 0.268 0.297 0.238 0.23 ...
  ..- attr(*, "dimnames")=List of 2
  .. ..$ : NULL
  .. ..$ : chr "theta"
  ..- attr(*, "mcpar")= num [1:3] 1001 11000 1
 - attr(*, "class")= chr "mcmc.list"
```

`coda.samples()` can also add "thin the chain", i.e. discard intermediate draws to decrease chain autocorrelation, with the `thin` argument.

Once the chain has been obtained, you can plot it with `plot()`, whose output is included in Figure 15.4.

```
plot(post_draws)
```

Figure 15.4 contains pristine plots. The trace plot (a), plots the sampled value of θ against the iteration of the chain. The "fat hairy caterpillar" strongly suggests convergence and efficient sampling; see [22] for a good practical reference. The estimated posterior density (b) looks just like the known Beta posterior. Similarly, we can use `autocorr.plot()`; this is Figure 15.5.

```
autocorr.plot(post_draws)
```

For more diagnostics, we can use the `coda` library, which stands for "Convergence Diagnosis and Output Analysis for MCMC" [23]. For example, the Gelman–Rubin and Geweke diagnostics are available as `gelman.diag()`, `geweke.diag()`. We illustrate the latter here as the former requires running multiple chains.

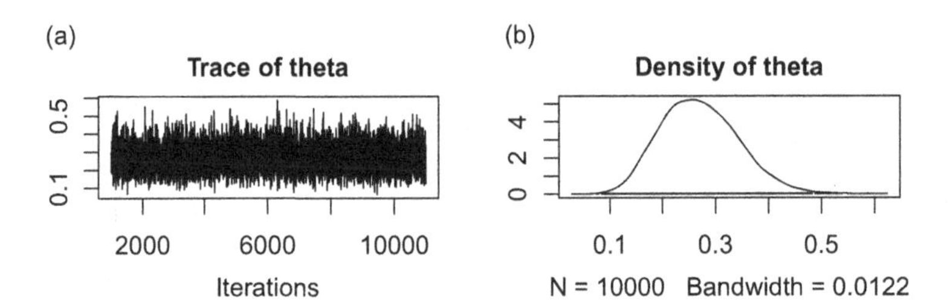

FIGURE 15.4
The trace plot (a) and estimated marginal posterior density of θ (b) of the sampled chain.

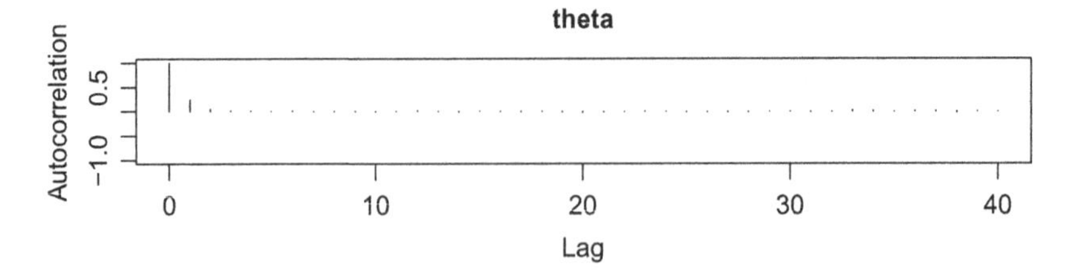

FIGURE 15.5
An autocorrelation plot of the chain, indicating efficient sampling.

```
geweke.diag(post_draws)

[[1]]

Fraction in 1st window = 0.1
Fraction in 2nd window = 0.5

  theta
0.02729
```

The Geweke diagnostic compares the mean of the first 10% of the chain draws to that of the last 50% (by default). This statistic is asymptotically standard normal distribution.

Note that we can use all of our R skills from earlier in this section to process the posterior draws. For example, the Bayes estimate and a 95% equal-tailed interval can be computed as follows:

```
mean(post_draws[[1]])

[1] 0.2694422

quantile(post_draws[[1]], c(.025, .975))

     2.5%     97.5%
0.1421511 0.4213925
```

These can be computed more easily using `summary()`:

```
summary(post_draws)


Iterations = 1001:11000
Thinning interval = 1
Number of chains = 1
Sample size per chain = 10000

1. Empirical mean and standard deviation for each variable,
   plus standard error of the mean:

          Mean             SD       Naive SE Time-series SE
     0.2694422      0.0726249      0.0007262      0.0009281

2. Quantiles for each variable:

  2.5%    25%    50%    75%  97.5%
0.1422 0.2168 0.2651 0.3179 0.4214
```

15.4 Basic Model Repository

In this section, we illustrate fitting three Bayesian models with JAGS. Each of these examples are models that can easily be adapted for different and more complex situations. The example code for posterior diagnostics or summaries can also be used for each example. Remember: start by loading rjags and coda:

```
library("rjags") # use jags 4.3 or higher
library("coda")
```

15.4.1 Normal Mean and Standard Deviation Model

We start this section with a bivariate prior problem, estimation of a normal mean and standard deviation, where we observe $n = 9$ normal random variables $\mathbf{y}$ and assume that they have a common mean μ and standard deviation σ. We use a standard non-informative prior for these unknown parameters. Specifically, we assume $y_i|\mu, \sigma^2 \overset{iid}{\sim} \mathcal{N}(\mu, \sigma^2)$ and $\pi(\mu, \sigma^2) \propto \frac{1}{\sigma}$. In JAGS, this is expressed as follows:

```
cat(file = "normal_model.txt", "
model{
  # set the likelihood
  for (i in 1:n) { y[i] ~ dnorm(mu, tau) }

  # set the priors
  mu ~ dnorm(0, .001)
  tau ~ dgamma(.01, .01)

  # convert precision to standard deviation
  sigma <- pow(tau, -.5)
}
")
```

The line `y[i] ~ dnorm(mu, tau)` indicates that each element of $\mathbf{y}$ has the same normal distribution. In JAGS as in BUGS, the normal distribution is parameterized in terms of its mean μ and precision $\tau = 1/\sigma^2$ instead of the variance or standard deviation. If you are interested in the standard deviation, you need to convert back by specifying a deterministic relation. Note: the `pow(x, a)` expresses x^a. JAGS does not allow you to use improper priors, but our prior can be approximated using a diffuse normal prior on μ and a diffuse gamma prior on τ. For more details on JAGS syntax and distribution conventions (names, parameterizations, etc.), see the JAGS user manual.

We now enter the data as a named list:

```
normal_data <- list("y" = c(4.5, 6.4, 7.8, 9.1, 9.9, 10.6,
                             12.5, 14.0, 15.4), "n" = 9)
```

Next, we compile and initialize the model with `jags.model()`. Note that unlike the previous example, here we run more than one chain:

```
normal_model <- jags.model(file = "normal_model.txt",
                           data = normal_data, n.chains = 3)

Compiling model graph
   Resolving undeclared variables
   Allocating nodes
Graph information:
   Observed stochastic nodes: 9
   Unobserved stochastic nodes: 2
   Total graph size: 18

Initializing model
```

After the model is compiled and ready to run, we use `coda.samples()` to do the sampling:

```
post_draws <- coda.samples(normal_model,
                           c("mu", "sigma", "tau"), n.iter = 1e4)
```

With the samples in hand, we can compute basic summary statistics of each parameter using `summary()`.

```
summary(post_draws)


Iterations = 1:10000
Thinning interval = 1
Number of chains = 3
Sample size per chain = 10000

1. Empirical mean and standard deviation for each variable,
   plus standard error of the mean:

         Mean      SD  Naive SE Time-series SE
mu    9.99702 1.36972 0.0079081      0.0078290
sigma 3.92539 1.16617 0.0067329      0.0077264
tau   0.07989 0.04015 0.0002318      0.0002601
```

```
2. Quantiles for each variable:

          2.5%     25%     50%     75%   97.5%
mu    7.28036 9.15995 9.99683 10.8280 12.7230
sigma 2.38298 3.12550 3.69798  4.4419  6.7749
tau   0.02179 0.05068 0.07313  0.1024  0.1761
```

And since we have several chains, we can compute the Gelman–Rubin statistic:

```
gelman.diag(post_draws, multivariate = FALSE)

Potential scale reduction factors:

      Point est. Upper C.I.
mu             1          1
sigma          1          1
tau            1          1
```

These support convergence of the chains. We can also plot the chains; this is shown in the following figure. The different colors in the trace plots are for different chains.

```
plot(post_draws, smooth = FALSE)
```

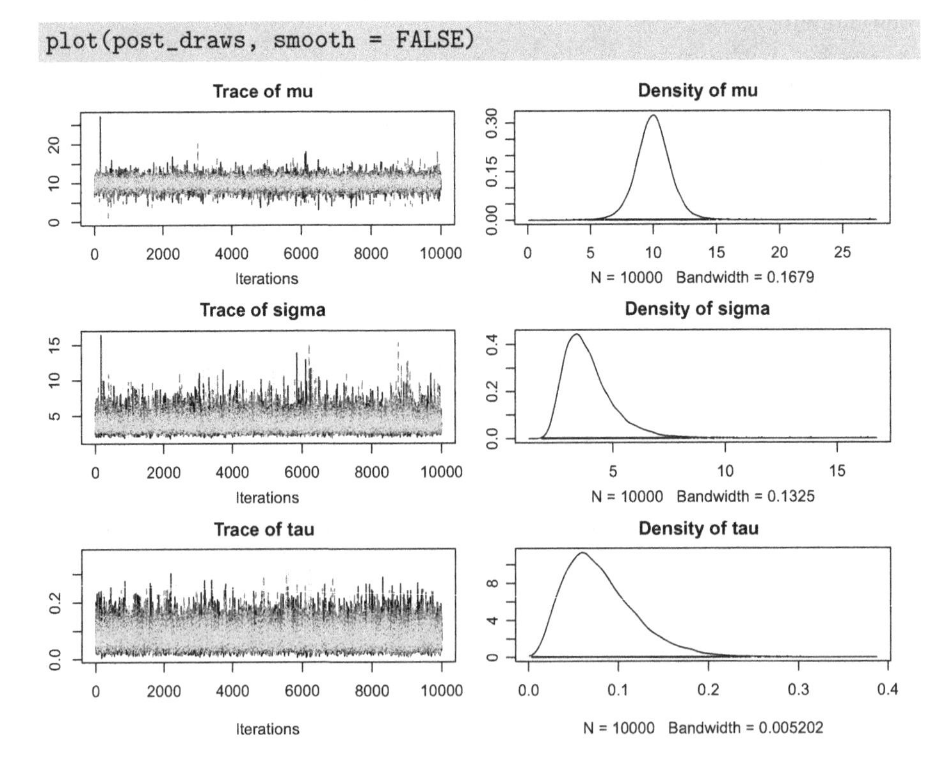

Note that for situations where a data frame is more helpful than an `mcmc.list`, you can convert the output of `coda.samples()` into a data frame:

```
post_draws_df <- post_draws %>% as.matrix() %>% as.data.frame()
head(post_draws_df)

         mu    sigma        tau
1 10.414745 3.507113 0.08130185
2  8.441247 5.652587 0.03129720
3  6.659325 3.446563 0.08418363
4 10.100777 3.677681 0.07393533
5 10.668015 2.934471 0.11612894
6 10.334409 2.208635 0.20499909
```

15.4.2 Binomial Random Effects Model

Hierarchical models (conjugate or not) can easily be implemented in `JAGS` using hyperpriors on the second-level parameters. Consider the binomial likelihood $r_i|p_i, n_i \sim \text{Bin}(n_i, p_i)$ independently with $\log(p_i) = \beta_i$ for $i = 1, 2, \ldots, n_{\text{study}}$. If we want the β_i to represent random effects, we can assume a second-level prior: $\beta_i|\mu, \sigma^2 \sim \text{Bin}(n_i, p_i)$ where μ and σ^2 represent the random effect's mean and variance, respectively. We can complete the prior specification with non-informative priors on the hyperparameters (third-level) $\pi(\mu, \sigma^2) \propto \frac{1}{\sigma}$.

A `JAGS` specification for this model is given below. The steps of creating a list of the data, running `jags.model()` and `coda.samples()`, along with the MCMC chain post processing can be applied as in the previous example.

```
cat(file = "logit_re_model.txt", "
model {
  for (i in 1:n_study) {
    # set the likelihood
    r[i] ~ dbinom(p[i], n[i])
    logit(p[i]) <- beta[i]

    # set the second level prior
    beta[i] ~ dnorm(mu, tau)
  }

  # set the hyper priors
  mu ~ dnorm(0, .001)
  tau ~ dgamma(.001, .001)
  sigma2 <- pow(tau, -1.0)
}")
```

15.4.3 Poisson Meta-Analysis Model

As a final example, we examine a version of the model in Example 2 of [24]. This is a relative effect based meta-analysis model for count data. Their dataset consists of ten studies comparing the effect of diet on cardiovascular (CV) mortality. Each study had a control arm as well as an experimental reduced fat diet arm (one study had two different reduced fat diets). For simplicity, we assume all experimental diets are the same. The studies reported the number of CV deaths r for each arm as well as the number of years of exposure E. Following [24], we use the Poisson model $r_{ik}|E_{ik}, \lambda_{ik} \sim \text{Pois}(\lambda_{ik} E_{ik})$ independently for $i = 1, 2, \ldots, 10$, where k indexes the arms of that study (up to 3), r_{ik} represents the number of CV deaths, D_{ik} is the person years represented, and λ_{ik} the parameter for the hazard in arm k of study i. This hazard is parameterized on the log scale with a relative effect structure: $\log(\lambda_{ik}) = \phi_i + \delta_{ik}$, where ϕ_i represents the "baseline" effect (also known as the control group or study effect) for study i and δ_{ik} is the relative treatment effect for arm k of study i.

In the tradition of meta-analysis, we use non-informative priors for the "baseline" effects, $\pi(\phi_i) \propto 1$, and random effects for the relative treatments, $\delta_{i1} = 0$ and $\delta_{ik} \sim \mathcal{N}(d_{trt_{ik}} - d_{trt_{i1}}, \sigma^2)$. In this example, there are only two treatments, but the model can be used for more than one treatment. The random effects have a mean which is the average change from the control group. These are given a standard non-informative prior. The standard deviation of the random effects, interpreted as a trial-to-trial variability in treatment effect, is often given a uniform prior on the standard deviation scale. The upper bound can be set based on scale of the data: $\sigma \sim \text{Unif}(0, 10)$.

The R code for this data and model is given below.

```
n_study <- 10

r <- cbind(
  c(113, 1, 24, 248, 31, 65,  3, 28, 177,  2),
  c(111, 5, 20, 269, 28, 48,  1, 39, 174,  1),
  c( NA, 3, NA,  NA, NA, NA, NA, NA,  NA, NA)
)

E <- cbind(
  c(1917, 43.6, 393.5, 4715, 715, 885, 87.8, 1011, 1544, 125),
  c(1925, 41.3, 373.9, 4823, 751, 895, 91.0,  939, 1588, 123),
  c(  NA, 38.0,    NA,   NA,  NA,  NA,   NA,   NA,   NA,  NA)
)

trt <- cbind(
  c( 1, 1,  1,  1,  1,  1,  1,  1,  1,  1),
  c( 2, 2,  2,  2,  2,  2,  2,  2,  2,  2),
  c(NA, 3, NA, NA, NA, NA, NA, NA, NA, NA)
```

```
)

n_trt <- 2
n_arms <- c(2, 3, 2, 2, 2, 2, 2, 2, 2, 2)

cat(file = "poisson_nma_model.txt", "
model {

  for(i in 1:n_study){
    # set the likelihoods
    for(k in 1:n_arms[i]) {
      r[i,k] ~ dpois(lambda[i,k] * E[i,k])
      log(lambda[i,k]) <- mu[i] + delta[i,k]
    }

    # set the second level priors
    mu[i] ~ dnorm(0, .001)
    delta[i,1] <- 0
    for (k in 2:n_arms[i]) { delta[i,k] ~ dnorm(d[t[i,k]]
                                 - d[t[i,1]],tau) }
  }

  # set the hyper priors
  d[1] <- 0
  for (k in 2:n_trt) { d[k] ~ dnorm(0, .001) }

  sigma ~ dunif(0, 10)
  tau <- pow(sigma, -2.0)
}")
```

15.5 Prior Elicitation Computations

Prior elicitation generally refers to the conversion of expert opinion into a prior distribution for use in a Bayesian context [25]. There are many excellent works on elicitation; we refer the reader to the summary article [26]. Properly performed prior elicitation typically involves several steps requiring a range of expertise and best-practice protocols. In this section, we describe how R can be used to approach one part of the elicitation process: the conversion of expert-specified quantities into the standard parameters of probability distribution.

All elicitation exercises ultimately result in the expert specifying information that is converted into the canonical parameters of a given family of probability distributions. Which quantities are chosen to be elicited varies from situation to situation but generally depends on how realistic it is for an expert to be able to reliably specify them. A typical example is provided by the so-called mode and percentile (MP) method, used to elicit two-parameter univariate probability distributions. Converting MP specifications into the (α, β) parameters of the beta distribution has attracted special interest, since beta distributions are used throughout Bayesian statistics, and several pieces of software exist that can perform the computations [27, 28, 29, 30]. In general, performing such computations requires solving a complicated system of nonlinear equations. For example, in the beta distribution case, if the percentile specified is the 90th percentile, the system that must be solved is

$$\begin{aligned} \text{mode} &= \arg\max_{\theta} \frac{\Gamma(\alpha+\beta)}{\Gamma(\alpha)\Gamma(\beta)} \theta^{\alpha-1}(1-\theta)^{\beta-1} \\ 0.90 &= \int_0^{\%\text{ile}} \frac{\Gamma(\alpha+\beta)}{\Gamma(\alpha)\Gamma(\beta)} \theta^{\alpha-1}(1-\theta)^{\beta-1}\, d\theta, \end{aligned}$$

where mode is the mode specified by the expert (given), %ile is the 90th percentile specified by the expert (given), and the system is to be solved for α and β. In this setting of the beta distribution, this resolves to

$$\begin{aligned} \text{mode} &= \frac{\alpha-1}{\alpha+\beta-2} \\ 0.90 &= \texttt{pbeta}(\%\text{ile},\ \texttt{shape1} = \alpha,\ \texttt{shape2} = \beta), \end{aligned}$$

where `pbeta()` is the (`R` implementation of the) CDF of the beta distribution. There are a number of ways to try to solve this system numerically; one way is to minimize the function

$$\begin{aligned} \text{error}(\alpha, \beta) = &\left(\text{mode} - \frac{\alpha-1}{\alpha+\beta-2}\right)^2 \\ &+ \Big(.90 - \texttt{pbeta}(\%\text{ile},\ \texttt{shape1} = \alpha,\ \texttt{shape2} = \beta)\Big)^2. \end{aligned}$$

Any (α, β) combination that yields $\text{error}(\alpha, \beta) = 0$ solves the desired system. How can we go about finding such a pair? This is where `R`'s built-in numerical optimization procedures come in.

Suppose we are interested in the specific problem where the expert thinks the likelihood of some event is 25% and is 90% sure it is less than 40%. We can perform the optimization with `optim()`:

```
# define the error function
error <- function(theta) {
  alpha <- theta[1]; beta <- theta[2]
  (.25 - (alpha-1)/(alpha+beta-2))^2
   + (.90 - pbeta(.40, alpha, beta))^2
}
```

```
# set initial values for routine and optimize
initial_values <- c("alpha" = 2, "beta" = 2)
o <- optim(initial_values, error)
str(o, 1)

List of 5
 $ par        : Named num [1:2] 5.84 15.52
  ..- attr(*, "names")= chr [1:2] "alpha" "beta"
 $ value      : num 1e-08
 $ counts     : Named int [1:2] 117 NA
  ..- attr(*, "names")= chr [1:2] "function" "gradient"
 $ convergence: int 0
 $ message    : NULL
```

By default, `optim()` seeks the minimum value of a given function starting at an initial value. `optim()` returns a named list containing (1) the value of the arguments where the minimum is obtained (a vector), (2) the value of the function at the minimum, (3) the number of times the function was evaluated, (4) a code communicating the result of the routine, and (5) a message. The convergence code `0` implies convergence. Thus, the values $\alpha = 5.84$ and $\beta = 15.52$ characterize a beta distribution whose mode is 20% and 90th percentile is 45%, consistent with the expert's specifications. We note that, in general, this conversion process is complicated and is not always one-to-one; nevertheless, `R`'s built-in functionality provides a wide range of resources for computations common to prior elicitation.

15.6 Bayesian Sample Size Determination

In this section, we provide a general description of Bayesian sample size determination as well as `R` code that can be used to carry it out. Bayesian sample size determination often involves two priors: one prior representing the prior that will be used during the analysis, called the *analysis prior*, and one that represents the prior belief about the parameters used to generate synthetic data, called the *design prior*. While our example is relatively simple, it is the same beta-binomial case as in Section 15.5, the concepts and code both generalize with surprisingly little modification. The exposition is intended to provide a feel for Bayesian sample size determination problems in general and the kinds of judgments that the Bayesian statistician must make in performing them as opposed to a rigorous treatment of the subject. For further reading in Bayesian sample size determination, many excellent treatments exist. We refer the reader to [31, 32, 33] and note that our approach is most similar to [34].

The sample size determination exercise in this section revolves around the following hypothetical scenario: A research physician is studying cyclic vomiting syndrome (CVS), a debilitating condition characterized by recurrent bouts of nausea, vomiting, and other gastrointestinal problems, and is interested in the antiemetic properties of a new compound on the condition. She is primarily interested in the proportion of patients θ whose symptoms are reduced by inhaling a specially designed vaporized preparation of the compound. To estimate θ, the physician intends to recruit several patients and determine the proportion of them whose symptoms are alleviated by the treatment.

Considering the problem from a theoretical perspective is helpful. Let $X_1, \ldots, X_n \overset{iid}{\sim}$ Bernoulli(θ), where each X_k represents a patient whose symptoms are reduced ($X_k = 1$) or not reduced ($X_k = 0$), and define $S_n = \sum X_k$, representing the total number of patients whose symptoms are reduced. Consider the sample size n as known. From elementary principles, we know that the expected number of patients whose symptoms are alleviated is $\mathbb{E}[S_n] = n\theta$; similarly, from sample to sample, the variability of the patients whose symptoms are alleviated is $\text{Var}[S_n] = n\theta(1-\theta)$. The standard frequentist approach estimates θ with $\hat{\theta} = \bar{X}_n = S_n/n$, the sample proportion; this can be obtained via maximum likelihood, for example. Using the previous facts, we have $\mathbb{E}[\hat{\theta}] = \theta$ and $\text{Var}[\hat{\theta}] = \theta(1-\theta)/n$. Recall that the standard error of an estimator is its standard deviation. The standard error of $\hat{\theta}$, denoted $s.e.(\hat{\theta})$, is therefore $s.e.(\hat{\theta}) = \sigma_{\hat{\theta}} = (\theta(1-\theta)/n)^{1/2}$. In most real applications, the standard error is not known exactly and must be estimated from data; the natural estimator of $s.e.(\hat{\theta})$ is $\widehat{s.e.}(\hat{\theta}) = (\hat{\theta}(1-\hat{\theta})/n)^{1/2}$. (Note: Some authors use the term standard error to refer to the estimate of the standard deviation of the estimator as well. We make the distinction explicit.) This estimator is well-known to have good asymptotic properties. Now, in this simple case, we can describe exactly the finite sample distribution of $\hat{\theta}$: it is a scaled Bin(n, θ) over the support $0, \frac{1}{n}, \frac{2}{n}, \ldots, 1$. However, it is more common in practice to use an approximating distribution since the exact distribution of statistics tend to be intractable; here, this simplifies calculations and eliminates challenges faced when considering discrete distributions. The central limit theorem (CLT) approximation to the distribution of $\hat{\theta}$ is given by $\hat{\theta} \approx \mathcal{N}(\theta, \theta(1-\theta)/n)$, and this provides the standard, albeit somewhat poorly performing, $(1-\alpha)100\%$ confidence interval $\hat{\theta} \pm z_{\alpha/2}\widehat{s.e.}(\hat{\theta})$.

Here is a practical problem: what sample size n is required so that the above interval places θ within 1% of $\hat{\theta}$? Mathematically, this demand is $|\theta - \hat{\theta}| \leq 0.01$, which is equivalent to $\hat{\theta} - 0.01 \leq \theta \leq \hat{\theta} + 0.01$ so that, from our interval, $0.01 = z_{\alpha/2}\widehat{s.e.}(\hat{\theta})$. There are two things to note about this equation. First, the right hand side is a random quantity that depends on the sample. Second, the equation may not be satisfiable: the right hand side is a discrete random variable assuming at most $n+1$ values, and .01 may not be one of those values. A common approach that resolves these problems is to swap $s.e.(\hat{\theta})$ for $\widehat{s.e.}(\hat{\theta})$ and then simply solve for n to obtain

$n = n(\theta) = (0.01)^{-2}z^2_{\alpha/2}\theta(1-\theta)$. For instance, if θ were 50%, we would need $n = n(0.50) = (0.01)^{-2}z^2_{0.05/2}(0.50)(1-0.50)$ observations to be 95% confident our estimate $\hat{\theta}$ is within 1% of the true value θ. Computing in `R` with `(.01)^(-2) * qnorm(1-.05/2)^2 *.50*(1-.50)` and rounding up, we determine this is $n = 9,604$ observations.

The sample size required depends on the unknown value of θ. Using `R`, we can visualize how this sample size changes as θ changes as follows; note that the code below requires `library("tidyverse")` to be loaded. For introductions to tidyverse packages, see the tidyverse webpage www.tidyverse.org.

```
freq_samp_size <- function(th) 10^4 * qnorm(1-.05/2)^2
                                  * th * (1-th)

data_frame(th = seq(.025, .975, .025),
           n = freq_samp_size(th)) %>%
  ggplot(aes(x = th, y = n)) + geom_point(size = .75) +
    geom_line() +
    labs(x = expression(theta), y = "Sample Size n")
```

Of course, in practice, θ is not known – that is the goal of the whole study. Since θ is unknown a priori, we need to address the fact that the expression that we have to compute sample sizes $n(\theta)$ contains it. One way to do that is to select the sample size that mitigates against all possible θ, the largest one. As a function of θ, $n(\theta) = 10^4 z^2_{\alpha/2}\theta(1-\theta)$ is maximized at $\theta = 0.50$; this is also evident from Figure 15.6. At that value, $\theta(1-\theta) = 0.25$ so that the formula reduces to $n = 2,500z^2_{\alpha/2}$. For a 95% confidence interval, this is $n = 9,604$. However, from Figure 15.6 it is clear that this quantity depends significantly on the unknown θ, so that if θ is actually 10%, the maximum

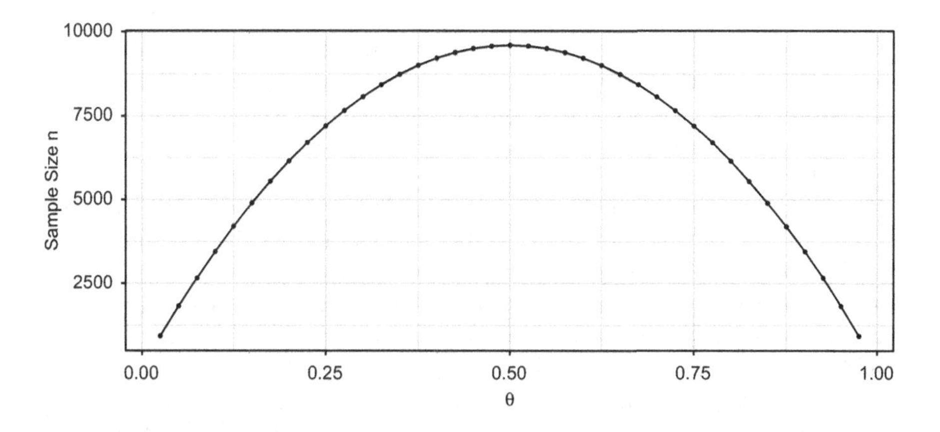

FIGURE 15.6
The sample size n required for a 95% confidence interval of length 2%.

value approach requires obtaining over 6,000 superfluous observations, triple the necessary sample size.

Suppose the physician believes that 25% of the patients will experience lessened symptoms and is quite sure the effect size is less than 40%. A *frequentist* strategy to incorporate this information may be to reason that, of the presumed possible values, the prescriptive sample size is maximized when $\theta = 0.40$, so plugging $\theta = 0.40$ into the sample size formula would give a reasonable answer. The previously determined formula can be used to determine a requisite sample size of $n = z^2_{0.05/2}\theta(1-\theta) \approx 9,220$.

How might a *Bayesian* incorporate this information into the above frequentist analysis to determine a reasonable sample size? Using the mode/percentile elicitation method described in Section 15.5, and assuming the 40% corresponds to the 95%th percentile, the physician's belief corresponds to a Beta(5.8, 15.5) distribution. If the correspondence between θ and the required sample size n is $n(\theta) = 10^4 z^2_{\alpha/2}\theta(1-\theta)$, then the beliefs concerning θ can be translated into beliefs concerning n, which we can determine easily using the following Monte Carlo strategy:

1. Sample N observations θ_k from the Beta(5.8, 15.5) distribution,
2. Transform each θ_k via the function $n(\theta) = 10^4 z^2_{\alpha/2}\theta(1-\theta)$, and
3. Compute a high percentile of the distribution of θ_k's.

The following code performs the computations in R using the 95th percentile of the induced distribution.

```
set.seed(1)
rbeta(1e6, 5.8, 15.5) %>% freq_samp_size %>% quantile(.95)

     95%
9446.965
```

This $n = 9,447$ can be reasonably compared to the $n = 9,220$ sample size determined using the frequentist method that incorporates the physician's information. This may seem strange: the Bayesian sample size is worse than the frequentist sample size in the sense that it demands a larger sample size. How is this possible? Aren't Bayesian methods supposed to incorporate expert information and thereby enable the use of smaller sample sizes? Generally yes, but not always. In this case, the Bayesian method admits the possibility that θ is larger than 40%, whereas the frequentist method forbids it. The fact that both are relatively close to the maximum bound ($n = 9,604$) is merely a consequence of 40% being close to the 50% bound; different circumstances would reduce the requisite sample size even further. Alternatively, if we used the 90th percentile of the distribution of θ_k's instead of the 95th, we would have obtained $n = 9,207$, which is much more consistent with the frequentist version. If we had instead used the expected number of observations needed, we would have obtained $n = 7,270$, but this is not recommended as the distribution is left-skewed, pulling the mean sample size down. Which should be used is the kind of judgment call Bayesians typically encounter, not unlike those of determining a reasonable confidence level or probability of Type I error in a frequentist analysis. In general, stable procedures are recommended in addition to forthright communication of the choices made to construct the design. Fortunately, the Bayes paradigm makes each of these very transparent.

The above considerations try to force a Bayesian approach to a demand for frequentist methods. We now turn to a more fully Bayesian strategy. The basic difference in a Bayesian design from what was done above is how the Bayesians create interval estimates; given a dataset $x_1, \ldots, x_n$, a Bayesian may use a $(1-\alpha)100\%$ equal-tailed credible interval as described in Section 15.2. Let $s_n = \sum x_k$, the total observed number of patients whose symptoms were alleviated. Presuming the data is to be analyzed with a Beta$(1, 1)$ prior, the Bayesian determines the posterior to be Beta$(1 + s_n, 1 + (n - s_n))$ and computes the equal-tailed interval using this distribution. The `bayes_ci()` function below computes a 95% equal-tailed credible interval using the Beta$(1, 1)$ prior for a given dataset `data` and demonstrates it on a dataset generated with $\theta = 25\%$.

```
bayes_ci <- function(data, alpha = 1, beta = 1) {
  n <- length(data); sn <- sum(data)
  qbeta(c(.025, .975), alpha + sn, beta + (n - sn))
}

set.seed(1)
rbinom(100, 1, .25) %>% bayes_ci

[1] 0.1927687 0.3646242
```

Note that if a different, non-conjugate (analysis) prior were used, or in a different modeling scenario, we would likely have run an MCMC subroutine in a probabilistic programming language such as `JAGS` inside `bayes_ci()`; that is, *every time* we create a credible interval. As we will see shortly, `bayes_ci()`, our main analytical computation, will itself be called thousands of times – or more – in our design computations. These are the kinds of computations that Bayesians must either be capable of performing, simplifying, or resolving in some other way, and hopefully this example serves as an indicator of how computationally intensive even simple Bayesian problems can ultimately be.

As in the frequentist case, the length of the interval varies from sample to sample (recall the exchange of $s.e.(\hat{\theta})$ for $\widehat{s.e.}(\hat{\theta})$). And, as in that setting, to simplify things we prefer to eliminate this dependence. For a hypothetical population with rate θ, one approach is to average over the sample space to determine a kind of expected interval length. Another, more conservative approach might be to take a high percentile, say the 95th percentile, of the distribution of lengths, and use that as the fixed "length" of the interval for a given θ independent of the sample. For example, suppose $\theta = 30\%$. If $n = 50$, we can use `rbinom()`, `replicate()`, `diff()` (which computes differences), and `bayes_ci()` to create a Monte Carlo estimate of how long the Bayesian equal-tailed intervals are on average. Recall that previously the intervals under consideration had length 0.02, and their interpretation was that the point estimate was within 1% of the true unknown value.

```
replicate(1e4, rbinom(50, 1, .30) %>% bayes_ci() %>% diff() )
  %>% mean()

[1] 0.2443712

replicate(1e4, rbinom(50, 1, .30) %>% bayes_ci() %>% diff() )
  %>% quantile(.95)

     95%
0.263016
```

Using `ggplot2`, we can plot how long the intervals are, on average, when $n = 50$ for values of θ ranging from 2.5% to 97.5%. This is shown in Figure 15.7. Note that the first line of the following function simply allows it to accept a vector of θ's.

```
avg_bayes_length <- function(th, n = 50, N = 1e4) {
  if(length(th) > 1) return(sapply(th, avg_bayes_length,
                                   n = n, N = N))
  replicate(N, rbinom(n, 1, th) %>% bayes_ci() %>% diff())
    %>% mean
}

data_frame(th = seq(.025, .975, length.out = 20),
                    length = avg_bayes_length(th)) %>%
  ggplot(aes(th, length)) + geom_point() + geom_line() +
    labs(x = expression(theta), y = "Average Interval Length") +
    stat_function(
      fun = function(th) 2 * qnorm(.975) * sqrt(th*(1-th)/50),
      linetype = 2
    )
```

Note that the curve in Figure 15.7 depends on the sample size n: as n increases, the whole curve gets pulled down as the interval becomes more precise. It is therefore analogous to $2z_{\alpha/2}s.e.(\hat{\theta}) = 2z_{\alpha/2}(\theta(1-\theta)/n)^{1/2}$, but it ignores the Type I error constraint.[2] Following the analogy, to determine a sample size we may choose to increase n until all the intervals are of expected length, at most 0.02, which would be roughly analogous to the bounding strategy used previously. If we instead use values of θ drawn from the physician's

[2] A more direct analogy would be $\mathbb{E}[2z_{\alpha/2}\widehat{s.e.}(\hat{\theta})]$.

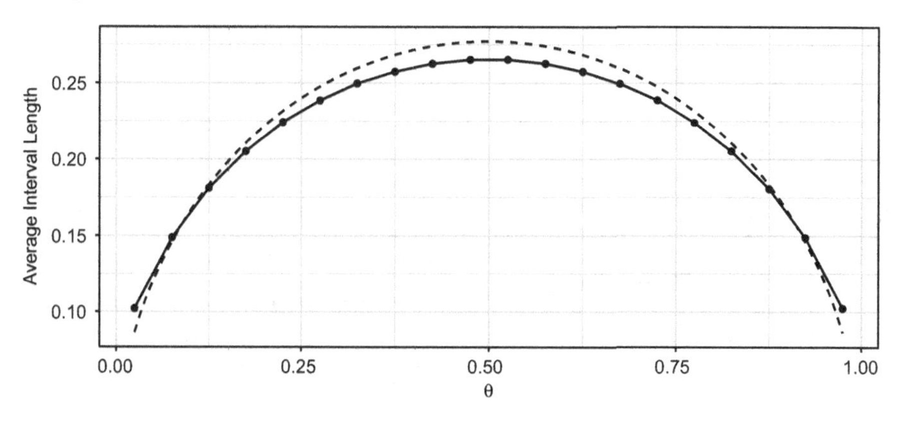

FIGURE 15.7
Average length of credible intervals when $n = 50$ as a function of θ (solid/dotted) superimposed onto the frequentist analogue (dashed, see text).

Beta(5.8, 15.5) prior, compute the average credible interval length for each for samples of size $n = 50$, and average the (average) lengths, we can obtain a Monte Carlo estimate for the expected length of a credible interval using a non-informative Bayesian analysis prior with respect to the physician's belief of what the true value is, the design prior.

```
rbeta(1e3, 5.8, 15.5) %>%
  sapply(avg_bayes_length, n = 50) %>%
  mean()

[1] 0.2314588
```

As n increases, the expected average length decreases monotonically. This can be illustrated by simply plotting it for a series of n's; this is displayed in Figure 15.8. A basic implementation that computes the expected average length of the interval, where the average refers to averaging over samples and the expected refers to averaging over the design prior, would be

```
expected_avg_bayes_length <- function(n, alpha = 5.8,
                                       beta = 15.5, N = 1e3){
  if(length(n) > 1) return(sapply(n, expected_avg_bayes_length))
  rbeta(N, alpha, beta) %>% sapply(avg_bayes_length, n = n)
    %>% mean()
}
```

However, even this computation is sufficiently cumbersome to prefer some basic speed enhancements. Here is a version of the same function that parallelizes the computations across all the cores of your local machine (including hyperthreading).

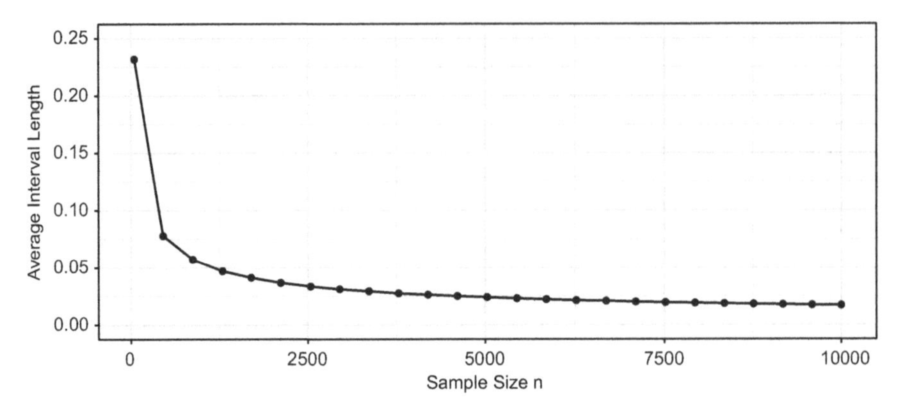

FIGURE 15.8
Average credible interval length, accounting for the physician's belief, as a function of the sample size.

```
library("parallel")

expected_avg_bayes_length <- function(n, alpha = 5.8,
                                      beta = 15.5, N = 1e3){
  if(length(n) > 1) return(sapply(n, expected_avg_bayes_length))
  cluster <- makeCluster(detectCores());
    on.exit(stopCluster(cluster))
  clusterExport(cluster, c(ls(.GlobalEnv), ls("package:dplyr")))
  rbeta(N, alpha, beta) %>%
    parSapply(cluster, ., avg_bayes_length, n = n) %>%
    mean()
}

# make range of ns
ns <- seq(50, 10000, length.out = 25) %>% round
expected_avg_lengths <- expected_avg_bayes_length(ns)

# make graphic
data.frame(n = ns, exp_avg_length = expected_avg_lengths) %>%
  ggplot(aes(n, exp_avg_length)) + geom_point() + geom_line() +
    labs(x = "Sample Size n") +
    scale_y_continuous("Average Interval Length",
                       limits = c(.00, .25))
```

We can use `uniroot()`, a numerical root finder, to more precisely determine the sample size needed by solving $w(\theta) = .02$ by looking for where $w(\theta) - .02 = 0$; here $w(\theta)$ is the function that computes the expected average length (width) of the Bayes intervals. Note that since the function involves a stochastic component, the number of iterations needs to be large to get a good estimate.

```
uniroot(
  f = function(n) expected_avg_bayes_length(round(n)) - .02,
  lower = 1000, upper = 10000
)$root

[1] 7212.392
```

Instead of sampling from the design prior and then the data model to construct data, it is more common in practice to sample directly from the design prior predictive distribution $\int f(x|\theta)\pi_d(\theta)d\theta$. In this case, the predictive distribution is a beta-binomial. Since this distribution is not built into R, we can create the sampler using the definition: $X \sim$ Beta-Bin(n, α, β) if $X|p \sim$ Bin(n, p) and $p \sim$ Beta(α, β). Thus,

```
rbetabinom <- function(n, size, alpha, beta)
                rbeta(n, alpha, beta) %>% rbinom(n, size, .)
```

We can use `rbetabinom()` to sample from the predictive distribution instead of sampling from the binomial for a fixed θ. Thus, using our collaborator's design prior, we can compute the expected average length of our Bayes intervals using a uniform analysis prior:

```
replicate(1e4,
  rbetabinom(50, 1, alpha = 5.8, beta = 15.5) %>% bayes_ci()
  %>% diff()
) %>% mean()

[1] 0.2377115
```

And we can dial the sample size n up until the average length is less than the desired .02:

```
expected_avg_bayes_length_2 <- function(n, N = 1e1) {
  if(length(n) > 1) return(sapply(n, expected_avg_bayes_length_2))
  replicate(N,
    rbetabinom(n, 1, alpha = 5.8, beta = 15.5) %>% bayes_ci()
    %>% diff()
  ) %>% mean()
}

uniroot(
  f = function(n) expected_avg_bayes_length_2(round(n)) - .02,
  lower = 1000, upper = 10000
```

```
)$root

[1] 7569.996
```

It should also be noted that, in addition to computing these using Monte Carlo techniques, numerical integration or even direct summation can be used can be used to compute expectations with respect to continuous or discrete distributions, and where possible these can dramatically expedite the process. See the references listed above for details.

References

[1] Christian P Robert and George Casella. *Monte Carlo Statistical Methods*. Springer: Berlin Heidelberg, 2005.

[2] David Smith. An updated history of R. http://blog.revolutionanalytics.com/2017/10/updated-history-of-r.html, 2017. Accessed: 2017-11-01.

[3] Hadley Wickham. *Advanced R*. Chapman and Hall/CRC Press, 2014.

[4] Douglas Bates and Martin Maechler. *Matrix: Sparse and Dense Matrix Classes and Methods*, 2017. R package version 1.2-11.

[5] Mordecai Ezekiel. *Methods of Correlation Analysis*. John Wiley & Sons: New York, 1930.

[6] Donald R. McNeil. *Interactive Data Analysis*. John Wiley & Sons: New York, 1977.

[7] David Kahle. mpoly: Multivariate polynomials in R. *The R Journal*, 5(1):162–170, 2013.

[8] Stan Development Team. RStan: the R interface to Stan, 2016. R package version 2.14.1.

[9] Hadley Wickham and Winston Chang. *devtools: Tools to Make Developing R Packages Easier*, 2017. R package version 1.13.3.

[10] Hadley Wickham. *ggplot2: Elegant Graphics for Data Analysis*. Springer-Verlag: New York, 2009.

[11] Hadley Wickham, Romain Francois, Lionel Henry, and Kirill Müller. *dplyr: A Grammar of Data Manipulation*, 2017. R package version 0.7.2.

[12] Hadley Wickham and Lionel Henry. *tidyr: Easily Tidy Data with 'spread()' and 'gather()' Functions*, 2017. R package version 0.7.1.

[13] Hadley Wickham. *tidyverse: Easily Install and Load 'Tidyverse' Packages*, 2017. R package version 1.1.1.

[14] Stefan Milton Bache and Hadley Wickham. *magrittr: A Forward-Pipe Operator for R*, 2014. R package version 1.5.

[15] Martyn Plummer. JAGS: A program for analysis of Bayesian graphical models using Gibbs sampling. In *Proceedings of DSC*, Vienna, Austria, vol. 2, pp. 1–10, 2003.

[16] David J Lunn, Andrew Thomas, Nicky Best, and David Spiegelhalter. WinBUGS – A Bayesian modelling framework: Concepts, structure, and extensibility. *Statistics and Computing*, 10(4):325–337, 2000.

[17] David Spiegelhalter, Andrew Thomas, Nicky Best, and Dave Lunn. OpenBUGS user manual, version 3.0.2. *MRC Biostatistics Unit, Cambridge*, 2007.

[18] Sibylle Sturtz, Uwe Ligges, and Andrew Gelman. R2winbugs: A package for running WinBUGS from R. *Journal of Statistical Software*, 12(3):1–16, 2005.

[19] Martyn Plummer. *rjags: Bayesian Graphical Models Using MCMC*, 2016. R package version 4-6.

[20] George Casella and Edward I George. Explaining the Gibbs sampler. *The American Statistician*, 46(3):167–174, 1992.

[21] Siddhartha Chib and Edward Greenberg. Understanding the Metropolis-Hastings algorithm. *The American Statistician*, 49(4):327–335, 1995.

[22] David Lunn, Chris Jackson, Nicky Best, Andrew Thomas, and David Spiegelhalter. *The BUGS Book: A Practical Introduction to Bayesian Analysis*. CRC Press: Boca Raton, FL, 2012.

[23] Martyn Plummer, Nicky Best, Kate Cowles, and Karen Vines. CODA: Convergence diagnosis and output analysis for MCMC. *R News*, 6(1):7–11, 2006.

[24] Sofia Dias, Nicky J Welton, Alex J Sutton, and Tony Ades. *NICE DSU Technical Support Document 2: A Generalised Linear Modelling Framework for Pairwise and Network Meta-Analysis of Randomised Controlled Trials*. Number TSD2 in Technical Support Document in Evidence Synthesis. National Institute for Health and Clinical Excellence, 8 2011.

[25] Anthony O'Hagan, Caitlin E Buck, Alireza Daneshkhah, J Richard Eiser, Paul H Garthwaite, David J Jenkinson, Jeremy E Oakley, and Tim Rakow. *Uncertain Judgements: Eliciting Experts' Probabilities*. John Wiley & Sons: New York, 2006.

[26] Paul H Garthwaite, Joseph B Kadane, and Anthony O'Hagan. Statistical methods for eliciting probability distributions. *Journal of the American Statistical Association*, 100(470):680–701, 2005.

[27] Yujun Wu, Weichung J Shih, and Dirk F Moore. Elicitation of a beta prior for Bayesian inference in clinical trials. *Biometrical Journal*, 50(2):212–223, 2008.

[28] David Kahle, James Stamey, Fanni Natanegara, Karen Price, and Baoguang Han. Facilitated prior elicitation with the Wolfram CDF. *Biometrics and Biostatistics International Journal*, 3(6):1–6, 2016.

[29] Chun-Lung Su. BetaBuster. Department of Medicine and Epidemiology, University of California, Davis.

[30] David E Morris, Jeremy E Oakley, and John A Crowe. A web-based tool for eliciting probability distributions from experts. *Environmental Modelling and Software*, 52:1–4, 2014.

[31] Pierpaolo Brutti, Fulvio De Santis, and Stefania Gubbiotti. Robust Bayesian sample size determination in clinical trials. *Statistics in Medicine*, 27(13):2290–2306, 2008.

[32] Giovanni Parmigiani and Lurdes Inoue. *Decision Theory: Principles and Approaches*, vol. 812. John Wiley & Sons: New York, 2009.

[33] Shein-Chung Chow, Jun Shao, Hansheng Wang, and Yuliya Lokhnygina. *Sample Size Calculations in Clinical Research*, 4th edn. Chapman & Hall: Boca Raton, FL, 2017.

[34] Lawrence Joseph, Roxane Du Berger, and Patrick Belisle. Bayesian and mixed Bayesian/likelihood criteria for sample size determination. *Statistics in Medicine*, 16(7):769–781, 1997.

16

Software for Bayesian Computation—An Overview of Some Currently Available Tools

Melvin Munsaka
AbbVie, Data and Statistical Sciences

Mani Lakshminarayanan
Statistical Consultant

CONTENTS

16.1 Introduction

The use of Bayesian methods has become increasingly popular in modern statistical analysis, with applications in a wide variety of scientific fields. Bayesian methods incorporate existing information, based on expert knowledge, past studies, and other information into current data analysis. This existing information is represented by a prior distribution, and the data likelihood is effectively weighted by the prior distribution as the data analysis results are computed. The main outcomes of a Bayesian analysis are the posterior distributions of a model's parameters, rather than point estimates and their standard errors. Access to a model's parameters' posterior distributions enables one to address scientific questions of interest directly, because once the model parameters are estimated, it is easy to compute the posterior distributions for any functions of the parameters or any quantities of interest. Bayesian methods can be computer intensive, and recently there have been enormous improvements in computational tools. There is currently evidence of competitive continual refinements and enhancements in a wide range of computational tools with newer and more efficient tools. For example, within the framework of Bayesian inference, Markov Chain Monte Carlo (MCMC) methods that comprise a class of algorithms for sampling from a probability distribution have continued to evolve from random walk proposals to Hamiltonian Monte Carlo, with both theoretical and algorithmic innovations opening new opportunities to practitioners, see for example Green et al. (2015) and Robert (2014). The previous chapter focused on R computational tools. In this chapter, we will provide a brief overview of some of the currently available tools for Bayesian computations including non-R tools. The discussion will include a brief overview of Bayesian methods which will be followed by brief discussion of MCMC. Available tools for Bayesian computing will be discussed. For discussion purposes, these have been divided into Open source Bayesian-specific software, Open Source General Bayesian Software Tools, and Commercial General Bayesian Software. Available functionality within each software tool will be discussed.

16.2 The Bayesian Framework

In Bayesian framework, there may be information about model parameters θ available to the user before data from a new study are available. This information is quantified by putting a probability distribution $p(\theta)$, called the prior on the parameters. Information regarding model parameters contained in the data $p(x)$ from the new study is expressed in the likelihood, which is proportional to the distribution of the observed data given the model parameters $p(x|\theta)$. The information in the likelihood and prior is combined to produce an updated probability distribution, the posterior distribution or $p(\theta|x)$, which is the basis of Bayesian inference. Proportional to the product of the prior and the likelihood, the posterior distribution theoretically is always available but sometimes not easily mathematically tractable. Expressed mathematically, we have

$$p(\theta|x) = \frac{p(x|\theta)p(\theta)}{p(x)}.$$

More generally, we have the following setting that encompasses various quantities that are involved in Bayesian inference:

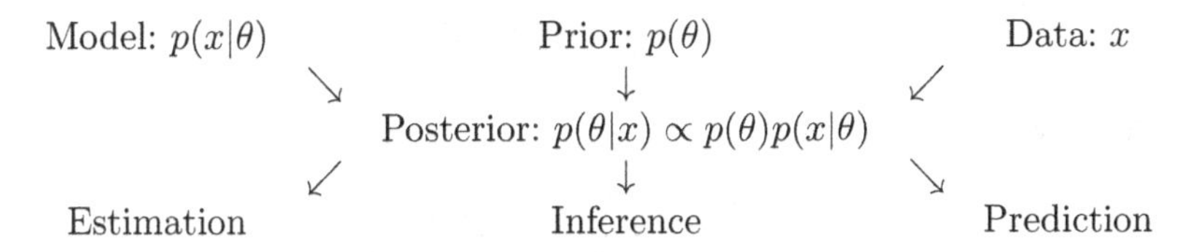

For complex models, the analytic computations often are intractable particularly, the required integration to obtain the normalizing constant of the joint posterior distribution and the calculations of posterior marginal distributions of individual parameters of interest. In fact for typical high-dimensional models, even standard numerical integration techniques are inadequate.

16.3 Use of Bayesian Methods in Practical Clinical Trial Applications

As noted earlier, Bayesian statistical methods have long been computationally out of reach because the analysis often requires integration of high-dimensional functions. Recent advancements in computational tools to apply MCMC methods make Bayesian inference accessible for use across many disciplines. Within the drug development setting, we can broadly identify two statistical methodologies that are applicable to the design and analysis of clinical trials, namely, frequentist and Bayesian. For the most part, traditional clinical trial designs rely heavily on frequentist statistics. In frequentist statistics, prior information is utilized formally only in the design of a clinical trial (e.g., assumptions made on the model parameters such as delta, clinically meaningful difference, that need to be detected, and the standard deviation, in the sample size calculation) but not in the analysis of the data. On the other hand, Bayesian statistics provide a formal mathematical framework for combining prior information with current information at the design stage, during the conduct of the trial, and at the analysis stage. With the availability of computing tools, there has been an increase in the use of Bayesian approaches in all aspects of drug development, including adaptive trial designs, dose-finding studies, proof-of-concept studies, seamless trials, postmarketing surveillance, and meta-analysis. The opportunities and advantages along with disadvantages of using Bayesian methods have been discussed widely in the literature, see, for example, Gupta (2012), Wijeysunder et al. (2009), Walley (2015), Bittl and He (2017), Hobbs and Carlin (2008), Kruschke and Liddell (2018), and Lazic et al. (2016).

The fundamental objective of Bayesian data analysis is to determine the posterior distribution:

$$p(\theta \mid X) = \frac{p(X \mid \theta)p(\theta)}{p(X)}$$

$$p(X) = \int d\theta p(X \mid \theta)p(\theta),$$

and hence we can write $p(\theta \mid X)$ as

$$p(\theta \mid X) = \frac{p(X \mid \theta)p(\theta)}{\int d\theta p(X \mid \theta)p(\theta)}.$$

Here, $p(X \mid \theta)$ is the likelihood, $p(\theta)$ is the prior, and $p(X)$ is a normalizing constant also known as the evidence or marginal likelihood.

The computational issue is the difficulty of evaluating the integral in the denominator. There are many ways to address this difficulty, including:

- In cases with conjugate priors (with conjugate priors, the posterior has the same distribution as the prior), use of closed form solutions
- Using numerical integration
- Approximating the functions used to calculate the posterior with simpler functions and show that the resulting approximate posterior is "close" to true posterior (variational Bayes)
- Use of Monte Carlo methods, of which the most widely used is MCMC.

For most problems of interest, Bayesian analysis requires integration over multiple parameters, making the calculation of a posterior intractable whether via analytic methods or standard methods of numerical integration. However, it is often possible to approximate these integrals by drawing samples from posterior distributions. For example, consider the expected value (mean) of a vector-valued random variable $\mathbf{x}$:

$$E[\mathbf{x}] = \int \mathbf{x}f(\mathbf{x})\mathrm{d}\mathbf{x}, \quad \mathbf{x} = \{x_1, \ldots, x_k\},$$

where k (dimension of vector $\mathbf{x}$) is perhaps very large. If we can produce a reasonable number of random vectors $\{\mathbf{x_i}\}$, we can use these values to approximate the unknown integral. This process is known as Monte Carlo integration. In general, Monte Carlo integration allows integrals against probability density functions

$$I = \int h(\mathbf{x}) f(\mathbf{x}) d\mathbf{x}$$

to be estimated by finite sums

$$\hat{I} = \frac{1}{n} \sum_{i=1}^{n} h(\mathbf{x}_i),$$

where $\mathbf{x}_i$ is a sample from f. This estimate is valid and useful because $\hat{I} \to I$ with probability 1 by the strong law of large numbers. Simulation error can be measured and controlled.

16.4 MCMC Computation Algorithms

Put simply, MCMC is just a method for generating random samples from a probability distribution. MCMC algorithms, sometimes called samplers, are numerical approximation algorithms (Gelman and Rubin, 1996; Robert and Casella, 2010; Hastings, 1970). The name is derived from the fact that the Monte Carlo is used to describe random sampling methods, and since the distribution of the next sample is dependent on the current sample, the set of samples forms a Markov chain. The benefit of using MCMC is that it lets one to replace an impossible integral with a simple summation. The summation is of course less precise than the integral, but given a large enough number of samples the difference is negligible. The tricky aspect of MCMC is how it generates samples from the distribution. There are a number of variations on MCMC, but almost all of them involve a random walk through the distribution space. The primary difference between these variations is in how they generate a potential step for the random walk. In almost all cases, once a potential step has been generated, it is accepted or rejected based on the ratio of its weight relative to the the current state's weight. When this ratio is greater than or equal to one, the state is automatically accepted, and if the ratio is less than one, the state is randomly accepted with probability equal to the ratio.

A key question is how one knows that the random walk has produced a set of samples that really approximate the distribution that is being estimated. That is, how do we determine when it is safe to stop sampling and use the samples to estimate characteristics of the distribution of interest and subsequently carry out statistical inference. There are a lot of answers to this, and unfortunately none of them is perfect. The simplest method is to look at the rate of change of the distribution's marginal variance for its different dimensions. This can be done by running the sample in batches of a few thousand iterations at a time, and between batches, and calculating marginal variances from a randomly selected subset of samples. Once the new marginal variance values are within some threshold of previous values, then the process is complete. Regardless of the algorithm, the goal in Bayesian inference is to maximize the unnormalized joint posterior distribution and collect samples of the target distributions, which are marginal posterior distributions, later to be used for inference. It turns out that there are many algorithms for performing MCMC, too numerous to discuss each one. We will briefly discuss a few of the common ones, including Gibbs, Metropolis, Metropolis–Hastings, and Hamilton algorithms.

16.4.1 Gibbs Sampling Algorithm

The Gibbs sampling algorithm is the simplest and most prevalent MCMC algorithm. The way it works is as follows: if a posterior distribution has some parameters that need to be estimated, one can condition each parameter on current values of the other $k-1$ parameters and sample from the resultant distributional form. This is then repeated on the other parameters in turn. This procedure generates samples from the posterior distribution. Note that we have now combined Markov chains (conditional independence) and Monte Carlo techniques (estimation by simulation) to yield MCMC. The Gibbs sampler takes an alternating approach for multiparameter problems, sampling one parameter given the values of the others and thus reducing a potentially high-dimensional problem to lower-dimensional conditional densities. It can be described based on its steps generally as follows:

- Select starting values for parameters: $\theta = [\theta_1^{(0)}, \theta_2^{(0)}, \ldots, \theta_k^{(0)}]$.
- Initialize counter $j = 1$.
- Draw the following values from each of the k conditional distributions:

$$
\begin{aligned}
\theta_1^{(j)} &\sim \pi(\theta_1 | \theta_2^{(j-1)}, \theta_3^{(j-1)}, \ldots, \theta_{k-1}^{(j-1)}, \theta_k^{(j-1)}) \\
\theta_2^{(j)} &\sim \pi(\theta_2 | \theta_1^{(j)}, \theta_3^{(j-1)}, \ldots, \theta_{k-1}^{(j-1)}, \theta_k^{(j-1)}) \\
\theta_3^{(j)} &\sim \pi(\theta_3 | \theta_1^{(j)}, \theta_2^{(j)}, \ldots, \theta_{k-1}^{(j-1)}, \theta_k^{(j-1)}) \\
&\vdots \\
\theta_{k-1}^{(j)} &\sim \pi(\theta_{k-1} | \theta_1^{(j)}, \theta_2^{(j)}, \ldots, \theta_{k-2}^{(j)}, \theta_k^{(j-1)}) \\
\theta_k^{(j)} &\sim \pi(\theta_k | \theta_1^{(j)}, \theta_2^{(j)}, \theta_4^{(j)}, \ldots, \theta_{k-2}^{(j)}, \theta_{k-1}^{(j)}).
\end{aligned}
$$

- Increment j and repeat until convergence occurs.

As we can see from the algorithm, each distribution is conditioned on the last iteration of its chain values, constituting a Markov chain as advertised. The Gibbs sampler has all of the important properties outlined in the previous section: it is aperiodic, homogeneous, and ergodic. Once the sampler converges, all subsequent samples are from the target distribution. This convergence occurs at a geometric rate. Advantages of Gibbs sampling include the fact that there is no need to tune proposal distributions and proposals are always accepted. Disadvantages include the need to be able to derive conditional probability distributions, the need to be able to draw random samples from conditional probability distributions, and it can be very slow if parameters are correlated because one cannot take "diagonal" steps.

16.4.2 Metropolis Algorithm

This algorithm as the name suggests is due to Metropolis (Metropolis et al., 1953; Robert 2016). It can be described based on its steps generally as follows:

- Begin the algorithm at the current position in parameter space (θ_{current}).
- Propose a "jump" to a new position in parameter space (θ_{new}).
- Accept or reject the jump probabilistically using the prior information and available data.
- If the jump is accepted, move to the new position and return to the first step.
- If the jump is rejected, stay where you are and return to the first step.
- After a set number of jumps have occurred, return all of the accepted positions.

The Metropolis algorithm uses a normal distribution to propose a jump. This normal distribution has a mean value μ which is equal to the current position and takes a "proposal width" for its standard deviation σ, see, for example, Quantcademy (2017). This proposal width is a parameter of the Metropolis algorithm and has a significant impact on convergence. A larger proposal width will jump further and cover more space in the posterior distribution but might miss a region of higher probability initially. However, a smaller proposal width won't cover as much of the space quickly and thus could take longer to converge. A normal distribution is a good choice for such a proposal distribution (for continuous parameters) as, by definition, it is more likely to select points nearer to the current position than further away. However, it will occasionally choose points further away, allowing the space to be explored. Once the jump has been proposed, we need to decide (in a probabilistic manner) whether it is a good move to jump to the new position. How do we do this? We calculate the ratio of the proposal distribution of the new position and the proposal distribution at the current position to determine the probability of moving, p,

$$p = P(\theta_{\text{current}}/\theta_{\text{new}}).$$

We then generate a uniform random number on the interval $[0, 1]$. If this number is contained within the interval $[0, p]$, then we accept the move; otherwise, we reject it. While this is a relatively simple algorithm, it is not immediately clear why this makes sense and how it helps us avoid the intractable problem of calculating a high-dimensional integral of the evidence, $P(D)$.

As Wiecki (2015) points out in his article on MCMC sampling, we are actually dividing the posterior of the proposed parameter by the posterior of the current parameter. Using Bayes' rule, this eliminates the evidence $P(D)$ from the ratio:

$$p = \frac{P(\theta_{\text{current}})}{P(\theta_{\text{new}})} = \frac{\frac{P(D|\theta_{\text{new}})P(\theta_{\text{new}})}{P(D)}}{\frac{P(D|\theta_{\text{current}})P(\theta_{\text{current}})}{P(D)}} = \frac{P(D|\theta_{\text{new}})P(\theta_{\text{new}})}{P(D|\theta_{\text{current}})P(\theta_{\text{current}})}.$$

The right hand side of the latter equality contains only the likelihoods and the priors, both of which we can calculate easily. Hence by dividing the posterior at one position by the posterior at another, we're sampling regions of higher posterior probability more often than not, in a manner which fully reflects the probability of the data. Note that the above acceptance ratio is the reason Metropolis works out and we get around the integration. We can show this by computing the acceptance ratio over the normalized posterior and seeing how it is equivalent to the acceptance ratio of the unnormalized posterior. Stated in other words, dividing the posterior of proposed parameter setting by the posterior of the current parameter setting gets canceled out. So, you can intuit that we are actually dividing the full posterior at one position by the full posterior at another position. This way, we are visiting regions of high posterior probability relatively more often than those of low posterior probability.

16.4.3 Metropolis–Hastings Algorithm

Recall that the key to success in applying the Gibbs sampler to the estimation of Bayesian posteriors is being able to specify the form of the complete conditionals of θ because the algorithm cannot be implemented without them. In practical applications however, the posterior conditionals cannot always be specified in straightforward manner. The Metropolis–Hastings algorithm in contrast generates candidate state transitions from an alternate distribution and accepts or rejects each candidate probabilistically.

The Metropolis algorithm is a special case of the general Metropolis–Hastings algorithm (Hoff, 2009). The main difference is that the Metropolis–Hastings algorithm does not have the symmetric distribution requirement. The Metropolis algorithm can be slow, particularly for those cases in which the initial starting point is way off target. By using the asymmetric proposal distributions via the more general Metropolis–Hastings algorithm can speed up the process.

Consider a simple Metropolis–Hastings algorithm for a single parameter, θ. We will use a standard sampling distribution, referred to as the proposal distribution, to produce candidate variables $q_t(\theta'|\theta)$. That is, the generated value $\theta^{'}$, is a possible next value for θ at step $t+1$. We also need to be able to calculate the probability of moving back to the original value from the candidate, or $q_t(\theta|\theta')$. These probabilistic ingredients are used to define an acceptance ratio:

$$a(\theta', \theta) = \frac{q_t(\theta'|\theta)\pi(\theta')}{q_t(\theta|\theta')\pi(\theta)}.$$

The value of $\theta^{(t+1)}$ is then determined by

$$\theta^{(t+1)} = \begin{cases} \theta' & \text{with probability } \min(a(\theta', \theta^{(t)}), 1) \\ \theta^{(t)} & \text{with probability } 1 - \min(a(\theta', \theta^{(t)}), 1). \end{cases}$$

This transition kernel implies that movement is not guaranteed at every step. It only occurs if the suggested transition is likely based on the acceptance ratio.

A single iteration of the Metropolis–Hastings algorithm proceeds as follows:

- Sample θ' from $q(\theta'|\theta^{(t)})$.
- Generate a Uniform$[0, 1]$ random variate u.

  If $a(\theta', \theta) > u$, then $\theta^{(t+1)} = \theta'$; otherwise, $\theta^{(t+1)} = \theta^{(t)}$.

The original form of the algorithm specified by Metropolis required that $q_t(\theta'|\theta) = q_t(\theta|\theta')$, which reduces $a(\theta', \theta)$ to $\pi(\theta')/\pi(\theta)$, but this is not necessary. In either case, the state moves to high-density points in the distribution with high probability and to low-density points with low probability. After convergence, the Metropolis–Hastings algorithm describes the full target posterior density, so all points are recurrent. A disadvantage of the Metropolis–Hastings algorithm is that the sampling of high-dimensional distributions becomes very inefficient in practice.

16.4.4 Hamiltonian Monte Carlo

A more efficient algorithm is the Hamiltonian Monte Carlo algorithm which in essence is a variant of the Metropolis–Hastings algorithm. The major differences compared to Metropolis–Hastings are as follows:

- Distances between successive generated points are typically large, so we need less iterations to get representative sampling.
- "Price" of a single iteration is higher, but Hamiltonian Monte Carlo is still significantly more efficient.
- Hamiltonian MC in most cases accepts new states.
- Still, Hamiltonian Monte Carlo has problems with sampling from distributions with isolated local minimums.

Algorithmically, the Hamiltonian Monte Carlo takes the parameter θ as collectively denoting the position of a particle in some space with momentum ϕ (of same dimension as θ). Both θ and ϕ are updated at each Metropolis step and jointly estimated, though we are only interested in θ. It can be described in the following steps:

- At iteration t, take a random draw of momentum ϕ from its posterior distribution.
- Update the position vector θ given current momentum; update ϕ given the gradient of θ.
- Calculate
$$r = \frac{p(\theta^* | y)p(\phi^*)}{p(\theta^{t-1})p(\phi^{t-1})}.$$
- Set $\theta^t = \theta^*$ with probability min(r,1), else $\theta^t = \theta^{t-1}$.

The overall process allows it to move quite rapidly through the parameter space, and it can work well where other approaches such as Gibbs might be very slow. However, it is limited in the sense that it only works in those applications where gradient exists and can be computed in reasonable time.

16.4.5 Other MCMC Algorithms, Variations, and Approximate Methods

There are many other MCMC algorithms that are available, some of which are modifications of the above algorithms. For example, within this Metropolis–Hastings approach, there are variations such as slice sampling, reversible jump, particle filtering, and so on. Also, one can reparameterize the model to help overcome some convergence issues if applicable. In addition, there exist many approximate methods such as variational Bayes, approximate Bayesian computation, and so on. The main point here is just to be familiar with what is out there in case it might be useful. Any particular method might be well suited to certain models, particularly those that are notably complex, or they may just be convenient for a particular case. It is also not uncommon for some software to employ more than one algorithm. Some of the various algorithms are included in the table below.

Adaptive Directional Metropolis-within-Gibbs	Multiple-Try Metropolis
Adaptive Griddy-Gibbs	No-U-Turn Sampler
Adaptive Hamiltonian Monte Carlo	Preconditioned Crank–Nicolson
Adaptive Metropolis	Oblique Hyperrectangle Slice Sampler
Adaptive Metropolis-within-Gibbs	Random Dive Metropolis-Hastings
Adaptive-Mixture Metropolis	Random-Walk Metropolis
Affine-Invariant Ensemble Sampler	Reflective Slice Sampler
Automated Factor Slice Sampler	Refractive Sampler
Componentwise Hit-And-Run Metropolis	Reversible-Jump
Delayed Rejection Adaptive Metropolis	Robust Adaptive Metropolis
Delayed Rejection Metropolis	Sequential Adaptive Metropolis-within-Gibbs
Differential Evolution Markov Chain	Sequential Metropolis-within-Gibbs
Elliptical Slice Sampler	Slice Sampler
Griddy-Gibbs	Stochastic Gradient Langevin Dynamics
Hamiltonian Monte Carlo with Dual-Averaging	Tempered Hamiltonian Monte Carlo
Hit-And-Run Metropolis	t-walk
Independence Metropolis	Univariate Eigenvector Slice Sampler
Interchain Adaptation	Updating Sequential Adaptive Metropolis-within-Gibbs
Metropolis-Adjusted Langevin Algorithm	Updating Sequential Metropolis-within-Gibbs
Metropolis-Coupled MCMC	Metropolis-within-Gibbs

16.5 Software for Bayesian Analysis

16.5.1 Open Source Bayesian-Specific Software

16.5.1.1 BUGs, WinBUGs, and OpenBUGs

BUGS

BUGS (Bayesian inference Using Gibbs Sampling) began as a project in 1989 at the Medical Research Council (MRC) Biostatistics Unit with the aim of making practical MCMC methods available to applied statisticians. It is a general purpose software for fitting arbitrarily simple and complex Bayesian models using MCMC methods. The idea was to develop a flexible software for Bayesian analysis for complex statistical models using MCMC methods. It was first released in 1993 and was the only program available that would allow users to fit Bayesian models easily. Some key options of BUGS include exact means of propagating uncertainty in graphical structures, understanding that simulation methods could be used for inference, and recognizing that object-oriented programming could be exploited to generalize the simulation algorithm. Although BUGS initially only used fairly specialized algorithms, efforts were made to expand the software's capabilities and at later stages additional functionality was implemented including non-linear models. The development of a stand-alone Windows version (WinBUGs discussed below) of the software gave it lots of momentum. In subsequent versions, a number of other challenging models were tackled, including spatial models, dynamic models (involving differential equations), and variable-dimension models (fitted using reversible jump MCMC). In BUGS, the user specifies a statistical model by simply stating the relationships between related variables. The software includes an "expert system", which determines an appropriate MCMC scheme based on the Gibbs sampler for analyzing the specified model. The user then controls the execution of the scheme and is free to choose from a wide range of output types.

WinBUGS

WinBUGS is the Windows version of the original BUGS which has facilitated the growth of Bayesian statistics by making these methods more accessible computationally. It is currently available as a free stable software with no further development. It is the most widely used Bayesian software tool with many published papers referencing it, and these cut across many applications including food safety, forestry, mental health policy, clinical trials, population genetics, pharmacokinetics, and other diverse fields. It can also be accessed from within other software including SAS, Stata, and R. It is highly recommended for both simple and complex Bayesian analyses, with the caveat that users require knowledge of both Bayesian methods and issues in MCMC WinBUGS syntax must be incorporated appropriately to obtain desired results. It is well documented with plenty of practical resources, see for example WinBUGS Home (2018), WinBUGS Examples A (2018), WinBUGS Examples B (2018), and WinBUGS Examples C (2018).

WinBUGS enables the user to specify a Bayesian model either by using code in R-like syntax or via drawing directed graphs called DoodleBUGS. Examples can be found on software website and the tool itself. WinBUGS syntax must be incorporated appropriately in a specific set of steps to obtain results for the desired model. The software then determines the transition kernel for a Markov chain to generate samples from the joint posterior distribution of the unknown quantities in the model using either a graphical user interface (GUI) or a script, the user specifies the number of parallel MCMC chains to be run, the number of iterations, the model unknowns to monitor for analysis and reporting, and the types of convergence assessment and output summaries. The final result is numeric and graphical summaries of the estimated univariate marginal posterior distributions of the requested model quantities. The choice of initial values may affect convergence substantially. Other decisions include choice to run a single Markov chain or several independent chains initialized at different values and the selection of the number of early iterations to discard before the sampler is judged to have converged closely enough to the target distribution to provide reasonable inference. Furthermore, because the samples produced by a Markov chain are correlated, a larger number of samples are required for a desired degree of accuracy in estimation than would be the case with independent samples.

OpenBUGS

All of the BUGS development is now centered around the open source OpenBUGS software (OpenBUGS Home, 2018). The development team of BUGS has now shifted their attention more toward OpenBUGS, thus making the direct access to WinBUGS available to the public indefinitely which no longer requires the users to access the free key as part of their annual registration process. WinBUGS and OpenBUGS somewhat diverged in some functionality, each with its own advanced features unavailable in the other. OpenBUGS has progressed from being an experimental to a stable and reliable package with ongoing development efforts. OpenBUGS therefore represents the future of the BUGS project whereas WinBUGS, on the other hand, is an established and stable, stand-alone version of the software, which will remain available but not further developed. The latest version of OpenBUGS has been designed to be at least as efficient and reliable as WinBUGS over a wide range of test applications. One of the main differences between OpenBUGS and WinBUGS is the way in which the expert system makes its decisions. WinBUGS defines one algorithm for each possible computation type whereas there is no limit to the number of algorithms that OpenBUGS can make use of for much greater flexibility and extensibility.

16.5.1.2 JAGS

JAGS (Just Another Gibbs Sampler) was developed as an open-source program for analysis and statistical inference of Bayesian hierarchical models and was first released in 2003. It has since seen several subsequent versions. It was created to be compatible with the BUGS software, but it is arguably more extensible, flexible, and user-friendly than the BUGS software. It was essentially developed using the dialect of Bayesian inference Using Gibbs Sampling. The initial goal of JAGS was to be able to analyze all of the examples provided with classic BUGS. There are several reasons provided for choosing classic BUGS, rather than WinBUGS, as a target. These included the fact that classic BUGS had not been developed for a while and hence was not a moving target. Additionally, classic BUGS came with a large collection of example scripts that could be run on Linux and which could therefore be used as a test suite, and a third reason for choosing classic BUGS concerned the BUGS language itself. JAGS is written in C++ allowing an object-oriented style, which is extremely useful in this context. At the same time, it allows existing software written in C and FORTRAN.

JAGS uses MCMC algorithms to sample from probability distributions. JAGS was initially developed to be a clone of WinBUGS and OpenBUGS, while allowing its users additional flexibility to directly modify the program as in BUGS and hence the use of the BUGS language. However, unlike WinBUGS, it is able to run on platforms other than Windows (e.g., Unix, Linux, or OS X). Additionally, JAGS users are able to create their own modules to extend the program's capabilities. A JAGS module is a general term that can encompass various functions, distributions, and samplers—the latter of which refers to specific sampling algorithms (e.g., the Gibbs sampler) used with the MCMC estimation algorithm. Several JAGS modules have been created for use with the R software. JAGS does not have a built-in GUI. Instead, it can be run directly from the command line. However, since the base program uses C++ language, existing software written in C or FORTRAN (e.g., R, MATLAB, Python, or Stata) can be easily used in conjunction with JAGS. JAGS is most commonly called from the R programming environment, and it is designed to work closely with the R language. This was highlighted in the previous chapter (Depaoli et al., 2016; Gurrin et al., 2013; McElreath, 2016; Plummer, 2003).

16.5.1.3 STAN

Named in honor of Stanislaw Ulam, pioneer of the Monte Carlo method, Stan is a probabilistic programming language for specifying statistical models and software for describing data and models for Bayesian inference. A Stan program imperatively defines a log probability function over parameters conditioned on specified data and constants. Current versions of Stan provide full Bayesian inference for continuous-variable models through MCMC methods and make the required computation automatic using state-of-the-art techniques including automatic differentiation, Hamiltonian Monte Carlo and No-U-turn Sampler (NUTS), and automatic variational inference (ADVI). Penalized maximum likelihood estimates are calculated using optimization methods such as the limited memory Broyden–Fletcher–Goldfarb–Shanno algorithm. It is also a platform for computing log densities and their gradients and Hessians, which can be used in alternative algorithms such as variational Bayes, expectation propagation, and marginal inference using approximate integration. In this regard, Stan is set up so that the densities, gradients, and Hessians, along with intermediate quantities of the algorithm such as acceptance probabilities, are easily accessible. Stan also provides extensive diagnostics for the inference. Models written in Stan language are compiled to C++, which makes the inference faster and also allows easy portability to other languages. Stan already has interfaces for to other commonly used software including R, Python, and Stata which extends many commonly used statistical modeling tools, such as generalized linear models, providing options to specify priors and perform full posterior inference. These interfaces support sampling and optimization-based inference with diagnostics and posterior analysis and access to log probabilities, gradients, Hessians, parameter transforms, and specialized plotting. Additional details about Stan can be seen in the documentation from the Stan Development Team (2018), Carpenter (2017), Clark (2016 and 2018), Core Development Team (2018), and Farrell and Myers-Smith (2018).

16.5.1.4 MAMBA

MAMBA (2014) is a Julia package designed for general Bayesian model fitting via MCMC. Julia itself was designed for high performance and excels at numerical computing (Julia, 2017). Like OpenBUGS and JAGS, Mamba supports a wide range of model and distributional specifications, and provides a syntax for model specification. But unlike those two, and like PyMC (to be discussed later), Mamba provides a unified environment in which all interactions with the software are made through a single, interpreted language. Any Julia operator, function, type, or package can be used for model specification; and custom distributions and samplers can be written in Julia to extend the package. Conversely, interactions with and extensions to OpenBUGS and JAGS can involve three different programming environments—R wrappers used to call the programs, their domain specific language (DSLs), and the underlying implementations in Component Pascal and C++. Advantages of a unified environment include more flexible model specification; tighter integration with supplied functions for convergence diagnostics and posterior inference; and faster development, testing, and debugging of extensions. Advantages of the BUGS domain-specific modeling language include more concise model specification and facilitation of automated sampling scheme formulation. Indeed, sampling schemes must be selected manually in the initial release of Mamba. Nevertheless, Mamba holds other distinct advantages over existing offerings. In particular, it provides arbitrary blocking of model parameters and designation of block-specific samplers; samplers that can be used with the included simulation engine or apart from it; and command-line access to all package functionality, including its simulation application programming interface. Likewise, advantages of the Julia language include its familiar syntax, focus on technical computing, and benchmarks showing it to be one or more orders of magnitude faster than R and Python. Finally, the intended audience for Mamba includes individuals interested in programming in Julia who wish to have low-level access to model design and implementation and in some cases are able to derive full conditional distributions of model parameters up to normalizing constants (Smith, 2015).

16.5.2 Open Source General Bayesian Software

16.5.2.1 R

R has many tools for Bayesian analysis, see, for example, Park (2018). Some of this was also seen in the previous chapter where a basic introduction to the R programming language and probabilistic programming language JAGS with practical case examples on common models and topics such as prior elicitation and sample size was provided. From a Bayesian computation perspective, R programs for Bayesian analysis can be broken down into four broad categories, including:

- User written R programs—usually limited to a specific application
- Wholly R Bayesian Packages for performing set of Bayesian analyses
- R packages for linking to Bayesian-specific programs like WinBUGS, JAGS, and Stan
- R package tools for post-Bayesian analysis
- R packages for educational purposes.

There are also numerous R Shiny-based Bayesian tools and programs that are currently emerging having self-contained GUIs and applications within R.

User Written Packages

User-written programs tend to range from basic simple R programs to complicated programs that are geared at a specific task. A simple program could, for example, be one that performs a simple task such as comparing two binomial proportions, π_1 and π_2. A complex model could be one that implements the Berry and Berry (2004) approach in the analysis of adverse events, see, for example, Johnson (2009).

Wholly R Bayesian Packages

Wholly R Bayesian Packages are packages in R designed to perform various Bayesian analyses. The table below summarizes some of these R packages.

Package Name	Descriptions
MCMCpack	Provides model-specific MCMC algorithms for a wide range of models commonly used in the social and behavioral sciences. It contains R functions to fit a number of regression models (linear regression, logit, ordinal probit, probit, Poisson regression, etc.), measurement models (item response theory and factor models), changepoint models (linear regression, binary probit, ordinal probit, Poisson, panel), and models for ecological inference. It also contains a generic Metropolis sampler that can be used to fit arbitrary models.

(*Continued*)

Package Name	Descriptions
BayesianTools (BT)	This package supports model analysis (including sensitivity analysis and uncertainty analysis), Bayesian model calibration, as well as model selection and multi-model inference techniques for system models.
c212	This software package implements a number of methods for the detection of safety signals in Clinical Trials based on groupings of adverse events by body-system or system organ class. The methods include an implementation of the Three-Level Hierarchical model for Clinical Trial Adverse Event Incidence Data of Berry and Berry (2017), an implementation of the same model without the point mass (Model 1a from Xia et al. (2011)), and extended Bayesian hierarchical methods based on system organ class or body-system groupings for interim analyses. The package also implements a number of methods for error control when testing multiple hypotheses, specifically control of the False Discovery Rate (FDR). The FDR control methods implemented are the Benjamini–Hochberg procedure, the Double FDR, the Group Benjamini–Hochberg and subset Benjamini–Hochberg methods. Also included are the Bonferroni correction and the unadjusted testing procedure.
NIMBLE	A system for writing hierarchical statistical models largely compatible with "BUGS" and "JAGS", writing nimbleFunctions to operate models and do basic R-style math and compiling both models and nimbleFunctions via custom-generated C++. "NIMBLE" includes default methods for MCMC, particle filtering, Monte Carlo Expectation Maximization, and some other tools. The nimbleFunction system makes it easy to do things like implement new MCMC samplers from R, customize the assignment of samplers to different parts of a model from R, and compile the new samplers automatically via C++ alongside the samplers "NIMBLE" provides. "NIMBLE" extends the "BUGS"/"JAGS" language by making it extensible: New distributions and functions can be added, including as calls to external compiled code. Although most people think of MCMC as the main goal of the "BUGS"/"JAGS" language for writing models, one can use 'NIMBLE' for writing arbitrary, other kinds of model-generic algorithms as well

(*Continued*)

Package Name	Descriptions
RBest	Tool-set to support Bayesian evidence synthesis. This includes meta-analysis, (robust) prior derivation from historical data, operating characteristics, and analysis (one and two sample cases).

Learning Tools

Package Name	Description
LearnBayes	This package provides a collection of functions helpful in learning the basic tenets of Bayesian statistical inference. It contains functions for summarizing basic one- and two-parameter posterior distributions and predictive distributions. It provides MCMC algorithms for summarizing posterior distributions defined by the user. It also contains functions for regression models, hierarchical models, Bayesian tests, and illustrations of Gibbs sampling. The package was written by Jim Albert and is computational companion to his book titled *Bayesian Computation with R* (Albert, 2009).
AtelieR	This package provides an interface for teaching basic concepts in statistical inference and doing elementary Bayesian statistics (inference on proportions, multinomial counts, means and variances).
BaM	This package is an R package associated with Jeff Gill's book, *Bayesian Methods: A Social and Behavioral Sciences Approach, Second Edition* (CRC Press, 2007).
BayesDA	This package provides R functions and datasets for *Bayesian Data Analysis, Second Edition* (CRC Press, 2003) by Andrew Gelman, John B. Carlin, Hal S. Stern, and Donald B. Rubin.
Bolstad	This package contains a set of R functions and data sets for the book *Introduction to Bayesian Statistics*, by Bolstad, W.M. (2007).

Linking to Other Programs or Assessment of MCMC Diagnostics

Package Name	**Descriptions**
bayesplot	This R package provides an extensive library of plotting functions for use after fitting Bayesian models (typically with MCMC). Currently bayesplot offers a variety of plots of posterior draws, visual MCMC diagnostics, as well as graphical posterior predictive checking. Additional functionality (e.g., for forecasting/out-of-sample prediction and other inference-related tasks) will be added in future releases. The plots created by bayesplot are ggplot objects, which means that after a plot is created, it can be further customized using the various functions for modifying ggplot objects provided by the ggplot2 package. The idea behind bayesplot is not only to provide convenient functionality for users, but also a common set of functions that can be easily used by developers working on a variety of packages for Bayesian modeling, particularly (but not necessarily) those powered by RStan.
R2WinBUGS	This package provides convenient functions to call WinBUGS from R. It automatically writes the data and scripts in a format readable by WinBUGS for processing in batch mode, which is possible since version 1.4. After the WinBUGS process has finished, it is possible either to read the resulting data into R by the package itself, which gives a compact graphical summary of inference and convergence diagnostics, or to use the facilities of the coda package for further analyses of the output. Examples are given to demonstrate the usage of this package (Sturtz et al., 2005).
coda	This package includes a number of functions for analyzing posterior draws including functions for summarizing and plotting the output from MCMC simulations, as well as diagnostic tests of convergence to the equilibrium distribution of the Markov chain, see, for example, Myles (2010).
RStan	RStan is the R interface to the Stan C++ library for Bayesian estimation.
rstanarm	Estimates previously compiled regression models using the RStan package. Users specify models via the customary R syntax with a formula and data.frame plus some additional arguments for priors, see Ali (2017).

(*Continued*)

Package Name	Descriptions
r2OpenBugs	Using this package, it is possible to call a BUGS model, summarize inferences and convergence in a table and graph, and save the simulations in arrays for easy access in R.
rjags	This package provides an interface to the JAGS MCMC library.
R2jags	This package provides wrapper functions to implement Bayesian analysis in JAGS. Some major features include monitoring convergence of a MCMC model using Rubin and Gelman Rhat statistics, automatically running an MCMC model till it converges, and implementing parallel processing of an MCMC model for multiple chains.
BRugs	Fully interactive R interface to the "OpenBUGS" software for Bayesian analysis using MCMC sampling. Runs natively and stably in 32-bit R under Windows. Versions running on Linux and on 64-bit R under Windows are in "beta" status and less efficient.
iBUGS	Opens a widget for WinBUGS.
rube	WinBUGS (or JAGS) Enhancer. Makes working with WinBUGS much easier. Built on top of the R2WinBUGS package. Works with either the WinBUGS or the JAGS.
ShinyStan	ShinyStan provides visual and numerical summaries of model parameters and convergence diagnostics for MCMC simulations. ShinyStan can be used to explore the output of any MCMC program (including but not limited to Stan, JAGS, BUGS, MCMCpack, NIMBLE, emcee, and SAS). ShinyStan is coded in R using the Shiny web application framework (The Joint Development Team, 2017)
tidybayes	Compose data for and extract, manipulate, and visualize posterior draws from Bayesian models ("JAGS", "Stan", "rstanarm", "brms", "MCMCglmm", "coda", etc.) in a tidy data format. Functions are provided to help extract tidy data frames of draws from Bayesian models and to generate point summaries and intervals in a tidy format. In addition, "ggplot2" "geoms", and "stats" are provided for common visualization primitives like points with multiple uncertainty intervals, eye plots (intervals plus densities), and fit curves with multiple, arbitrary uncertainty bands.

(Continued)

Package Name	Descriptions
brms	Fits Bayesian generalized (non-)linear multivariate multilevel models using "Stan" for full Bayesian inference. Supports a wide range of distributions and link functions, allowing users to fit—among others—linear, robust linear, count data, survival, response times, ordinal, zero-inflated, hurdle, and even self-defined mixture models all in a multilevel context. Further modeling options include non-linear and smooth terms, auto-correlation structures, censored data, meta-analytic standard errors, and quite a few more. In addition, all parameters of the response distribution can be predicted in order to perform distributional regression. Prior specifications are flexible and explicitly encourage users to apply prior distributions that actually reflect their beliefs. Model fit can easily be assessed and compared with posterior predictive checks and leave-one-out cross-validation. For more details, see Burkner (2017).
MCMCVis	Performs key functions for MCMC analysis using minimal code—visualizes, manipulates, and summarizes MCMC output. Functions support simple and straightforward subsetting of model parameters within the calls, and produce presentable and "publication-ready" output. MCMC output may be derived from Bayesian model output fit with JAGS, Stan, or other MCMC samplers (Youngflesh, 2018).

R Shiny Tools

There is additionally a host of numerous R Shiny applications that have been developed to perform various tasks from Bayesian study design and analysis of both efficacy and safety data and for benefit risk. There are way too many of these applications for a comprehensive discussion within the current discussion.

16.5.2.2 Python

Like R, Python has many tools for Bayesian analysis and possessed these before Stan came around. Among the more prominent were those that allowed the use of BUGS (e.g., r2OpenBugs), one of its dialects JAGS (rjags), and packages like PyMC3 that allowed for customized approaches, further extensions, or easier implementation. Other packages might be regarded as being a specific type or family of models (e.g., bayesm), but otherwise be mostly Python R-like in specifying the model (e.g., MCMCglmm for mixed models). It is now easy to conduct standard and more complex models using Stan while staying within the usual framework of R-style modeling and, in some cases, may not require writing Stan code directly (Fonnesbeck, 2015; Wiecki, 2015).

Package Name	Description
PyMC3 Library	PyMC3 is a Python package for Bayesian statistical modeling and Probabilistic Machine Learning focusing on advanced MCMC and variational inference (VI) algorithms. Its flexibility and extensibility make it applicable to a large suite of problems (Salvatier et al., 2016, 2017).
BayesPy	BayesPy provides tools for Bayesian inference with Python. The user constructs a model as a Bayesian network, observes data, and runs posterior inference. The goal is to provide a tool which is not only efficient, flexible, and extendable enough for expert use but also accessible for more casual users. Currently, only variational Bayesian inference for conjugate-exponential family (variational message passing) has been implemented. Future work includes variational approximations for other types of distributions and possibly other approximate inference methods such as expectation propagation, Laplace approximations, MCMC, and other methods. It is based on the variational message passing framework and supports conjugate exponential family models. By removing the tedious task of implementing the variational Bayesian update equations, the user can construct models faster and in a less error-prone way. Simple syntax, flexible model construction, and efficient inference make BayesPy suitable for both average and expert Bayesian users. It also supports some advanced methods such as stochastic and collapsed variational inference (Luttinen, 2017).
Pystan	PyStan provides an interface to Stan. For more details, see the documentation at PyStan (2017).
SurvivalStan	SurvivalStan is a library of Survival Models written in Stan (Novik, 2017).
KCBO	This library implements some common Bayesian tests: the t-test for (independent groups and for an individual group), binomial test, Chi-square test, Poisson, Exponential, One-way ANOVA, Multifactor ANOVA, Linear Regression (Simple and Multiple), Logistic Regression, Ordinal Regression, and Power and Sample size planning (Hammond, 2017).

16.5.3 Commercial General Software with Bayesian Functionality

16.5.3.1 SAS

Bayesian computation with SAS can be done by using one of the following approaches: user written task-specific programs which range from simple to complex, use of SAS procedures with functionality for Bayesian analysis, use of the PROC MCMC procedure, and linking to Bayesian specific programs such as WinBUGS from within SAS. A simple example of a simple user-written program to perform Bayesian analysis can be seen in Nieto et al. (2011) applied in the context of hypothesis testing and a complex user-written program which can be seen in Gemperli (2010) who implemented the Berry and Berry (2004) hierarchical model for the analysis of frequency counts of adverse events. SAS procedures that include Bayesian functionality are GENMOD, LIFEREG, and PHREG where one can use the Bayes statement to obtain Bayesian analysis (see for example, Stokes et al. (2014). These procedures perform frequentist (or likelihood-based) analyses as a default, but the Bayes statement can be used to request a Bayesian analysis following the frequentist analysis using Gibbs sampling (or other MCMC sampling algorithms, with the default sampling method depending on the distribution of the data and model type). The default prior distribution for these fixed-effect parameters is the uniform distribution. However, in the GENMOD procedure, a special kind of uniform distribution is used in which the value of the distribution (i.e., density function) is 1 at all values of the parameters also known as the improper prior distribution because when a true (i.e., proper) distribution is integrated over all values, the result is a value equal to 1. With an improper distribution, the integration across all values results in a value that is greater than 1 (and usually equal to infinity). Two other procedures in SAS that are used for Bayesian analysis include the MCMC procedure which is a general purpose Bayesian simulation procedure that uses MCMC techniques to fit a wide range of Bayesian models, the BCHOICE procedure which fits Bayesian discrete choice models, and the FMM procedure which fits statistical models to data for which the distribution of the response is a finite mixture of distributions. The various SAS procedures are briefly discussed below.

Procedure Name	**Descriptions**
PROC GENMOD	This provides Bayesian analysis for the following distributions: binomial, gamma, inverse-Gaussian, negative binomial, normal, and Poisson and can work with the following link functions: identity, log, logit, probit, complementary log-log, and power. The procedure currently does not provide Bayesian analysis for multinomial distributions and does not include links for

(*Continued*)

Procedure Name	Descriptions
	cumulative complementary loglog (CCLL), cumulative logit (CLogit), and probit. The model parameters are the regression coefficients and dispersion (or the precision or scale) parameter, if the model has one. This procedure provides Bayesian analysis for distributions like binomial, gamma, Gaussian, normal, and Poisson. It also provides Bayesian analysis for links like identity, log, logit, probit, etc. In a Bayesian analysis, the model parameters are treated as random variables, and inference about parameters is based on the posterior distribution of the parameters, given the data.
PROC LIFEREG	This procedure provides analyses for parametric lifetime models for the following distributions: exponential, three-parameter gamma, log-logistic, log-normal, logistic, normal, and Weibull. The procedure does not currently provide Bayesian analysis for the binomial distribution. The model parameters are the regression coefficients and dispersion (or the precision or scale) parameter, if the model has one. The procedure fits parametric models to data that can be uncensored, right censored, left censored, or interval censored. The models for the response variable consist of a linear effect composed of the covariates and a random disturbance term. Bayesian analysis of parametric survival models can be requested by using the Bayes statement in the LIFEREG procedure.
PROC PHREG	This procedure provides analysis for Bayesian semiparametric survival models for the Cox regression models and uses the partial likelihood as the likelihood (Sinha et al., 2003), time-independent and time-dependent, all TIES= methods and can also handle piecewise exponential models. The Bayesian functionality in PROC PHREG currently does not fit models with certain data constraints, for example, data that include recurrent events. Model parameters are the regression coefficients and hazards (piecewise exponential models). This procedure performs survival analysis of data. Many types of models have been used for survival data. Two of the more popular types of models are the accelerated failure time model (Kalbfleisch and Prentice, 2002) and the Cox proportional hazards model (Cox, 1972). The PHREG procedure performs a regression analysis of survival data based on the Cox proportional hazards model.
PROC MCMC	This is a general purpose Bayesian simulation procedure that uses MCMC techniques to fit a wide range of Bayesian models (Chen, 2015). It is relatively new in SAS and requires

(Continued)

Procedure Name	Descriptions
	specification of a likelihood function for the data and a prior distribution for the parameters. It enables analysis of data that have any likelihood or prior distribution as long as they are programmable using SAS DATA step functions. Using this procedure, one can declare the parameters in the model and assign the starting values for the Markov chain with the PARMS statements and can specify prior distributions for the parameters with the PRIOR statements. One can also specify the likelihood function for the data with the MODEL statements. It is also flexible and capable of handling more general models, including single-level or multilevel (hierarchical) models, linear or nonlinear models, such as regression, mixture, survival, ordinal multinomial, and so on. Outputs include the posterior mean, standard deviation, percentiles, equal-tail and highest posterior density (HPD) intervals, covariance-correlation matrices, and deviance information criterion (DIC). From the MCMC procedure, one can get Markov chain convergence diagnostics, including Geweke test, Heidelberger-Welch stationarity and half-width tests, Raftery-Lewis test, Posterior sample autocorrelations, Effective sample size (ESS), and Monte Carlo standard error (MCSE). The MCMC procedure also includes functionality for visualization, including graphical display of the posterior samples, trace plots (with optional smoothed mean curve), autocorrelation plot, and kernel density plot (with optional fringe plot). This procedure is a general purpose simulation procedure that uses MCMC techniques to fit Bayesian models. PROC MCMC draws samples from a random posterior distribution and uses these samples to approximate the data distribution. You need to specify only parameters, prior distributions, and a likelihood function.
PROC FMM	This procedure fits statistical models to data for which the distribution of the response is a finite mixture of distributions; that is, each response is drawn with unknown probability from one of several distributions. It enables the user to describe data with mixtures of different distributions so that one can account for underlying heterogeneity and address overdispersion. PROC FMM offers a wide selection of continuous and discrete distributions, and it provides automated model selection to help you choose the number of components. Bayesian techniques are also available for many analyses.

(*Continued*)

Procedure Name	Descriptions
PROC BCHOICE	This procedure fits Bayesian discrete choice models by using MCMC methods. The procedure's capabilities include the following model fits: multinomial logit, multinomial probit, nested logit, multinomial logit with random effects, and multinomial probit with random effects. This procedure performs Bayesian analysis for discrete choice models. Discrete choice models are derived under the assumption of utility-maximizing behavior by decision makers. When individuals are asked to make one choice among a set of alternatives, they usually determine the level of utility that each alternative offers.

Accessing Other Programs from SAS

It is also possible to call WinBUGS from within a SAS session to perform Bayesian analysis. For example, Zhang et al. (2008) provided a SAS macro that enables SAS to implement Bayesian analysis with WinBUGS as part of a standard set of SAS routines. Zhang provided examples of how to implement this by fitting a multiple regression model and a linear growth curve model. An additional application that they provided also demonstrates how to iteratively run WinBUGS inside SAS for Monte Carlo simulation studies. Their SAS code can be extended to accommodate many other models with only slight modification. This interface is of practical benefit in many aspects of Bayesian methods as it allows SAS users to benefit from the implementation of Bayesian estimation and it also allows the WinBUGS user to benefit from the data processing routines available in SAS.

Smith and Richardson (2007) also presented a SAS macro for remote execution of WinBUGS. This macro does the data handling and input and output from WinBUGS via SAS. It produces a column format data file and also writes a list format data file for constants. It then writes a script file to the WinBUGS directory referencing appropriate datafile, model file, init file, and log file names. The script then runs WinBUGS in batch mode and reads in the node statistics block from the log file. The user needs to specify the input and output file names and directory path as well as the statistics to be monitored in WinBUGS. The code works best for a model that has already been set up and checked for convergence diagnostics within WinBUGS. The obvious use of this macro is for running simulations where the inputs and output files all have the same name, but all that differs between simulation iterations is the input dataset.

16.5.3.2 STATA

Bayesian estimation in Stata is similar to standard estimation. To implement Bayesian analyses, one prefixes the estimation command with "bayes". The bayes prefix combines Bayesian features with regression models, including, logistic, ordered probit, multinomial logistic, Poisson, generalized linear, conditional logistic, zero-inflated, sample-selection, multilevel models, and others. One can also select from many prior distributions for model parameters or use default priors or even define new priors. Additionally, Stata can also link to WinBUGS and Stan (via the StataStan function). Additional details of Stata functionality for Bayesian analysis can be seen in the Stata reference document (STATA, 2016; Grant et al., 2016; Marchenko, 2016; Stan Development Team, 2017; Thompson, 2014; Thompson et al., 2016).

16.5.3.3 SPSS

Starting with Version 25, SPSS software implemented support for some Bayesian analysis (Peck, 2017). These included the following: One Sample and Pair Sample T-tests, Binomial Proportion tests, Poisson Distribution Analysis, Related Samples, Independent Samples T-tests, Pairwise Correlation (Pearson), Linear Regression, One-way ANOVA, and Log-Linear Regression. Additional details can be seen in the documentation at SPSS (2017).

16.5.3.4 MLwiN

MLwiN also includes functionality for Bayesian analysis. By default, MLwiN sets diffuse priors which can be used to approximate maximum likelihood estimation. There are two procedures that are used by MLwiN. These include Gibbs sampling that is used with Normal responses and the Metropolis–Hastings algorithm with normal or binary/proportion responses. Estimation is controlled using options that appear on the toolbar when MCMC is selected. For additional details of the functionality within MLwiN for Bayesian analysis, see Browne (2015).

16.6 Summary and Discussion

Computations are part and parcel of any statistical analysis and inference especially if it involves computational steps in extracting information, for estimation or prediction or something else, from raw data. We certainly have come a long way from the 19th century when Karl Pearson (1894: Contribution to the mathematical theory of evolution: Proc Trans R soc A, 185, 71-110) used some painstaking approach in estimating ratio in deriving the moment estimates of a mixture of two normal distributions to where we are today when information gets synthesized instantaneously.

Introduction of simulation in 1940s did lead to invention of faster and better computers. The use of probabilistic simulations predated the existence of a computer. The introduction of MCMC methods became part of the Bayesian toolbox that was preceded by computer-dependent tools such as bootstrap or the expectation-maximization (EM) algorithm. For most problems of interest,

Bayesian analysis requires integration over multiple parameters, making the calculation of a posterior intractable whether via analytic methods or standard methods of numerical integration. However, it is often possible to approximate these integrals by drawing samples from posterior distributions. Prior to the introduction of MCMC in Bayesian analysis, a portfolio of quadrature methods was in practice in handling high-dimensional integrals. Integration generally involves probability distributions in the integrand, which thereby suggests the simulation methods to be employed. Such pervasive use of simulation methods in science persists even today despite the well-known asymptotic of deterministic approaches to integration.

MCMC was invented soon after ordinary Monte Carlo at Los Alamos, one of the few places where computers were available at the time. Its appeal came from the fact that it is a stochastic procedure, unlike deterministic maximum-likelihood algorithms, that repeatedly generates random samples that characterize the distribution of parameters of interest.

MCMC methods have been in existence in the literature for almost as long as Monte Carlo techniques, but their impact on Statistics had not been truly felt until the publication by Gelfand and Smith (1990). In this pioneering paper, Gelfand and Smith were able to create new interest in the Gibbs sampler by providing its potential in plethora of statistical problems. In their follow-up paper (1992), Smith and Gelfand provided easier interpretation for Gibbs sampling by showing that one can draw samples from any distribution, provided one can provide all the conditional distributions of the joint distribution analytically.

The timelines of these papers can be appreciated more when the BUGS software was introduced to the public in 1991 at the Valencia meeting (2011 Statistical Science, Robert and Casella), which provided another compelling argument to adapting MCMC algorithms at a much larger scale. So, it is almost universal to say that the computing aspect of Bayesian calculations started with the introduction of BUGS software. With the introduction of R as a vehicle for newly developing methods of interactive data analysis in the early 1990s, Bayesian computations and its use had taken off in growth in various fields by multitude of practitioners. Over the past 25 years, Bayesian modeling and inferential techniques have been applied successfully to thousands of problems across a wide range of application domains. During this time, as described in previous sections, the research community has seen an explosion of computing tools for Bayesian calculations. As was stated in the previous sections, before deciding on what tool is appropriate for the application in hand, it is crucial to consider

- model design/interpretation (iterative nature of modeling)
- whether one should write their own code for posterior computing for convenience
- whether you have surveyed software that does automatic posterior inference.

As each application can bring its own constraints in terms of model dimensionality and complexity, data, inference, accuracy, and computing times, it is prudent and time-saving to have a clear goal to what needs to be accomplished before deciding on a specific software. With future in mind, it is also worthwhile to study how adaptable the computing methodologies and tools are and how well those tools can combine simulation, variational approximations, and optimization. Choice of software for Bayesian computations, as it stands now, may depend on

- stage of model development (debugging or mass production)
- scale of analysis
- documentation and online community
- R or Python as the primary data processing language.

In the Bayesian research community, we have gone past answering the standard question from others, "Where does the prior come from?", and we are now currently enjoying an explosion of computing tools that are developed to meet the challenges arising out of various scientific fields. With programming languages such as Stan and Python at our disposal, in the midst of R as an integrated suite of software facilities for data manipulation, calculation, and graphical display, the future looks very bright for Bayesian approach to hit prime time in various applications.

References

Albert, J. 2009. *Bayesian Computation with R.* Springer: New York.

Ali, I. 2017. A brief introduction to RStanARM. http://imadali.net/bayesian-statistics/rstanarm/intro/.

Berry, S. M. and D. A. Berry. 2004. Accounting for multiplicities in assessing drug safety: A three-level hierarchical mixture model. *Biometrics* **60**: 418–26.

Bittl, J. A. and Y. He. 2017. Bayesian analysis: A practical approach to interpret clinical trials and create clinical practice guidelines. *Circ. Cardiovasc. Qual. Outcomes* doi:org/10.1161/CIRCOUTCOMES.117.003563.

Bolstad, W. M. 2007. *Introduction to Bayesian Statistics.* John Wiley & Sons: Hoboken, NJ, ISBN 978-1-118-09156-2.

Browne, W. J. 2015. MCMC estimation in MLwiN. www.bris.ac.uk/cmm/media/software/mlwin/downloads/manuals/2-32/mcmc-web.pdf.

Burkner, P.-C. 2017. brms: An R package for Bayesian multilevel models using Stan. *J. Stat. Software* **80**: 1–28.

Carpenter, B., A. Gelman, M. D. Hoffman, D. Lee, B. Goodrich, M. Betancourt, M. A. Brubaker, J. Guo, P. Li, and A. Riddell. 2017. Stan: A probabilistic programming language. *J. Stat. Software* **76**: 1–32.

Chen, F. 2015. Bayesian computation using PROC MCMC. www.bayes-pharma.org/wp-content/uploads/2014/10/FangChen_MCMC.pdf.

Clark, M. 2016. Bayesian basics. https://m-clark.github.io/docs/IntroBayes.html#HMCexample.

Clark, M. 2018. Bayesian basics. https://m-clark.github.io/bayesian-basics/.

Core Development Team. 2018. Stan: Probabilistic programming language. http://andrewgelman.com/wp-content/uploads/2015/07/darpa-ppaml-2015.pdf.

Cox, D. R. 1972. Regression models and life-tables. *J. R. Stat. Soc.* **34**:187–220.

Depaoli, S., J. P. Clifton, and P. R. Cobb. 2016. Just Another Gibbs Sampler (JAGS): Flexible software for MCMC implementation. *J. Educ. Behav. Stat.* **41**: 628–49.

Farrell, M. and I. Myers-Smith. 2018. Introduction to Stan: Getting started with Bayesian Farrell Myers-Smith. https://ourcodingclub.github.io/2018/04/17/stan-intro.html.

Fonnesbeck, C. 2015. Computational methods in Bayesian analysis in Python. https://plot.ly/ipython-notebooks/computational-bayesian-analysis/.

Gelfand, A. E. and Smith, A. F. M. 1990. Sampling-based approaches to calculating marginal densities. *J. Am. Stat. Assoc.* **85**: 398–409.

Gelman, A. and D. B. Rubin. 1996. Markov chain Monte Carlo methods in biostatistics. *Stat. Meth. Med. Res.* **5**: 339–55.

Gemperli, A. 2010. Safety analysis using Bayesian simulation methods in SAS 9.2. *Pharma. Programming* **3**: 29–36.

Grant, R. L., B. Carpenter, D. Furr, and A. Gelman. 2016. Introducing the StataStan interface for fast, complex Bayesian modeling using Stan. http://robertgrantstats.co.uk/papers/StataStan.pdf.

Green, P. J., K. Latuszynski, M. Pereyra, and C. P. Robert. 2015. Bayesian computation: A summary of the current state, and samples backwards and forwards. *Stat. Comput.* **25**: 835–62.

Gupta, S. 2012. Use of Bayesian statistics in drug development: Advantages and challenges. *Int. J. Appl. Basic Med. Res.* **25**: 3–6.

Gurrin, L., B. Carstensen, S. H. Jsgaard, and C. Ekstrm. 2013. Practical data analysis with JAGS using R. http://bendixcarstensen.com/Bayes/Cph-2012/pracs.pdf.

Hammond, H. 2017. KCBO – A Bayesian data analysis toolkit. http://blog.henryhhammond.com/kcbo-a-bayesian-data-analysis-toolkit/.

Hastings, W. K. 1970. Monte Carlo sampling methods using Markov chains and their applications. *Biometrika* **57**: 97–109.

Hobbs, B. P. and B. P. Carlin. 2008. Practical Bayesian design and analysis for drug and device clinical trials. *J. Biopharm. Stat.* **18**: 54–80.

Hoff, P. 2009. *A First Course in Bayesian Statistical Methods.* Springer Science and Business Media: New York.

Johnson, J. 2009. My implementation of Berry and Berry's hierarchical Bayes algorithm for adverse events. www.r-bloggers.com/my-implementation-of-berry-and-berrys-hierarchical-bayes-algorithm-for-adverse-events/.

The Joint Development Team. 2017. ShinyStan: Analysis and visualization GUI for MCMC. http://mc-stan.org/users/interfaces/shinystan.

Julia. 2017. The Julia programming language. https://julialang.org/.

Kalbfleisch, J. D. and R. L. Prentice. 2002. *The Statistical Analysis of Failure Time Data.* Wiley: New York.

Kruschke, J. K. and T. M. Liddell. 2018. The Bayesian new statistics: Hypothesis testing, estimation, meta-analysis, and power analysis from a Bayesian perspective. *Psychon. Bull. Rev.* **25**: 178–206.

Lazic, S. E., N. Edmundsa, and C. E. Pollard. 2016. Predicting drug safety and communicating risk: Benefits of a Bayesian approach. *Toxicol Sci.* **162**: 89–98 doi:10.1101/193169.

Luttinen, J. 2017. BayesPy – Bayesian Python. www.bayespy.org/.

Mamba. 2014. The Mamba package website. https://mambajl.readthedocs.io/en/latest/intro.html.

Marchenko, Y. 2016. Bayesian analysis using Stata. www.stata.com/meeting/germany16/slides/de16_marchenko.pdf.

McElreath, R. 2016. *Statistical Rethinking: A Bayesian Course with Examples in R and Stan.* CRC Press: Boca Raton, FL.

Metropolis, N., A. W. Rosenbluth, M. N. Rosenbluth, A. H. Teller, and E. Teller. 1953. Equations of state calculations by fast computing machines. *J. Chem. Phys.* **21**: 1087–92.

Myles, J. 2010. MCMC diagnostics in R with the coda package. www.johnmyleswhite.com/notebook/2010/08/29/mcmc-diagnostics-in-r-with-the-coda-package/.

Nieto, A., S. Extremera, and J. Gomez. 2011. Bayesian hypothesis testing for proportions. www.lexjansen.com/phuse/2011/sp/SP08.pdf.

Novik, J. 2017. Introducing SurvivalStan. www.hammerlab.org/2017/06/26/introducing-survivalstan/.

OpenBUGS Home. 2018. OpenBUGS homepage. www.openbugs.net/w/FrontPage.

Park, J. H. 2018. CRAN Task view – Bayesian inference. https://cran.r-project.org/web/views/Bayesian.html.

Peck, J. K. 2017. SPSS: The new SPSS statistics version 25 Bayesian procedures. https://developer.ibm.com/predictiveanalytics/2017/08/18/new-spss-statistics-version-25-bayesian-procedures/.

Plummer, M. 2003. JAGS: A program for analysis of Bayesian graphical models using Gibbs sampling. www.ci.tuwien.ac.at/Conferences/DSC-2003/.

PyStan. 2017. PyStan: The Python interface to Stan. https://pystan.readthedocs.io/en/latest/.

Quantcademy. 2017. Markov chain Monte Carlo for Bayesian inference – The Metropolis algorithm. www.quantstart.com/articles/Markov-Chain-Monte-Carlo-for-Bayesian-Inference-The-Metropolis-Algorithm.

Robert, C. P. 2014. Bayesian computational tools. *Annu. Rev. Stat. Appl.* **1**: 153–57.

Robert, C. P. 2016. The Metropolis-Hastings algorithm. https://arxiv.org/pdf/1504.01896.pdf accessed June 19, 2018.

Robert, C. P. and G. Casella. 2010. Introducing Monte Carlo Methods with R, use R: Chapter 6: Metropolis–Hastings Algorithms. DOI 10.1007/978-1-4419-1576-4_6.

Salvatier, J., T. V. Wiecki, and C. Fonnesbeck. 2016. Probabilistic programming in Python using PyMC3. *Comput. Sci.* doi:10.7717/peerj-cs.55.

Salvatier, J., T. V. Wiecki, and C. Fonnesbeck. 2017. Getting started with PyMC3. http://docs.pymc.io/notebooks/getting_started.

Sinha, D., M.-H. Chen, and J. G. Ibrahim. 2003. Bayesian inference for survival data with a surviving fraction. In *Crossing Boundaries: Statistical Essays in Honor of Jack Hall*, edited by J. E. Kolassa and D. Oakes, Vol. 43, pp. 117–38. Lecture Notes–Monograph Series. Institute of Mathematical Statistics: Beachwood, OH. doi:10.1214/lnms/1215092394.

Smith, A. F. M. and Gelfand, A. E. 1992. Bayesian statistics without tears: A sampling-resampling perspective. *Am. Stat.* **46**: 84–88.

Smith, B. J. 2015. Mamba.jl documentation. https://media.readthedocs.org/pdf/mambajl/julia-0.4/mambajl.pdf.

Smith, M. K. and H. Richardson. 2007. WinBUGSio: A SAS macro for the remote execution of WinBUGS. *J. Stat. Software* **23**: 1–10.

SPSS. 2017. Bayesian statistics. https://www.ibm.com/support/knowledgecenter/en/SSLVMB_25.0.0/statistics_mainhelp_ddita/spss/advanced/idh_bayesian.html.

STATA. 2016. Stata Bayesian analysis reference manual release 15. www.stata.com/manuals/bayes.pdf.

Stokes, M., F. Chen, and F. Gunes. 2014. An introduction to Bayesian analysis with SAS/Stat. https://support.sas.com/resources/papers/proceedings14/SAS400-2014.pdf.

Sturtz, S., U. Liggesy, and A. Gelman. 2005. R2WinBUGS: A package for running WinBUGS from R. *J. Stat. Software* **12**: 1–16.

Stan Development Team. 2017. StataStan: The Stata interface to Stan. http://mc-stan.org.

Stan Development Team. 2018. RStan: The R interface to Stan. http://mc-stan.org/.

Thompson, J. 2014. Bayesian analysis with STATA: Application to neonatal mortality in the UK. http://fmwww.bc.edu/repec/usug2014/thompson_uksug14.pdf.

Thompson, J., T. Palmer, and S. Moreno. 2016. Bayesian analysis in Stata using WinBUGS. http://citeseerx.ist.psu.edu/viewdoc/download?doi=10.1.1.110.7503&rep=rep1&type=pdf.

Walley, R. J., C. L. Smith, J. D. Gale, and P. Woodward. 2015. Advantages of a wholly Bayesian approach to assessing efficacy in early drug development: A case study. *Pharma. Stat.* **14**: 205–15.

Wijeysundera, D. N., P. C. Austin, J. E. Hux, W. S. Beattie, and A. Laupacis. 2009. Bayesian statistical inference enhances the interpretation of contemporary randomized controlled trials. *J. Clin. Epid.* **62**: 13–21.

Wiecki, T. 2015. While my MCMC gently samples Bayesian modeling, data science, and Python. https://twiecki.github.io/blog/2015/11/10/mcmc-sampling/.

WinBUGS Examples A. 2018. Volume 1. www.mrc-bsu.cam.ac.uk/wp-content/uploads/WinBUGS_Vol1.pdf.

WinBUGS Examples B. 2018. Volume 2. www.mrc-bsu.cam.ac.uk/wp-content/uploads/WinBUGS_Vol2.pdf.

WinBUGS Examples C. 2018. Volume 3. www.mrc-bsu.cam.ac.uk/wp-content/uploads/WinBUGS_Vol3.pdf.

WinBUGS Home. 2018. WinBUGS Homepage. www.mrc-bsu.cam.ac.uk/software/bugs/the-bugs-project-winbugs/.

Xia, H. A., H. Ma, and B. P. Carlin. 2011. Bayesian hierarchical modeling for detecting safety signals in clinical trials. *J. Biopharm. Stat.* **21**: 1006–29.

Youngflesh, C. 2018. MCMCvis: Tools to visualize, manipulate, and summarize MCMC output. *J. Open Source Software* **3**: 640–50.

Zhang, Z., J. J. McArdle, L. Wang, and F. Hamagami. 2008. A SAS interface for Bayesian analysis with WinBUGS. *Struct. Equ. Model. Multidisciplinary J.* **15**: 705–28.

17

Considerations and Bayesian Applications in Pharmaceutical Development for Rare Diseases

Freda Cooner

Sanofi US

CONTENTS

17.1 Background

Since the establishment of Orphan Drug Act[1] in 1983, the FDA (Food and Drug Administration) Office of Orphan Products Development (OOPD) has been tasked to "advance the evaluation and development of products (drugs, biologics, devices, or medical foods) that demonstrate promise for the diagnosis and/or treatment of rare diseases or conditions" by providing incentives for sponsors to develop products for such diseases.[2] It significantly changed the landscape of rare disease clinical programs. In the 2017 New Drug Therapy Approvals Report[3] from FDA Center for Drug Evaluation and Research (CDER), it is reported that 18 out of 46 novel drugs (39%) were approved to treat rare or "orphan" diseases (Figure 17.1).

The OOPD provides orphan designation (or sometimes "orphan status") to drugs and biologics which are defined as those intended for the safe and effective treatment, diagnosis, or prevention of rare diseases/disorders that affect fewer than 200,000 people in the United States, or that affect more than

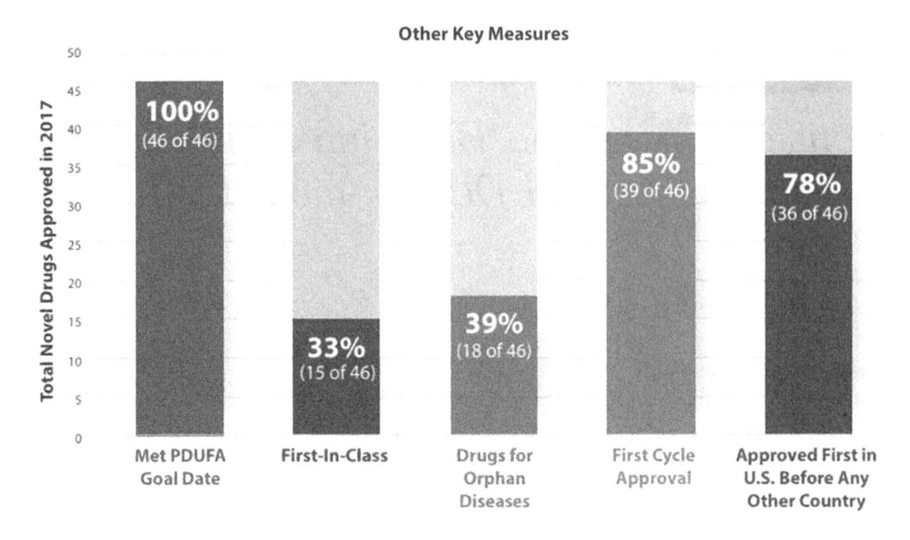

FIGURE 17.1
Key measures of FDA CDER Novel Drug Approvals in 2017. (Source: 2017 New Drug Therapy Approvals Report (FDA CDER).[3])

200,000 persons but are not expected to recover the costs of developing and marketing a treatment drug. A sponsor seeking orphan designation for a drug must submit a request for designation to OOPD with the information required in 21 CFR[4] 316.20 and 316.21. The granting of an orphan designation request does not alter the standard regulatory requirements and process for obtaining marketing approval. Safety and effectiveness of a drug must be established through adequate and well-controlled studies.

Although rare diseases patient population consists of roughly 70% pediatric patients,[5] rare diseases and pediatric patient populations are usually so distinct that different considerations should be taken for trial designs. It is worth mentioning that under Section 529 to the Federal Food, Drug, and Cosmetic Act (FD&C Act[6]), FDA awards priority review vouchers to sponsors of rare pediatric disease product applications that meet certain criteria. On September 30, 2016, the Advancing Hope Act of 2016[7] amended Section 529 of the FD&C Act. Among the changes, the term "rare pediatric disease" has been revised to mean a disease that meets each of the following criteria:

A. The disease is a serious or life-threatening disease in which the serious or life-threatening manifestations primarily affect individuals aged from birth to 18 years, including age groups often called neonates, infants, children, and adolescents.

B. The disease is rare disease or conditions, within the meaning of Section 526.

The Act changed the original definition: "The disease primarily affects individuals aged from birth to 18 years, including age groups often called neonates, infants, children, and adolescents."

These regulations and incentive programs have enabled the development and marketing of over 600 drugs and biologic products for rare diseases since 1983. However, it is unclear where the standard holds for the approval of orphan products, especially for rare diseases that affect far fewer people than 200,000. A few natural challenges for a clinical development program in rare disease population are small number of patients available for enrollment, lack of understanding of the disease, and site recruitment rate differences. Historical data collection, Bayesian framework, and innovative trial designs along with continuously refined efficient simulation tools can significantly alleviate these concerns, albeit there are no perfect solutions.

The statutory requirement for all drug approvals is the demonstration of substantial evidence of effectiveness in treating or preventing the condition and evidence of safety for that use. Substantial evidence of effectiveness should be obtained from one or more adequate and well-controlled studies, each convincing on its own. FDA acknowledges the difficulties for such requirements implementation in rare diseases and provides flexibility because of the many types and intended uses of drugs. In particular, FDA "exercise[s] its scientific judgment" in determining the kind and quantity of data a sponsor is required to provide for individual drug development programs.[8] This flexibility extends from early phase of development to design of adequate and well-controlled clinical studies required to demonstrate safety and effectiveness to support marketing approval.

Before considering detailed aspects of trial designs suitable for rare diseases, careful attention should be paid to endpoints selection and trial population identification.

- Choosing endpoints can be especially difficult for rare diseases due to lack of knowledge of disease progression. The endpoints for the clinical trial should be both feasible and clinically meaningful. They can be decided based on knowledge of either the nature of a rare disease derived from a natural history study or the pharmacological effects and mechanisms of action of the investigational drugs or interventions. Surrogate endpoints[9,10] that can be accurately measured earlier in follow-up than clinical endpoints and are thought to be reasonably likely to predict clinical benefit are often used in rare disease trials. Trial designs should be suitable for the endpoints of interest.

- It is well-understood that all clinical trials are somewhat enriched in terms of trial population selection. This is particularly critical to include the right population in the trial that could demonstrate the effect of investigational medicinal products. On the other hand, the limited patient population sets boundaries on how selected a trial could be in order to maintain trial integrity and to collect sufficient data forreasonable statistical comparisons. The FDA published a guidance document specifically on enrichment strategies,[11] and some of the designs, such as randomized withdrawal studies, have been implemented in rare disease clinical programs.[12]

Other regulatory health agencies, such as European Medicines Agency,[13] have published guidance document on this topic as well. Although the definition of rare diseases may be different, the general guidelines are remarkably similar across different nations.

17.2 Trial Designs

The limited number of patients in a rare disease population makes the size of the clinical trials conducted in that population inevitably small and that in turn restricts the usage of inferential statistics. With the usual requirements for significance level and study power, small population trials with a conventional design may only be able to detect very large intervention effects. It is therefore necessary to practice flexibility in establishing evidence of effectiveness and safety in rare diseases, and there are many ways this can be accomplished.[14] Historically, FDA has allowed the following program options to provide development of a rare disease therapy[15]: (1) one adequate and well-controlled trial with supporting evidence from pharmacological data or other sources, (2) a single-arm, non-randomized, open-label trial, and (3) a single adequate and well-controlled trial with the significance level relaxed somewhat. Although these may be the only feasible approaches, careful consideration should be paid to the consequences, and other options should be explored when possible.

A control group is critical to the ability to estimate the true treatment effect of an investigational therapy.[16] There are three types of control groups typically considered for use in rare disease trials: historical controls, patient self-controls, and concurrent controls.

The source of historical control data may be scarce on its own for rare diseases. Patient-level data are often not available or not of the same quality as data from prospectively designed clinical trials FDA usually requires to support approval. Moreover, some historical data are from a different group of patients whose disease status or diagnoses cannot be verified. The standard of care may vary significantly at different times and different locations. It should also be noted that patients under any standard of care may not represent the natural course of the disease. All these elements present difficulties in the use of patient registry data or natural history studies as a comparator.

Using patient-self as its control means either comparing the outcome to the baseline value for the same patient or comparing the outcomes from two periods of a cross-over design. Changes from baseline values do not account for changes unrelated to treatment. One should also take into account the impact of regression to the mean, especially when only a more severe stage of the disease is represented in a trial. Without a proper control or knowledge of the natural history of the disease, accurately measuring the treatment effect could be challenging and/or spontaneous improvement may be misinterpreted as evidence of a treatment effect.

The benefit of a concurrent control group is that its use in comparative analyses eliminates the potential for bias due to differences in patient populations, time periods, locations, and regression to the mean associated with the usage of historical or patient self-controls. Furthermore, it allows consistent reporting of prognostic factors in data collection. Although the benefit is obvious, it is universally recognized that including a concurrent placebo control group may not always be ethical for rare disease trials. Other options, such as the use of a very low dose of the investigational treatment or standard of care as a concurrent control, may be more practical.

Once the control group is selected, some clinical trial features, such as randomization and blinding, should also be weighed against ethical and logistical considerations in rare disease clinical trials to ensure interpretability of the trial results. Additional attention should be paid to the data collection process to ensure data quality because of the limited sources of data on rare diseases. Alternative trial designs[17] and more effective operational aspects[18] should be considered.[19]

Adaptive and Bayesian trial designs[20] have caught much attention since the 21st Century Act (Cures Act)[21] signed into law on December 13, 2016, which designated a section for modern trial design and evidence development that discussed utilizing both novel clinical trial designs and real-world evidence. The PDUFA VI[22,23] also initiated a pilot program[24] "for highly innovative trial designs." While adaptive designs are among the most popular novel trial designs, Bayesian framework can readily incorporate real-world evidence in data evaluation. Moreover, with Bayesian predictive probabilities and proper simulation tools, one can envision the course of trials at the design stage to more readily solicit and incorporate inputs from clinical and pharmacologic groups.

17.3 Trial Conduct

With limited subjects' availability and sometimes a more complicated trial design, holding trial conduct to a rigorous standard[25] is critical for small sample size trials, where achieving a high level of accuracy is essential. First and most important component in trial conduct is to maximize data quality. This goal requires obtaining accurate and precise measurements of key variables through the implementation of standardized data collection processes. The overarching goal is to minimize bias in data ascertainment and reduce the measurement errors of key variables in order to increase the precision of estimates. Other aspects that may aid trial conduct are training and monitoring. Use of a well-organized data monitoring committee (DMC)[26,27] to implement interim analyses can enhance trial conduct and efficiency. Moreover, central training of clinic personnel on data acquisition and data entry and clinical monitoring and auditing of data collection can reduce measurement errors and variabilities.

17.4 Data Analysis

Once data are collected according to an appropriate trial design and standardized procedures, the small sample sizes make it essential to optimize data analysis strategies. Efficient statistical models and analysis methods should be considered for effect estimation and hypothesis testing. For example, use of repeated measurements models may have advantages over analyses that are based only on one-time observation in a trial. Similarly, time-to-event analyses may be considered in lieu of analyses of event proportions. If the primary outcome is based on a continuous measurement and not the occurrence of a discrete event, then analyzing the outcome with a linear model (e.g., ANCOVA) will provide efficiencies compared to determining a threshold for response and then conducting a responder analysis. Covariate adjustment in a statistical model may also minimize variability of study outcomes and increase the ability to detect a clear signal of efficacy. The covariates to be included in the model should be both statistically and clinically meaningful and pre-specified in the protocol.

Sensitivity analyses should be planned to assess the impact of assumptions required for the analysis to be valid on the trial's results. In particular, assumptions that are difficult to verify, e.g., assumptions about the missing data mechanism, should be evaluated with appropriate sensitivity analyses. Different efficacy endpoints should also be explored as a part of sensitivity analyses consideration, which will benefit future clinical programs in rare diseases.

For trials with small sample sizes, standard statistical methods based on large-sample theory often do not apply, thus making the results and conclusions of the analysis less reliable. Also, modern trials often involve several subgroups, such as gene-mutation subgroups, with subgroup sample sizes even smaller. In such situations, Bayesian analysis methods may be appropriate as they allow borrowing information from homogeneous subgroups, while discounting information from heterogeneous subgroups, regardless of subgroup sample sizes.[28] In addition, when historical or external information is available; for example, through historical controls, early phase trials, or some observational studies, Bayesian methods naturally lead to the incorporation of the historical/external information through the application of Bayes theorem in a way that allows for some of the historical/external data that are obsolete or less useful to be discounted.[29]

With that said, it is crucial to keep in mind that early and frequent conversation with the regulatory health authorities could be the key to the final approval of effective medicinal products. While it is preferable to conduct many exploratory analyses at the end of a trial in order to better understand rare diseases progresses, patient populations, and mechanisms of action of the investigational drugs, one should always pre-specify a primary efficacy analysis, however difficult it is, for orphan product clinical programs. Regulatory interactions should consult the draft FDA guidance[30] on common issues in drug development for rare diseases.

17.5 Bayesian Application

Bayesian framework and statistics can be implemented throughout the lifecycle of a medicinal product. During compound development period, Bayesian statistics can offer a reasonable summary of the existing real-world data or external information, and provide predictive probabilities of trial success rate to assist the critical go/no-go decision.[31] At the proof-of-concept and dose ranging/dose selection stage, Bayesian framework can provide a more efficient trial in terms of duration and size.[32,33] Once decisions are made to proceed to registry confirmatory clinical program, Bayesian framework can incorporate all the data collected in a clinical program and provide a more robust treatment effect estimate. After the product approved for market use, Bayesian statistics can better aid safety evaluations with continuous data collection.[34] It should be recognized that Bayesian framework is an efficient venue to maximize the available information systematically. Hence, it should be considered for all rare disease pharmaceutical development programs, including many oncology programs,[35] where efficiency is imperative.

There are two main aspects of a clinical trial where Bayesian statistics have advantages over traditional frequentist statistics. One is to augment the current trial by historical data through prior distribution or other mechanism, such as hierarchical modeling. Augmenting placebo or control group[36] and increasing estimates' precision have a wider acceptance than leveraging external active treatment group information and shifting treatment effect away from the null hypothesis where no treatment effect is assumed. The other aspect is to utilize Bayesian statistics during a trial, typically in an adaptive design, to modify trial features, such as randomization ratio,[37] target patient population defined by biomarkers,[33] or reducing number of treatment groups. The second implementation is often practiced when historical data are insufficient or sometimes non-existing. Besides these two main aspects, Bayesian estimates are being considered for treatment effect as they can morereadily take subgroup effects into account[27] and design simulation for operational characteristics demonstration.

17.6 Case Examples

1. Natural History Control

Utilizing natural history control is the most conventional option and deservedly has been widely implemented for orphan product clinical programs. Recently, the FDA approved the first treatment (Brineura; cerliponase alfa) for a specific form of Batten disease, neuronal ceroid lipofuscinosis type 2 (CLN2) disease, based on a non-randomized, single-arm dose escalation clinical trial with a natural history control.[38]

Batten disease is a fatal rare disease of the nervous system that typically has onset of symptoms in childhood and causes worsening problems with vision, movement, and thinking ability.[39,40] It is usually referring to a group of disorders known as neuronal ceroid lipofuscinoses (NCLs), to which CLN2 disease belong. CLN2 disease is also known as tripeptidyl peptidase-1 (TPP1) deficiency. In the late infantile form of the CLN2 disease, signs and symptoms typically begin between ages 2 and 4. The initial symptoms usually include language delay, recurrent seizures (epilepsy), and difficulty coordinating movements (ataxia). Affected children also develop muscle twitches (myoclonus) and vision loss. CLN2 disease affects essential motor skills, such as sitting and walking. Individuals with this condition often require the use of a wheelchair by late childhood and typically do not survive past their teens. Batten disease collectively is relatively rare, occurring in an estimated two to four of every 100,000 live births in the United States.[41]

Brineura[42] is an enzyme replacement therapy developed by BioMarin Pharmaceutical Inc. Its active ingredient (cerliponase alfa) is a recombinant form of human TPP1. Brineura is administered into the cerebrospinal fluid (CSF) by infusion via a specific surgically implanted reservoir and catheter in the head (intraventricular access device). It is the first approved treatment to slow loss of walking ability (ambulation) in symptomatic pediatric patients 3 years of age and older with late infantile CLN2, with recommended dose of 300 mg administered once every other week, followed by an infusion of electrolytes. The complete Brineura infusion, including the required infusion of intraventricular electrolytes, lasts approximately 4.5 h. Pre-treatment of patients with antihistamines with or without antipyretics (drugs for prevention or treatment of fever) or corticosteroids is recommended 30–60 min prior to the start of the infusion.

The efficacy of Brineura[43] was established in a non-randomized, single-arm, open-label, dose escalation clinical study in 23 symptomatic pediatric patients with CLN2 disease who completed the study and compared to 42 untreated patients with CLN2 disease from a natural history cohort (an independent historical control group), extracted from DEM-CHILD database, who were at least 3 years old and had motor or language symptoms. A two-item version of the CLN2 rating scale (i.e., Motor and Language domains) was the instrument to collect the primary efficacy data of clinician-rated CLN2 scores in both clinical study and natural history cohort. The originally proposed primary efficacy endpoint was proportion of patients with an absence of an unreversed (sustained) 2-point rate (slope) of decline or a score of 0 in the Motor–Language total score over 48 weeks. It was later rejected by the FDA, and instead, the data on the Motor domain only over a longer period of 72–96 weeks were used as the basis of the statistical considerations. As usual, the data quality of the natural historical control and its comparability to the current trial were under the microscope. The data issues were the main reason for the FDA to seek a more reliable and readily interpretable efficacy endpoint. After full deliberation, taking into account age, baseline walking ability, and genotype, the FDA was able to confirm that Brineura-treated patients demonstrated fewer declines in walking ability compared to untreated patients in the natural history cohort.

This application adopted traditional usage of historical control group in a clinical trial and frequentist methods. Bayesian framework could be introduced to alleviate the concerns surfaced during >>FDA review. First of all, the data collected on the Language domain can be easily bridged in the efficacy assessment through a proper utility function or Bayesian model assuming language domain data as external or additional information. Then, as the historical data quality is questionable, Bayesian framework could more systematically summarize all available historical data instead of selecting "matching" subjects from the historical data pool. Last but not least, using Bayesian modeling and predictive probabilities, one can potentially simulate a control group using historical data as prior information and adjusting the degree of borrowing by different data comparability assumptions. Such proposals can be generalized to any single-arm trial comparing to a historical control group or objective performance criteria.

The safety of Brineura was evaluated in 24 patients with CLN2 disease aged three to 8 years who received at least one dose of Brineura in clinical studies. The safety and effectiveness of Brineura have not been established in patients less than 3 years of age.[44]

2. Innovative Trial Design

Facing many regulatory challenges, innovative trial designs have been slowly pushed to the main stage to tackle the resources issues prevalent in drug development programs for rare diseases. Dominantly Inherited Alzheimer Network Trials Unit (DIAN-TU) program[45] is one of very few adaptive platform trials that have been implemented in registry clinical programs. Observational trials have been initiated since 2009 to better understand a rare form of Alzheimer's disease, dominantly inherited Alzheimer. Many pharmaceutical companies and research institute have joined this program, and DIAN-TU has launched the world's first prevention trial for this disease.

Alzheimer's is a brain disease that causes a slow decline in memory, thinking, and reasoning skills.[46,47] It is a progressive disease and the most common cause of dementia. Although Alzheimer's greatest known risk factor is aging, it is not just a disease of old age. There is currently no cure for Alzheimer's, and there are several treatments available for the symptoms. Researches are ongoing for this deliberating disease. Dominantly inherited Alzheimer's disease, also known as autosomal-dominant Alzheimer's disease, represents less than 1% of all cases of Alzheimer's and typically manifests before the age of 60.[48]

The DIAN-TU Pharma Consortium is a public–private partnership formed in 2010 to facilitate collaborative among industry, patient community, research institutes, and health agencies in order to address key Alzheimer's prevention trial design challenges in a population almost certain to develop Alzheimer's disease—the autosomal-dominant Alzheimer's disease population. This Alzheimer's population was chosen for clinical trial researches due to its almost certain risk of developing Alzheimer's disease and the predictability of the age at symptom onset. However, there is still much unknown in both biomarkers and mechanisms of potential treatments.

A seamless phase 2/3 double-blind, randomized, parallel-group 2-year biomarker trial began in 2012 by testing two drugs, from two different pharmaceutical companies, against a pooled placebo group that has transitioned to a 4-year Adaptive Prevention Trial in 2014. As these two treatment groups' enrollment concluded in 2015, the DIAN-TU platform launched two new drug arms as the Next Generation (NexGen) Prevention Trial,[49] which is a phase 3 trial with similar design but 4-year treatment period. The DIAN-TU trial is now operational internationally. The early trials provided critical information in primary endpoint (a cognitive composite) selection, primary analysis model (a disease progression model[50]) validation, and valuable database of dominantly inherited Alzheimer's disease population to potentially identify predictive biomarker for Alzheimer's disease.

The disease progression model used in this program is a mixed-effects model through a decline function of estimated years from symptom onset (EYO) as the predictive variable. The potential treatment effect on the cognition decline rate was to be estimated through the declined function that combines the natural decline function and the effect of the intervention. The effect of the intervention was measured through a log-progression rate. The complexity of this model demands the usage of simulation for parameters estimation as opposed to analytical approaches. Bayesian framework was introduced to streamline the simulation process that not only to provide treatment effect estimate through posterior distributions but also to demonstrate the operational characteristics required by the regulatory health agencies. Prior distributions used in the simulation were non-informative, and historical data were only used to develop the disease progression model (Figure 17.2).

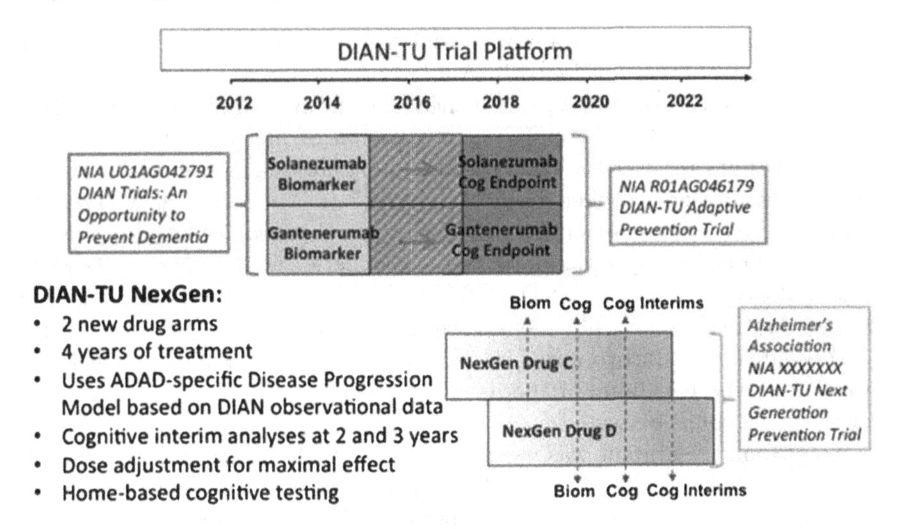

FIGURE 17.2
Existing and proposed structure of the DIAN-TU trial platform including the Next Generation Prevention Trial. (Source: Bateman RJ, Benzinger TL, Berry S, et al. The DIAN-TU Next Generation Alzheimer's prevention trial: Adaptive design and disease progression model.[49]) Light grey indicates enrollment of the first two arms, medium grey the biomarker phase, and dark grey denotes graduation to a cognitive endpoint.

Besides platform design with a pooled placebo group, another critical component of DIAN-TU trials is the adaptive feature where dosing was to be adjusted with interim combined safety/biomarker-target analyses to maximize biomarker-target engagement. Formal interim analyses were also proposed for both early efficacy stopping and futility assessment. One could assume that Bayesian predictive probabilities were to be implemented in interim analyses to assist decision-making.

The DIAN-TU platform trials are dedicated to Alzheimer's disease researches with biomarker adaptations and many innovative clinical trial design aspects, including novel endpoints and analysis models. These trials have been and will continue to benefit from these operational and analytical innovations in saving both time and cost.

3. Bayesian Methodology

Bayesian statistics naturally and systematically borrowing strength from previously collected data. However, it has been slowly adopted into registry clinical programs due to clinical and statistical inertia. Through the last two case examples, we have seen the potential of using Bayesian statistics to maximize the information at hand and simplify simulation process for trial design. Because of Ebola virus disease (EVD) outbreak in West Africa 2014–2016,[51,52] a much needed more efficient adaptive trial design was used in the investigation of a new drug ZMapp developed by Mapp Biopharmaceutical.[53]

The PREVAIL II trial was launched within weeks after the trial drug became available. Study population is patients of any age who had positive test results for EBOV infection on a polymerase-chain-reaction (PCR) assay. Patients were enrolled from March 2015 to November 2015. The primary objective was to determine whether the combination of ZMapp plus the current standard of care was both safe and superior to the current standard of care alone in managing EVD. A feature of this adaptive design was that an investigational agent that was subsequently shown to have activity against EVD could then be incorporated into an evolving standard of care, which was provided as the backbone of therapy in each trial group and against which newer agents could be tested.

The primary endpoint was mortality at day 28. It was estimated that 100 patients per group would be needed for the trial to have 88% power to detect a 50% relative difference in mortality between the two groups, assuming that the 28-day mortality in the group receiving the current standard of care alone was 40%.

Interim and final analyses were performed with the use of a Bayesian approach. In particular, a skeptical prior distribution, which is in favor of smaller effect versus larger ones and assumes equal results between two treatment groups, was formulated for the treatment effect. Imposing this prior on the trial data, a posterior probability distribution for treatment effect was then assessed for the probability that ZMapp plus the current standard of care result in lower 28-day mortality than the current standard of care alone. A posterior probability of 97.5% or more was required to establish efficacy. Key analyses were summarized with both posterior probabilities and 95% credible intervals.

Out of the 72 patients (36 per treatment group) enrolled into the trial, 71 patients completed the trial and were evaluated. An overall case fatality rate of 30% was observed per 21 death occurred. There were 13 deaths out of 35 patients who received the current standard of care alone and 8 deaths out of 36 patients who also received ZMapp. The crude estimates of 28-day mortality were 37% and 22%, respectively. The Bayesian estimate of the absolute difference in mortality between the ZMapp group and the group that received the standard of care alone was −14%, and the relative difference was −38%. These mortality differences gave a 91.2% posterior probability that ZMapp plus the current standard of care was superior to the current standard of care alone, which was below the pre-specified probability threshold of 97.5%. The 95% credible interval for the absolute difference in mortality was −34% to 6%, and the 95% credible interval for the relative risk of death was 0.29–1.24 (Figures 17.3 and 17.4).

An independent data and safety monitoring board convened seven times during the relatively short duration of this trial. Due to the increasing success of numerous public health measures in reducing and aiming to extinguish

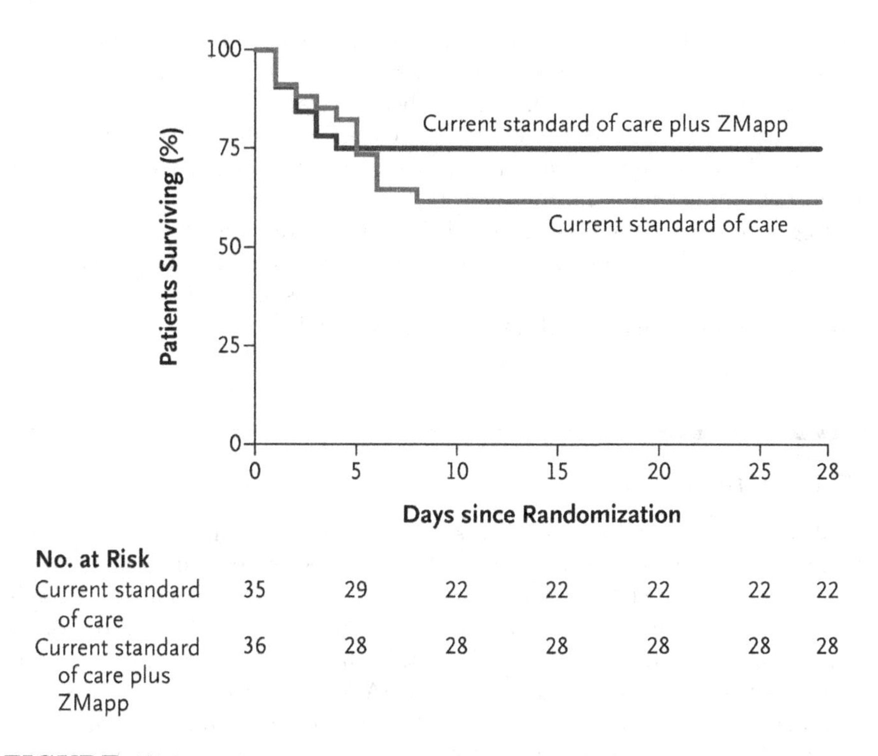

FIGURE 17.3
Kaplan–Meier plot of survival, according to the two assigned treatment groups. (Source: The PREVAIL II Writing Group. A Randomized, Controlled Trial of ZMapp for Ebola Virus Infection.[53])

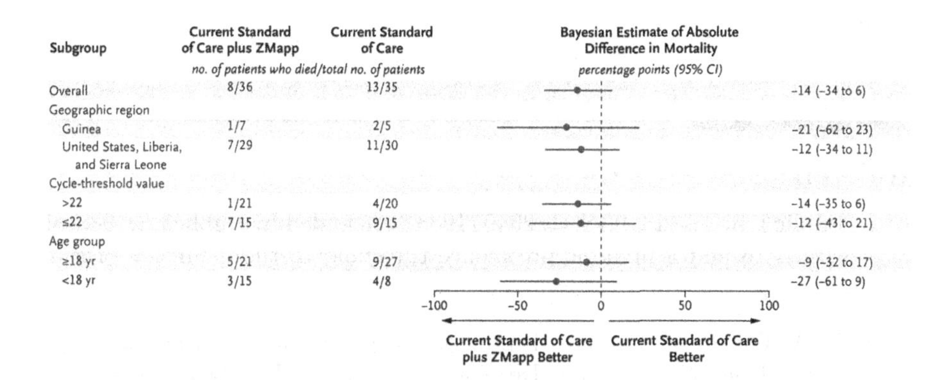

FIGURE 17.4
Forrest plot of absolute difference between groups in 28-day mortality, overall and according to subgroup. (Source: The PREVAIL II Writing Group. A Randomized, Controlled Trial of ZMapp for Ebola Virus Infection.[53])

EVD, the board endorsed a plan for stopping the trial if the already low incidence of EVD did not increase in late 2015. On January 29, 2016, after Liberia, Sierra Leone, and Guinea had been declared nearly Ebola-free, the trial was closed and the data were unblinded. As a result, the trial was unable to attain the originally planned 100 patients. The findings of this trial, albeit promising, are arguably inconclusive.

Regardless, PREVAIL II trial established the advantage of Bayesian framework in conducting a more efficient randomized controlled clinical trial in the face of challenges in population size, time, and costs. It set a precedent for Bayesian trial designs in confirmatory clinical trials.

17.7 Summary

Small population trial in rare diseases poses unique challenges in both trial designs and data analyses, while it also provides opportunities for innovative designs and methods. The FDA not only provides regulatory flexibilities but also has led great efforts in research and practice for rare disease clinical programs. Recently, the FDA established complex innovative designs pilot program as an action following PDUFA VI,[18] calling for sponsor's early submissions and discussions with the FDA regarding innovative clinical trial designs. It is reasonable to anticipate that more trials utilizing unconventional designs and Bayesian methodologies will surface. There remains unexplored area to discover and develop in small population trial designs and analyses; understanding a paradigm shift such as a new clinical trial landscape is never easy. After all, the progress is encouraging and with an increasing pace especially in orphan product clinical programs with regulatory health agencies' support.

References

[1] Public Law 97-414, 96 Stat. 2049 (1983). Amended by Public Law 98-551 (1984) to add a numeric prevalence threshold to the definition of rare diseases.

[2] FDA Office of Orphan Products Development (OOPD) webpage www.fda.gov/ForIndustry/DevelopingProductsforRareDiseases Conditions/default.htm.

[3] 2017 New Drug Therapy Approvals Report (CDER) www.fda.gov/downloads/AboutFDA/CentersOffices/OfficeofMedicalProductsand Tobacco/CDER/ReportsBudgets/UCM591976.pdf.

[4] Electronic Code of Federal Regulation (CFR Title 21) www.ecfr.gov/cgi-bin/text-idx?SID=3ee286332416f26a91d9e6d786a604ab&mc=true&tpl=/ecfrbrowse/Title21/21tab_02.tpl.

[5] Stanford Research Pediatric and Rare Diseases webpage www.sanfordresearch.org/researchgroups/pediatricsandrarediseases/.

[6] FDA Federal Food, Drug, and Cosmetic Act (FD&C Act) webpage www.fda.gov/regulatoryinformation/lawsenforcedbyfda/federalfood drugandcosmeticactfdcact/default.htm.

[7] Advancing Hope Act of 2016 (Public Law No: 114-229) September 29, 2016 www.congress.gov/114/plaws/publ229/PLAW-114publ229.pdf.

[8] 21 CFR 314.105.

[9] FDA Drugs webpage (Table of Surrogate Endpoints That Were the Basis of Drug Approval or Licensure) www.fda.gov/Drugs/Development ApprovalProcess/DevelopmentResources/ucm613636.htm.

[10] FDA Drugs webpage (Surrogate Endpoint Resources for Drug and Biologic Development) www.fda.gov/Drugs/DevelopmentApproval Process/DevelopmentResources/ucm606684.htm.

[11] Guidance for Industry—Enrichment Strategies for Clinical Trials to Support Approval of Human Drugs and Biological Products (December 2012).

[12] Biaggioni I, Freeman R, Mathias CJ, Low P, Hewitt LA, Kaufmann H. Randomized withdrawal study of patients with symptomatic neurogenic orthostatic hypotension responsive to droxidopa. *Hypertension.* 2015; 65(1): 101–107. doi:10.1161/HYPERTENSIONAHA.114.04035.

[13] Committee for Medicinal Products for Human Use (CHMP)—Guideline on Clinical Trials in Small Populations (July 2006) www.ema.europa.eu/documents/scientific-guideline/guideline-clinical-trials-small-populations_en.pdf.

[14] Hilgers RD, König F, Molenberghs G, Senn S. Design and analysis of clinical trials for small rare disease populations. *Journal of Clinical Research on Rare Diseases.* 2016; 1(3): 53–60.

[15] Mitsumoto J, Dorsey ER, Beck CA, Kieburtz K, Griggs RC. Pivotal studies of orphan drugs approved for neurological diseases. *Annals of Neurology.* 2009; 66(2): 184–190. doi:10.1002/ana.21676.

[16] ICH E10 Choice of Control Group in Clinical Trials (2001) www.ema.europa.eu/docs/en_GB/document_library/Scientific_guideline/2009/09/WC500002925.pdf.

[17] Abrahamyan L, Diamond IR, Johnson SR, Feldman BM. A new toolkit for conducting clinical trials in rare disorders. *Journal of Population Therapeutics and Clinical Pharmacology.* 2014; 21(1): e66–e78.

[18] Whicher D, Philbin S, Aronson N. An overview of the impact of rare disease characteristics on research methodology. *Orphanet Journal of Rare Diseases.* 2018; 13(1): 1. doi: 10.1186/s13023-017-0755-5.

[19] Institute of Medicine. 2001. *Small Clinical Trials: Issues and Challenges.* Washington, DC: The National Academies Press. doi:10.17226/10078.

[20] Facey K, Granados A, Guyatt G, Kent A, Shah N, van der Wilt GJ, Wong-Rieger D. Generating health technology assessment evidence for rare diseases. *International Journal of Technology Assessment in Health Care.* 2014; 30(4): 416–422. doi: 10.1017/S0266462314000464. Epub 2014 Nov 19.

[21] Cures Act (Public Law 114-255) December 13, 2016 www.congress.gov/114/plaws/publ255/PLAW-114publ255.pdf.

[22] PDUFA VI (PDUFA Reauthorization Performance Goals and Procedures Fiscal Years 2018 through 2022) www.fda.gov/downloads/forindustry/userfees/prescriptiondruguserfee/ucm511438.pdf.

[23] FDA PDUFA VI webpage www.fda.gov/forindustry/userfees/prescription druguserfee/ucm446608.htm.

[24] Federal Register on 08/30/2018 for Complex Innovative Designs Pilot Meeting Program [Docket No. FDA-2018-N-0049] https://federalregister.gov/d/2018-18801.

[25] Poolman RW, Hanson B, Marti RK, Bhandari M. Conducting a clinical study: A guide for good research practice. *Indian Journal of Orthopaedics.* 2007; 41(1): 27–31. doi: 10.4103/0019-5413.30522.

[26] Guidance for Clinical Trial Sponsors—Establishment and Operation of Clinical Trial Data Monitoring Committees (March 2006) www.fda.gov/downloads/RegulatoryInformation/Guidances/ucm127073.pdf.

[27] Yao B, Zhu L, Jiang Q, and Xia HA. Safety monitoring in clinical trials. *Pharmaceutics.* 2013; 5(1): 94–106. doi: 10.3390/pharmaceutics5010094.

[28] Jones HE, Ohlssen DI, Neuenschwander B, Racine A, Branson M. Bayesian models for subgroup analysis in clinical trials. *Clinical Trials.* 2011; 8(2): 129–143. doi: 10.1177/1740774510396933. Epub 2011 Jan 31.

[29] Viele K, Berry S, Neuenschwander B, et al. Use of historical control data for assessing treatment effects in clinical trials. *Pharmaceutical Statistics.* 2014; 13(1): 41–54. doi:10.1002/pst.1589.

[30] Guidance for Industry—Rare Disease: Common Issues in Drug Development (Draft; August 2015) www.fda.gov/downloads/Drugs/Guidance ComplianceRegulatoryInformation/Guidances/ucm458485.pdf

[31] Ibrahim JG, Chen MH, Lakshminarayanan M, Liu GF, and Heyse JF. Bayesian probability of success for clinical trials using historical data. *Statistics in Medicine.* 2015; 34(2): 249–264. doi:10.1002/sim.6339.

[32] Basu C, Ahmed MA, Kartha RV, et al. A hierarchical Bayesian approach for combining pharmacokinetic/pharmacodynamic modeling and Phase IIa trial design in orphan drugs: Treating adrenoleukodystrophy with Lorenzo's oil. *Journal of Biopharmaceutical Statistics.* 2016; 26(6): 1025–1039. doi: 10.1080/10543406.2016.1226326.

[33] Miller F, Zohar S, Stallard N, et al. Approaches to sample size calculation for clinical trials in rare diseases. *Pharmaceutical Statistics*. 2018; 17(3): 214–230. doi: 10.1002/pst.1848. Epub 2018 Jan 10.

[34] Sharrar RG, Dieck GS. Monitoring product safety in the postmarketing environment. *Therapeutic Advances in Drug Safety.* 2013; 4(5): 211–219. doi:10.1177/2042098613490780.

[35] Bogaerts J, Sydes MR, Keat N, et al. Clinical trial designs for rare diseases: Studies developed and discussed by the International Rare Cancers Initiative. *European Journal of Cancer.* 2015; 51(3): 271–281. ISSN 0959-8049, doi:10.1016/j.ejca.2014.10.027.

[36] Hampson LV, Whitehead J, Eleftheriou D, Brogan P. Bayesian methods for the design and interpretation of clinical trials in very rare diseases. *Statistics in Medicine.* 2014; 33(24): 4186–4201. doi:10.1002/sim.6225.

[37] Williamson SF, Jacko P, Villar SS, Jaki T. A Bayesian adaptive design for clinical trials in rare diseases. *Computational Statistics & Data Analysis.* 2017;113:136–153. doi:10.1016/j.csda.2016.09.006.

[38] FDA News Release (April 27, 2017) www.fda.gov/NewsEvents/Newsroom/PressAnnouncements/ucm555613.htm.

[39] Brain Foundation—Batten Disease http://brainfoundation.org.au/disorders/batten-disease.

[40] Batten Disease Support and Research Association website www.bdsra.org.

[41] NIH Batten Disease Fact Sheet www.ninds.nih.gov/Disorders/Patient-Caregiver-Education/Fact-Sheets/Batten-Disease-Fact-Sheet.

[42] Brineura website www.brineura.com/.

[43] FDA reviews for Brineura (cerliponase alfa) Injection (Application No. 761052; Approval Date April 27, 2017) www.accessdata.fda.gov/drugsatfda_docs/nda/2017/761052Orig1s000TOC.cfm.

[44] Brineura (cerliponase alfa) Injection Label (April 2017) www.accessdata.fda.gov/drugsatfda_docs/label/2017/761052lbl.pdf.

[45] The Dominantly Inherited Alzheimer Network website https://dian.wustl.edu/.

[46] Alzheimer's Association website www.alz.org/alzheimers-dementia/what-is-alzheimers.

[47] NIH Alzheimer's Disease Fact Sheet www.nia.nih.gov/health/alzheimers-disease-fact-sheet.

[48] Bateman RJ, Aisen PS, De Strooper B, et al. Autosomal-dominant Alzheimer's disease: A review and proposal for the prevention of Alzheimer's disease. *Alzheimer's Research & Therapy.* 2011; 3(1): 1. doi:10.1186/alzrt59.

[49] Bateman RJ, Benzinger TL, Berry S, et al. The DIAN-TU Next Generation Alzheimer's prevention trial: Adaptive design and disease progression model. *Alzheimer's & Dementia: The journal of the Alzheimer's Association.* 2017; 13(1): 8–19. doi:10.1016/j.jalz.2016.07.005.

[50] Wang G, Berry S, Xiong C, et al. A novel cognitive disease progression model for clinical trials in autosomal-dominant Alzheimer's disease. *Medicine.* 2018; 37: 3047–3055.

[51] Report of an International Commission. Ebola haemorrhagic fever in Zaire, 1976. *Bulletin of the World Health Organization.* 1978; 56(2): 271–293.

[52] Piot P, Muyembe JJ, Edmunds WJ. Ebola in West Africa: From disease outbreak to humanitarian crisis. *Lancet Infectious Diseases.* 2014; 14: 1034–1035.

[53] The PREVAIL II Writing Group. A Randomized, Controlled Trial of ZMapp for Ebola Virus Infection. *New England Journal of Medicine* 2016; 375:1448–1456. doi: 10.1056/NEJMoa1604330.

18

Extrapolation Process in Pediatric Drug Development and Corresponding Bayesian Implementation for Validating Clinical Efficacy

Margaret Gamalo-Siebers and Simin Baygani

Eli Lilly & Co.

CONTENTS

18.1 Introduction

Legislations in the United States (US) and Europe require, monitor, and encourage the conduct of pediatric studies. In the US, for example, the US Congress enacted the Pediatric Research Equity Act (PREA) [1] requiring pediatric assessment whenever a sponsor submits an application to market a new active ingredient, new indication, new dosage form, new dosing regimen, or new route of administration [2]. Similarly, the European Medicines Agency (EMA) Paediatric Regulation (EC) No 1901/2006 [3] requires clinical studies for pediatric drug development in the European Union (EU) for all new products and line extensions (new indications, new formulation, new dosage form, etc.) for existing products. Other regulations also provide incentives to sponsors for doing studies in pediatric patients, i.e., a 6-month extension of market exclusivity (per Section 505(A) of Food and Drug Administration Modernization Act (FDAMA) or Best Pharmaceuticals for Children Act (BPCA) [4]) or supplementary protection certificate (SPC) (per EC No 1901/2006 [3]) as a reward for the expense of conducting studies in pediatric patients, even if the study results do not support an indication (see Penkov et al. [5]).

While the enactment of the aforementioned pediatric regulations addressed the economic challenges of pediatric drug development, they have not provided sponsors with sufficient headway to overcome many other challenges. Completion of many paediatric studies required under the Paediatric Regulation are generally delayed [6]. In a recently published article in the journal *Pediatrics*, researchers from Harvard Medical School and Boston Children's Hospital estimated that up to half of pediatric clinical trials are abandoned or never published based on a retrospective, cross-sectional study of 559 pediatric randomized clinical trials (RCTs) registered in ClinicalTrials.gov from 2008 to 2010 that were completed by 2012 [7]. The authors also found that nearly one out of five trials (104) ended early, primarily because investigators could not attract enough children to participate in the first place (e.g., 36 trials were withdrawn before recruitment began). While lack of enrollment is often a problem with adult trials, pediatric investigations face unique challenges. There simply are not as many children as adults. Just 19% of the US population is younger than 14 years of age [8]. Furthermore, children are less likely than adults to suffer serious illnesses making most childhood diseases rare.

On the methodological aspects (e.g., technical, ethical, and logistical considerations), even when data on drugs successfully tested on adults are available, there may still be lack of information on the dosing, safety, and/or efficacy of these drugs in children. Previous research demonstrates that children may not respond to medications in the same way as adults (see, e.g., Selevan et al. [9] and Miller et al. [10]). Differences between children and adults include ways in which medicines are absorbed, distributed, metabolized, and excreted (ADME) by the body, as well as what the medicines do to the body. There is also the ethical requirement which precludes pediatric research that poses higher risks and does not offer a prospect of direct benefit see [11,12]. This implies that for "higher risk" interventions, clinical investigations (e.g., proof-of-concept studies in adults) must have been conducted to show anticipated direct benefits to the child. Ethics also has implications on the logistics of implementing the trial. For example, there might be no active control available for pediatric use while withholding treatment may pose risk of serious harm, vaccination schedules might preclude them from trial participation, or the disease may have low incidence in children. Different drug formulations in children need to be stable (temperature, reliable drug release) with proof of bioequivalence and with an acceptable palatability. Regardless of the difficulty of patient enrollment, drugs for children are subject to the same statistical evidentiary standards for efficacy and safety as drugs in adults (see 21CFR 314.50 [13] and PHS Act, 505(d) [14]), and that Institutional Review Boards (IRBs) should find that "the research is likely to generate vital knowledge about the children's disorder or condition" [2]. Both of these statements indicate enrolling a sufficient number of patients to answer the scientific question of interest adequately. These challenges diminish the feasibility of conducting clinical trials of the size necessary to demonstrate statistical significance by traditional means and contribute to the long time lapse between initial adult label and pediatric label updates, which on average is 9–10 years between the initial adult approval and pediatric approval [15].

Under the law, children are considered a vulnerable group and as such are granted additional protection as research subjects [16]. In particular, and as stated earlier, research is permitted if the level of risk is no greater than minimal, regardless of whether there is a prospect of direct benefit to the child. Furthermore, if the answer to the scientific question can be obtained by enrolling adult subjects in a clinical trial, pediatric subjects should not be exposed to those research risks (cf. *scientific necessity* [17]). Hence, the use of extrapolation reduces "the amount of, or general need for, additional information (types of studies, design modifications, number of patients required) needed to reach conclusions" [11]. Thus, there is a moral obligation to build the foundation for the use of pediatric extrapolation and related innovative analytical strategies with appropriately designed adult clinical trials. Harmonization of methodologies and strategies into overall drug development plans is needed to improve the speed of access to new drugs for pediatric patients.

This chapter is structured as a combination of pediatric drug development process and statistical methodology. Section 2 covers extrapolation strategy, extrapolation process, including the plan, concept, and validation, and extrapolation types. Section 3 discusses the applicability of Bayesian methodology within the framework of extrapolation and the incorporation of a "validative" approach within process. Section 3 further provides a discussion on related references, e.g., Gamalo-Siebers et al. [18] for a review of Bayesian methods to the extrapolation of adult data to support drug approvals in a pediatric population. This review mentions both hierarchical models Gelman et al. [19], power prior methods [20], and provide two worked out examples of the methods.

In addition, the review also mentions the hybrids of the above two methods (see [21]), as well as the commensurate prior [22] which in the case of a single prior study can be shown to be equivalent to using a hierarchical model to construct a prior. Schoenfeld et al. [23] proposed a variation of a hierarchical model where the adult treatment response and pediatric treatment response are considered exchangeable. In their variation, there could be several adult studies which are assumed *exchangeable*, and whose treatment effects come from a common population with population mean effect exchangeable with the treatment effect from the pediatric population. Loosely, *exchangeability* means that one could not distinguish the studies only by looking at the study clinical efficacy results because there is nothing known a priori that would imply a certain rank ordering of treatment effects. In general, there could be several pediatric studies which are assumed exchangeable among themselves as well. Section 4 discusses regulatory implications and ways of optimizing pediatric drug development programs within a investigational/study plan. Lastly, Section 5 provides a short conclusion to tie up the arguments in the manuscript.

18.2 Extrapolation

One way to maintain feasibility of conducting trials in children while achieving both efficacy requirements and reducing risks to pediatric trial subjects and/or spare children from unnecessary clinical studies is to extrapolate efficacy conclusion about the drug and the disease indication from adults or older age pediatric groups. In particular, *extrapolation* from adult is an approach to provide evidence in support of effective and safe use of drugs in the pediatric population when it can be assumed that the course of the disease and the expected response to a medicinal product would be sufficiently similar in the pediatric and reference population (e.g., adult or other pediatric). The EMA defines extrapolation formally as "extending information and conclusions available from studies in one or more subgroups of the patient population (source population(s)), or in related conditions or with related medicinal products, to make inferences for another subgroup of the population (target population), or condition or product, thus reducing the need to generate additional information (types of studies, design modifications, number of patients required) to reach conclusions for the target population, or condition or medicinal product" [24]. While the FDA does not define extrapolation nor the process of extrapolation explicitly, it gives a set of rules on the extent of pediatric development work needed. This relies on fundamental assumptions such as (1) that the course of the disease is similar in adults, (2) the expected response to therapy is similar to adults, and (3) the effective pediatric dose can be established. When these assumptions are scientifically justified, the FDA may conclude that safety and effectiveness may be supported by effectiveness data in adults together with additional data such as dosing, pharmacokinetic (PK), and safety data in pediatric patients [25]. From these definitions, extrapolation provides rational interpretation of the limited evidence, i.e., it maximizes use of existing information and avoid unnecessary studies and in so doing, it accelerates timely access to safe and effective medications by increasing the speed and efficiency. Hence, a concept that is center to extrapolation is the generation of just the right amount of data to fill the knowledge gap to target (pediatric) population, i.e., the development of a framework for reduction of the required evidence generated in the target population in accordance with the predicted degree of similarity to the source population [24].

In all circumstances, the regulatory agency considers that the pharmaceutical companies (henceforth called sponsors) should bring forward the evidence supporting the extrapolation concept for any disease and target population, instead of the agency determining beforehand whether such extrapolation is generally endorsed. The burden of evidence gathering to justify whether that process (concept, plan, validation, and mitigation of risks) is thus transferred to all applicants individually and to have early discussions about the pediatric plan with respective agencies to reduce uncertainty in the process. But while a reduction in the amount of data may be agreed, it is still possible that extrapolation may not be the correct decision at the end of the trial when the results do not validate the assumption. The approval of extrapolation is a regulatory decision whether there is sufficient information to warrant the conduct of a "leaner" development program in the target population and the transfer of conclusion and communicated through an initially agreed Pediatric Investigational Plan (PIP), Pediatric Study Plan (PSP), or an issued Written Request (WR). In fact, decisions about the extent of extrapolation, for example with the FDA, have evolved overtime further illustrating why there are disease indications with full/partial/no extrapolation that have been submitted and agreed as a development program. For example, epilepsy and pain went from partial to full extrapolation [26]. Over the last decade, the FDA has tested its assumptions about extrapolation and modified its approaches as knowledge and experience increased. Products moved from no extrapolation to partial extrapolation and *vice versa* which illustrates a fluid process where both the experiences and new science inform variations in approach to leveraging of information from the source population.

18.2.1 Extrapolation Process

The EMA Extrapolation Reflection paper [24] and ICH E11 (R1) give a process for extrapolation to provide a systematic approach on the use of extrapolation in pediatric drug development, including the framework for extrapolation [27]. While there might be differences in the application of this approach across competent regional authorities, it is desirable to harmonize acceptable techniques to the extent possible to develop a framework that

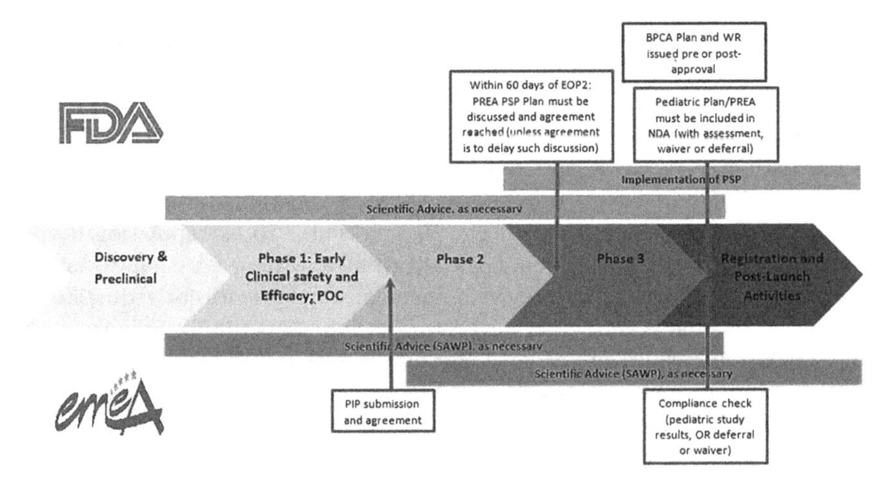

FIGURE 18.1
Submission timeline of required pediatric plans in the EU and US.

provides the basis for an explicit and systematic approach to extrapolation to support pediatric drug development in instances where target populations fulfill extrapolation criteria. This framework would consist of several key steps, to be considered in sequence across the drug development program.

Note that, in the US, the FD&C Act requires sponsors to submit an initial PSP (iPSP) within 60 days after the completion of the Phase 2 study (see Figure 18.1 for when these documents are required in relation to the adult product development). In addition, there is no extrapolation process in the US but an expectation of what type of data package will be required to extend label to the target population based on certain understanding of the disease process and response to therapy. Sponsors may amend the extent of the proposed data package in the iPSPs at any time prior to approval of the new drug application (NDA) [28]. On the other hand, in the EU, sponsors are required submit proposals for PIPs to the European Medicines Agency's Paediatric Committee (PDCO) "no later than the completion of adult human pharmacokinetic studies." The development plan can be modified at a later stage as knowledge increases or if the applicant encounters difficulties with the implementation of a PIP, which render it unworkable or no longer appropriate [29].

18.2.1.1 Extrapolation Concept

Extrapolation starts with the hypothesis or prediction that there is similarity in disease progression and response to intervention in the target population being studied with the source population from which efficacy conclusion is extrapolated from. However, this assumption is not automatic and requires initial substantiation and subsequent validation. In substantiating the assumption, there are a number of questions that must be answered which include whether (1) there is similarity in disease pathogenesis or underlying cause and in tissue findings (histopathologic); (2) there is similarity in presentation or manifestation of the disease (signs and symptoms) characteristics; (3) there is similarity in the criteria for disease diagnosis and classification; and (4) there is similarity in the response to therapy among all the drugs, including standard of care, that provide therapeutic benefit in the indication. To answer these questions and make predictions in the target population, it is helpful to look at all available data sources relevant to their product, the condition it treats, the populations affected, and what information can it provide toward the target population. In fact, initial PK, pharmacodynamics (PD) and or PK/PD predictions through modeling can be characterized through these data. In addition, a systematic assessment and synthesis of available data from clinical trials and potentially other relevant evidence from clinical practice (e.g., from other pediatric age groups, related pediatric indications, adult indication for (similar) pediatric indication, real world evidence or historical or placebo controls (pediatric)) can also provide context for extrapolation. In particular, the review can provide information on the response of the drug and of placebo or standard of care. Source information to justify extrapolation concept on clinical response may also come from products to treat the same indication, other formulations of same active ingredient, surrogate endpoints and data from other clinical trials and observational studies. Safety information on the drug observed from adults may provide benchmark safety understanding but with consideration that different age cohorts might have varying ADME properties that will affect safety response in that cohort.

18.2.1.2 Extrapolation Plan and Types

Given information obtained from the initial substantiation, a plan is then needed to identify and address the knowledge gaps or the level of precision that is desired to make a robust decision to extend efficacy conclusion to the target population. It is helpful to clearly define what are the objectives of the plan that aligns with the knowledge gaps or precision identified. For example, a general objective might be (1) to present data supporting the assumption that the outcome of treatment is likely to be similar in pediatric subsets by age and by any other relevant characteristics compared to adults; and (2) to model and to evaluate PK, PD, response and efficacy data in adults and PK data in children as well as data on the maturational profile of these parameters and using results from a physiology-based model. Additional clinical trial simulations based on data obtained from the exposure-response study may be helpful to determine whether the results obtained from these exposure-response study rule will already rule out conclusion that the investigational drug is inefficacious in the target population.

As stated earlier, in the US, the FDA has established a pediatric decision tree to help identify the studies required to extrapolate adult efficacy conclusion towards the pediatric population, but also data between different pediatric age groups. The decision tree includes a series of questions or decision points to determine if an information from an existing clinical data is suitable for extrapolation to support pediatric drug effectiveness. The decision tree classifies three types of extrapolation: *full extrapolation*, *partial extrapolation*, and *no extrapolation*. Full extrapolation implies that only exposures in the source and target population need to match while no extrapolation implies that a full development program has to be conducted in children. Partial extrapolation approaches range along a continuum and to date, the following development programs have been used to support efficacy of a drug that is already approved in adults, to be approved in children: single adequate well controlled efficacy and safety trial and PK study, single uncontrolled efficacy and safety trial and PK study, single exposure-response trial, PK and safety study, PK/PD study and an uncontrolled efficacy and safety trial, PK/PD and safety study [26]. Recently, there has been a push to re-think this decision tree towards the consideration of the non-binary nature of similarity. In the EU, however, no such types exist because a PIP is required to be submitted no later than upon completion of the human PK studies or by the end of adult Phase I trials for new medicinal products where, at this stage, little is known about the PK, PD, efficacy and safety of the experimental drug (see Figure 18.1). Furthermore, the development plan for a drug in the PIP can be modified at a later stage as knowledge increases and thereby making extrapolation plan an iterative process. However, it is important to look at the extrapolation types and how it can guide the extrapolation plan. In the US, for example, there are indications where partial extrapolation is known to be appropriate and that the use of a PK/PD study and safety study is sufficient. Given this goal, what data needs to be provided for the drug in the indication to maximize acceptability of the approach. This is where the EU concept of an extrapolation plan becomes more relevant since it is asking for what data is available and what knowledge gaps need to be addressed (Figure 18.2).

While the extrapolation plan attempts to identify and address the knowledge gaps in the pediatric population, the question of when, to what extent, and how extrapolation can be applied and validated consistently and quantitatively is not yet clear. Also, it is not clear what quantitative data would need to be provided. In fact, the use of quantitative methods may not always be feasible. Nevertheless, extrapolation is the place where the use of Bayesian methodology may be more appropriate as there is a need to project information within data from different sources and populations so that it's value in the new population can be ascertained approximately depending on the similarity of the disease between the populations. From this, it is then possible to optimize pediatric clinical trials in the specific pediatric population the investigational drug is being studied. The relevant Bayesian methodologies will be discussed in subsequent sections.

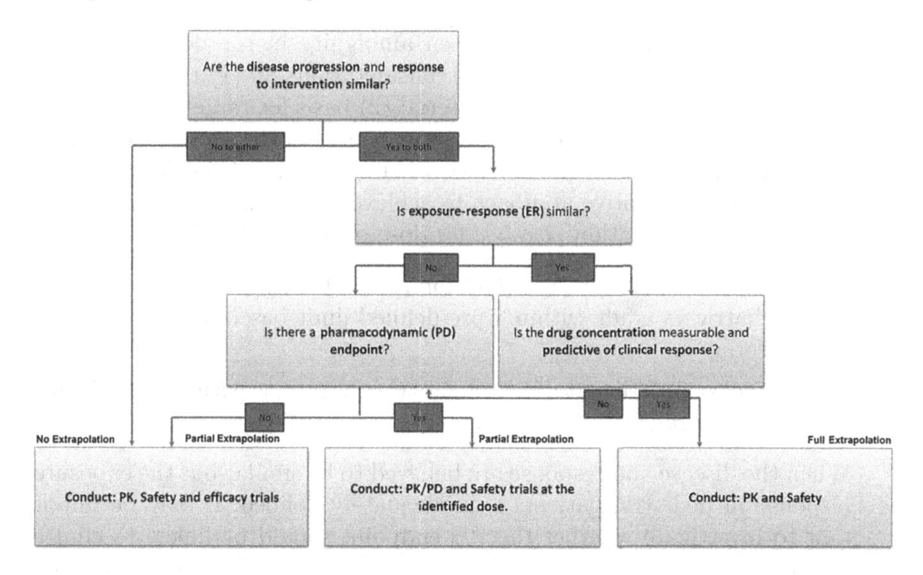

FIGURE 18.2
US FDA's extrapolation decision tree.

18.2.1.3 Validation of Assumptions

There are three considerations for validating the extrapolation assumptions with respect to the extrapolation plan.

i. **Confirming similarity of PK, PD, and/or PK/PD**: The EMA reflection paper stipulates that the extrapolation concept should result in explicit predictions of differences in PK, PK/PD, the nature of disease (manifestation, severity, progression, etc.), and clinical response to treatment in the target population as compared to the source population and stressing that these predictions "should be quantified to the greatest degree possible." For example, when there is sufficient data on similarity (disease, response, and exposure-response) between adults and children, a PK study in relevant age groups may be sufficient to identify dose. Initial dose selection strategy may be obtained from physiologically-based PK (PBPK) modeling and/or allometric scaling to define the starting dose. In the modeling, careful consideration should be given to identifying the key PK parameter (e.g., maximum concentration (C_{max}), minimum concentration (C_{min}), area under the curve (AUC), etc.) as well as target exposure range and acceptance criteria used to determine whether matching exposures is achieved. Pediatric-formulation related challenges such as differences in bioavailability (in the adult and pediatric formulation) and dosing accuracy/flexibility need to be considered. Because there are variable methods for assessing similarity of systemic exposures, there may be a need for an objective approach in the context of the drug, indication, age group, and formulation to lessen ambiguity in the decision making process. Such an approach requires considerations on (1) target exposure range and statistical acceptance criteria; (2) basis for target criteria based on therapeutic range of the drug and risk benefit of the product for a given indication; (3) simulations of doses when planning pediatric trials; and (4) need for an adaptive approach to achieve target exposure versus using modeling and simulation *post-hoc* for dose optimization. For example,

a. $X\%$ confidence interval (CI) for ratio of mean exposure metric in pediatric vs adult within a pre-defined limit based on defined target criteria;
b. $X\%$ of population at different age/weight groups within a pre-defined exposure range.

When the disease and response are believed to be similar but the exposure-response in pediatric patients is inadequately defined, it may be important to investigate whether the PD endpoint should be linked to clinical response. Dose range should account for observed differences in response between adults and the pediatric population. For example, one may require exposures to be greater than the approved adult dose (exposure) provided that exposure-response and safety data justify such an exposure. Distinctly different ranges of exposures are desirable for exposure-response analysis and dose optimization. PK/PD understanding will sometimes be sufficient but other times may not. The better the understanding of the link between PD and clinical response, the more weight can be given to PD; and hence, potential assessment of the relationship between PD and clinical response, and the related decision on the evidence to be generated, is made in committees with a broad expertise.

The analysis of data generated from these PK, PD, and/or PK/PD studies are needed to review and revise uncertainties and assumptions made in the extrapolation concept and the actual extrapolation step is required to be based on emerging data and pre-specified methodology. Mitigation of uncertainty and risk (including sensitivity analyses) through calibration of extrapolation step are also needed in case data do not validate assumptions. Furthermore, similarity in PK/PD can reduce Phase 3 trial size or even eliminate the need for an efficacy trial if there is a clear understanding of a PK/PD and clinical response relationships. In some cases, a formal similarity measure can be obtained to measure the degree of uncertainty quantitatively and which can be used to size how much data is needed for the Phase 3 trial (see, e.g., Basu et al. [30]).

ii. **Confirming efficacy and similarity of clinical response**: This step is more relevant in the case when the extrapolation plan is related to partial extrapolation, i.e., the disease/condition and/or response to intervention are similar but there is uncertainty about the strength of the assumptions. Usually, this extrapolation plan includes a framework for reduction of the required evidence generated in the target population in accordance with the predicted degree of similarity to the source population. If the similarity in exposure-response curves is confirmed quantitatively, the requirement for a clinical trial and its size should be sufficient to address any remaining uncertainty in efficacy. For example, the plan can include a proposal for additional clinical trials or observational studies in the target population. As mentioned in Section 18.2.1.2, the following development programs have been used to support efficacy of a drug that is already approved in adults, to be approved in children, e.g., single adequate well controlled efficacy and safety trial and PK study, single uncontrolled efficacy and safety trial and PK study, single exposure-response trial, PK and safety study, PK/PD study and an uncontrolled efficacy and safety trial, PK/PD and safety study [26]. However, it is not clear how much clinical data is required to support pediatric safety and effectiveness given the degree of similarity of exposures or exposure-response curves. This will be addressed in the statistical proposals in the next section. In addition, since safety is not extrapolated, how much clinical data is needed to assess safety in pediatrics is needed.

If the disease process is unique to children, its progression undefined or dissimilar to that in adults, and no pertinent biomarker to predict effectiveness, the extrapolation plan for confirming clinical efficacy may include one or more clinical efficacy studies usually evaluating more than one dose. The approach for dose selection is similar to dose selection in partial extrapolation with a single efficacy study. Examples of disease indications following this plan include major depressive disorders (MDD), attention deficit hyperactivity disorder (ADHD) [31]. There are also diseases that may appear similar in pediatric patients and adults, but underlying physiology suggests a difference for which many failed trials. Examples of diseases following this scenario include neurologic/psychiatric conditions such as selective serotonin reuptake inhibitors (SSRI's) and antidepressants, in general. Furthermore, this may include areas with numerous failed trials, for example, migraine, hypertension, and diabetes [31]. Bayesian procedures for this kind of setting will be described in the succeeding section as well.

iii. **Mitigation of risks**: As stated in Section 18.2.1.3 part (i.), knowing that there is uncertainty and risk of not validating the extrapolation assumption, a mitigation strategy is needed (including sensitivity analyses) through the calibration of extrapolation step. The mitigation may include analysis where extrapolation of conclusions from the source data are not used in the analysis for the data in the target population. This will be further described in the next section. In particular, we will describe a Bayesian strategy that leverages information when consistency of responses or efficacy is warranted and wholly reject borrowing if the consistency is not achieved in a streamlined manner. In fact, the Bayesian strategy is most suitable in this case as it naturally leverages information from the source data but can be further restricted by additional statistical guardrails to ensure that borrowing is only used when certain clinical response similarity conditions are met.

The use of the extrapolated source data will also have consequences in the understanding of the risk/benefit profile. As well as these data sources are considered with an awareness of the benefit risk profile in the intended indication. Note that because children may display developmentally unique differences ADME and resultantly exhibit variable patterns of drug sensitivity when compared to adults, the decision tree does not permit for extrapolation of PK and safety data between adults and children. Caution is needed as there would be few or no long-term pediatric studies for most products.

Furthermore, the sample size used to confirm clinical efficacy in children is usually small to gather sufficient information on the safety of the drug at the dose. Further steps can be done in terms of surveillance if the therapeutic product is approved for its intended use. Real world Evidence (see Section 18.3.3) can be conditionally pre-specified in the extrapolation approach as a mitigation strategy when results do not support extrapolation. Ultimately, the success of the extrapolation can only be extablished in post-marketing surveillance when sufficient number of children have been exposed to the drug and the purported label update in children provided information toward appropriate dosing.

18.3 Statistical Confirmation of Efficacy and Similarity of Response

The previous thinking about extrapolation heavily relies on conducting PK or PK/PD studies in the pediatric population to determine how the dosage regimen should be adjusted to achieve approximately the same level of systemic exposure that is safe and effective in adults. Some examples include antihypertensives [32] and clopidrogel (Plavix) for prevention of shunt thrombosis (Plavix FDA Clinical Review). Although these approaches are still relevant, the current pediatric drug development has branched out to three quantitative approaches toward systematic review of existing data as well as the data that need to be generated. These approaches maximize the use of existing data, increase the efficiency of development programs, and limit the number of children required for enrollment in clinical trials. First, quantitative methods to establish confirmatory evidence that structurally incorporates the systematic synthesis of evidence, e.g., Bayesian approaches in conjunction with as meta-analysis of clinical trial data or cross-design synthesis to integrate evidence from both clinical trials and observational studies. Results of similarity in exposure response can also guide how much information is needed in the clinical trial in conjunction with the data obtained in other age cohorts (say, adults). Furthermore, this formal way of incorporating information can be done in a "validative" way. This is important particularly in cases where these methods are usually not acceptable in adult drug development programs for these purposes. Second, quantitative methods to justify decreased alpha levels for statistical testing and collection of post-market safety data. Note that in this case, the tolerance is over how much error can be accepted is probably relevant in diseases where there is established efficacy of the drug. As a second example, there will be a necessity to define suitable success criteria. If the totality of data (based on extent of the similarity of disease and response to the therapy) permits the move away from success criteria like "$p < 0.05$" towards new paradigms, an explicit upfront agreement between regulatory agencies and sponsors will be critical. Third, the feasibility of using clinical evidence in place of clinical trials for the purpose of pediatric extrapolation. Given recent interest in the use of evidence from clinical experience (i.e., observational studies, registries, and therapeutic use) for regulatory purposes in the 21st–century Cures Act [33], some discussion of the feasibility of using such evidence in place of clinical trials for the purpose of pediatric extrapolation is suggested.

18.3.1 Bayesian Approach to Extrapolation

Since the extrapolation plan addresses what data needs to be generated to reach a level of certainty that warrants extension of efficacy conclusion toward the target population, the Bayesian methodology appears as the most reasonable approach. The structure of borrowing of information in is a form of approximation of projected information from the source toward the target population which is done in a quantitative way (which, henceforth, we call *Bayesian extrapolation*). In *Bayesian extrapolation*, the prior information (in the form of a density) is used as the approximate to the projected source information in the target population, then we collect appropriately sized data in the target population which forms a likelihood. We compute the posterior which is generally more precise that the likelihood (i.e., posterior mean and its associated credible interval [CrI]). Note that the Bayesian extrapolation approach is intuitive in the sense that it formalizes what pediatricians do when they combine the results from large adult trials with the results of smaller pediatric trials to make treatment decisions as Schoenfeld et al. [23] notes in their paper in 2009. Alternatively, Hlavin et al. [34] propose calibrating the design of a frequentist trial in pediatrics with a Bayesian interpretation of the trial in mind, acknowledging that the pediatric results will be interpreted in light of the adult data.

18.3.1.1 Projection of Information from Adult to Pediatric Population through the Use of Informative Priors

Because there is preponderance that there is similarity in disease intervention and response to therapy, then it makes sense that information from the adult or source information on both the treatment and placebo can be projected toward the target population. The general question is that how much does the information in the adult or source population contribute into the adolescent or target population. Consider a simple pediatric (adolescent) trial with two treatments: experimental treatment (E) and control (C). Let $D_i = (y_i, z_i, \boldsymbol{x}_i)$, $i = 1, \ldots, n$, denote the data from subject i, where y_i is its treatment response, z_i is its treatment group indicator with $z_i \in \{1, 0\}$ and $\boldsymbol{x}_i = (x_{i1}, \ldots, x_{ip})$ is a vector of covariates. Let θ_{z_i} denote the response rate of treatment z_i, which, in the dichotomous endpoint setting ($n_i = 1$ for all i), is the proportion of events, with its maximum likelihood estimator $\hat{\theta}_z = Y_z/N_z$ where $Y_z = \sum_{i=1}^{n} I_{y_i}(1) I_{z_i}(z)$ and $N_z = \sum_{i=1}^{n} I_{z_i}(z)$, within the treatment group where the observation belongs. Furthermore, suppose that a larger value of $\hat{\theta}_z$ means that the treatment is more efficacious, e.g., when $\hat{\theta}_z$ denotes the proportion of recovery or meeting primary endpoint. Consider the trial comparing $\theta_E (= \theta_1)$ and $\theta_C (= \theta_0)$, then the hypothesis of interest is

$$H_0 : \theta_E - \theta_C \leq -\lambda \quad \text{v.s.} \quad H_1 : \theta_E - \theta_C > -\lambda, \tag{18.1}$$

where $\lambda \in (-1, 1)$. When $\lambda = 0$, then the interest is to test whether E is superior to C; while when $\lambda > 0$, then the interest is to test whether E is non-inferior to C by a margin of λ. The decision for this hypothesis can be made by either performing the hypothesis test or constructing a confidence interval.

If y_i is the treatment response, with sampling distribution $y_i|\theta_{z_i} \sim F(\theta_{z_i}; n_i)$, then the instances with which the θ_{z_i}, $i = 1, \ldots, n$, occur are *exchangeable*. In addition, there exists a distribution G such that the transformed parameters $\vartheta_{z_i} = g(\theta_{z_i})$ are a random sample from G. This can be hierarchically represented as follows:

$$y_i|\theta_{z_i} \sim F(\cdot|\theta_{z_i}; n_i), \quad \vartheta_{z_i} = g(\theta_{z_i})|z_i, \boldsymbol{\zeta} \sim G_{z_i}, \quad \boldsymbol{\zeta} \sim H, \tag{18.2}$$

where $g(\cdot)$ can be logit function and ϑ_{z_i} can take the form of $b_0 + b_1 z_i$, for instance. Then, $\boldsymbol{\zeta} = (b_0, b_1)$ in this case. The second stage prior distribution (or *hyperprior*) H describes the initially available information about $\boldsymbol{\zeta}$ [35], and n_i is the sample size associated with y_i. If y_i are Bernoulli random variables, then $n_i = 1$ for all i. Note that an assessment of exchangeability of the θ's is fundamental because it is "borrowing of strength," or combination of information from subjects in the treatment group. From the specification above, if $\boldsymbol{D} = (D_1, \ldots, D_n)$, $\boldsymbol{Y} = (y_1, \ldots, y_n)$, $\boldsymbol{X} = (\boldsymbol{x}_1, \ldots, \boldsymbol{x}_n)$, and $\boldsymbol{x}_i = (1, z_i)^\top$, the posterior distribution is:

$$q(\boldsymbol{\zeta}|\boldsymbol{D}) \propto L(\boldsymbol{\zeta}|\boldsymbol{X}, \boldsymbol{Y})\pi(\boldsymbol{\zeta}) \\ = \left(\prod_{i=1}^{n} f(y_i|\boldsymbol{x}_i, \boldsymbol{\zeta})\right)\pi(\boldsymbol{\zeta}). \tag{18.3}$$

In the later discussions, we will consider two major model assumptions: Model 1 under the generic logistic model framework and Model 2 under the beta-binomial conjugacy property. Under the assumptions of Models 1, $g(\cdot)$ is the logit function and $\vartheta_{z_i} = \boldsymbol{x}_i^\top \boldsymbol{b}$,

$$f(y_i|\boldsymbol{x}_i, \boldsymbol{\zeta}) = f(y_i|\boldsymbol{x}_i, \boldsymbol{b}) = \exp\left(y_i(\boldsymbol{x}_i^\top \boldsymbol{b}) - \log(1 + \exp(\boldsymbol{x}_i^\top \boldsymbol{b}))\right). \tag{18.4}$$

The distribution G_{z_i} may not have a closed form expression, depending on the prior specifications or model assumptions. Under Model 1, the distribution G_{z_i} of ϑ_{z_i} is $\mathcal{N}(b_0 + b_1 z_i, \tau_0^{-1} + z_i^2 \tau_1^{-1})$. The hyperprior H of $\boldsymbol{\zeta}$ is characterized by the joint distribution, $\pi(\boldsymbol{b}) = \pi(b_0)\pi(b_1)$, i.e., $b_0 \perp b_1$. In Bayesian data augmentation, $\pi(b_0)$ and $\pi(b_1)$ come from the marginal posterior of b_0 and b_1, given by $\hat{p}(b_0|\boldsymbol{D}_0)$ and $\hat{p}(b_1|\boldsymbol{D}_0)$, respectively, and determined from the source data $\boldsymbol{D}_0 = (y_{0,i}, z_{0,i}, \boldsymbol{x}_{0,i})$, $i = 1, \ldots, n_0$. Unless these marginal posteriors can be expressed in closed form, an MCMC-based approximation of $\hat{p}(b_z|\boldsymbol{D}_0)$, $z \in 0, 1$ can be used. Schmidli et al. [36] suggest approximating the marginal posterior as a finite mixture of prior densities [37],

$$\hat{p}(b_j|\boldsymbol{D}_0) = \sum_{i=1}^{m} p_i \phi(b_z|\eta_{ji}, \tau_{ji}^2), \quad p_i > 0, \quad \sum_{i=1}^{m} p_i = 1 \tag{18.5}$$

with the value of m chosen to be small, and where ϕ is a density of the same kernel as the density of G. The selection of the number of components is guided by a measure of divergence so that the approximation is close to $p(b_z|\boldsymbol{D}_0)$ [38].

The assumption behind borrowing information from adult or source population is that $\boldsymbol{b}_{z,0}$ in the source data $\boldsymbol{D}_0$ is exchangeable with the counterpart $\boldsymbol{b}_z$ in the adolescent or target population $\boldsymbol{D}$, i.e., the treatment response in either arms is exchangeable or consistent between the adult and pediatric trial and that $\boldsymbol{b}_z = \boldsymbol{b}_{z,0}$. But more importantly is an ascertainment that the disease progression in $\boldsymbol{D}_0$ and in $\boldsymbol{D}$. If this assumption is not entirely certain, it is important to accommodate potential variations in responses between $\boldsymbol{D}_0$ and $\boldsymbol{D}$ that would diminish the projected information from $\boldsymbol{D}_0$ to $\boldsymbol{D}$. That can be accomplished by enhancing robustness in the prior for which a common technique is to add a vague distribution. For instance, if $\boldsymbol{b} = (b_0, b_1)$, a vague prior $p(b_j)$, $j = 0, 1$, can be added into the posterior distribution of b_0 from the source data, i.e.,

$$\hat{\pi}(b_j) = (1 - w)\hat{p}(b_j|\boldsymbol{D}_0) + wp(b_j), \tag{18.6}$$

where $0 < w < 1$ (cf. Berger and Berliner [39]). In general, heavy-tailed distributions have been shown to produce an approximation of the projection of $\boldsymbol{D}_0$ unto $\boldsymbol{D}$ by calibrating the disparity between responses in $\boldsymbol{D}_0$ and $\boldsymbol{D}$, typically by favoring one source of information over another. The less favored source is wholly or partially rejected as the conflict becomes increasingly extreme [40]. The issue here, however, is how to allocate the weight between the two distributions. In the absence of information over the optimal weight, a value of 0.5 is generally used. Alternatively, an expert elicitation can be obtained on how the diseases in the two populations are similar. An additional validation check can always be added to make sure that borrowing only happens if the pediatric data somehow is consistent wit the adult data by some pre-specified measure (see Section 18.3.1.4).

Finally, given this prior, the desired posterior arises from Equation 18.4 and 18.6 as $q(\boldsymbol{b}|\boldsymbol{D}) \propto L(\boldsymbol{b}|\boldsymbol{D})\tilde{\pi}(\boldsymbol{b})$. Since, the conditional posteriors cannot be expressed in closed form, one needs to sample for the posterior distribution through the Metropolis–Hastings algorithm to obtain the posterior mean of the parameters and then transform them back to $\hat{\boldsymbol{\theta}}$ and the estimate of the treatment effect $\hat{\Delta}_{EC}$. Furthermore, any inference on Equation (18.1) is determined whether $\mathbb{P}(\Delta_{EC}+\lambda > 0) < \alpha/2$, where λ is a pre-specified margin.

Note that in the specification of Equation (18.2) a link function is used to describe how θ_z depends on the treatment assignment as predictor variable; and thus, provides a convenient way of handling multiple predictors. However, if the interest is only on the treatment assignment and no other covariate is required, the treatment responses for each arm can be modeled separately. For instance, if Y_C, Y_E are the number of events in treatment group C and E, respectively, then $Y_z \sim \text{Binomial}(N_z, \theta_z)$ and the prior for θ_z can be expressed as $\theta_z \sim \text{Beta}(\kappa_z\mu_z, \kappa_z(1-\mu_z))$. While prior for θ_z can be can be constructed using the approximation in Equation 18.5 with Beta distributions, a direct way of incorporating the source data as a prior information for θ_z in Equation (18.3), e.g., power priors [20]. What is attractive with this parameter is that it can be heuristically interpreted as the coefficient of projection between the source and target population in the sense that if these populations are planes in space, then the coefficient or projection is the *cosine* of angle between the two planes.

In other settings where multiple source data or populations are available, it is more reasonable to use study-specific weights because the projection of the information from each source may not be the same. For example, when the source data has a slightly different design in which a patient can access background therapy as needed whereas the current study strictly uses monotherapy design, then the value of the information coming from the source information may be less than when the information comes from a similar monotherapy trial. In this case, an assessment of exchangeability of the θ_z's is all the more fundamental since there still be some rationale that responses obtained in these two types of trials are still indistinguishable. The information about the similarity of the θ_z informs inferences about the distribution of the θ_z across other sources. Thus, inferences for the source-level parameters θ_z reflect not just the information in $Y_{z,0,k}$ itself, $k = 1, \ldots, K$, but, via the hierarchical model, will also draw on relevant information in the other sources.

The estimates of these weights can be data-driven by assuming that study-specific weights are independently distributed with joint distribution $\pi(\boldsymbol{\alpha}_0)$, $\boldsymbol{\alpha}_0 = (\alpha_{0,1}, \ldots, \alpha_{0,k})$, with k historical studies $\mathscr{D}_0 = (\boldsymbol{D}_{0,1}, \ldots, \boldsymbol{D}_{0,k})$. This gives the joint power prior

$$\pi(\theta_z, \boldsymbol{\alpha}_0 | \mathscr{D}_0) \propto \prod_{i=1}^{k} L(\theta_z | \boldsymbol{D}_{0,i})^{\alpha_{0,i}} \pi(\theta_z) \pi(\boldsymbol{\alpha}_0), \tag{18.7}$$

and the joint posterior is obtained as:

$$q(\theta_z, \boldsymbol{\alpha}_0 | \boldsymbol{D}, \mathscr{D}_0) \propto L(\theta_z | \boldsymbol{D}) \pi(\theta_z, \boldsymbol{\alpha}_0 | \mathscr{D}_0). \tag{18.8}$$

When a closed form expression for $p(\theta_z | \boldsymbol{D}, \mathscr{D}_0) = \int_{\boldsymbol{\alpha}_0} q(\theta_z, \boldsymbol{\alpha}_0 | \boldsymbol{D}, \mathscr{D}_0) \boldsymbol{\alpha}_0 d\boldsymbol{\alpha}_0$, an MCMC-based approximation can be done by fitting a mixture of beta distributions (see Equation (18.5) for comparison). This approach and the whole power prior specification can be readily extended to Model 1 as well.

Example 18.1: Following the extrapolation process outlined in Section 18.2.1, suppose sufficient information has been gathered on the similarity of disease progression and response to therapy between adults (or generic source) and adolescent (or generic target) population. The initial doses were then chosen based on modeling and simulation which predicted that the adolescents will have similar exposures to adults. While no exposure-response results in children is currently available, as this is incorporated in a PK lead-in to the single phase 3 study, there is a prediction that the selected doses will also yield similar clinical responses observed in the adult trials. Given these predictions, a trial will be conducted to confirm these predictions. The question then is how large a clinical trial is needed so that there is sufficient confidence in the eventual decision to expand labeling to adolescents. In addition, how much lower could that adolescent response be compared to adults before any borrowing will no longer be helpful given sufficient conditions to warrant borrowing information?

The treatment response rate observed in adults is about 30% while placebo is about 10%. Given sufficient similarity of the disease in adults and children, there is rationale to assume that the response rate in adolescents is about the same. In particular, the adult trial has one active dose and placebo and that each treatment arm has 100 patients. Figure 18.3 shows the simulated power when adults and adolescents have the same response as a function of sample size. This figure gives insight on what would the sample size per arm in the adolescent trial be to achieve a particular power assuming that the adolescent trial will borrow a certain amount information from the adult trial with a certain rate. For example, in Figure 18.3a, shows the power using power prior approach where the power prior is chosen conditionally at 0.5, 0.65, and 0.85. Even borrowing only about 50% of the adult information, a sample size of 30 patients per arm appears sufficient. Figure 18.3b, on the other hand, shows the power using robust prior

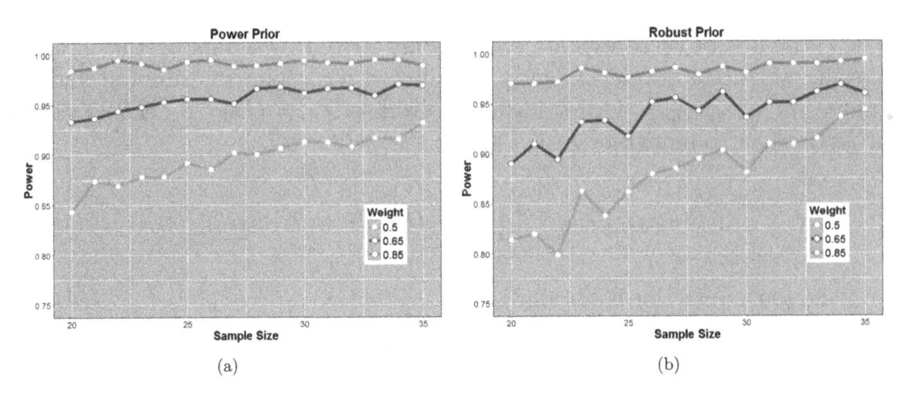

(a) (b)

FIGURE 18.3
Power as a function of sample size and weight. (a) For power priors, the weight corresponds to the power prior parameter and (b) while for robust priors, the weight corresponds to the mixture parameter.

with mixture weight of 0.5, 0.65, and 0.85. Again, this figure shows that a sample size of 30 patients per arm appears sufficient. Note that robust priors is used when there is knowledge that the trials are similar but not sure whether the result between the two cohorts will end up similar. Power prior is applicable when there is preponderance of dissimilarity in the population and trial and want to preempt from borrowing fully regardless of whether the responses turn out to be similar or not. This is because it effectively down-weights the information in the likelihood whereas robust priors just offer an "off-ramp." Given that a sample size of 30 appears sufficient, further investigation is conducted how sensitive this power is with respect to changes in the treatment response assumption in adolescents. This is shown in Figure 18.4a where if the responses observed in adolescent patients matches that of adult patients, then there is approximately 80% probability of making a conclusion of an efficacious dose. This diminishes gradually when the responses in adolescent patients becomes increasingly less than the responses observed in older patients. At the very least, if the treatment response in adolescents is about 25%, the power is still sufficient as long as the placebo response remains the same.

Figure 18.4b shows the simulated type-I error rate using robust priors as a function of sample size. Here, it is assumed that the treatment and placebo response rate in adults is 30% and 10%, respectively, but there is no treatment effect observed in adolescents, i.e., treatment and placebo response rate in adolescents are both 10%. In this figure, if the mixture weight is set at most 0.5, then the simulated type-I error rate is kept below 20%. In terms of congruence with the extrapolation concept, a lower mixture weight implies that there is less a priori confidence that there will be consistency between adults and adolescents. Hence, reliance

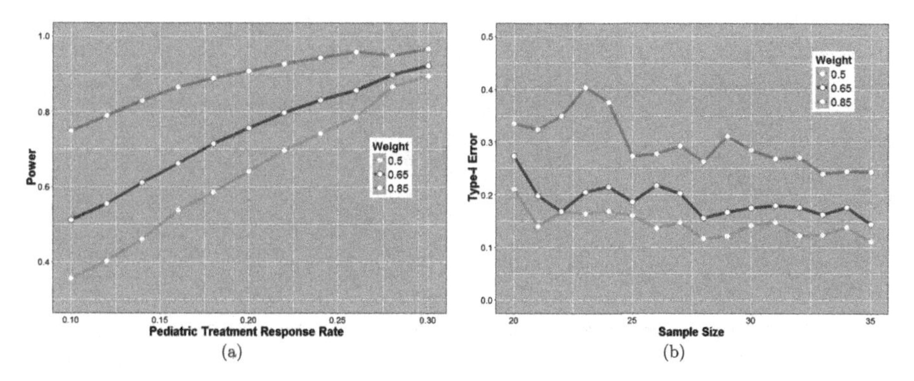

FIGURE 18.4
(a) Power using robust prior as a function of pediatric treatment response rate and weight. The pediatric placebo response rate is kept at 0.10. (b) Type-I error using robust prior as a function of sample size and weight or mixture parameter.

on the informative prior is severely handicapped so that it will less likely be chosen during the estimation process. Figure 18.5a shows the type-I error rate as a function of the adult treatment response. That is, if the adult placebo response is ket at 10% while the adult treatment response is allowed to vary from 20% to 30% the figure attempts to show how this variation affects the type I error rate for detecting superiority in adolescents. As expected, the maximal increase in type-I error occurs when the adult treatment response rate is about 17%–25% and starts to diminish beyond 25% which at that point, the robust prior will most likely choose a vague prior over an informative prior. Lastly, Figure 18.5b shows the simulated type-I error when the null response in adults is allowed to vary but the adult placebo and treatment response are fixed at 10% and 30%, respectively. Similar to Figure 18.5a, the type-I error rate is maximally increased when the robust prior seems indifferent whether to use the informative or non-informative prior because the treatment or placebo response is neither too close or too far from the adult placebo or treatment response. If the null response is about 15%, then the placebo arm will borrow more often from the adult placebo arm while the treatment arm will borrow more from the adult vague prior because it is the adult treatment response rate is farther in distance from the adolescent treatment response rate than adult placebo response rate with adolescent placebo response.

Note that the fixed hypothesis described in Equation 18.1 can always be tested in a group sequential way to minimize pediatric patients exposed to ineffective treatments and for timely access of effective drugs. In addition, a group sequential framework can be pursued regardless of whether an extrapolation strategy is taken by calibrating the prior information appropriately, i.e., if

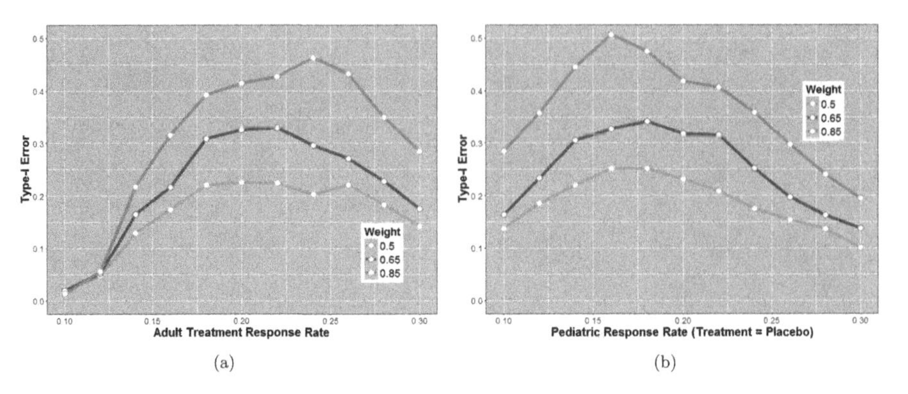

FIGURE 18.5
(a) Type-I error using robust prior as a function of adult treatment response and weight or mixture parameter. The adult placebo response is kept at 0.10. (b) Type-I error using robust prior as a function of pediatric null response (i.e., treatment response is equal to placebo response) and weight or mixture parameter.

extrapolation is used then the prior can take the form of a power or robust prior as discussed previously. If extrapolation is not pursued, the informative part of the prior is wholly abandoned for a non-informative prior.

Consider the comparisons of θ_E and θ_C, where the primary hypothesis of interest at the end of stage j is:

$$H_{0,j} : \theta_{E_j} - \theta_{C_j} \leq -\lambda \quad \text{v.s.} \quad H_{1,j} : \theta_{E_j} - \theta_{C_j} > -\lambda, \tag{18.9}$$

where $\lambda \in (-1, 1)$. The decision for this hypothesis is based on the posterior probability for the treatment effect exceeding a threshold $s_{1,j}$ given the clinical trial data at stage j. If $\Delta_j = \theta_{E_j} - \theta_{C_j}$ is specified in terms of improvement over the control treatment (i.e., positive values are used to express the benefit of the experimental treatment over the control), success and futility criteria for the Bayesian approach has the form:

$$\mathbb{P}(\Delta_j > s_{1j}|D) \geq p_{s_{1j}} \quad \text{and} \quad \mathbb{P}(\Delta_j < f_{1j}|D) \geq p_{f_{1j}}, \tag{18.10}$$

respectively. Here, s_{1j} and f_{1j} are user-specified effect thresholds, $p_{s_{1j}}$ and $p_{f_{1j}}$ are user-specified probability thresholds. There are cases when an additional success criteria may be added, e.g.,

$$\mathbb{P}(\delta_j > s_{2j}|D) \geq p_{s_{2j}}$$

which could describe an expectation that the treatment effect must be greater than what is observed from standard of care to be commercially viable. Usually, the probability thresholds $p_{s_{1j}}$ and $p_{f_{1j}}$ are chosen to inherently adjust for the amount of information yet to be collected, while controlling the overall false-positive rate and power as in the frequentist setting. On this note, the probability thresholds proposed in [41], O'Brien and Fleming [42], Pocock [43], and optimal Hwang et al. [44] group sequential rules have still been widely used. There are other probability thresholds derived through predictive probabilities, i.e., stop trials for efficacy if they currently show superiority and are likely to maintain superiority after remaining data are collected.

18.3.1.2 Externally Controlled Trials and Threshold Crossing

Among clinical trial designs, the conduct of randomized clinical trials (RCTs) have become the gold standard for estimating the effects of treatments on outcomes related to efficacy [45,46]. There is reason why this is so owing to the fact that randomization, when properly conducted, prevents patient selection bias and by the very nature of chance, random assignment tends to balance both observed or unobserved patient covariates so that the treatment status of the patient is not confounded [47,48]. Hence, the effect of treatment on outcomes can be estimated by comparing outcomes directly between treated and untreated subjects [49] or that any difference detected is not due to selection or other confounding covariates but attributable to the treatment alone. Furthermore, as there is no need to control for confounding factors, the analysis is simpler.

While there is preference for RCTs, their conduct may be unfeasible in many in pediatric indications because of their size, duration, cost, patient preference, or in some cases ethical constraints (see Gehan and Freireich [50] for reasons against randomization). For instance, when evidence already exists showing the superiority of a new treatment over the standard-of-care, it might be unethical for a RCT to assign patients to a potentially inferior treatment, e.g., thoracoscopic repairs of congenital diaphragmatic hernia in neonates [51], or parent/patient preference may altogether hinder accrual of pediatric patients making trials longer and more costly. In this case, data derived from external control (in which controls are not part of the same randomized study as the group receiving the investigational drug) could be preferable to the *status quo* of not being able to complete the trial by which any risk incurred by children is not justifiable. In fact, the use of external controls has been discussed in ICH E10 and endorsed in certain situations, particularly when enrollment is challenging (e.g., rare diseases) or when randomizing to placebo is unethical [52]. There are multiple options when considering the use external controls, including:

i. Placebo or active comparator patients from adult trials of the same treatment

ii. Placebo or active comparator patients from pediatric trials of similar treatments

iii. Patients receiving relevant comparator treatments in real-world disease state registries

iv. Natural history of untreated pediatric patients from real-world disease state registries

For example, adult and pediatric trial data were used to establish similar exposure-response relationships for eight anti-epileptic agents with complex partial seizures, thereby supporting full extrapolation of efficacy in future submissions [53]. Remicade® (infliximab) used adult and pediatric trial data to establish similar exposure-response relationships for infliximab (Remicaid) and the adult placebo response as external control to the pediatric trial to support partial extrapolation of efficacy and pediatric labeling in ulcerative colitis [54,55]. Pediatric patients from a real-world registry were used as a historical control comparison in the approval of Brineura® (cerliponase alfa) for a specific form of Batten disease [56].

The efficient use of historical data from the placebo-treated adult patients (source population) not only permits the reduction in the sample sizes of the pediatric studies and shortening the time from development to approval, but it also reduces the operational burden in conducting the trial since the use of placebo in pediatric trials has always proved to deter parents from enrolling their children [57] or that there are regional authorities prohibiting the use of placebo for conducting research in children. However, external controls should be carefully selected to minimize bias when comparing outcomes to patients receiving active treatment in the open-label study. As pointed out in ICH E10, the estimate of external control group outcomes should always be made conservatively, and the use of more than one external control may be advisable providing that the analytic plan specifies conservatively how each will be used in drawing inferences (e.g., requiring study group to be substantially superior to the most favorable control to conclude efficacy). For example, instead of using the posterior distribution of the treatment response $q(\theta_0|\mathscr{D}_0) \propto L(\theta_0|\boldsymbol{D}_0)\pi(\theta_0)$, the posterior predictive distribution

$$q(\boldsymbol{D}_1|\mathscr{D}_0) \propto \int_{\theta_0} f(\boldsymbol{D}_1|\theta_0) f(\theta_0|\mathscr{D}_0) d\theta_0 \tag{18.11}$$

may be better suited because it incorporates the "sampling uncertainty" from drawing a new data value. Note that there is no placebo in the trial and so the strategy reflects the prediction of what would placebo responses look like had placebo been a treatment arm in the trial. Then, a conservative threshold can be set at the 95% upper credible limit $\gamma = q_{0.975}(\boldsymbol{D}_1|\mathscr{D}_0)$. Alternatively, if $\hat{\Delta}_{EC}$ is the effect of treatment over placebo and $\theta_0 = \theta_C$ is the placebo response rate in $\mathscr{D}_0$, then a conservative threshold can also be defined as $\gamma = \hat{\theta}_C + f\hat{\Delta}_{EC}$, where f is a certain fraction in the interval $(0, 1)$. In fact, such a threshold is akin to establishing a non-inferiority margin wherein a certain effect of the investigational drug is preserved to ensure assay sensitivity and avoid declaring an ineffective drug effective in the absence of a placebo control [58]. In addition, the fraction f needs to preserve a certain clinically relevant level of effect of investigational treatment. Either of these thresholds, however, provide sufficient power when the treatment response in adults and pediatrics are similar and that power diminishes under significantly lower results in pediatrics. This is discussed in the example below.

Alternatively, because of concerns of interpretability of the results when combining projected adult and pediatric treatment response and comparing it back again with the adult placebo response, a logical objective for decision that still adheres with the principle of extrapolation is to

i. compare treatment responses from treated paediatric patients to treated adults which can be called (**confirmatory check**); **and**,

ii. compare response in treated paediatric patients to the threshold derived from the placebo response in adults which can be called an **efficacy check**.

Here, treated adult data will not be borrowed to supplement the treated paediatric data. Furthermore, the placebo response in adults is actually not extrapolated or "leveraged" in the usual sense of the word because there really is no quantified information that is incroporated into the likelihood from the pediatric data. Rather, the use of adult placebo is only to "project" what would be an extreme placebo response by which if we see an treatment response greater than this conservative placebo response threshold then that treatment is probably efficacious. Hence, the decision criteria can be defined as:

Confirmatory check: Consistency in treatment response will be confirmed if ξ, describing treatment response in paediatric patients is higher than the lower bound of the $(1-\alpha/2)100\%$ CI of the adult treatment response.

Efficacy check: Efficacy will be confirmed if ξ in paediatric patients is higher than the upper bound of $(1-\alpha/2)100\%$ of the treatment response in placebo-treated adults. The upper bound of the CI from the placebo response in adults should serve as a conservative benchmark for placebo response.

The value of ξ can be set conservatively as the lower bound of the $(1-\alpha/2)100\%$ CI of the mean treatment response, or in cases of serious unmet need, can be set as the mean treatment response itself. Both of these checks are in line with the EMA Reflection Paper on extrapolation in paediatric patients (EMEAextrapolation) which states that "specific ... methodological approaches should be proposed ... to generate evidence that strengthens and ultimately, based on success criteria, validates the extrapolation concept. This validation confirms whether regulatory decisions can rely on the initial, or revised, predictions for the expected effects of treatment in the target population or if more data needs to be generated."

Example 18.2: As mentioned above, Remicade® (infliximab) is one successful case example on the use of external control to validate clinical efficacy of a drug in children. Remicade® (infliximab) is an FDA-approved drug to treat Crohn's disease, ulcerative colitis (UC), rheumatoid arthritis, and other conditions. Its maker sought a meeting with a gastrointestinal (GI) FDA advisory panel for the purpose of expanding the drug's labeling to include pediatric ulcerative colitis. As is common in such settings, extrapolation from adult data was not permitted for dosing or safety assessment in children, but the panel did allow the sponsors to argue for extrapolation of efficacy from two existing adult studies (placebo-controlled), ACT 1 and ACT 2. The results of this trial (based on published summary data) are shown in Table 18.1. The results of the pediatric study, T72, are also shown under three different endpoints. The primary endpoint for the pediatric study is clinical response at week 8. Although this endpoint is not the primary endpoint of ACT 1 and ACT 2, they were collected and measured at Week 8 in that study as well.

Figure 18.6 shows the histograms of the results of clinical response in the Adult (a) and pediatric (b) trials. The light grey colored bars are the treatment responses under infliximab and the dark grey colored bars are for placebo. Since the T72 trial is an uncontrolled open label trial, it

TABLE 18.1
Study-Level Endpoint Data, Remicade UC Studies in Adults (ACT 1 and ACT 2) and Pediatrics (T72)

	ACT 1 (Adult)	**ACT 2 (Adult)**	**T72 (Pediatric)**
Endpoint	**Infliximab 5 mg/kg $n = 121$**	**Infliximab 5 mg/kg $n = 121$**	**Infliximab 5 mg/kg $n = 60$**
Clinical response	84 (69.4%)	78 (64.5%)	44 (73.3%)
Clinical remission	47 (38.8%)	41 (33.9%)	24 (40.0%)
Mucosal healing	75 (62.0%)	73 (60.3%)	41 (68.3%)

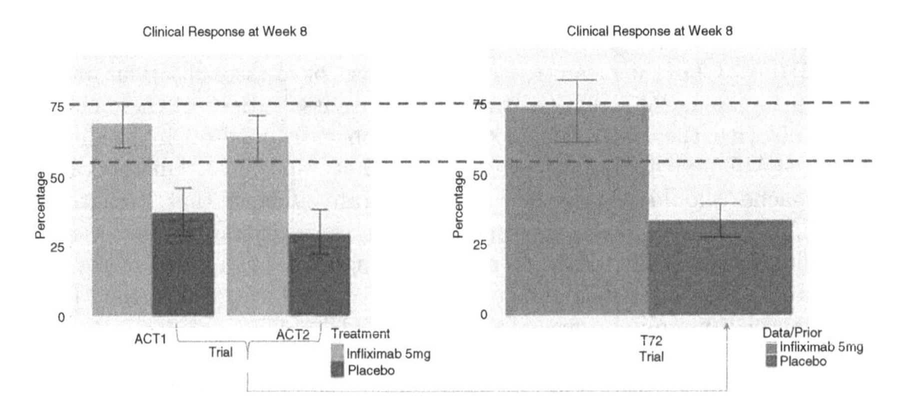

FIGURE 18.6
Infliximab: clinical response rates in adult and pediatric studies.

was assumed that the placebo response in adults is the same as placebo response in pediatrics. With that said, the pooled placebo response in adults is used as an objective threshold to determine whether infliximab is efficacious in pediatric ulcerative colitis. Let $\hat{p} + z_{\alpha/2}\hat{\sigma}_{\text{Adults}}$ be the upper 95% confidence limit of the estimate of the placebo response in adults, where $\hat{\sigma}_{\text{Adults}}$ is the standard error for the estimate of $\hat{p}$ in adults and $z_{1-\alpha/2}$ is the $1001 - \alpha/2$th quantile of the standard normal distribution. Pediatric clinical response was also compared *viz-a-viz* the confidence limits of the clinical responses in adults to make sure that the pediatric clinical response is within reasonable range of adult clinical response. The formal incorporation of historical controls is possible, but inherently introduces further uncertainties to such comparisons. The historical controls should match the treated pediatric population as closely as possible. Lastly, efficacy is deemed validated if the hypothesis of interest given by:

$$H_0 : \theta_E \geq \gamma \quad \text{v.s.} \quad H_1 : \theta_E < \gamma, \tag{18.12}$$

is rejected. The value of γ is chosen as $\hat{p} + z_{0.975}\hat{\sigma}_{\text{Adults}}$. The decision for this hypothesis can be made by either performing the hypothesis test or constructing a confidence interval. In particular, the combined treatment response under placebo is assumed to hold in the T72 trial. Then, a visual check is used to determine whether the observed treatment response in the T72 trial is within the range observed in the adult trials. Here that range is denoted by the two dashed bars. Since that seems to be true, extrapolation of efficacy from adults is warranted although not quantitatively used. We will call this as a "cursory" extrapolation. Efficacy of infliximab in pediatrics is concluded when the lower bound of the observed treatment response in pediatrics is greater than the upper bound of the assumed placebo response.

Suppose a new investigational drug is required to be explored in pediatric patients 2–18 years old. The study in 2–6 years old is deferred until after the the study in 6–18 years old is completed. Furthermore, an initial exposure-response study was conducted in a small number of children and was found out that the response in children is generally higher than the response in adults at the same exposure range using the doses that were studied in adults. The sponsor then plans to conduct a similar trial to the one done with Remicade® (infliximab) in pediatric patients 6-18 years of age. However, instead using a "cursory" look whether responses of children and adults are similar, a Bayesian extrapolation is planned, where the investigational drug treatment response in adults are projected into the pediatric trials to increase precision in the results. For example, for the investigational drug information from the adults are used as priors towards the estimation of the treatment response in pediatrics (cf. Equations (18.6) and (18.7) and its normalized form). A conservative placebo threshold, on the other hand, is created from observed adult placebo response. A decision criteria for declaring efficacy is then established as:

Confirmatory check: Consistency in treatment response will be confirmed if $\xi = \hat{\theta}_E + q_{0.975}(\theta_E|\boldsymbol{D}_1, \mathscr{D}_0)$ is higher than $\hat{\theta}_E + q_{0.975}(\theta_E|\mathscr{D}_0)$.

Efficacy check: Efficacy will be confirmed if ξ in paediatric patients is higher than $\hat{\theta}_C + f q_{0.975}(\theta_E|\mathscr{D}_0)$, where $f = 0.4$.

Figure 18.7a,b shows the simulated power for decision to be able to declare whether an investigational treatment is efficacious in children or not using either the **Efficacy Check** only (otherwise, called *Unconstrained*) or both the **Confirmatory** and **Efficacy Check** (otherwise called, *Constrained*). Generally, the *Unconstrained* decision is more powerful than the *Constrained* criterion. However, since the assumptions for the treatment response for pediatrics used in the simulations are about 6% higher than the treatment response in adults, the two approaches appear to have the same power.

Figure 18.8a,b shows the simulated type-I error rate for the *Unconstrained* and *Constrained* decision criteria, respectively. Note that because the placebo threshold is set at a conservatively extreme level, the error rate is low. Hence, the placebo threshold has to be chosen carefully so as not to be extremely conservative that a potentially efficacious treatment in children could be missed. On the other hand, in relation to the frequentist approach, the type-I error rate of either Bayesian approaches is higher but still within reasonable levels.

Figure 18.9a,b shows the type-I error rate as a function of the pediatric treatment response (i.e., pediatric treatment response = adult placebo

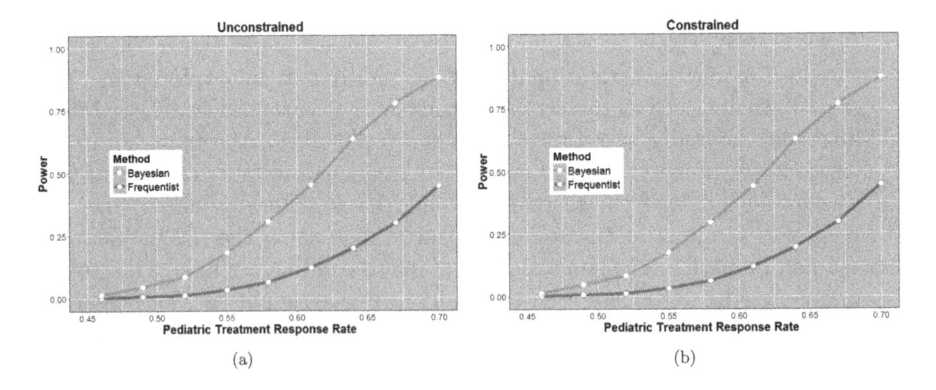

FIGURE 18.7
(a) Power as a function of pediatric treatment response without similarity check. (b) Power as a function of pediatric treatment response with similarity check. The adult treatment response is set as pediatric treatment response—0.03. The sample size for the adult treatment group is 150, while the sample size for the pediatric treatment group is 60.

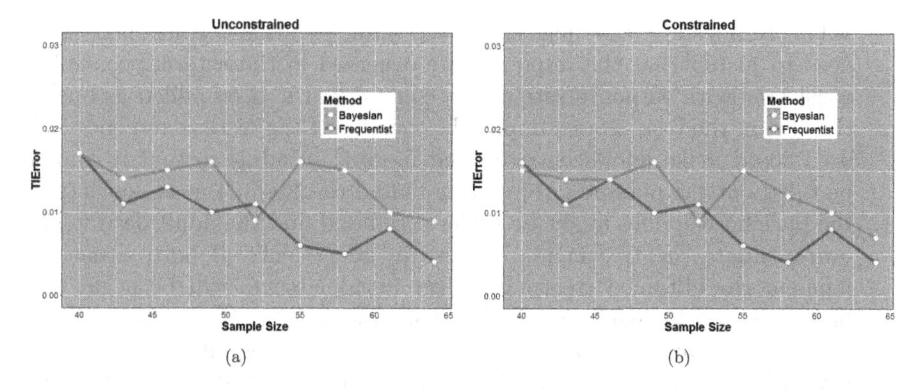

(a) (b)

FIGURE 18.8
(a) Type-I error as a function of sample size without similarity check. (b) Type-I error as a function of sample size with similarity check. The pediatric null response is set at 0.45, i.e., pediatric treatment response rate = adult placebo response rate = 0.45. The sample size for the adult treatment group is 150, while the sample size for the pediatric treatment group is 60.

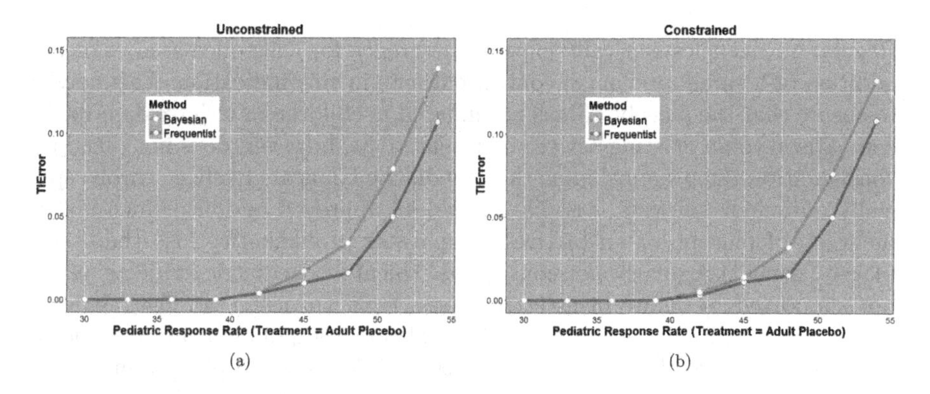

(a) (b)

FIGURE 18.9
(a) Type-I error as a function of sample size without similarity check. (b) Type-I error as a function of sample size with similarity check. The pediatric null response is variable over a range of null responses, i.e., pediatric treatment response rate = adult placebo response rate. The sample size for the adult treatment group is 150, while the sample size for the pediatric treatment group is 60.

response) in both *Unconstrained* and *Constrained* decision criteria, respectively. Note here that when the null response rate is low, then the simulated type-I error rate is almost zero. This is probably just because of the nature of binomial distributions where there is more variability when the responses are in the middle of the (0,1) interval.

> While Remicade® case has been successful, externally controlled trials need to justify that the response rate observed in the external control is consistent with response rate of the control had it been added as one of the arm in the trial. For example, the severity of the disease and endpoint used could make adolescents respond more than adult so that it cannot be ruled out the possibility of getting biased results for the adolescents as the placebo response might be under-estimated and the adult data might over-estimate the treatment response in adolescents. If both situations happen, the obtained treatment effect in adolescents will be negatively biased. Note that regulatory advice evolves with new understanding of the disease, mechanisms of action, etc. What was a successful strategy before may not necessarily be applicable in a particular setting no matter how similar. For example, similarity in PK/PD may need to be demonstrated if a PD marker exists. Otherwise exposure-response may need to be shown to be similar. Usually externally controlled trial needs justification for not having a control—either it is impossible to have patients to be randomized to placebo because of serious irreversible morbidity or there are operational barriers as demonstrated by other placebo controlled trials in the same indication.

Regulatory agencies generally express preference for controlled studies or the addition of a small amount of control patients in this indication. This maybe to ensure that the placebo rate is not inflated in children or to have descriptive comparison to an existing active treatment. This, however, does not take into consideration the practical issue that paediatric trials are likely to enroll especially refractory subjects, who have failed all approved options, which makes inclusion of placebo or an (active) comparator problematic. Also the use of placebo arms is regularly not considered ethical in paediatric subjects and is not supported by many investigators (see, e.g., Turner et al. [59]). Indeed, one of the potential benefits of the use of an extrapolation approach is to avoid the exposure of children to ineffective comparators or placebo in studies particularly when it is known from adult studies that there is direct prospect of benefit. Alternatively, an external control group (e.g., existing registry or new registry) could also be generated in parallel or as an addendum to the conduct of the trial in the treated adult population. This option has various advantageous over a historical control (e.g., current Standard of Care (SoC) and latest end point (EP) assessments), while it may not carry the difficulty of a blinded treatment with a not necessarily well-established efficacy, or worse, the potential for an intended placebo control.

18.3.1.3 Hierarchical Model for Incorporation of Multiple Cohorts

Given that a new investigational drug is generally studied over multiple trials in different types of patients, background therapies or age cohorts under the same the same indication, a hierarchical structure of analysis maybe more efficient. A typical hierarchical model incorporating children might have two levels, a patient level and an age-cohort level. In a two-level structure, cohorts have different but related treatment effects or means. The cohort treatment effects are then related through a common distribution. This relationship is the heart of what is meant by "exchangeable cohorts." Practically speaking, it means that one could not distinguish the cohorts only by looking at the cohort clinical outcome results because there is nothing known a priori that would imply a certain rank ordering of treatment effects. For a two-level hierarchical model, cohort treatment effects or means are exchangeable, and patients are exchangeable within studies. To say that patients are exchangeable means that one could not distinguish two patients from the same study only by looking at the patients' respective results on the endpoint of interest, because there is nothing known a priori that would imply a certain rank ordering of patient results.

In some instances, there might be differences across patient populations or age cohorts that affect the performance of the drug and thus prevent the assumption of general exchangeability across cohorts. If these differences can be identified and measured, then it is straightforward to account for them in a model. Essentially, the model will dictate that the studies are exchangeable, except for measured differences on certain variables. The differences may be related to the size or growth of a patient. For example, a drug whose effect is related to hormone level may have very different magnitudes of effect for adults than for adolescents, because adults and adolescents have different average hormone levels. If hormone level is highly associated with the effect of the drug, then cohorts run on adults are likely to have patient-level information on the amount of the hormone in adults. Patient-level information in adolescents would allow construction of a model that relates hormone level to outcome, and thus conditions on hormone level in order to assume exchangeability between the adult and adolescent populations. That is, except for hormone level, there are no known (and measured) differences between adults and adolescents that would allow one to identify an outcome as belonging to either an adult or adolescent. If there are, then those measured covariates should also be added to the model.

Since the cohorts form just one dataset, they will just be called subgroups henceforth. In regression models, subgroup treatment effects are determined by fitting a single model which includes both the main effects and interactions specified by the covariates, simultaneously. Let $D_i = (y_i, z_i, s_i, \boldsymbol{x}_i)$, $i = 1, \ldots, n$, denote the data from subject i, where y_i is its treatment response with sampling distribution $y_i|\vartheta_i \sim F(\vartheta_i; s_i)$, z_i is the treatment assignment with $z_i \in \{1, \ldots, J\}$, s_i is the subgroup indicator, and $\boldsymbol{x}_i = (x_{i1}, \ldots, x_{ip})$ is a vector of relevant covariates, i.e., if $\boldsymbol{x}'_i$ contains all the covariates, it is possible to work with a subset $\boldsymbol{x}_i \subseteq \boldsymbol{x}'_i$ so that, without loss of generality, $\boldsymbol{x}_i$ denotes the set of all relevant covariates or transformed versions of relevant covariates. When $\theta_{z_i} \equiv g(\vartheta_{z_i})$ there exists a prior distribution G such that $\{\theta_{z_i}\}$, are *exchangeable* as G, i.e., $\{\theta_{z_i}, i = 1, \ldots, n_z\}$ is a random sample from

G_{z_i} [35]. When G_{z_i} depends on a hyper-parameter $\boldsymbol{\eta}_{js} = (\alpha_{js}, \boldsymbol{b}_{js})$, we can write the Bayesian model as:

$$\theta_{z_i}(s_i, \boldsymbol{x}_i; \boldsymbol{\zeta}_i) = (\alpha_{js} + u(\boldsymbol{x}_i)^\top \boldsymbol{b}_{js}) I_{z_i}(j) I_{s_i}(s),$$
$$\boldsymbol{\zeta}_{js} | \boldsymbol{\eta}_j \sim G(\boldsymbol{\eta}_j), \quad \boldsymbol{\eta}_j | \xi \sim P, \quad \xi \sim Q, \tag{18.13}$$

where when $z_i = j$ we can drop the subscript on G_{z_i} as it is already dependent on $\boldsymbol{\eta}_j$, $j = 1, \ldots, J$. In this model α_{js} denotes the overall response of the sth subgroup in the jth treatment arm, $\boldsymbol{b}_{js}$ is the incremental effect of the baseline combination of covariates $u(\boldsymbol{x})$. For subgroup s, the treatment arm has a different intercept and covariate coefficients than the control arm such that $b_{0,s} \neq b_{1,s}$ and $b_{j,1} \neq b_{j,2}$. Such a model is similar to the *Full Interactions Model.* The function u is a defined subset of $\boldsymbol{x}$ such as, $u(\boldsymbol{x}) = (x_1, x_2, x_1 x_2)^\top$ for $\boldsymbol{x} = (x_1, x_2)^\top$. This makes the existence of different orders of interactions to depend on the membership of combinations in each subset. Shrinkage of the parameter estimates is facilitated by assuming $\boldsymbol{\zeta}_{js} \equiv (\alpha_{js}, \boldsymbol{b}_{js}) | \sigma^2, \sigma^2_{\alpha_j}, \boldsymbol{\Sigma}_j \sim G(\cdot | \sigma^2, \sigma^2_{\alpha_j}, \boldsymbol{\Sigma}_j)$, $\alpha_{js} \perp\!\!\!\perp \boldsymbol{b}_{js}$, and $\sigma^2_{\alpha_j} \times \boldsymbol{\Sigma}_j \sim P$, where $\boldsymbol{z}_i = u(\boldsymbol{x}_i^\top)$. What this implies is that cohorts have different but related treatment effects or means and are then related through a common distribution over the entire treatment arm.The variance parameters σ^2, σ^2_α, $\boldsymbol{\Sigma}_j$ are for the sampling distribution F, variance for the hyperprior of α_{js}, and covariance for the hyperprior of $\boldsymbol{b}$. The distributions G and P are priors and hyperpriors, respectively. The distribution G can be assumed as the product of the following prior specification for all the parameters, including the interaction term, i.e., $G = \pi(\alpha_{js} | \sigma^2_{\alpha_j}) \times \pi_{2p+1}(\boldsymbol{b}_{js} | \boldsymbol{\Sigma}_j)$, where $\pi(\alpha_{js} | \sigma^2_{\alpha_j}) = \mathrm{N}(0, \sigma^2_{\alpha_j})$ a diffuse prior, and $\pi_{2p+1}(\boldsymbol{b}_{js} | \boldsymbol{\Sigma}_j) = \pi_{2p+1}(\boldsymbol{b}_{js}^\top | \boldsymbol{\Sigma}_j) = \mathrm{N}_{2p+1}(0, \omega_j^2 \boldsymbol{I}_{(2p+1)\times(2p+1)})$, with $\boldsymbol{\Sigma} = \omega_j^2 \boldsymbol{I}_{(2p+1)\times(2p+1)}$; and P is the distribution of the hyper-parameter ω^2, which is taken as $\mathrm{IG}(c_0, d_0)$, $\pi(\sigma^2_{\alpha_j}) = 1$, and $\pi(\sigma^2)$ which is taken to be another inverse-gamma distribution.

There are situations when the effect of a treatment on a pediatric population is of a lower (or higher) magnitude than on an adult population, due to the pediatric population having a generally lower or higher mean level of a covariate that influences the treatment effect. Furthermore, there may be several covariates that interact with treatment, and a model could assume several treatment by covariate interactions to account for them. In this case, a proportionally interactions model (PIM) may be also used which assumes that some of these interactions are the same proportionally across covariates, leading to a more parsimonious model (see, e.g., Kovalchik et al. [60]). Any model with interactions or proportional interactions can be easily extended to a hierarchical model with partially exchangeable studies.

As mentioned earlier, this type of model relies on assumption of *exchangeability* so that leveraging of information only happens among patients whose outcomes exhibit some degree of similarity. In fact, the assumption is that cohorts have different but related treatment effects or means and are then related through a common distribution. While this may be sometimes true in diseases that manifest in both adults and children, the assumption does not provide latitude when responses in some age cohorts are consistent only to a few but not all. The pediatric population is many be heterogeneous from neonates to adolescents with major developmental cognitive and physiological changes ending with puberty and significant differences in PD and PK. Hence, the value of the information of one cohort onto another may differ and one may be lead to believe that cohorts are similar to either even if the outcomes of the patients are too different. In this case, the standard *exchangeability* assumption bears the risk of too much shrinkage and excessive borrowing for extreme strata. For this reason, a prior distribution that allows some cohorts to be very different from other cohorts or to have groups of subjects that cluster close together is desirable. There models that can accommodate clustering are more relevant without going beyond the *a priori* exchangeability assumption. One such approach is described by Neuenschwander et al. [61] who proposed an exchangeability-nonexchangeability (EX-NEX) approach as a robust mixture extension of the standard exchangeability approach. In particular, clustering is accommodated by using a mixture distribution $G_\alpha = \pi(\alpha_{js}|\sigma^2_{\alpha_j}) = (1 - w_j)\mathrm{N}(\mu_j, \sigma^2_{\alpha_j}) + w_j\mathrm{N}(m_j, s^2_{\alpha_j})$, i.e., with chosen probability $1 - w_j$ a particular cohort has overall mean following $\mathrm{N}(\mu_j, \sigma^2_{\alpha_j})$ but is non-exchangeable with other cohorts with probability w_j and in that case follows $\mathrm{N}(m_j, s^2_{\alpha_j})$. If it is suspected that a particular cohort behaves systematically different from the others, then the value of w_j can be increased.

As a form of generalization to the EXNEX model, in the absence of parametric knowledge of G_{z_i}, one should choose a prior for G_{z_i} with support on the set of distributions on the real line, with this prior effectively corresponding to a distribution over distributions. One popular choice proposed by Bush and MacEachern [62] is the Dirichlet process [63–65]. A Dirichlet Process, denoted as $\mathrm{DP}(\kappa, B_0)$, can be characterized by two parameters: G_0, the base measure, and κ, the precision parameter. The base measure specifies the expectation $\mathbb{E}(G(A)) = G_0(A)$, which is the prior guess of $G(A)$, for any set A in the sample space. The precision parameter κ specifies the uncertainty in the guess, i.e., the larger the value, the more we expect a realization of DP to be from G_0. Sethuraman [66] and Sethuraman and Tiwari [67] gave a constructive definition of DP through the stick-breaking implementation described as:

$$G(\cdot) = \sum_{l=1}^{\infty} w_l \delta_{\zeta_l}(\cdot), \quad \zeta_l \overset{i.i.d.}{\sim} G_0 \quad \text{and}$$
$$w_l = v_l \prod_{j=1}^{l-1}(1 - v_j) \quad \text{with} \quad v_l \overset{i.i.d.}{\sim} \mathrm{Beta}(1, \kappa), \tag{18.14}$$

where $\delta_A(\cdot)$ is a Dirac delta function denoting a point mass at A. Under this construction, θ_i has the induced distribution from the distributions of α_i and $\alpha_i + \mu_j|\alpha_i$, namely, $G \sim \pi^J(\alpha_i + \mu_j|\alpha_i) \times H$, $\alpha_i + \mu_j|\alpha_i \sim \pi(\alpha_i +$

$\mu_j|\alpha_i) = \mathrm{N}(\alpha_i, \omega^2)$. In addition, κ is commonly assigned a gamma hyper-prior to allow the data to inform more strongly about clustering in the data and the extent to which H is similar to H_0, with Gamma$(2,1)$ providing a commonly used choice as it has mean equal to 2 and variance equal to 2. Another choice for the prior for κ is the uniform distribution $\mathrm{U}(0.3, 10)$ [68], where the lower bound of 0.3 was specified to avoid computational difficulties caused by small w_l's. Completing the prior specification, τ_0 and τ, respectively, the precisions associated with σ^2 and ω^2 are given a gamma distribution Gamma(c_0, d_0) while $\alpha \sim \mathrm{N}(0, 10^5)$. Since DP can be represented as an infinite mixture as seen in Equation (18.14), Ishwaran and Zarepour [69] suggested truncating the number of allocations at a large integer L, taking $L = \sqrt{n}$ or $\log(n)$ for large n, and $L = n$ for small n. This reduces $H(\cdot)$ into a finite dimensional form as $H = \sum_{l=1}^{L} w_l \delta_{\zeta_l}(\cdot)$.

Example 18.3: A Phase 3 placebo-controlled study to evaluate the safety and efficacy of an investigational treatment in adult and adolescent patients with moderate-to-severe atopic dermatitis (AD) who have responded inadequately to or who are intolerant to topical therapy. If the Phase 2 study demonstrates proof-of-concept in reducing disease activity in adults with AD and has an acceptable safety profile, pediatric patients ≥12 years will be eligible to enroll in a combined phase 3 adult/adolescent study. The adolescents will participate in the combined adult/adolescent phase 3 study through an addendum which will include an open-label PK lead-in to confirm exposure-matching to the dose(s) planned for the adults ($\tilde{1}0$ patients), followed by a 52-week placebo controlled portion ($\tilde{1}60 - 240$ adolescents). By an addendum, if the adolescent cohort does not enroll as fast as the adult cohort, the adult and adolescent cohorts will be analyzed independently. The doses for the adolescents will be identified through a PK modeling and simulation approach using PK data from adults, where the impact of bodyweight on PK will be evaluated. The primary endpoint for will be the proportion of patients achieving an IGA score of 0 or 1 (clear or almost clear skin) at Week 16 (end of induction period) with a ≥2-point improvement from baseline. Assuming at least 70 patients per arm complete the 16-week primary endpoint, the proposed sample size will ensure >80% power to detect an absolute difference of 20% between an LY treatment group and the placebo treatment group, assuming a 10% placebo response rate for the primary endpoint using a two-sided alpha of 0.05. Patients who complete the study will be eligible to enter a year long-term extension study. The primary analysis for the adult cohort will be done using traditional frequentist means. The primary analysis will be conducted through a hierarchical model (which will be described further below) that analyzes all cohorts, including adolescents, together. In addition, a multi-center, placebo-controlled trial to evaluate an investigational treatment in mono-therapy for pediatric patients 6 months to <12 years old with moderate to severe AD is proposed once the safety and efficacy of LY in mono-therapy has been demonstrated in adults and pediatric patients ≥12 years with AD. The entry criteria will be similar to criteria used for pediatric patients 12 to <18 years old. This study will include a PK lead-in (1 dose with 4 weeks follow up), the number of doses and dose range to be used in PK lead-in will be informed by initial results from the study in patients 12 to <18 years old. Similar to the adolescents, the doses will be identified through a PK modeling and simulation approach using data from older children and adults, where the impact of bodyweight on PK will be evaluated. The PK lead-in phase will be followed by a 16-week placebo-controlled phase (unequal randomization).The primary endpoint will be the proportion of patients achieving an IGA score of 0 or 1 (clear or almost clear skin) at Week 16 with a ≥2-point improvement from baseline. This study will enroll at least 120 children. Given that there is preponderance of similarity in disease progression and response to therapy in these populations as determined by an expert industry advisory board (see [70]), a Bayesian hierarchical model is proposed for the analysis of the primary efficacy outcome at 16 weeks for the adolescent, children with ages 6–12 years old, and children with ages 6 months to 12 years old cohorts (see Figure 18.10). In addition, a long-term extension is planned to monitor long-term efficacy and safety.

Simulations were done to explore the operating characteristics of the hierarchical model which follows a simple form as given in Equation 18.13. Results of the explorations show that the type-I error rate is generally well-controlled (minimal increase in type-I error) as shown in Figures 18.11

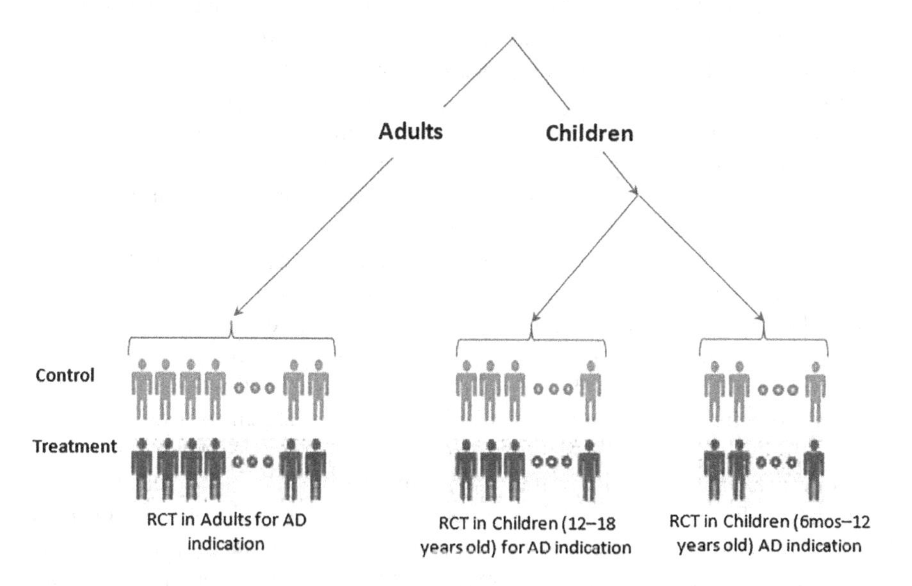

FIGURE 18.10
Hierarchical model for Example 18.3.

and 18.12. In these figures, the response rate in adults and adolescent or adults and children are fixed in order to see what would be the type-I error rate over varying responses in children and adolescents, respectively. Figures 18.13 and 18.14 show the corresponding power curves when the response rate in adults and adolescent or adults and children are fixed. It is clear that the Bayesian approach is more powerful with minimal increase in type-I error.

The hierarchical model is the most flexible method to incorporate different but related studies. It is a reasonable analysis when studying multiple age cohorts or populations (e.g., with background therapy, standardization periods, etc.). Furthermore, this analysis can still be done even if studies are not done in parallel but as long as there is patient level data in each of the studies for the relevant model parameters. It can also be used for both controlled and uncontrolled cohorts. See Sections 18.3.1.1 and 18.3.1.2.

18.3.1.4 Additional Validation Criterion

Note that the Bayesian approach may still borrow empirically despite there is preponderance that similarity in the response for intervention may not be biologically supported. To mitigate this problem, an additional validation for the extrapolation assumption with respect to clinical efficacy can be placed by incorporating the following in the testing procedure.

(1) Comparison of adult and children's clinical response to treatment: Compare whether the treatment response is greater than the lower limit of the 95% CrI of the estimate of the posterior mean of treatment in the adult studies. The lower limit can be replaced by any clinically meaningful margin related to equivalence. This is to verify that the response to treatment between adults and children is numerically similar or at least the response in children is not significantly worse than the adults. In controlled trials, this is an important step to effective extrapolation, i.e., provide rules toward proper extrapolation aside from the empirical calibration of extrapolation which can be implemented through the Bayesian methodology. Furthermore, it can be implemented in the borrowing procedure explicitly by adding an indicator of whether the response in children is within the credible bands of the response in adults. For example, borrowing can be warranted if $\hat{\theta}$ is in the 95% CrI of the corresponding estimate of the parameter in the adult or source population. See Example 18.4.

(2) Comparison of adult and children's response to placebo or relative comparator: For controlled trials, this may not be relevant. However, it does provide some clue into the heterogeneity of disease progression and response to therapy in children. For externally controlled trials, this comparison needs to be established and perhaps the most logical way to demonstrate efficacy absent placebo control. For example, the adult

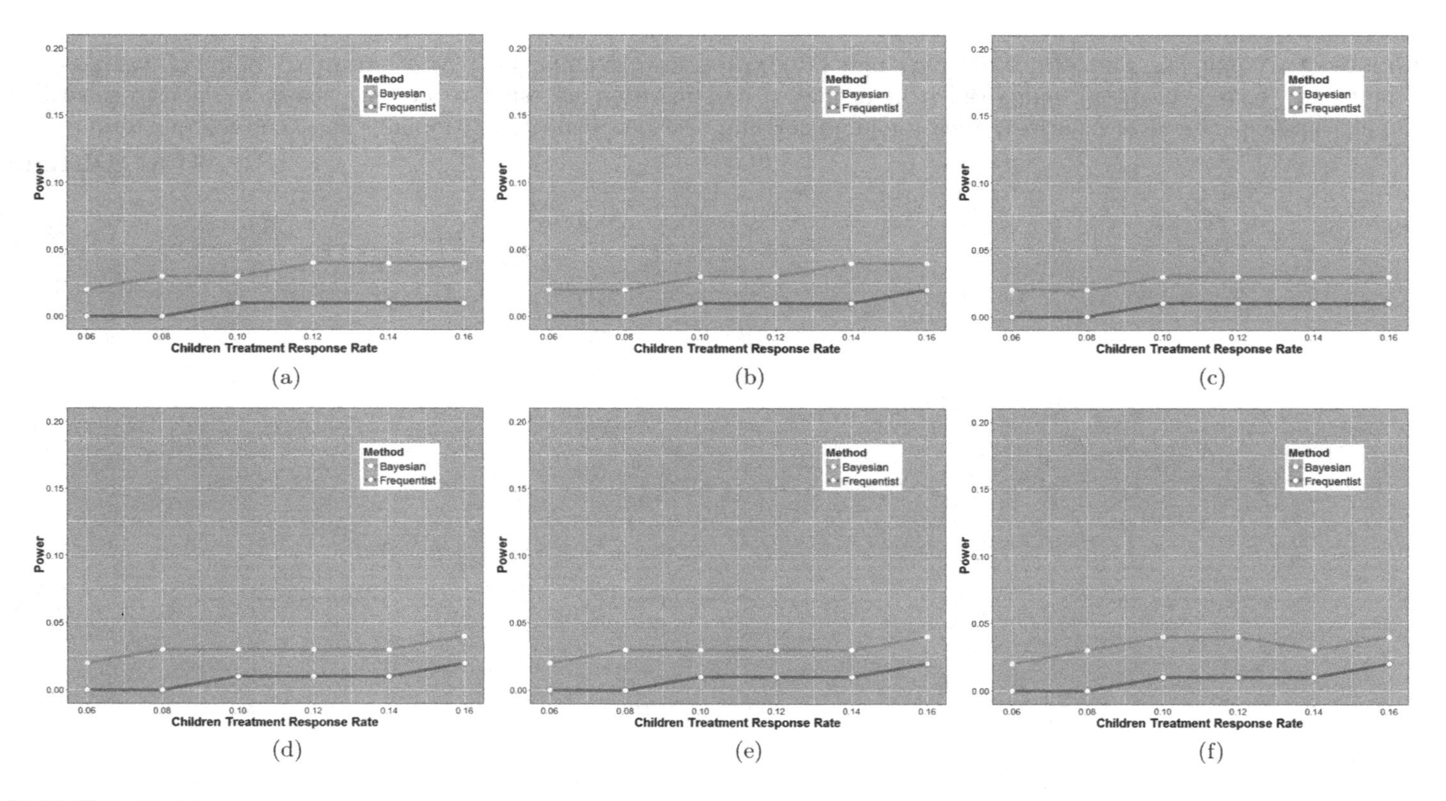

FIGURE 18.11
Type-I error rate for detecting superiority in Children as a function of children's treatment response when the adolescent's treatment response is fixed. Placebo response for adolescents is kept at 0.10. (a) Adol Trt Resp = 0.06, (b) Adol Trt Resp = 0.08, (c) Adol Trt Resp = 0.10, (d) Adol Trt Resp = 0.12, (e) Adol Trt Resp = 0.14, and (f) Adol Trt Resp = 0.16.

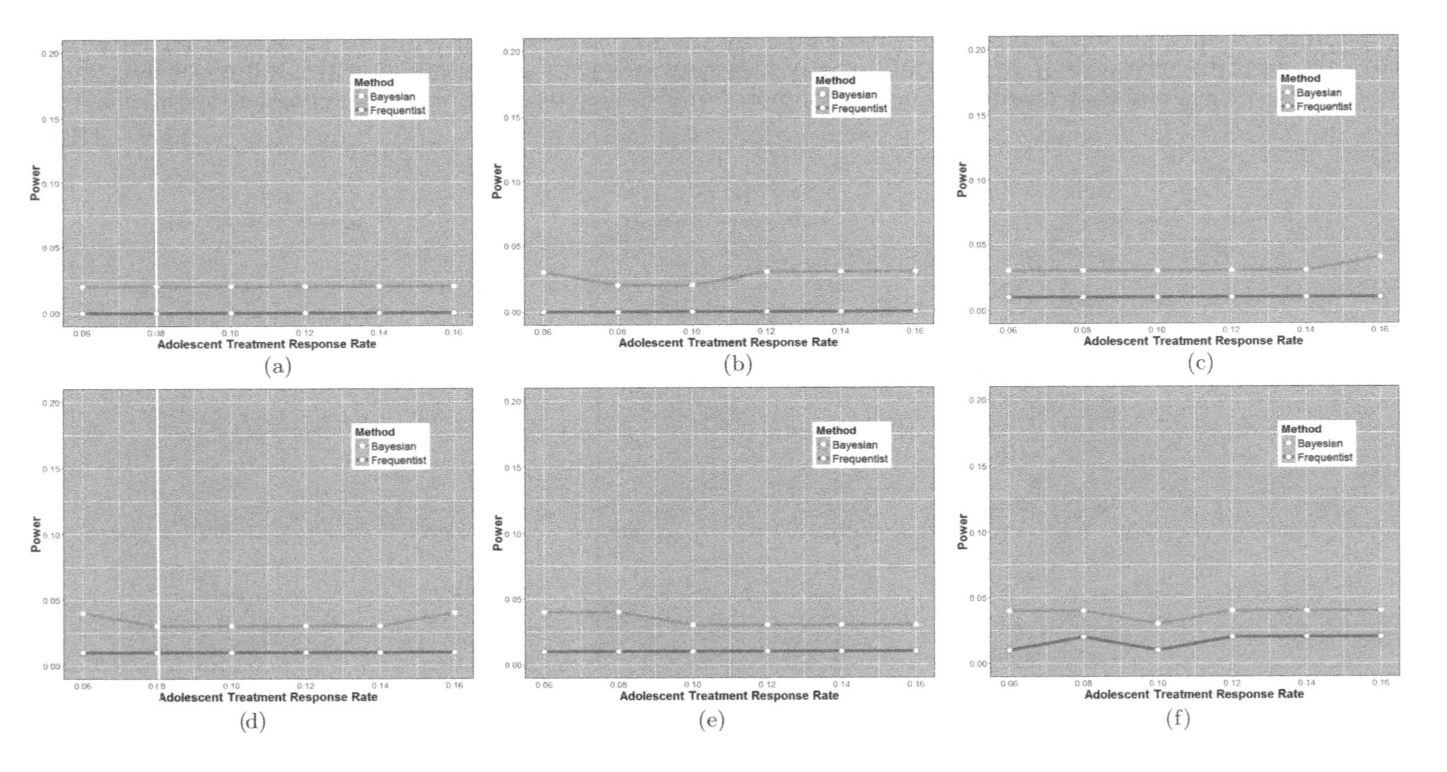

FIGURE 18.12
Type-I error rate for detecting superiority in Adolescents as a function of adolescents' treatment response when the children's treatment response is fixed. Placebo response for both children is kept at 0.10. (a) Child Trt Resp = 0.06, (b) Child Trt Resp = 0.08, (c) Child Trt Resp = 0.10, (d) Child Trt Resp = 0.12, (e) Child Trt Resp = 0.14, and (f) Child Trt Resp = 0.16.

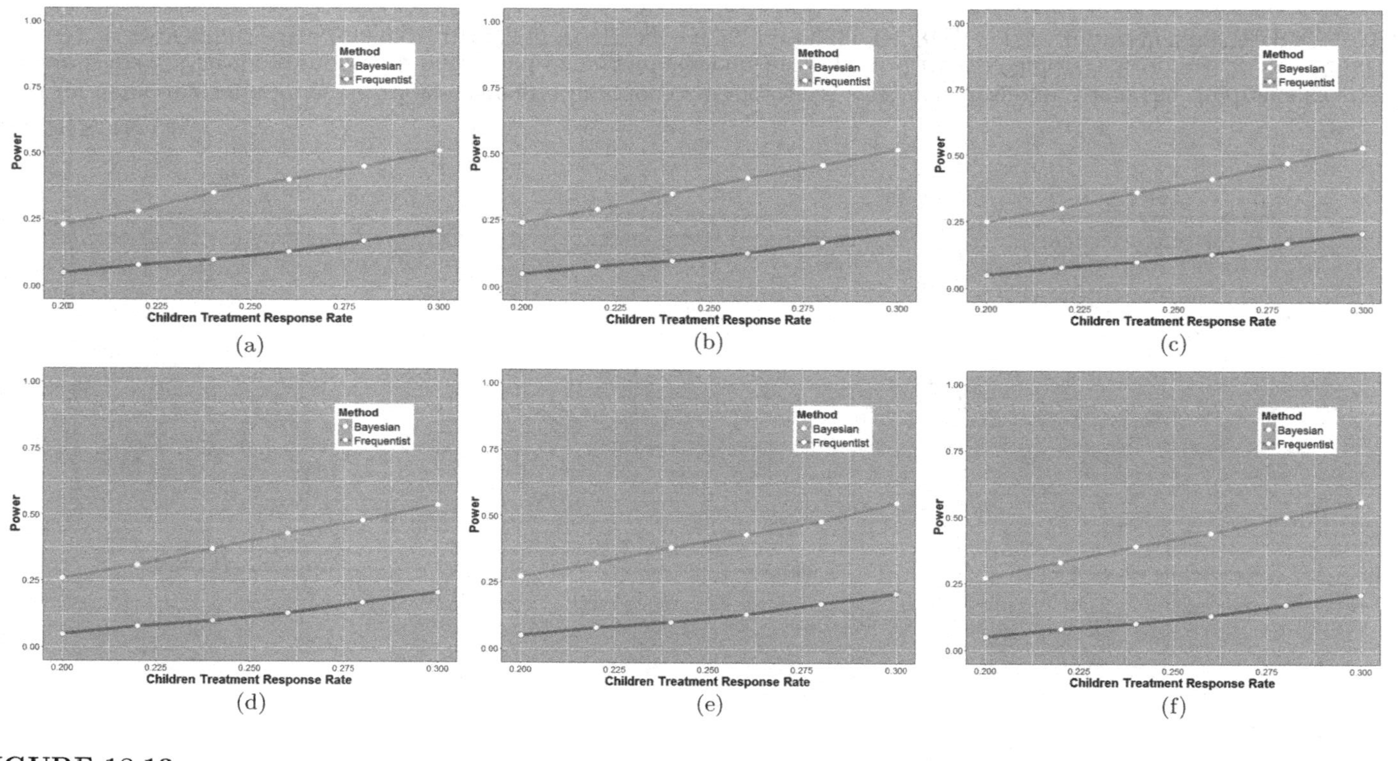

FIGURE 18.13
Power for detecting superiority in Children as a function of children's treatment response when the adolescent's treatment response is fixed. Placebo response for both children and adolescents are kept at 0.10. (a) Adol Trt Resp = 0.20, (b) Adol Trt Resp = 0.22, (c) Adol Trt Resp = 0.24, (d) Adol Trt Resp = 0.26, (e) Adol Trt Resp = 0.28, and (f) Adol Trt Resp = 0.30.

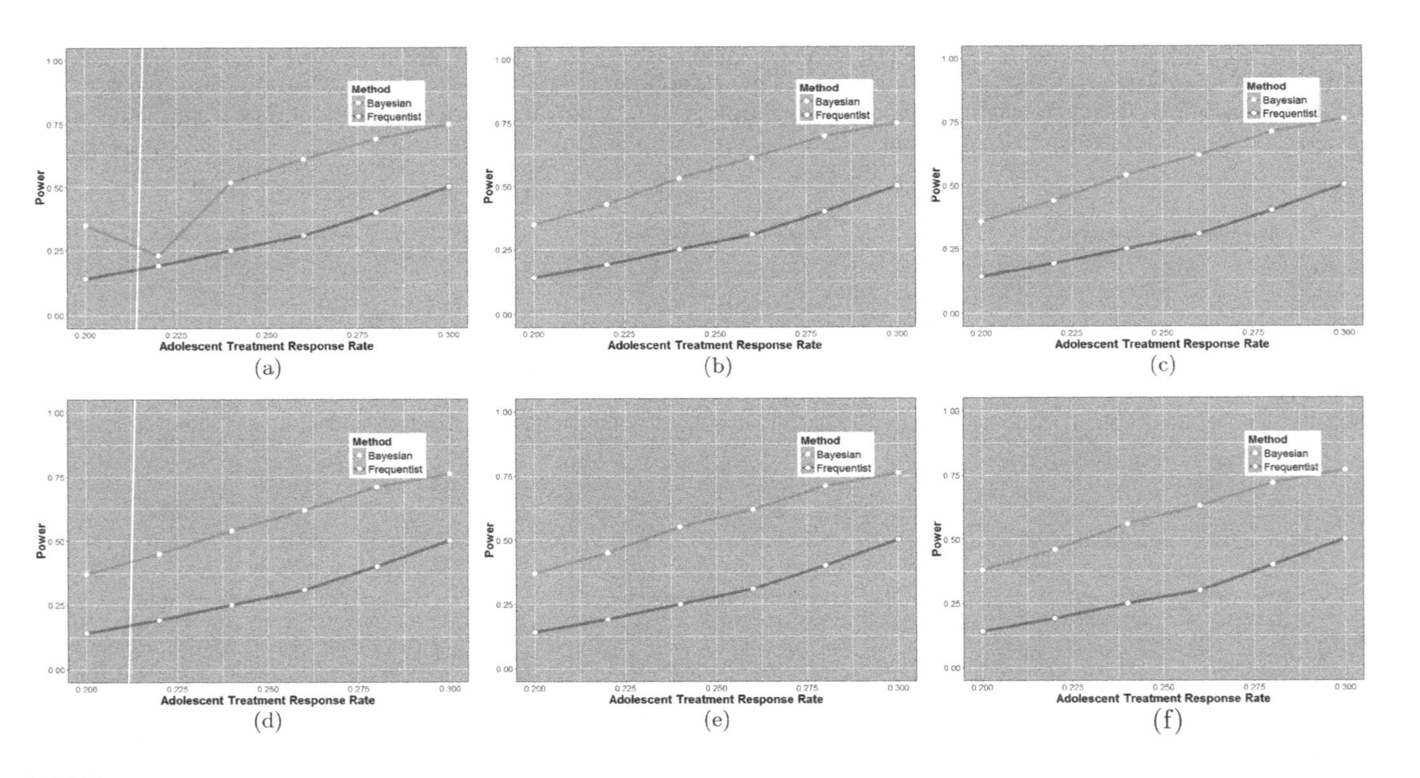

FIGURE 18.14
Power for detecting superiority in Adolescents as a function of adolescents' treatment response when the children's treatment response is fixed. Placebo response for both children and adolescents are kept at 0.10. (a) Child Trt Resp = 0.20, (b) Child Trt Resp = 0.22, (c) Child Trt Resp = 0.24, (d) Child Trt Resp = 0.26, (e) Child Trt Resp = 0.28, and (f) Child Trt Resp = 0.30.

placebo or relative comparator response rate might be compared with literature data on response rate in the target population. Ultimately, the question is about whether the borrowing of information will lead to a decision of approving a drug that does not offer benefit. Hence, the high hurdle that needs to be put in place for that decision to be warranted.

In general, for controlled studies, one can argue that establishing similarity in response to the drug between the source and target population is sufficient, i.e., establishing similarity in response to placebo may not be relevant. Because of the existence of the control, internal validity is always present as given by the measure of clinical. For uncontrolled studies, both conditions need to be satisfied. For multiple cohorts, response to treatment in the target cohort needs to be consistent with cluster it belongs.

Example 18.4: To validate the assumption of similar expected results in adults and adolescent patients, one can explicitly include a consistency check on the prior information in Equation 18.6 by restricting the leveraging of information from adult to adolescent patients only to situations when the responses for specific treatment arms in the adolescents patients are consistent with the responses observed in the respective treatment arms in older paediatric patients, i.e.,

$$\hat{\pi}(b_j) = [(1-w)\hat{p}(b_j|\boldsymbol{D}_0) + wp(b_j)]I_{[c_{z,0},c_{z,1}]}(\hat{\theta}_z), \tag{18.15}$$

$z \in \{0,1\}$. Specifically, leveraging of information in respective treatment arms is allowed only when the following conditions are met:

(i) For the investigational treatment arm: Leveraging will occur only if the mean treatment response of treated adolescent patients is above the lower bound of the 95% confidence limit for the mean treatment response of the corresponding treated adults, i.e., $c_{1,0} = \hat{\theta}_1 + q_{0.975}Var(\hat{\theta}_1)^{1/2}$ and $c_{1,1} = \infty$.

(ii) For placebo: Leveraging will occur only if the mean treatment response for placebo in adolescent patients is below the upper bound of the 95% confidence limit for the placebo mean responses in adult patients, i.e., $c_{0,0} = -\infty$ and $c_{0,1} = \hat{\theta}_1 - q_{0.025}Var(\hat{\theta}_1)^{1/2}$.

The amount of weight that is put on data from adults will vary depending on results of the consistency check, i.e., if the treatment response in adolescents is consistent with the treatment response in adolescents, the maximum amount of weight that would be applied is w; while no information will be leveraged between age cohorts if substantially differing treatment responses are observed between the cohorts. If the consistency check is not satisfied, the informative part of the prior is wholly abandoned and only a vague (non informative prior) will be used. Alternatively, the consistency check can be based on what is clinically meaningful from previous adult studies, e.g., at least half of the treatment effect based on the adult Phase 2 study.

Another criterion that can be used if the goal of the analysis is to ensure consistency of response between analogous doses of the two cohorts determined by whether the treatment effect in the younger paediatric population meets the minimum of the following:

(i) preserves at least half of the treatment effect (investigational treatment minus placebo) observed in adult patients (this criterion is aligned with criteria typically applied for regional subgroup analyses for multi-regional studies [71,72]), OR

(ii) treatment effect (investigational treatment minus placebo) in adolescent patients is at least 10% (i.e., at least half of the treatment effect based on the adult Phase 2 study);

AND

(iii) to determine whether there is sufficient separation by the high dose from placebo in the younger paediatric patients as indicated by superiority over placebo.

The threshold used is consistent with assessments of subgroups of populations which are typically set at 0.10 two-sided level. A dose is deemed efficacious in younger paediatric patients if these two success criteria are satisfied. If the consistency check is not achieved, the usual frequentist testing strategy with 0.05 (one-sided) alpha will be used. Hence, the type-I error will always be at the nominal 0.05 level.

In any of these proposals, a simulation is generally needed to describe the operating characteristics of the decision criterion for declaring whether an investigational treatment is efficacious.

It is important to note that if the goal in extrapolation is to validate an assumption of similarity (e.g., paediatric efficacy versus efficacy predicted from the source), then it is important to apply confirmatory standing to the situation of validation, e.g., narrow clinically relevant equivalence margin perhaps at a higher nominal level than 5% to reflect the justified assumption supporting similarity. This is the proposition in Equation 18.15 above. In fact, one could argue that both validation and generation of pivotal evidence should be considered similarly with a confirmatory-like criterion (equivalence limit for the validation and significant difference for confirmatory) and, for both validation and pivotal evidence, a reduced nominal alpha level maybe proposed to reflect the justified assumption supporting similarity. A reduced alpha level is also arguably justifiable given the stringent nature of a joint decision criterion.

18.3.2 Use of Two-Sided "$p > 0.05$" or One-Sided Probability Threshold Less Than 0.975

With the usual requirements for significance level and study power, pediatric trials employing a conventional design may only be able to detect very large treatment effects. A related challenge is to design a trial with an acceptable compromise between (1) level of scientific evidence, in accordance with similarity of the disease between, say, adult and adolescents and the level of available supporting information from adult or source data, and (2) feasibility in terms of trial size and duration, recognizing that a trial that cannot be completed in a timely manner does not have social value in terms of addressing public health question either. One solution is to increase α, as was done in the alternating design trial of itraconazole by Gallin et al. [73]. The type I error is the probability of wrongly rejecting a null hypothesis (H_0); erroneously concluding the investigational treatment is efficacious, active or interesting. This is traditionally 5% i.e., 1/20. The type II error is the probability of erroneously accepting the null hypothesis; missing an interesting treatment. In effect, one might decrease the required sample size by accepting a higher type I error rate more. In the context of extrapolation, where the diseases and response to treatment are believed to be similar in adults and pediatrics, the phenomenon of "*chance*" finding of positive effect of an ineffective treatment, which by increasing the significance level to 0.10 implies a finding of 1/10 replications, does not happen quite often if the drug has been established to be effective either in the adult indication or other related indications. For this reason, the context of a strict control of type-I error in extrapolation appears ill-construed.

Another solution is to conduct the underpowered study and incorporate the results into a prospectively planned meta-analysis. This approach is used in EORTC-1206-HNCG of salivary gland carcinomas (SGC), a heterogeneous group of rare tumours [74]. This can also be an attractive option especially when placebo has been studied extensively in head-to-head trials with other compounds in the same pediatric indication and can leverage the information on placebo in a frequentist network meta-analysis [75]. Alternative Bayesian formulation of the estimation can be used as well (see e.g., [76]). A third option is to incorporate the results into a Bayesian framework as discussed in Sections 18.3.1.1–18.3.1.3. Many authors, including Lilford [77], recommend this approach for trials in pediatric diseases in which the individual trials are unlikely to result in a definitive answer but each can change the level of certainty around the clinical question. As discussed previously, the Bayesian approach uses all available data from the trial and other sources-to calculate probabilities that a particular treatment is effective. These probabilities can then be applied to clinical practice. In fact, uncertainties can still be addressed within a Bayesian framework through informative priors.

There may be arguments that reducing the alpha level might dilute interpretation of evidence with respect to clinical relevance and should be restricted only to cases where clinically justified. In the context of extrapolation, this is most likely clinically justified. Furthermore, in the light of the principle of *scientific necessity*, there are references from Neyman as well as from Fisher that the level of significance has to be chosen based on the whole context (scientific, economic, aims, limitations ...). In the Neymanian philosophy a "conventional level" makes no sense because the fixed constraint is the cost/benefit ratio of the research, and there is anyway no possibility to sensibly decide on an acceptable level *a posteriori*. In the Fisherian philosophy there is no cost/benefit ratio, and the criteria to select a level are made not so explicit. It may be helpful, however, if such a reduced level of significance is combined with other measures to see consistency across varying endpoints. Furthermore, reporting the effect size, confidence intervals, and p-value itself rather than the dichotomy of significance or non-significance maybe more informative.

18.3.3 Use of Alternative Data When Extrapolation Is not Supported or Validated

Real-world data (RWD) consisting of data that are routinely generated or collected in the course of health care delivery or otherwise [78], they could also have potential to support clinical development or potential source of evidence for regulatory consideration [79]. Potentially, real world evidence (RWE), defined as the clinical evidence regarding the usage, and potential benefits or risks, of a medical product derived from analysis of RWD [79], can potentially reduce the cost and duration of clinical trials in the process of evaluating the safety and effectiveness of medicinal products. On this note, there has been is substantial enthusiasm to generate real-world evidence (RWE) for regulatory decision making with the wide availability of various RWD. In fact, public-private partnerships, health technology assessment organizations, either in the EU or US, have launched major initiatives to address the concerns and consideration in the use of real world evidence (RWE) to inform regulatory decision making. Recent publications from regulatory leadership (cf. [78,80]), draft guidance for use of RWE for devices [79], many medical & regulatory related conference discussions and white papers–such as from the Duke Margolis conference and NAM (National Academy of Medicine), have put forth the challenges and presented potential scenarios where RWE may be considered as part of the evidence package in support of regulatory decisions.

The use of RWD in support of pediatric clinical trial is in line with the principle of *scientific necessity*. However, because the decision to potentially extend labeling to children can be based on these data, it is important that the data and how it is synthesized is well characterized. For instance, as RWD are not collected or organized with the goal of supporting research nor optimized for such purposes, discussion must be informed by a clear understanding of the assembly of the data as well as the methods selection of appropriate analytic approaches will need to be determined using the key dimensions of study design, including the use of prospectively planned interventions and randomization [80,81]. Previous research from OMOP (Observational Medical

Outcomes Partnership) raised challenging questions regarding the ability of observational research to accurately detect causal effects and avoid false findings. To improve the operating characteristics - and lead to confidence in decision making from such research, improvements are needed in (1) quality and linking of data sources; (2) research designs; and (3) statistical methodology. It is also important to adopt a rigorous approach to the analysis and to how it weighs the evidence from RWD in arriving at a decision. Note that implicit in the decision to combine data across studies and RWD sources is the assumption that the observations are similar or exchangeable, i.e., patients in the studies and or alternative data sources are similar and measuring the same outcome [82]. While randomization is maintained within each study, it no longer holds across the pooled data included as patients are not randomized to different trials nor when their observations are collected in RWD. As a result, there are systematic differences in study characteristics (interventions, outcomes and measurements, study design elements, etc.) or the distribution of patient characteristics across data sources. If these characteristics are effect modifiers (i.e., they only influence the treatment effects), then there are systematic differences in treatment effects across studies or between-study heterogeneity [83]. Varying levels of measured and unmeasured effect modifiers in each of the studies manifests as heterogeneity that needs to be explained.

There are several analytic approaches that can be employed to leverage patient-level data from RWD sources. Study entry criteria can be applied to exclude non-comparable patients from the historical control arm. A staged approach to incorporation of data can also be planned to avoid bias. In this approach, outcomes of the RWD sources are masked prior to matching (see Figure 18.15). Comparison of treatment and control can then be conducted by first stratifying patients on important covariates and weighting strata-specific outcomes based on cohort differences to generate a valid overall control outcomes estimate. Alternatively, a matched-pair control cohort can be created using propensity scores and outcomes compared (cf. [84]). Both of these approaches were used to compare outcomes in adult patients with B-precursor Ph-negative relapsed/refractory acute lymphoblastic leukemia in the recent blinatumomab (Blincyto) submission [85,86]. A further extension of these methods could involve the use of external data to set success and futility thresholds to inform a "threshold crossing study" [87] or using the information from the matched observations as a prior to ensure exhangeability of populations from where the prior is obtained [88]. Results from patients treated in a single-arm open-label study can then result in one of three outcomes: (1) the product is judged "effective" and initial approval granted if outcomes exceed the efficacy threshold; (2) the product is deemed "ineffective" and development terminated if outcomes are below the futility threshold; or (3) given intermediate outcomes, the product is deemed "promising" and study continues in an adaptive manner with either addition of a randomized control arm or with a subsequent single-arm threshold crossing study. The success of any such approach requires that the aforementioned concerns about data

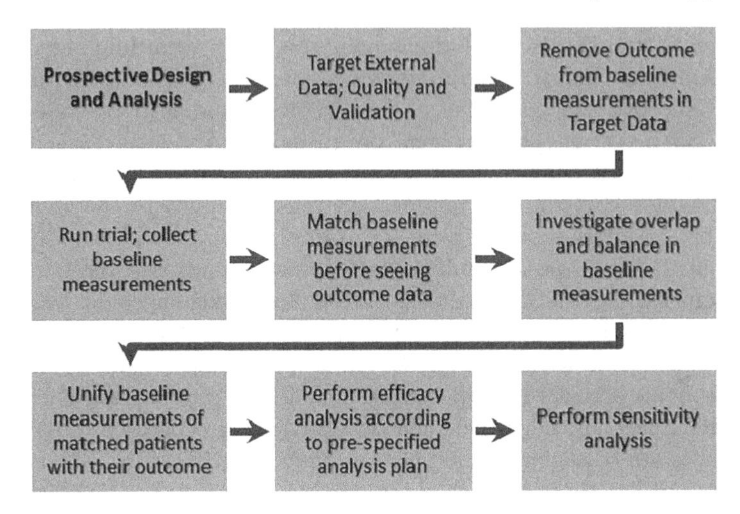

FIGURE 18.15
Proposal for incorporating RWD in support of clinical trial data.

availability, quality, and completeness are adequately addressed. However, all have the potential to obviate or at least decrease the size of the control arm (and potentially the treatment arm) in a randomized study, and to increase study power relative to traditional randomized trials.

Ultimately, these data sources and approaches should not be used in a one-size-fits-all manner, but rather in a thoughtful, fit-for-purpose approach depending on the novel treatment of interest, benefit-risk, and unmet need. For example, use of multiple historical real-world controls and a more rigorous analytic design might be required when assessing treatment with a novel mechanism of action, whereas pediatric placebo data and/or adult treatment data might be used to inform a threshold-crossing study when assessing the efficacy of an additional treatment option in an already crowded drug class.

18.4 Commentary and Discussion

18.4.1 Important Considerations to Extrapolation

Although extrapolating efficacy from adult or other data to the pediatric population provided a path toward a streamlined pediatric drug development, one must be aware of some important considerations including the following:

i. In using pediatric extrapolation, early planning during adult development is necessary to generate the required data to extrapolate to the overall pediatric population or to pediatric subgroups so that adult trials can be designed with awareness of how this can support pediatric labeling in the future. In particular, this ensures that adult development can be supportive of the pediatric drug development (e.g., innovative trial design, endpoint selection and measurement). Typically, early-phase development for investigational drugs is conducted in adult healthy volunteers and adult patients as part of the proof-of-concept evaluation. For certain diseases and population subsets (adults and adolescents), there may be little difference in renal capacity or hepatic enzyme expression, leading to a high degree of congruence on dosing and allowing for the inclusion of some pediatric subsets within the adult Phase II and III program [89] as long as the prospect of direct benefit has been established wither through exposure-response assessment or Phase 2 adult studies. This may lead to the inclusion of some pediatric subsets within the indication for use statements of newly approved therapies, facilitating earlier pediatric access to new medicines. Note also that pediatric drug development has a long sequence of steps including the following:

- Toxicology in juvenile animals where appropriate;
- Formulations: ease of administration and flexibility in dosing;
- Natural history and disease assessment;
- Modeling through PBPK and systems biology and clinical trial simulations;
- Exposure-Response assessment and modeling;
- Innovative study designs and alternative data: prior-augmented designs, adaptive designs, "threshold-crossing" open label studies, enrichment strategies, withdrawal designs, etc.

Many of these steps can be optimized within the overall clinical development plan.

ii. The identification of the correct pediatric dose(s) in pediatric trials is important as incorrect dose has been shown to be the primary reason for failure in pediatric trials (see Momper et al. [90]). On this note and to use extrapolation effectively, the goal is to use doses that have comparable exposure-response to adults as proof of concept (particularly, in establised MOAs) and (2) to facilitate potential borrowing of source clinical data. As uncertainty often exists relating to pediatric dose selection, a range of doses should generally be tested. At the higher end of the dose range, a dose should be used that achieves drug exposure at least as high as established with an efficacious response in adults. Testing more than one dose provides valuable information regarding dose response relationships, which are critical to selecting the optimal dose. Exploring a broad range of tolerated doses may also be useful to determine if drug exposure beyond that which is efficacious in adults is necessary. Pediatric dose ranges should account for observed differences in PK, PD, and/or clinical response between adults and the pediatric. It may also be beneficial to introduce specific clinical study design elements in the adult trials (e.g., additional timepoints, biomarkers, wider distribution of body weight) to inform modeling of doses in target cohorts. Note that it will be difficult to leverage efficacy information if the doses or exposures do not match between adults and children, or at least exposures achieved by these doses do not match between the source and target population. Techniques such as extrapolation and Bayesian statistics cannot be used unless the adult E-R is sufficiently characterized; must be done in early phase adult program [91].

iii. Emphasis should be given on sufficient quality of data from adult population in terms of study design, data collection, and measurement. Practice of medicine has changed since the drug was initially approved to such an extent that historical data would likely be different than prospectively collected data. A crucial factor in designing pediatric efficacy studies is that the clinical endpoints must be age and developmentally appropriate and validated for use in the pediatric age group under study. If adult and pediatric endpoints are different, this also casts shadow on the ability to leverage existing information. In this case, it maybe helpful to ensure that there is a common measure that will help correlate the two endpoints. Alternatively, one can do Bayesian modeling on the common endpoint that would help support establish efficacy in the primary endpoint. If there is no similar endpoint or modeling to project information contained in source endpoint toward target endpoint is mathematically intractable then perhaps alternative designs can be used, e.g., adaptive designs to determine efficacy sooner when evidence of superiority is achieved at a pre-specified timepoint.

18.4.2 Regulatory Considerations with Respect to Innovative Designs and Analysis or Use of RWD

As mandated by regional regulations, every new medicinal product intended for pediatric use should be prospectively agreed with authorities in context of the PSP or PIP. In the EU, for instance, there is a series of procedural and regulatory questions that need to be addressed prior to the agreement of the PIP. Such questions may revolve around how exactly should extrapolation concepts and plans should be handled, or what are the procedural aspects to get agreement on extrapolation plans according to the proposal in the reflection paper. Note that in the EMA reflection paper it is stated that "Based on the extrapolation concept, the specification of key scientific questions of interest and specific trials listed with objectives, key design elements and criteria for success that can inform the size of the trial should be presented using the extrapolation framework in regulatory procedures at e.g., PDCO, Scientific Advice Working Party (SAWP) or Committee for Medicinal Product for Human Use (CHMP)."

Early discussion with regulators may involve ascertaining the extent to which extrapolation can be used and the pharmacokinetic or pharmacodynamics considerations that are needed to warrant this approach. Generally, the assessment of disease similarity is not a simple dichotomy (i.e., yes or no) but an assessment with different degrees of uncertainty. Critically important is the understanding disease mechanism and pathophysiology and the level of confidence in this understanding. These considerations may effect the specification of key scientific questions of interest, and the specific trials and objectives, key design elements and the criteria for success that inform the size of the trial. For example, it may be helpful to seek feedback on the possibility to borrow placebo data from adults and adolescents as a comparator in a study in young pediatric patients or to borrow both placebo and investigational drug data from adults and adolescents to use in a study of young paediatric patients, and what scientific evidence would support these proposals. Moreover, should techniques such as Bayesian statistics be used only when the adult E-R is sufficiently characterized [91] as proof-of-concept (in established MOAs). The agreements for these design and analysis could also take into consideration prior information, its weight, comparability of studies from which the prior information would derive, covariates, etc. Because these details can sometimes be less understood, thorough discussions with regulatory agencies and numerous simulations may be required to understand operating characteristics prior to the adoption or agreement of the methodology. Furthermore, such requests may sometimes be needed at a timely manner, particularly towards the final stages of Pediatric Investigational Plan (PIP) discussions.

Another dimension that needs to be considered when proposing an innovative design is the awareness that these discussions are currently done separately with each regional authority for which the chances for discrepancies are greater. While there might be an avenue for optimization through the parallel scientific advice procedure, this procedure might be limited due to resource constraints at the agencies' sides.

18.5 Conclusion

The conduct of pediatric clinical trials is legally required, monitored, and encouraged in major geographic areas such as the United States and Europe. However, because pediatric patients are considered vulnerable populations, they should only be enrolled as research subjects in a clinical trial if enrolling adult subjects will not be able to the answer the scientific question related to health and welfare of children. Thus, there is an ethical obligation to build the foundation for the use of pediatric extrapolation and related innovative analytical strategies with appropriately designed pediatric and adult clinical trials to reduce the amount of, or general need for, additional information needed from children to reach conclusions. This manuscript has described applications of innovative clinical trial designs, analytic strategies to more efficiently leverage prior information, and modeling approaches that impact on the data required to determine efficacy of an investigational drug in pediatrics. The timely planning of pediatric trials and regulatory interactions related to required pediatric studies and expectations for innovative analytics are also discussed.

References

[1] US Congress. Pediatric Research Equity Act of 2003. *Public Law*, 108 (155):117, 2003.

[2] Richard E Behrman, Marilyn J Field, and Institute of Medicine (US) Committee on Clinical Research Involving Children. *Ethical Conduct of Clinical Research Involving Children.* National Academies Press: Washington, DC, 2004.

[3] B Regulation. No 1901/2006 of the European Parliament and of the council of 12 December 2006 on medicinal products for paediatric use and amending Regulation (EEC) no 1768/92. Technical report, Directive 2001/20/EC, Directive 2001/83/EC and Regulation (EC) No 726/2004 (Official Journal L 378, 27/12/2006 p. 1–19). Available at: http://ec.europa.eu/health/documents/eudralex, 2006.

[4] US Food Drug Administration, Best Pharmaceuticals for Children Act. *Public Law*, 107–109, 2002.

[5] Dobromir Penkov, Paolo Tomasi, Irmgard Eichler, Dianne Murphy, Lynne P Yao, and Jean Temeck. Pediatric medicine development: An overview and comparison of regulatory processes in the European Union and United States. *Therapeutic Innovation and Regulatory Science*, 51 (3):360–371, 2017.

[6] Thomas J Hwang, Paolo A Tomasi, and Florence T Bourgeois. Delays in completion and results reporting of clinical trials under the Paediatric Regulation in the European Union: A cohort study. *PLoS Medicine*, 15 (3):e1002520, 2018.

[7] Natalie Pica and Florence Bourgeois. Discontinuation and nonpublication of randomized clinical trials conducted in children. *Pediatrics*, 138:e20160223, 2016.

[8] Population ages 0-14 (% of total). https://data.worldbank.org/indicator/SP.POP.0014.TO.ZS. Accessed: 2018-02-04.

[9] Sherry G Selevan, Carole A Kimmel, and Pauline Mendola. Identifying critical windows of exposure for children's health. *Environmental Health Perspectives*, 108(Suppl 3):451, 2000.

[10] Mark D Miller, Melanie A Marty, Amy Arcus, Joseph Brown, David Morry, and Martha Sandy. Differences between children and adults: Implications for risk assessment at California EPA. *International Journal of Toxicology*, 21(5):403–418, 2002.

[11] Michelle Roth-Cline, Jason Gerson, Patricia Bright, Catherine S Lee, and Robert M Nelson. Ethical considerations in conducting pediatric research. In: *Pediatric Clinical Pharmacology*, Seyberth HW, Rane A, and Schwab M (eds) pp. 219–244. Springer, Berlin Heidelberg, 2011.

[12] Robert M Nelson. The scientific and ethical path forward in pediatric product development, 2010. https://www.slideserve.com/jacinda/the-scientific-and-ethical-path-forward-in-pediatric-product-development-applying-21-cfr-50-subpart-d link and accessed 07/14/2019

[13] US Food Drug Administration. CFR-Code of Federal Regulations Title 21. *Current Good Manufacturing Practice for Finished Pharmaceuticals Part*, 211, 2017.

[14] Part D. Miscellaneous Provisions Relating to Substance Abuse, Public Health Service Act.

[15] Gerold T Wharton, Dianne Murphy, Debbie Avant, John V Goldsmith, Grace Chai, William J Rodriguez, and Eric L Eisenstein. Impact of pediatric exclusivity on drug labeling and demonstrations of efficacy. *Pediatrics*, 134(2):e512–e518, 2014.

[16] National Commission for the Protection of Human Subjects of Biome Beha Resea and Kenneth John Pres Ryan. *The Belmont Report: Ethical Principles and Guidelines for the Protection of Human Subjects of Research-the National Commission for the Protection of Human Subjects of Biomedical and Behavioral Research.* US Government Printing Office, 1978.

[17] Education Department of Health. The Belmont Report. Ethical principles and guidelines for the protection of human subjects of research. *The Journal of the American College of Dentists*, 81(3):4, 2014.

[18] Margaret Gamalo-Siebers, Jasmina Savic, Cynthia Basu, Xin Zhao, Mathangi Gopalakrishnan, Aijun Gao, Guochen Song, Simin Baygani, Laura Thompson, H Amy Xia, et al. Statistical modeling for Bayesian extrapolation of adult clinical trial information in pediatric drug evaluation. *Pharmaceutical Statistics*, 16:232–249, 2017.

[19] Andrew Gelman, John B Carlin, Hal S Stern, David B Dunson, Aki Vehtari, and Donald B Rubin. *Bayesian Data Analysis*, vol. 2. CRC Press: Boca Raton, FL, 2014.

[20] Joseph G Ibrahim and Ming-Hui Chen. Power prior distributions for regression models. *Statistical Science*, 15:46–60, 2000.

[21] Brian P Hobbs, Bradley P Carlin, Sumithra J Mandrekar, and Daniel J Sargent. Hierarchical commensurate and power prior models for adaptive incorporation of historical information in clinical trials. *Biometrics*, 67(3):1047–1056, 2011.

[22] Brian P Hobbs, Daniel J Sargent, and Bradley P Carlin. Commensurate priors for incorporating historical information in clinical trials using general and generalized linear models. *Bayesian Analysis*, 7(3):639, 2012.

[23] David A Schoenfeld, Hui Zheng, and Dianne M Finkelstein. Bayesian design using adult data to augment pediatric trials. *Clinical Trials*, 6(4): 297–304, 2009.

[24] European Medicines Agency. Concept paper on extrapolation of efficacy and safety in medicine development. www.ema.europa.eu/docs/en_GB/document_library/Scientific_guideline/2013/04/WC500142358.pdf, 2013.

[25] Food Drug Administration. Regulations requiring manufacturers to assess the safety and effectiveness of new drugs and biological products in pediatric patients: Final rule. *Federal Register*, 63:66631–66672, 1998.

[26] Julia Dunne, William J Rodriguez, Dianne Murphy, B Nhi Beasley, Gilbert J Burckart, Jane D Filie, Linda L Lewis, Hari C Sachs, Philip H Sheridan, Peter Starke, et al. Extrapolation of adult data and other data in pediatric drug-development programs. *Pediatrics*, 2011. doi: 10.1542/peds.2010-3487.

[27] ICH Harmonised Tripartite Guideline. Addendum to ICH E11: Clinical investigation of medicinal products in the pediatric population (R1). Current Step, 4, 2017.

[28] Marilyn J Field, Thomas F Boat, et al. *Safe and Effective Medicines for Children: Pediatric Studies Conducted under the Best Pharmaceuticals for Children Act and the Pediatric Research Equity Act*. National Academies Press: Washington, DC, 2012.

[29] Paediatric Investigation Plans. www.ema.europa.eu/ema/index.jsp?curl=pages/regulation/general/general_content_000608.jsp. Accessed: 2018-02-05.

[30] Cynthia Basu, Mariam A Ahmed, Reena V Kartha, Richard C Brundage, Gerald V Raymond, James C Cloyd, and Bradley P Carlin. A hierarchical Bayesian approach for combining pharmacokinetic/pharmacodynamic modeling and Phase IIa trial design in orphan drugs: Treating adrenoleukodystrophy with Lorenzo's oil. *Journal of Biopharmaceutical Statistics*, 26(6):1025–1039, 2016.

[31] Lily Mulugeta. What do we look for to support pediatric dosing? Pediatric Clinical Investigator Training, 2014. www.fda.gov/ downloads/NewsEvents/MeetingsConferencesWorkshops/ UCM415211.pdf.

[32] Daniel K Benjamin, Phillip B Smith, Pravin Jadhav, Jogarao V Gobburu, Dianne Murphy, Vic Hasselblad, Carissa Baker-Smith, Robert M Califf, and Jennifer S Li. Pediatric antihypertensive trial failures: Analysis of end points and dose range. *Hypertension*, 51(4):834–840, 2008.

[33] US Senate and House of Representatives. 21st Century Cures Act. www.congress.gov/bill/114th-congress/house-bill/34/, 2015.

[34] Gerald Hlavin, Franz Koenig, Christoph Male, Martin Posch, and Peter Bauer. Evidence, eminence and extrapolation. *Statistics in Medicine*, 35(13):2117–2132, 2016.

[35] José M Bernardo. The concept of exchangeability and its applications. *Far East Journal of Mathematical Sciences*, 4:111–122, 1996.

[36] Heinz Schmidli, Sandro Gsteiger, Satrajit Roychoudhury, Anthony O'Hagan, David Spiegelhalter, and Beat Neuenschwander. Robust meta-analytic-predictive priors in clinical trials with historical control information. *Biometrics*, 70(4):1023–1032, 2014.

[37] Mike West. Approximating posterior distributions by mixture. *Journal of the Royal Statistical Society, Series B*, 55:409–422, 1993.

[38] Jose M Bernardo and Adrian FM Smith. *Bayesian Theory.* Wiley: New York, 1994.

[39] James Berger and Mark Berliner. Robust Bayes and empirical Bayes analysis with ε-contaminated priors. *The Annals of Statistics*, 14:461–486, 1986.

[40] Anthony O'Hagan, Luis Pericchi. Bayesian heavy-tailed models and conflict resolution: A review. *Brazilian Journal of Probability and Statistics*, 26(4):372–401, 2012.

[41] Peter F Thall and Kyle J Wathen. Bayesian designs to account for patient heterogeneity in phase II clinical trials. *Current Opinion in Oncology*, 20:407, 2008.

[42] Peter C O'Brien and Thomas R Fleming. A multiple testing procedure for clinical trials. *Biometrics*, 35:549–556, 1979.

[43] Stuart J Pocock. Interim analyses for randomized clinical trials: The group sequential approach. *Biometrics*, 38:153–162, 1982.

[44] Irving K Hwang, Weichung J Shih, and John S De Cani. Group sequential designs using a family of type I error probability spending functions. *Statistics in Medicine*, 9(12):1439–1445, 1990.

[45] FDA. Guidance for industry: Providing clinical evidence of effectiveness for human drug and biological products. Rockville, MD, 1998.

[46] Carl C Peck, Donald B Rubin, and Lewis B Sheiner. Hypothesis: A single clinical trial plus causal evidence of effectiveness is sufficient for drug approval. *Clinical Pharmacology and Therapeutics*, 73(6):481–490, 2003.

[47] Austin Bradford Hill. *Controlled Clinical Trials*. Blackwell Scientific Publications: Oxford, UK, 1960.

[48] David P Byar, Richard M Simon, William T Friedewald, James J Schlesselman, David L DeMets, Jonas H Ellenberg, Mitchell H Gail, and James H Ware. Randomized clinical trials. Perspectives on some recent ideas. *The New England Journal of Medicine*, 295(2):74–80, 1976.

[49] Sander Greenland, Judea Pearl, and James M Robins. Causal diagrams for epidemiologic research. *Epidemiology*, 10:37–48, 1999.

[50] Edmund A Gehan and Emil J Freireich. Non-randomized controls in cancer clinical trials. *The New England Journal of Medicine*, 290(4):198–203, 1974.

[51] David Cho, Sanjay Krishnaswami, Julie C Mckee, Garret Zallen, Mark L Silen, and David W Bliss. Analysis of 29 consecutive thoracoscopic repairs of congenital diaphragmatic hernia in neonates compared to historical controls. *Journal of Pediatric Surgery*, 44(1):80–86, 2009.

[52] ICH Harmonised Tripartite Guideline. Choice of control group and related issues in clinical trials E10. *Choice*, E10, 2000.

[53] Sarah Owens. FDA analysis supports extrapolation of data on aeds for adults with partial-onset seizures to children. *Neurology Today*, 2016.

[54] Omoniyi J Adedokun, Zhenhua Xu, Lakshmi Padgett, Marion Blank, Jewel Johanns, Anne Griffiths, Joyce Ford, Honghui Zhou, Cynthia Guzzo, Hugh M Davis, et al. Pharmacokinetics of infliximab in children with moderate-to-severe ulcerative colitis: Results from a randomized, multicenter, open-label, phase 3 study. *Inflammatory Bowel Diseases*, 19(13):2753–2762, 2013.

[55] Jeffrey Hyams, Lakshmi Damaraju, Marion Blank, Jewel Johanns, Cynthia Guzzo, Harland S Winter, Subra Kugathasan, Stanley Cohen, James Markowitz, Johanna C Escher, et al. Induction and maintenance therapy with infliximab for children with moderate to severe ulcerative colitis. *Clinical Gastroenterology and Hepatology*, 10(4):391–399, 2012.

[56] Anthony Markham. Cerliponase alfa: First global approval. *Drugs*, 77(11):1247–1249, 2017.

[57] Sandeep B Bavdekar. Pediatric clinical trials. *Perspectives in Clinical Research*, 4(1):89, 2013.

[58] James Hung, Sue-Jane Wang, Yi Tsong, John Lawrence, and Robert T O'Neil. Some fundamental issues with non-inferiority testing in active controlled trials. *Statistics in Medicine*, 22(2):213–225, 2003.

[59] Dan Turner, Sibylle Koletzko, Anne M Griffiths, Jeffrey Hyams, Marla Dubinsky, Lissy de Ridder, Johanna Escher, Paolo Lionetti, Salvatore Cucchiara, Michael J Lentze, et al. Use of placebo in pediatric inflammatory bowel diseases: A position paper from ESPGHAN, ECCO, PIBDnet, and Canadian children ibd network. *Journal of Pediatric Gastroenterology and Nutrition*, 62(1):183–187, 2016.

[60] Stephanie A Kovalchik, Sara De Matteis, Maria Teresa Landi, Neil E Caporaso, Ravi Varadhan, Dario Consonni, Andrew W Bergen, Hormuzd A Katki, and Sholom Wacholder. A regression model for risk difference estimation in population-based case–control studies clarifies gender differences in lung cancer risk of smokers and never smokers. *BMC Medical Research Methodology*, 13(1):143, 2013.

[61] Beat Neuenschwander, Simon Wandel, Satrajit Roychoudhury, and Stuart Bailey. Robust exchangeability designs for early phase clinical trials with multiple strata. *Pharmaceutical Statistics*, 15(2):123–134, 2016.

[62] Christopher A Bush and Steven N MacEachern. A semiparametric Bayesian model for randomised block designs. *Biometrika*, 83(2):275–285, 1996.

[63] Thomas S Ferguson. A Bayesian analysis of some nonparametric problems. *The Annals of Statistics*, 1:209–230, 1973.

[64] Charles E Antoniak. Mixtures of Dirichlet processes with applications to Bayesian nonparametric problems. *The Annals of Statistics*, 2:1152–1174, 1974.

[65] David Blackwell and James B MacQueen. Ferguson distributions via Pólya urn schemes. *The Annals of Statistics*, 1:353–355, 1973.

[66] Jayaram Sethuraman. A constructive definition of dirichlet priors. *Statistica Sinica*, 4:639–650, 1994.

[67] Jayaram Sethuraman and Ram C Tiwari. Convergence of dirichlet measures and the interpretation of their parameter. In: *Statistical Decision Theory and Related Topics II*, Gupta SS and Berger JO (eds), pp. 305–315, Academic Press: Cambridge, MA 1982.

[68] David I Ohlssen, Linda D Sharples, and David J Spiegelhalter. Flexible random-effects models using Bayesian semi-parametric models: Applications to institutional comparisons. *Statistics in Medicine*, 26(9):2088–2112, 2007.

[69] Hemant Ishwaran and Mahmoud Zarepour. Exact and approximate sum representations for the dirichlet process. *Canadian Journal of Statistics*, 30(2):269–283, 2002.

[70] FDA. Guidance for industry: Developing drugs for treatment of atopic dermatitis in pediatric patients (draft). Accessed: 2018-09-05.

[71] Hui Quan, Peng-Liang Zhao, Ji Zhang, Martin Roessner, and Kyo Aizawa. Sample size considerations for Japanese patients in a multiregional trial based on MHLW guidance. *Pharmaceutical Statistics*, 9(2): 100–112, 2010.

[72] Hiroyuki Uesaka. Sample size allocation to regions in a multiregional trial. *Journal of Biopharmaceutical Statistics*, 19(4):580–594, 2009.

[73] John I Gallin, David W Alling, Harry L Malech, Robert Wesley, Deloris Koziol, Beatriz Marciano, Eli M Eisenstein, Maria L Turner, Ellen S DeCarlo, Judith M Starling, et al. Itraconazole to prevent fungal infections in chronic granulomatous disease. *New England Journal of Medicine*, 348(24):2416–2422, 2003.

[74] Jan Bogaerts, Matthew R Sydes, Nicola Keat, Andrea McConnell, Al Benson, Alan Ho, Arnaud Roth, Catherine Fortpied, Cathy Eng, Clare Peckitt, et al. Clinical trial designs for rare diseases: Studies developed and discussed by the international rare cancers initiative. *European Journal of Cancer*, 51(3):271–281, 2015.

[75] Gerta Rücker and Guido Schwarzer. Ranking treatments in frequentist network meta-analysis works without resampling methods. *BMC Medical Research Methodology*, 15(1):58, 2015.

[76] Junjing Lin, Margaret Gamalo-Siebers, and Ram Tiwari. Non-inferiority and networks: Inferring efficacy from a web of data. *Pharmaceutical Statistics*, 15(1):54–67, 2016.

[77] Richard J Lilford. The ethics of underpowered clinical trials. *JAMA*, 288(17):2118–2119, 2002.

[78] Jonathan P Jarow, Lisa LaVange, and Janet Woodcock. Multidimensional evidence generation and FDA regulatory decision making: Defining and using "real-world" data. *JAMA*, 318(8):703–704, 2017.

[79] FDA. Guidance for industry: Use of real-world evidence to support regulatory decision-making for medical devices. Rockville, MD, 2017.

[80] Rachel E Sherman, Steven A Anderson, Gerald J Dal Pan, Gerry W Gray, Thomas Gross, Nina L Hunter, Lisa LaVange, Danica Marinac-Dabic, Peter W Marks, Melissa A Robb, et al. Real-world evidence: What is it and what can it tell us. *The New England Journal of Medicine*, 375(23):2293–2297, 2016.

[81] Jessica M Franklin and Sebastian Schneeweiss. When and how can real world data analyses substitute for randomized controlled trials? *Clinical Pharmacology and Therapeutics*, 102(6):924–933, 2017.

[82] Rongwei Fu, Gerald Gartlehner, Mark Grant, Tatyana Shamliyan, Art Sedrakyan, Timothy J Wilt, Lauren Griffith, Mark Oremus, Parminder Raina, Afisi Ismaila, et al. Conducting quantitative synthesis when comparing medical interventions: AHRQ and the Effective Health Care Program. *Journal of Clinical Epidemiology*, 64(11):1187–1197, 2011.

[83] Jeroen P Jansen, Bruce Crawford, Gert Bergman, and Wiro Stam. Bayesian meta-analysis of multiple treatment comparisons: An introduction to mixed treatment comparisons. *Value in Health*, 11(5):956–964, 2008.

[84] Junjing Lin, Margaret Gamalo-Siebers, and Ram Tiwari. Propensity score matched augmented controls in randomized clinical trials: A case study. *Pharmaceutical Statistics*, 175:593–606, 2018.

[85] Nicola Gökbuget, Michael Kelsh, Victoria Chia, Anjali Advani, Renato Bassan, Hervè Dombret, Michael Doubek, Adele K Fielding, Sebastian Giebel, Vincent Haddad, et al. Blinatumomab vs historical standard therapy of adult relapsed/refractory acute lymphoblastic leukemia. *Blood Cancer Journal*, 6(9):e473, 2016.

[86] FDA. FDA Blincyto drug approval package. www.accessdata.fda.gov/drugsatfda_docs/nda/2014/125557Orig1s000TOC.cfm. Accessed: 2018-02-05.

[87] Hans-Georg Eichler, Brigitte Bloechl-Daum, Peter Bauer, Frank Bretz, Jeffrey Brown, Lisa Victoria Hampson, Peter Honig, Michael Krams, Hubert Leufkens, Robyn Lim, et al. "Threshold-crossing": A useful way to establish the counterfactual in clinical trials? *Clinical Pharmacology and Therapeutics*, 100(6):699–712, 2016.

[88] Junjing Lin, Margaret Gamalo-Siebers, and Ram Tiwari. Propensity-score-based priors for bayesian augmented control design. *Pharmaceutical Statistics*, 18(2), 223–238.

[89] Christina Bucci-Rechtweg. Enhancing the pediatric drug development framework to deliver better pediatric therapies tomorrow. *Clinical Therapeutics*, 39(10):1920–1932, 2017.

[90] Jeremiah D Momper, Yeruk Ager Mulugeta, and Gilbert J Burckart. Failed pediatric drug development trials. *Clinical Pharmacology and Therapeutics*, 98(3):245–251, 2015.

[91] Lisa V Hampson, Ralf Herold, Martin Posch, Julia Saperia, and Anne Whitehead. Bridging the gap: A review of dose investigations in paediatric investigation plans. *British Journal of Clinical Pharmacology*, 78(4): 898–907, 2014.

19

A Brief Guide to Bayesian Model Checking

John W. Seaman, Jr., James D. Stamey, and David J. Kahle
Baylor University

Somer Blair
JPS Health Network

CONTENTS

19.1 Overview

This brief guide covers critical features of Bayesian model building.[1] Checking and documenting such features facilitates evaluation of the models, rendering potential problems with assumptions, convergence, and inferential accuracy more transparent to the user.

The items to be checked are arranged in six sections. Section 19.2 contains a "quick kill" list allowing rapid identification of a model with convergence problems or other "fatal" flaws. The next three sections cover convergence, posterior accuracy, prior justification, and prior-to-posterior sensitivity. We conclude with brief sections on model characteristics and model performance.

This document is not meant to be a tutorial on model checking but rather a quick guide. We have included several references for the user to pursue

[1]Although written with Markov chain Monte Carlo (MCMC) methods in mind, much of the document is applicable regardless of how the posterior distribution is obtained.

further details. Thorough, textbook-level treatments of model checking can be found, for example, in Carlin and Louis (2009), Chapter 4, and Lesaffre and Lawson (2012), Chapter 10. Another excellent source is Spiegelhalter et al. (2004), Chapter 5. An extensive collection of diagnostic tools are available in the boa and coda R packages from the Comprehensive R Archive Network at http://CRAN.R-project.org/.

19.2 Rapid Identification of Problems in Models: "Quick Kill" Items to Check

A thorough examination of each of the items in Sections 19.3–19.7 is necessary for a full assessment of the approximations afforded by iterative simulation methods such as MCMC. However, in many cases, a problematic model can be identified rather rapidly by examining these "quick kill" items. If any of the following are missing or fail to hold, then failure to converge to a stable distribution, model identifiability, or other problems should be suspected. See the main entries in following sections for more detail.

1. The posterior kernel density plots should be relatively "smooth". These should be examined for all parameters of interest, as well as for variance components. See 2, Section 19.3.
2. Prior-to-posterior comparisons (as in 2, Section 19.3) should reveal change from the prior to the posterior. That is, the prior should be "updated". (Models that are purposefully over-parameterized, such as those with measurement error or unmeasured confounding components, are exceptions.)
3. For priors with compact support, say on an interval $[a, b]$, the corresponding posterior density plot should not exhibit mass "piling up" near either endpoint without reason.
   (a) If a posterior for a probability concentrates mass near zero, and it is reasonable that the event in question is "rare", then there is no need for alarm.
   (b) By contrast, if a uniform prior on $(0, B], B > 0$, is used for a standard deviation, and the posterior "piles up" at B, then B was evidently too small. See Figure 19.8.
4. History plots should exhibit good "mixing". See 3, Section 19.3.
5. Autocorrelation plots should dampen quickly. See 4, Section 19.3.
6. Brooks–Gelman–Rubin diagnostic plots should approach 1. See 6, Section 19.3. This assumes the modeler employed at least two chains. In fact, multiple chains are recommended, starting from "overdispersed" initial values.

7. Interval estimates for the main parameters of interest should be sufficiently narrow to be of practical use. See 3, Section 19.4.
8. Prior choices should be documented, whether based on expert opinion or historical data. If the prior predictive probability of success is "too large", or the prior effective sample size is unreasonable compared to the study sample size, then the analysis is suspect. See Section 19.5 for further detail.
9. Where "diffuse" or "non-informative" priors are used, induced priors should be examined, as they may be unintentionally informative. This is easily accomplished using simulation. See Section 19.5 for more detail.
10. Model characteristics such as the use of fixed vs. random effects, exchangeability assumptions, and identifiability should be noted and justified. See Section 19.6.
11. The model should be tested using simulated data representing various truth scenarios. For more detail, see Section 19.7.

19.3 Convergence

Before MCMC samples can be used for inference, the chains from which they were drawn must be checked for convergence. To help assess convergence, there are several things to consider.

1. Number of chains, burn-in length, and subsequent chain length should be noted, along with initial values and how the latter were chosen. For more on this, see Lunn et al. (2013), p. 70ff, or Carlin and Louis (2009), pp. 158–159.[2] Any thinning or the use of over-relaxation should also be indicated. As for thinning, see item 4 below.
2. Prior and posterior densities for all parameters of interest, including variance components, should be provided and plotted on the same graph.[3] In some cases, functions of model parameters (e.g. hazard ratios or ED50s) will require computation of induced priors via separate simulation. Note that the posterior densities should be relatively "smooth", as illustrated in Figure 19.1. (Of course, "smoothness" is a subjective assessment and will

[2]See also Gelman and Shirley (2011). For an alternate view on the use of multiple chains, see, for example, Geyer (2011), who advocates the use of single, very long chains.

[3]Such plots may be problematic if the prior is extremely diffuse compared to the posterior; that is, the relative maxima of prior and posterior density may make plotting difficult.

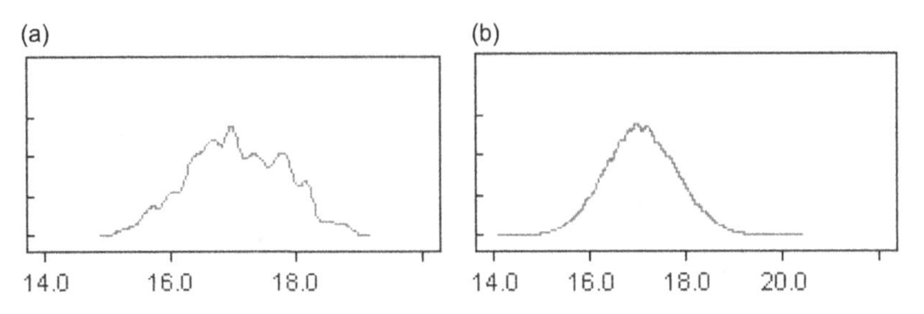

FIGURE 19.1
Kernel density plots for posteriors from OpenBUGS for similar chain lengths. (a) Convergence not attained and (b) convergence attained.

depend on graphics choices, such as features of a kernel density estimator.)

3. History plots including all chains should be provided for each parameter of interest as well as variance components. Like the smoothness of the kernel density plots, these "trace" plots can indicate problems with convergence. Note that requisite chain lengths may well vary across parameters and starting values. Ultimately, the chain lengths you utilize should be at least that required by the most "demanding" parameter. Several examples for single chains are shown in Figure 19.2. Illustrations with two chains are shown in Figure 19.3. With multiple chains, an indication that there is a sufficient number of iterations for convergence is that values from the different chains are similar—they "mix". Visually, the plots of the individual chains intermingle and "mix" together after convergence.

4. Autocorrelation plots should be provided for all parameters of interest as well as variance components. Note that, as autocorrelation increases, chain lengths required for convergence and subsequent inference also increase. That is, the effective sample size from the posterior is reduced as autocorrelation in the chain increases. Thinning may be required to alleviate this problem. It is often helpful to center continuous covariates. See Figure 19.4 for some illustrations.

5. Bivariate plots for each pair of the parameters of interest should be provided in order to gauge posterior dependencies between parameters. In OpenBUGS, this can be done for individual pairs using the Correlations "button" under the inference menu. A better method would be to export the chains to R and use the pairs() function in the GGally package, which employs ggplot2 graphics. An example

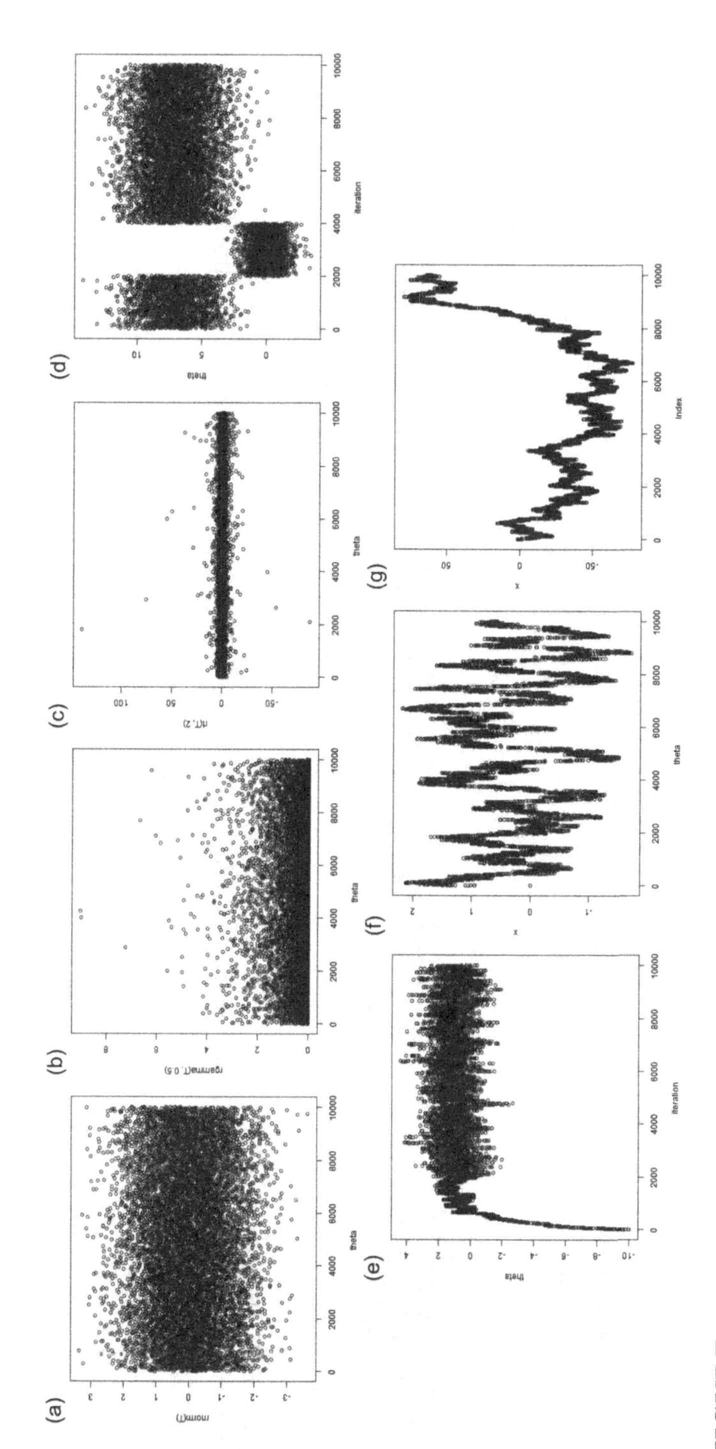

FIGURE 19.2
History (trace) plots for a hypothetical parameter. (a)–(c) exhibit no evidence of a lack of convergence. They appear to be from symmetric, skewed, and heavy tailed posteriors, respectively. Plot (d) is likely from a bimodal posterior, (e) suggests there are burn-in iterations that should be removed, with possibly poorly chosen starting values, (f) and (g) indicate heavily correlated chains.

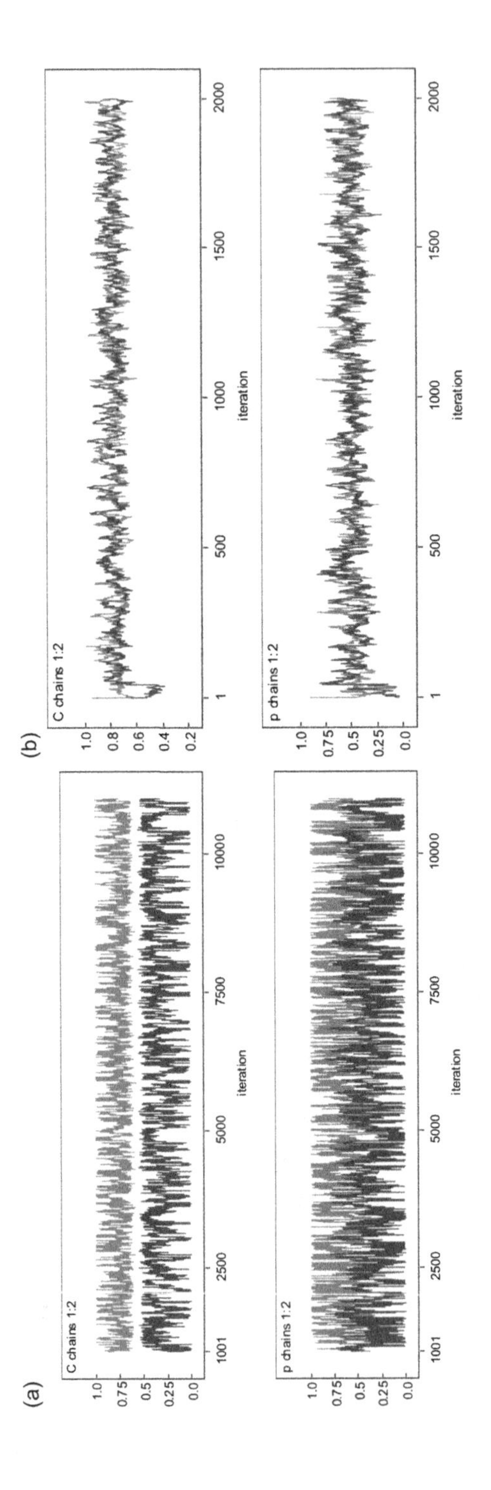

FIGURE 19.3
History (trace) plots for two chains from the posterior of a hypothetical parameter. The plots on the left exhibit poor mixing. Those on the right exhibit good mixing.

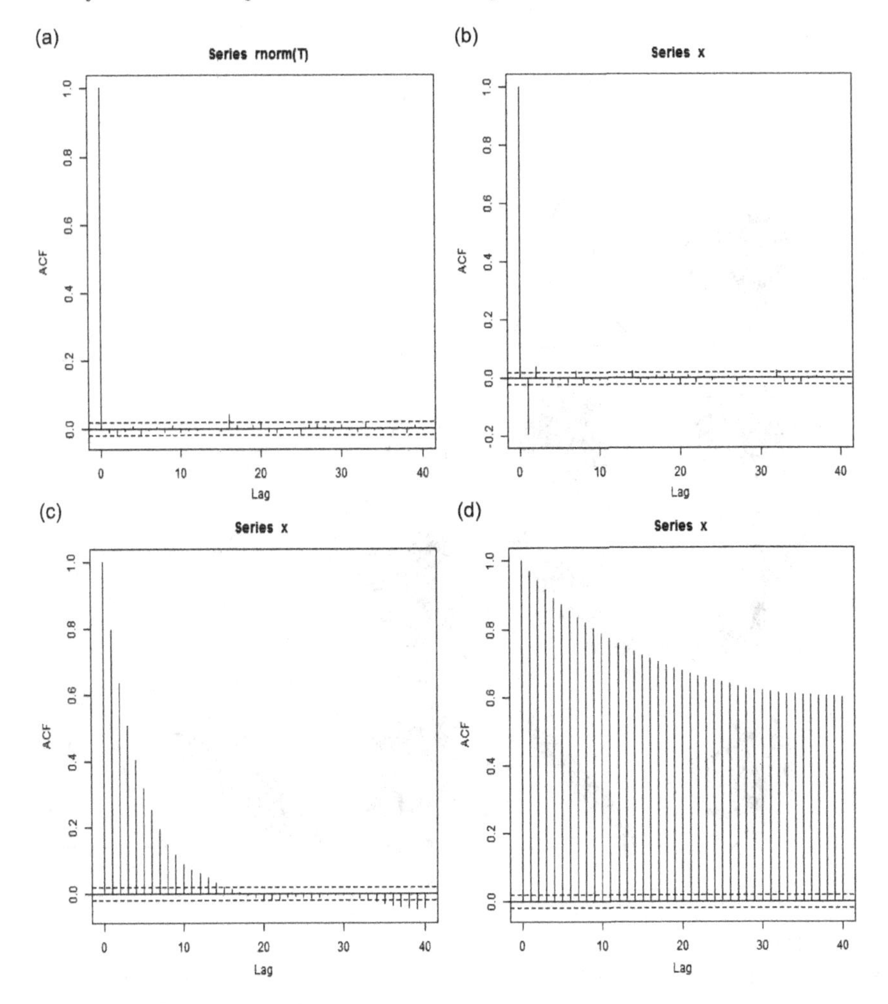

FIGURE 19.4
Autocorrelation function (ACF) plots. The plots (a) and (b) are "good", indicating little or no autocorrelation in the chain. The plots (c) and (d) are "bad". The (d) plot, in particular, exhibits large correlation even for high lag values. The effective sample size from the posterior will be greatly reduced in the lower two cases.

appears in Figure 19.5. Should a pair of parameters be highly dependent, marginal posterior inference on either alone is problematic. See Carlin and Louis (2009), p. 159.

6. Brooks–Gelman–Rubin (BGR) diagnostic plots should be provided. This diagnostic is available in OpenBUGS but requires the use of at least two chains. The BGR statistic compares the variance of

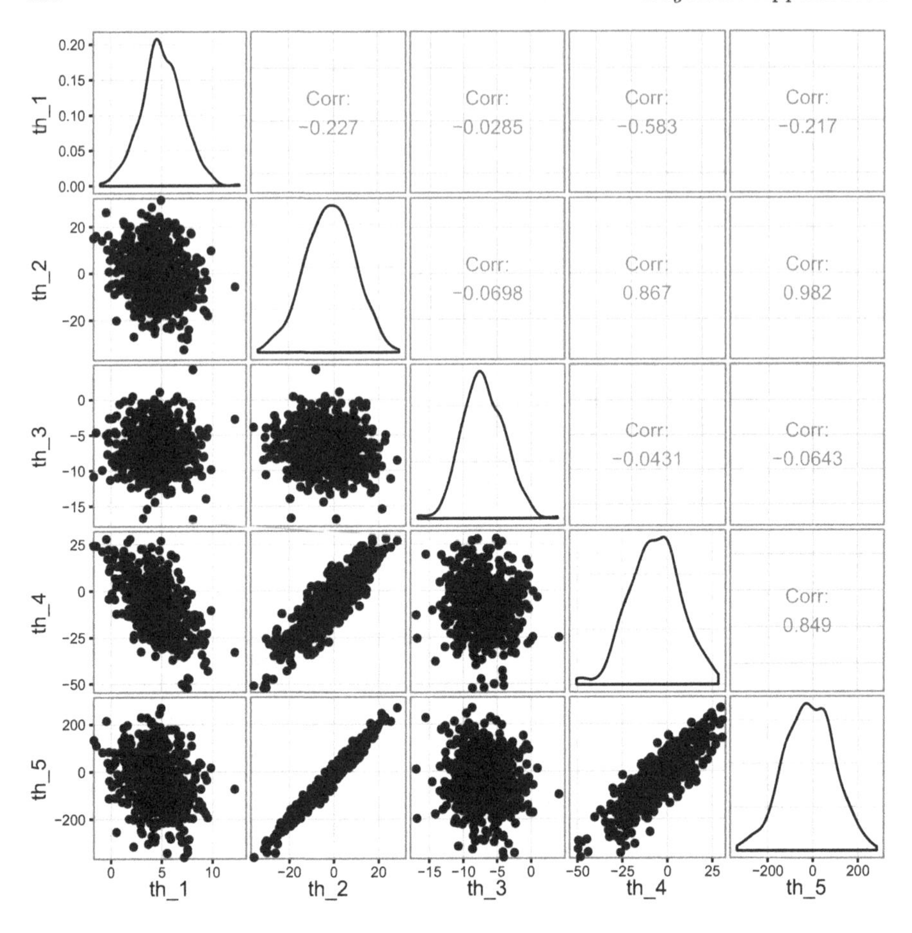

FIGURE 19.5
Scatter plots of posterior samples from pairs of five parameters. Parameter pairs $(\theta_2, \theta_4), (\theta_2, \theta_5)$, and (θ_4, θ_5) appear to be confounded.

the values sampled by each chain separately with the variance of the pooled chains. If convergence has been attained, then the two variances should be similar. See Figure 19.6 for illustrations. In OpenBUGS, this is done by comparing a pooled posterior variance approximation (light gray in the plots) with the average within-sample variance (dark gray in the plots) and their ratio (black in the plots).[4] The red curve should tend to 1, and the others should stabilize.

[4]BUGS actually uses the width of 80% credible sets for these purposes. Details can be found, for example, in Lunn et al. (2013), p. 75ff, or Ntzoufras (2009). See also Lesaffre and Lawson (2012), p. 183ff.

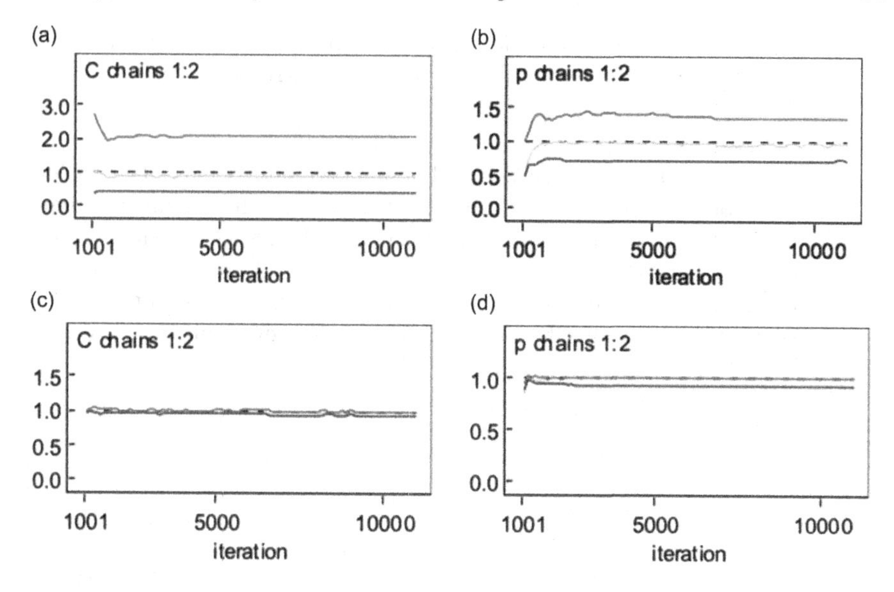

FIGURE 19.6
Brooks–Gelman–Rubin plots. Plots (a) and (b) exhibit poor chain performance. Plots (c) and (d) exhibit good chain performance.

19.4 Posterior Accuracy and Suitability

Chain length leading to convergence is one aspect of the compound approximation that is inherent in Bayesian posterior inference using iterative simulation methods such as MCMC. Once convergence is attained, posterior inference must be based on sufficiently large samples from that posterior.

1. To this end, the Monte Carlo standard error (MCSE) and the posterior standard deviation should be included for each parameter of interest, as well as variance components. A reasonable rule of thumb is that the chain should extend long enough beyond convergence to render the MCSE no more than 5% of the posterior standard deviation for the parameter. See Cowles (2013) pp. 137–138 and Lunn et al. (2013), p. 77ff.[5]

2. An alternative is to plot posterior summary values (mean, SD, percentiles) as a function of post burn-in chain length. Once all such summaries have stabilized within prescribed boundaries, accuracy can be said to have been attained. See, for example, Lunn et al. (2009), p. 79. One can choose a stability boundary, say a small

[5] A more extensive discussion is provided in Flegal et al. (2008).

$\delta > 0$, and require that the posterior summary of interest (such as a posterior mean) be within $\pm\delta$. This is illustrated in Figure 19.7.

3. Interval estimates for parameters of interest should be sufficiently narrow for practical purposes. For a fixed sample size, this need must be carefully balanced against the justifiability of informative priors. Priors can usually be made sufficiently informative as to render credible sets arbitrarily narrow. However, doing so may result in unreasonably high prior probability of study success, or unrealistically large prior effective sample size. For more on justification of priors, see Section 19.5.

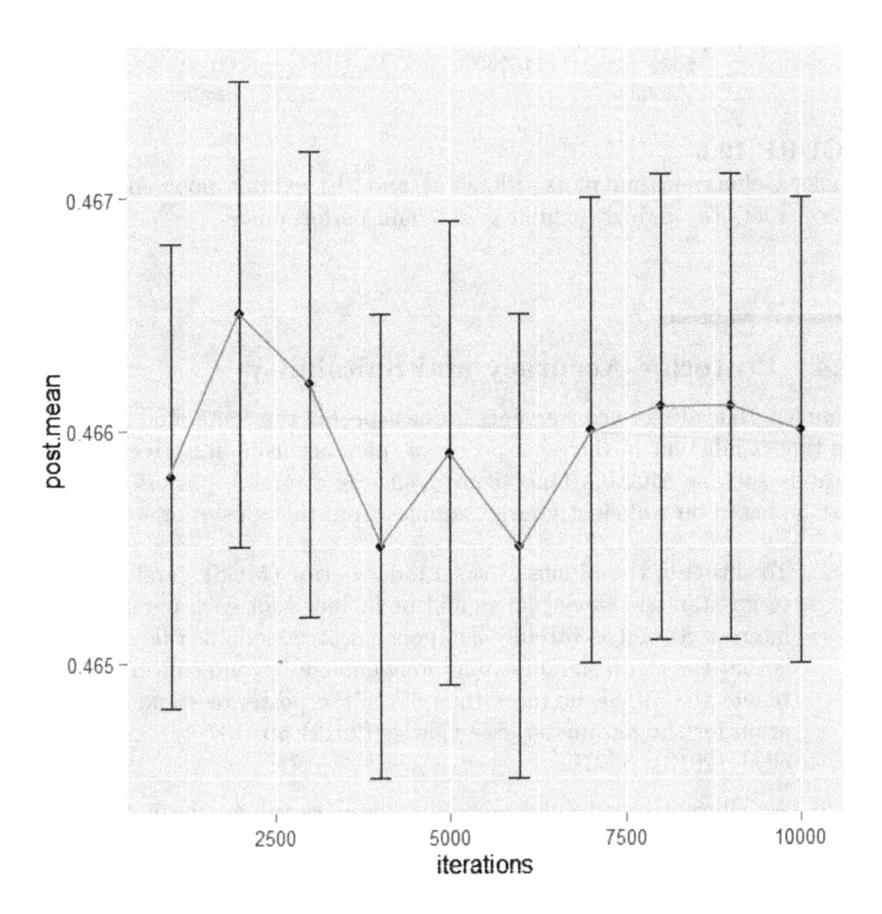

FIGURE 19.7
Posterior means plotted as a function of iterations. This shows the "tail" of the chain, in which all the plotted means are within 0.001 of each other.

19.5 Prior Justification and Prior-to-Posterior Sensitivity Analysis

The posterior distribution clearly depends on both the choice of data model and the joint prior distribution. The former is an issue for any approach to modeling.[6] The latter is unique to Bayesian methods. Priors should be carefully justified, even if using relatively non-informative priors. Furthermore, posterior results should be reasonably robust with respect to prior choices. That is, "small" changes in the joint prior structure should not produce "large" changes in posterior results. Thus, *checking the sensitivity of focal posterior results to changes in the prior structure is a necessary task in Bayesian model building.* The list below provides guidance for prior-to-posterior sensitivity analyses.[7]

1. The choice of each prior distribution should be justified. Note that it is not sufficient to say that the default in a given tool, such as FACTS, was chosen.[8] The default itself must be justified for the problem at hand.

   (a) Justification is needed even for relatively non-informative (diffuse) priors on parameters, especially if functions of those parameters are of interest, as their induced priors may in fact be informative. See 7 below.

   (b) Furthermore, priors intended to be relatively non-informative can be chosen to be unnecessarily diffuse, sometimes resulting in convergence problems. For example, if a relatively non-informative prior is desired for the mean change from baseline for, say, systolic blood pressure in mm/Hg, then a normal prior with a standard deviation of 100 mm/Hg is surely overly diffuse. For more on this, see 3c and 3d below.

2. Potential dependencies among parameters should be reflected in the joint prior structure if possible, although this can be very challenging. For example, it would be unreasonable to claim that the mean change in A1c from baseline is independent of the coefficient for a body weight covariate. If parameters are thought to be independent, justification should be provided. Of course, *a priori* independence need not yield *a posteriori* independence. Indeed, looking at

[6]See Section 19.6 for some comments on this.

[7]More formal approaches have been considered in the literature. Such methods are beyond the scope of this chapter. For a good brief overview of formal methods, see Roos et al. (2015). In addition, they offer a relatively practical approach to formal sensitivity analysis that has the virtue of reproducibility.

[8]This is a package designed by Berry Consultants for fixed and adaptive clinical trial simulation. See www.berryconsultants.com/software/.

posterior plots of pairs of parameters can be revealing.[9] Strong posterior bivariate relationships suggest that modeling dependency in the joint prior might have afforded more efficient use of the data. Against that, the difficulty of modeling prior dependencies (e.g., covariance structures for joint priors) must be considered as, again, such tasks are typically difficult.

3. Prior-to-posterior sensitivity with respect to both location and scale of marginal priors should be considered.
   (a) A check for posterior sensitivity to small changes in prior parameters should be made. Slight changes in location or scale should not result in decision-altering changes in the posterior.
   (b) One approach regarding choices for marginal prior location is to perform the Bayesian analysis using several prior points of view. For example, it can be useful to compare posterior results under vague ("relatively non-informative" or "diffuse"), skeptical, enthusiastic, clinical (i.e., expert-elicited), and just-significant (i.e., barely meeting success criteria). See Lunn et al. (2011), p. 97ff, for an illustration. Lesaffre and Lawson (2012) provide examples of such "archetypal" priors. They illustrate use of skeptical priors in two clinical trials and an enthusiastic prior for another. Spiegelhalter et al. (2004) discuss the use of skeptical priors in various clinical contexts at length.
   (c) Posterior results are often critically dependent on assumptions made about variance components. The suitability of families used for priors on variance components is highly problem-dependent. In general, use of Gamma(a, b) priors, with a and b small, is not recommended. Use of alternatives, such as uniform or half-normal distributions on standard deviations is often preferable. If support for a prior is bounded, posterior probability mass should not "pile up" at the boundaries. This problem can occur, for example, when placing a uniform distribution with support $(0, B], B > 0$, on a standard deviation. One can immediately rule out a value of B as being too small if it results in posterior mass that "piles up" near B. See Figure 19.8 for an illustration.
   (d) Justification of priors on variance components should be done in terms of standard deviations, arguing from operational scales.
      i. For example, if a uniform distribution on $(0, B], B > 0$, is used for a standard deviation of a random variable, X, then B should be justified in terms of what is known about X. If X is, say, systolic blood pressure in mm/Hg, then $B = 100$ is probably unreasonable.

[9]See 5, Section 19.3.

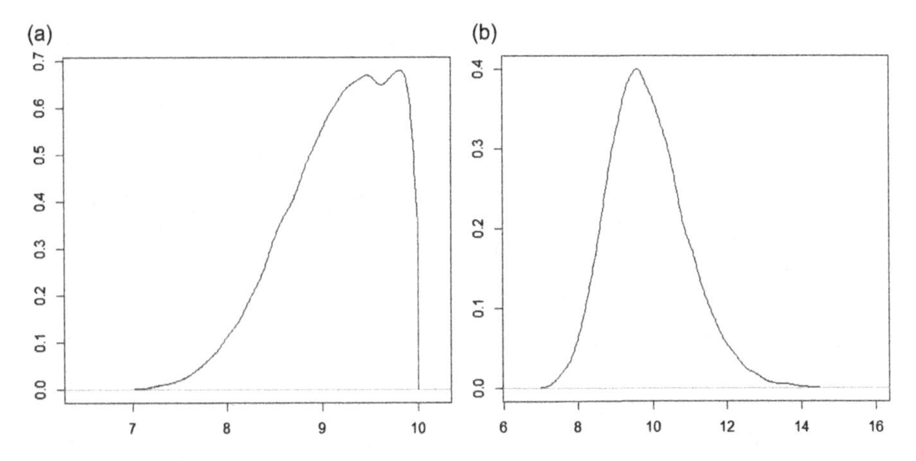

FIGURE 19.8
Posterior plots for a standard deviation given a $U(0, B)$ prior. (a) $B = 10$, resulting in a "pile-up" of probability mass at that bound. (b) $B > 14$, resulting in a reasonable posterior plot.

   ii. One approach to choosing bounds that do not influence posterior inference is a form of prior-to-posterior sensitivity analysis. One can plot a posterior quantity of interest, such as the posterior probability of study success, or the width of a 95% credible set, as a function of B. The bound should be chosen so as to avoid undue influence on that posterior quantity of interest. Figure 19.9 illustrates a case where $B = 2.5$ or 3 seems reasonable.[10]

4. Prior-to-posterior comparisons may reveal that the marginal posteriors for some parameters are little changed from their priors. This could be an indication of an overly informative prior or an identifiability problem. (This may be an unavoidable feature of the model as with, for example, some measurement error problems.)

5. Some indication of prior informativeness should be provided. This can be assessed in a number of ways, including prior effective sample size (ESS). See Morita et al. (2008, 2010, 2012) and the R code referred to therein. In effect, this method provides the sample size that would be necessary to achieve similar posterior results if a vague prior structure had been used. At a minimum, per FDA (2010) guidance, the prior probability of study success should be carefully considered.

[10]While we are using the data (via the posterior) to determine a feature of the prior, we are doing so to *avoid* unintentional prior influence. This makes it clear that priors on variance components have not been chosen to yield "more favorable" posterior conclusions.

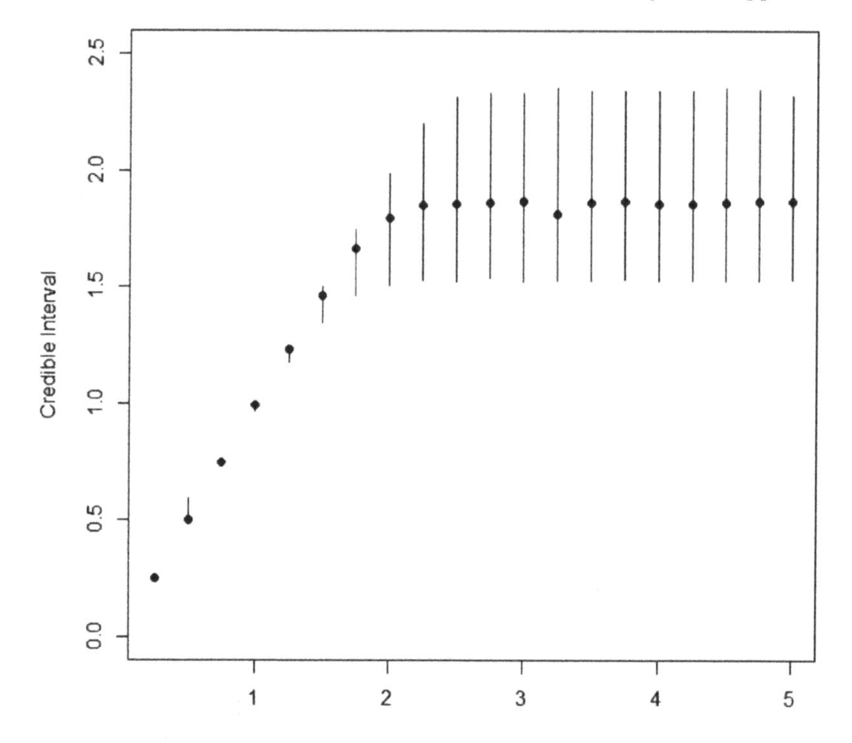

FIGURE 19.9
Median values and credible intervals from the posterior on a parameter as a function of B, the upper bound on a uniform distribution used as a prior for a standard deviation related to the parameter. Each median and interval corresponds to a separate "run" of the model given a value of B.

6. If historical data are used for prior construction, great care must be used in considering the resulting prior variance. Large historical sample sizes can yield inappropriately small prior variances with correspondingly high posterior influence. Methods such as power priors can be used to attenuate this influence. Use of these methods must be carefully detailed in advance. For example, when using power priors, the power parameter (usually denoted by a_0) must be very carefully justified. Placing a prior on a_0 is not recommended. One can plot the prior probability of study success (PSS) against values of a_0 to assess the power parameter's influence. Computing the prior ESS as a function of a_0 is another approach. See Figure 19.10 for illustrations. Again see the papers by Morita et al. See also FDA (2010), pp. 26, 39.
7. If diffuse priors are placed on parameters that are components of some function of interest, then care must be taken with the induced prior on that function. The latter may be highly informative, in

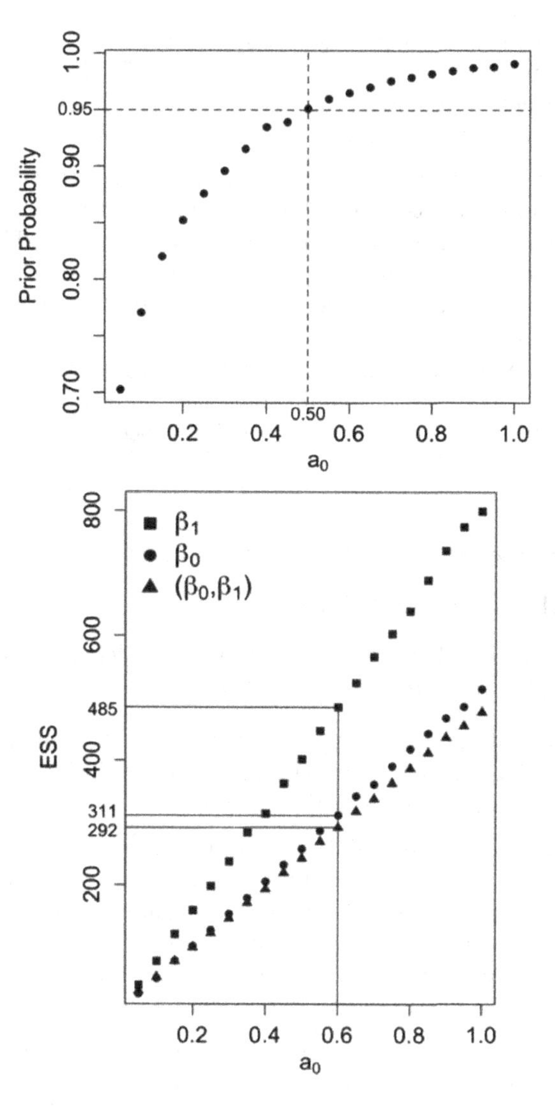

FIGURE 19.10
Two methods for justifying the choice of a fixed power parameter, a_0, when using power priors. In the upper plot, the prior probability of success for some model is plotted against values of the power parameter, a_0. The value of a_0 can be chosen so as to achieve a prior probability of success no larger than some specified value. In this case, all values of a_0 yield relatively large values for the prior PSS. In the lower plot, we have illustrated selection of a_0 with respect to prior ESS. Here, prior ESS is computed for each parameter, as well as the joint parameter, for an hypothetical two-parameter model. Selection of $a_0 = 0.6$ corresponds to a prior ESS of approximately 292 for the vector parameter.

contrast to the joint prior placed on its arguments. This may result in unintended influence on the posterior for the function, especially when sample sizes are small and the function is nonlinear. See Seaman et al. (2012).

8. There is a difference between conducting a sensitivity analysis and model selection. As noted above, the former concerns the sensitivity of the posterior to small changes in the model. This typically focuses on changes to the prior structure. Model choice involves deciding which among several models is most appropriate.[11] This means contrasting candidates often differing in both their data models (likelihood functions) and corresponding prior structures. The candidate models considered should each have been subjected to a sensitivity analysis *before* the selection process.

19.6 Model Characteristics

While model choice is beyond the scope of this brief guide, there are some model features that require elaboration.

1. Models that can be fit using either fixed or random effects require justification of either choice. (Of course, this is a concern in frequentist modeling as well.)
2. Some meta-analysis models require baseline modeling. How this is carried out should be carefully described, especially if such modeling is to be done separately as, for example, recommended when constructing network meta-analysis models using guidelines from the National Institute for Health and Care Excellence (Dias et al., 2013).
3. Hierarchical Bayesian models require assumptions regarding exchangeability and conditional exchangeability. The reasonableness of these assumptions should be addressed. For example, in device studies, exchangeability is frequently not a reasonable assumption. See FDA (2010), p. 17ff. See also Gelman et al. (2014).
4. Identifiability may be an issue if it is observed that there is little updating of the prior to the posterior. This may indicate the need for a reparameterization (perhaps the addition of constraints) or employment of more informative priors. This is typical in models containing measurement error or misclassification components, as well as with other partially identified models. See Gustafson (2015).

[11] Or the use of more sophisticated methods such as Bayesian model averaging.

19.7 Model Performance

A thorough treatment of model performance analysis is also beyond the scope of this guide, but here are some brief comments.

1. A Bayesian model requires specification of a likelihood function, $l(\boldsymbol{\theta}|\mathbf{x})$, $\boldsymbol{\theta} \equiv (\theta_1, \ldots, \theta_p)$, and a joint prior structure, $\pi(\boldsymbol{\theta})$. The joint probability distribution, $f(\mathbf{x}|\boldsymbol{\theta})$, associated with the likelihood function can be used to simulate data sets, with parameter values generated using a joint "design prior", $\pi_D(\boldsymbol{\theta})$, analogous to Bayesian sample size determination.[12] Location of the design prior is chosen to reflect θ_i values of interest. Dispersion parameters for $\pi_D(\boldsymbol{\theta})$ are chosen to reflect uncertainty about which values of θ_i will be

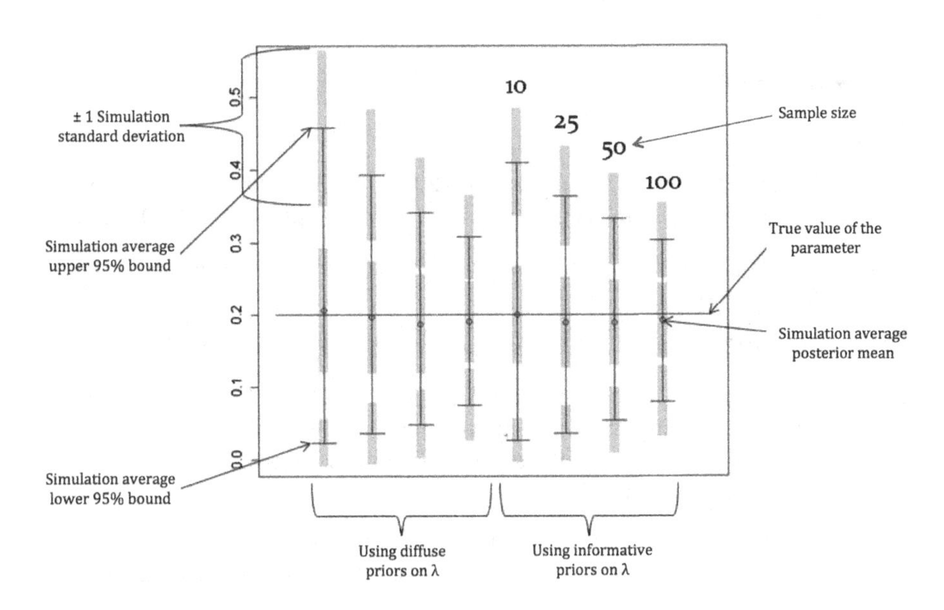

FIGURE 19.11
Simulation summary plot for 500 samples simulated to study model performance under a single truth scenario, indicated by the horizontal line at 0.2. In this case, the plot compares model performance for a diffuse prior structure to an informative prior structure for an hypothetical parameter, λ, at various simulated sample sizes. Each vertical bar summarizes the results of multiple simulations. Included are the average of 500 posterior means, 500 upper and lower 95% credible set bounds, and associated ± 1 simulation standard deviation bars (in gray). In some cases, simulation and/or posterior medians or other measures may be preferred to simulation averages.

[12]See, for example, Brutti et al. (2008) for more detail.

obtained in a trial. For each generated data set, the model is fit using the joint prior, $\pi(\boldsymbol{\theta})$. More than one design prior can be chosen to represent different scenarios, e.g., optimistic or skeptical.

2. Testing a model under various scenarios (values of $\boldsymbol{\theta}$) in the fashion described above can result in hundreds or even thousands of MCMC runs. Clearly not all can be checked for convergence, but we recommend that you do so for a small subset of such runs, chosen to represent the range of parameters generated by the design prior.
3. Summarizing the results of testing a model across multiple model characteristics (diffuse priors vs. informative priors, for example) and/or truth scenarios can be done compactly with graphical tools. A particularly useful graphical summary can be constructed like that illustrated in Figure 19.11.

References

Brutti, P., De Santis, F., Gubbiotti, S. (2008) Robust Bayesian sample size determination in clinical trials, *Statistics in Medicine*, **27**, 2290–2306.

Carlin, B. and Louis, T. (2009) *Bayesian Methods for Data Analysis*, 3rd edn. CRC Press: Boca Raton, FL.

Cowles M. (2013) *Applied Bayesian Statistics.* Springer: New York.

Dias, S., Sutton, A., Ades, A., and Welton, N. (2013) Evidence synthesis for decision making 2: a generalized linear modeling framework for pairwise and network meta-analysis of randomized controlled trials, *Medical Decision Making*, **33**, 607–617.

Food and Drug Administration (2010) Guidance for the use of Bayesian statistics in medical device clinical trials. www.fda.gov/MedicalDevices/ucm071072.htm

Flegal, J., Haran, M. and Jones, G. (2008) Markov chain Monte Carlo: Can we trust the third significant figure? *Statistical Science*, **23**(2), 250–260.

Gelman, A., Carlin, J., Stern, H., Dunson, D., Vehtari, A., and Rubin, D. (2014) *Bayesian Data Analysis*, 3rd edn. CRC Press: Boca Raton, FL.

Gelman, A. and Shirley, K. (2011) Inference from simulations and monitoring convergence. In *Handbook of Markov Chain Monte Carlo*, edited by Brooks, S., Gelman, A., Jones, G., and Meng, X. CRC Press: Boca Raton, FL, pp. 163–174. (This is Chapter 6 in the book. As of 13 December 2018, it is available free of charge at www.mcmchandbook.net/HandbookChapter6.pdf)

Geyer, C. (2011) Introduction to Markov chain Monte Carlo. In *Handbook of Markov Chain Monte Carlo*, edited by Brooks, S., Gelman, A., Jones, G., and Meng, X. CRC Press: Boca Raton, FL, pp. 3–48. (This is Chapter 1 in the book. As of 13 December 2018 it is, available free of charge at www.mcmchandbook.net/HandbookChapter1.pdf)

Gustafson, P. (2015) *Bayesian Inference for Partially Identified Models.* CRC Press: Boca Raton, FL.

Lesaffre, E. and Lawson, A. (2012) *Bayesian Biostatistics.* Wiley: New York.

Lunn, D., Jackson, C., Best, N., Thomas, A., and Spiegelhalter, D. (2012) *The BUGS Book.* CRC Press: Boca Raton, FL.

Morita, S., Thall, P., and Müller, P. (2008) Determining the effective sample size of a parametric prior, *Biometrics*, **64**(2), 595–602.

Morita, S., Thall, P., and Müller, P. (2010) Evaluating the impact of prior assumptions in Bayesian biostatistics, *Statistics in Biosciences*, **2**, 1–17.

Morita, S., Thall, P., and Müller, P. (2012) Prior effective sample size in conditionally independent hierarchical models, *Bayesian Analysis*, **6**(3), 591–614.

Ntzoufras, I. (2009) *Bayesian Modeling Using WinBUGS.* Wiley: New York.

Roos, M., Martins, T., Held, L., and Rue, H. (2015) Sensitivity analysis for Bayesian hierarchical models, *Bayesian Analysis*, **10**(2). 321–349.

Seaman, J. III, Seaman, J. Jr., and Stamey, J. (2012) Hidden dangers of specifying noninformative priors, *The American Statistician*, **66**(2), 77–84.

Spiegelhalter, D., Abrams, K., and Myles, J. (2004) *Bayesian Approaches to Clinical Trials and Health-Care Evaluation.* Wiley: Chichester, England.

Index

C

D

E

F

N

O

P

S

Printed in the United States
by Baker & Taylor Publisher Services